Y0-BCR-543

Using the Engineering Literature

The field of engineering is becoming increasingly interdisciplinary, and there is an ever-growing need for engineers to investigate engineering and scientific resources outside their own area of expertise. However, studies have shown that quality information-finding skills often tend to be lacking in the engineering profession. *Using the Engineering Literature* is a guide to the wide range of resources in all fields of engineering. The information age has made a great impact on the way engineers find information. While print is still important, resources are increasingly being made available in electronic format and the Web is now a major resource. Engineers impact upon almost all aspects of our lives, and it is vital that they find the right information at the right time for better products and processes.

The book takes an engineering sub-discipline approach, detailing those resources that are most important for the practicing engineer and the librarians who work in engineering. Each chapter provides a short history and description of the discipline, then lists the most important resources by format: handbooks, dictionaries, texts, journals, websites, etc. Most references include a short annotation. The authors of each chapter are well-known, experienced librarians or faculty in the appropriate engineering discipline, sharing their expertise and experiences with engineering information.

Using the Engineering Literature is a guide to often unknown resources to the practicing engineer, a textbook for the library school student or new engineering librarian, and a time-saving handbook to the engineering librarian. The arrangement of materials provides easy and logical access to evaluated resources in engineering and supporting disciplines, providing a tool that is useful in reference services and collection development.

Bonnie A. Osif is Engineering Reference and Instruction Librarian at the Pennsylvania State University. Her research interests include the value of information, transportation information and instructional methodologies.

Routledge studies in library and information science

1 **Using the Engineering Literature**
Edited by Bonnie A. Osif

Previous titles to appear in *Routledge studies in library and information science* include:

Using the Mathematics Literature
Edited by Kristine K. Fowler

Electronic Theses and Dissertations
A sourcebook for educators: students, and librarians
Edited by Edward A. Fox

Global Librarianship
Edited by Martin A. Kesselman

Using the Financial and Business Literature
Edited by Thomas Slavens

Using the Biological Literature
A practical guide
Edited by Diane Schmidt

Using the Agricultural, Environmental, and Food Literature
Edited by Barbara S. Hutchinson

Becoming a Digital Library
Edited by Susan J. Barnes

Guide to the Successful Thesis and Dissertation
A handbook for students and faculty
Edited by James Mauch

Electronic Printing and Publishing
The document processing revolution
Edited by Michael B. Spring

Library Information Technology and Networks
Edited by Charles Grosch

Using the Engineering Literature

T
10.7
U85
2006
W03

Edited by Bonnie A. Osif

LONDON AND NEW YORK

First published 2006
by Routledge
2 Park Square, Milton Park, Abingdon, Oxon OX14 4RN

Simultaneously published in the USA and Canada
by Routledge
270 Madison Ave, New York, NY10016

Routledge is an imprint of the Taylor & Francis Group, an informa business

© 2006 Selection and editorial matter Bonnie A. Osif; individual chapters, the contributors

Typeset in Garamond by Wearset Ltd, Boldon, Tyne and Wear
Printed and bound in Great Britain by TJI Digital, Padstow, Cornwall

All rights reserved. No part of this book may be reprinted or reproduced or utilized in any form or by any electronic, mechanical, or other means, now known or hereafter invented, including photocopying and recording, or in any information storage or retrieval system, without permission in writing from the publishers.

British Library Cataloguing in Publication Data
A catalogue record for this book is available from the British Library

Library of Congress Cataloging in Publication Data
A catalog record for this book has been requested

ISBN10: 0-8247-2964-1

ISBN13: 978-0-8247-2964-6

This book is dedicated to present and future engineering librarians who strive to provide quality services and promote the value of evaluated information. And to my father, Earl William Baldwin, who encouraged me to always be curious and keep learning.

Contents

Contributors

Michael Chrimes is Head of Library and Information Services at the Institution of Civil Engineers.

Thomas W. Conkling is Head of the Engineering Library at the Pennsylvania State University. From 1975 to 1981, he was Assistant Librarian at the Princeton University Plasma Physics Laboratory Library. He has a BS in physics and math from the State University of New York at Stony Brook and an MLS from Queens College of the City University of New York. His research interests include the production and use of technical information. He is an active member of the American Society for Engineering Education (ASEE) Engineering Libraries Division. He was a co-recipient of the 1995 Engineering Information/Special Libraries Association Engineering Librarian of the Year Award, and he also received the Homer I. Bernhardt Distinguished Service Award in 2000 from the ASEE Engineering Libraries Division.

Mel DeSart is currently Head of the Engineering Library at the University of Washington. He is liaison to the Departments of Aeronautics and Astronautics, Industrial Engineering, and Mechanical Engineering. He has served in previous engineering-related positions at the Universities of Kansas and Illinois. He received his BA (1982) in classical civilizations and his MS in library and information science (1987), both from Illinois. He is a current member of American Library Association and Association of College and Research Libraries (Science and Technology Section and University Libraries Section), but is most active in the Engineering Libraries Division (ELD) of the American Society for Engineering Education (ASEE). He initiated and moderates the ELDNET-L (1991) and ELD-L (1998) e-mail discussion lists, and has served ELD in a variety of positions, including Secretary/Treasurer, Program Chair, and Division Chair. His professional interests include initiating and supporting positive change in the current system of scholarly communication (he currently serves as Chair of the UW Libraries Scholarly Communication Steering Committee), and in the training and mentoring of new sci-tech librarians. His non-professional interests include a variety of college and professional sports, in particular baseball, and delighting in the long-standing hunt for the perfect microbrewed beer.

Rita Evans has been Reference Librarian at the Harmer E. Davis Transportation Library at the Institute of Transportation Studies at the University of California, Berkeley since 2001. Most of her career has been spent in corporate libraries, including thirteen years managing Dolby Laboratories' technical library in San Francisco. She also started and organized an engineering library for a division of Westinghouse Corporation and worked for a number of years at Gulf Oil's headquarters in Pittsburgh. She has been a consultant for taxonomy and intranet projects for banking and software companies. Active in the Special Libraries Association, Rita is currently on the board of the Transportation Division and served previously as President of the San Francisco Bay Region Chapter. She received her MLS from the University of Pittsburgh.

Kathy Fescemyer is the Life Sciences Librarian in the Life Sciences Library at Penn State. She is a graduate of the University of Illinois with a BS in zoology and an MS in library and information science. She also has a MS in entomology from Louisiana State University. She is active in the American Library Association, Council on Botanical and Horticultural Libraries, and United States Agricultural Information Network.

Aleteia Greenwood is a Science and Engineering Librarian at the University of British Columbia, whose subject responsibilities include civil and mechanical engineering, mining and mineral process engineering, mathematics and statistics, and the integrated sciences undergraduate programs. Two areas of interest are mentoring librarians-to-be and inspiring patrons to enjoy using the library while becoming sophisticated researchers. She has a BA in English literature and an MLIS, both from UBC. Eschewing television, she devotes her off-work hours to creating innovative and colorful paintings as well as print-making. Having not yet become a famous artist, she plans to remain an engineering librarian for some years to come.

Godlind Johnson is Head of Science and Engineering Library, Stony Brook University. She has an undergraduate library science degree from Berlin Library School/Berlin, Germany, and an MLS from George Peabody College/now Vanderbilt University, Nashville, Tennessee. She was Science and Engineering Reference Librarian at Vanderbilt University from 1984 to 1988 and has been at Stony Brook University since 1988. She has been active in the Engineering Libraries Division of the American Society for Engineering Education since 1990. She has edited several guides in the series *Selective Guides to the Literature on...* published by ASEE/ELD, compiled the guide on *Advanced Ceramics* and co-compiled the guide on *Applied Optics.*

Jerry Kowalyk has been a Public Services Librarian for the Cameron Science and Technology Library at the University of Alberta in Edmonton, Alberta, Canada since 1992. Currently, he is responsible for library liaison, bibliographic instruction, and collection development in support of teaching and research in the Faculty of Engineering Departments for Civil, Mining and Petroleum, and Electrical and Computer Engineering. Prior to 1992, He was the Slavic literature and language materials cataloguer in the Bibliographic Services Division of the University Library for ten years.

Mary Frances Lembo is an Information Specialist for the Hanford Technical Library at the Pacific Northwest National Laboratory where she provides in-depth reference services for laboratory and Hanford site staff. She coordinates demos of library products and services and collaborates in the teaching of a two-day workshop on presentation skills. She provides support for the library collection in the areas of chemistry, materials science and physics. She holds a Masters in library science from the University of Washington. The Pacific Northwest National Laboratory is operated by Battelle for the US Department of Energy.

Linda Martinez is currently Head of the Vesic Library for Engineering, Mathematics and Physics at Duke University and the subject specialist in biomedical, civil, electrical, mechanical engineering and computer science at Duke since 1996. Prior to coming to Duke, she was at the Massachusetts Institute of Technology for more than ten years, in various positions, such as the Subject Specialist in Electrical Engineering and Computer Science, and the Assistant Librarian for Core Information Competencies. She holds a MLS from Simmons College Graduate School of Library and Information Science and an Ed.M. from Harvard University Graduate School of Education.

Renée McHenry has more than twenty years of experience in corporate, special and academic libraries, specializing in business and transportation-related research. She has worked at Northwestern University and is currently working as an independent contractor.

Barbara Opar is Architecture and French Librarian at Syracuse University. She holds a BA in French and English, and an MA and MLS, all from Syracuse. She has been employed at Syracuse University since 1975. Barbara has been a member of the Art Libraries Association of North America and of Western New York (ARLIS/WNY), and of the Association of Architectural School Librarians. She has been President of both AASL and ARLIS/WNY. She prepares the booklist for the Society of Architectural Historians.

Bonnie A. Osif has been Engineering Reference and Instruction Librarian in the Engineering Library at the Pennsylvania State University since 1991. Prior to that she was a physical sciences librarian at Penn State and managed the biology library at Temple University. She holds a BS in biology from Penn State, an MS in information science from Drexel, and an Ed.D. in science education from Temple University. She was co-recipient of the SLA Engineering Librarian of the Year Award in 1995. She is active in the Special Libraries Association, the Transportation Research Board, and is a columnist for the American Library Association's *Library Administration and Management.* She is the co-author of *TMI 25 Years Later.*

Mary A. Osorio holds a BA degree in English (minor in history) from the State University of New York (SUNY) at Buffalo and an MLS from the SUNY at Geneseo. Currently, Mary is employed at Messenger Public Library, North Aurora, IL. Previous positions held include working for the Information Center of the Centro Colombo Americano and as instructor of English at Universidad del Norte both in Barranquilla, Colombia.

Nestor L. Osorio received a degree in mathematics and physics from the Universidad del Atlantico, Colombia; he also holds a Masters in physics and a Masters in library and information science from the State University of New York. He joined Northern Illinois University, DeKalb, Illinois in 1982 and is currently a Professor and Subject Specialist for Science and Engineering as well as a senior member of the Graduate Faculty. Professor Osorio is an active member of the American Society for Engineering Education, The American Society for Information Science and Technology, the Association for College and Research Libraries, and the Society for the History of Technology. He is a board member of the electronic journal *Issues in Science and Technology Librarianship* (ISTL), and a referee board member of the journal *Science and Technology Libraries.* An author of numerous scholarly papers and bibliographies, he has presented a number of papers at national and international conferences. His areas of interest include technical information, the use and development of digital collections, content analysis, and bibliometrics.

Andrew W. Otieno is an Assistant Professor of Manufacturing Engineering Technology at Northern Illinois University's Department of Technology. He received his Ph.D. in mechanical engineering from Leeds University, United Kingdom, in 1994. He has worked in various institutions in different capacities including University of Nairobi, Kenya and University of Missouri-Rolla. His research and teaching interests are in the areas of manufacturing engineering, computer applications in manufacturing, finite element modeling structural health monitoring and damage detection, and machining. He has published several research and pedagogical articles in various journals and has made numerous conference presentations. He is also a member of the Society of Manufacturing Engineers, American Society of Engineering Education, National Association of Industrial Technology, and Pi Epsilon Tau. He is currently involved in teaching and research in the area of manufacturing automation, and tool-wear imaging and analysis in micro-machining applications.

Jean Z. Piety is Head of the Science and Technology Department of the Cleveland Public Library in Cleveland, Ohio. She received a Bachelor of Arts degree from Ohio University (1955) and a Masters in library science from Western Reserve University (1957) (now CASE). She is active in SLA (Special Libraries Association), and is a member of ASIST (American Society for Information Science and Technology) and SES (Standards Engineering Society). She has written articles in *Science and Technology Libraries*, in *Library Hi-Tech News*, and has spoken at conferences and lectured on environmental issues and on science and technology, especially on running a large subject department in a public library. Recently, she spoke on the complexity of identifying and locating historical industrial standards. Other activities include being the Archivist for the Cleveland Technical Societies Council, a consortium in the Greater Cleveland area, where she represents the local chapters of ASIST and SLA. In 2002, the Council presented her with its Visionary Award. Both she and her husband enjoy the Cleveland Orchestra, museum touring, and the theater. They are board members of the Ensemble Theater, a Northeastern Ohio theater group.

John Piety has a BA in anthropology from the University of Arizona and a MLS from the University of Oklahoma. He started special library work at General Atomic in San Diego, and was the Collections Officer who built the University of Wisconsin Green Bay Library when it was known as Ecology U. His years at Pan American University Library ended when he married Jean and became Library Director at John Carroll University. He has published, especially with the Special Libraries Association where he has been active with the Science and Technology Division. He is now retired and loving every minute of it.

Jill H. Powell is Reference and Instruction Coordinator for the Engineering Library, Cornell University. She received a BA degree in German and economics from Cornell University in 1982. She worked for Cornell University Press in Ithaca, New York and Mayfield Publishing Company in Palo Alto, California before obtaining an MLS degree from Syracuse University in 1985. In 1986 she started working as a reference librarian in the Engineering Library, Cornell University. In 1992 she became the Reference and Instruction Coordinator.

Hema Ramachandran worked for a large construction company in London, England for six years doing reference, interlibrary loans and ordering publications. After moving to the United States and while studying for an MLS at Florida State University, she worked at the University of Florida's NASA/STAC (an information center), conducting online searches and document delivery for fee-paying clients. After being in charge of the University of New Hampshire's Chemistry Library for three years, She joined Northwestern University's Transportation Library as Reference/Bibliographic Instruction Librarian. She left Northwestern University in 1997 and – after a short stop at San Jose State University – she took up the position as one of three engineering librarians at the California Institute of Technology specializing in computer science, electrical engineering and mathematics. She has recently taken up a new position as Access Services Librarian at Pasadena City College.

Carol Reese has over sixteen years of experience handling civil engineering information for the American Society for Civil Engineers (ASCE). She first worked in the Publications Division where her main responsibilities included the production of indexes for all journal and book publications, the development of ASCE's Web-based Civil Engineering Database and the Society's full-text online publications. With these projects accomplished, she then became ASCE's Cybrarian and Archivist where she serves the civil engineering community by providing research assistance to those who need it. As Archivist, she is responsible for organizing and maintaining the history of the Society. In addition, she has been active in Special Libraries Association and has held several positions. She holds a Master of Science degree in library science from Florida State University and worked for over fifteen years in reference at Brookdale Community College in New Jersey.

Randy Reichardt is Public Services Librarian (Engineering) at the University of Alberta's Science and Technology Library. His subject areas include chemical and materials engineering, mechanical engineering, engineering management, and nanotechnology. He maintains the weblog, "The SciTech Library Question" (stlq.info), which features occasional postings of interest to sci-tech librarians. His recent publications include articles on the application of weblogs in science and technology libraries, and on finding chemical and petroleum pricing information. He loves NYC and movies, and is a professional guitarist, but has yet to hit it big in the entertainment industry.

Dana Roth has a BS in chemistry from UCLA, an MS in chemistry from Caltech and an MLS in library service from UCLA. Employed at Caltech since 1965, he served as Head of Science and Engineering Libraries from 1983 to 1995 and is currently the Chemistry Librarian. His other interests are politics and history, especially the World War II era.

Helen Smith is the Agricultural Sciences Librarian at Penn State where she provides reference, instructional, and collection development services in the Life Sciences Library with a special focus on the food sciences, animal sciences, and agricultural and biological engineering. She is active in the American Library Association, the United States Agricultural Information Network, and has published in the areas of integration of library services into course management software, and has evaluated several databases for national publications.

Mary D. Steiner is Head, Engineering Library, and Coordinator, Physical Sciences and Engineering Libraries, at the University of Pennsylvania. She is responsible for providing services and collections to support the research and educational goals of the School of Engineering and Applied

Science, which includes the Department of Bioengineering. Mary is involved with collection development and management issues, scholarly communication, and institutional repositories. She has served as Chair of the SLA Engineering Division, President of the SLA Princeton-Trenton Chapter, and Director for the Engineering Libraries Division of the American Society for Engineering Education. Mary holds a BSE degree in aerospace engineering from the University of Michigan, and worked in that field for several years before obtaining her MLS degree from Rutgers University in 1994.

Larry Thompson is College Librarian, Engineering, at the Virginia Polytechnic Institute and State University. He is an active member of the Engineering Libraries Division of the American Society for Engineering Education.

Linda Vida is currently Director of the Water Resources Center Archives, located on the University of California, Berkeley campus. She has held this position for over thirteen years. Her previous experience includes work as the Head Librarian at the US Environmental Protection Agency Library in San Francisco (Region 9) under contract to Labat-Anderson Inc. and a reference librarian for the firm of Earl & Wright Engineering in San Francisco. She was awarded her Masters in library and information sciences (MLIS) from UC Berkeley. She is active in the San Francisco Bay Region Chapter of the Special Libraries Association (SLA) and also in the Environment and Resource Management Division of SLA.

Lois Widmer is currently Manager of the Science Library at Brandeis University, Waltham, MA. Her previous experience includes work as a reference librarian at the National Institute of Environmental Health Sciences and as Director of Information Services for the Center for Transportation and the Environment, a federally funded research program at North Carolina State University. She holds a Master of science in the Library Science Program (MSLS) from the University of North Carolina at Chapel Hill. She has been an active member of the Environment and Resource Management Division of Special Libraries Association.

Preface

"Libraries? I don't need libraries. The facts I need are in my personal collection of books and journals in my office. If I don't find what I need, my colleague in the next office will have it. Anyway, most everything is on the Web now."

These words, or those of a similar meaning, are the bane of all information specialists. While it is probable that to some extent, people from every discipline truly believe this sentiment, it could be argued that some of the strongest supporters of the "I don't need libraries" mindset are engineers. Various articles have been written on the subject of engineers and information use. In an article promoting information instruction, England notes that "many engineers lack skills in accessing and retrieving information. Yet the ability to monitor, access, retrieve, evaluate, use and communicate information will be critical in a global information society characterized by rapid technological change" (England, 1995: 1). Holland and Powell (1995) concur in their study of engineering graduates and information instruction. Pinelli (1991) has published a number of articles on the information skills of engineering and the role of information. These studies all support the observations that engineers do not perceive libraries and information resources as a high priority, and are not conversant with or strong users of the range of resources available to them.

Tenopir and King have published a book that provides a fascinating and eye-opening study of the information-seeking behaviors of engineers. *Communication Patterns of Engineers* is a longitudinal study of engineers from the 1960s to the present day. A number of observations are reviewed and documented by years of communication use data, many of which substantiate anecdotal observations of the long-term engineering librarian and the previously cited research. For example, "easy access is an engineer's top information priority" (Tenopir and King, 2004: 58) and the monetary value of reading the literature is documented. They find that "nearly all reading by university scientists and engineering (95 percent) resulted in some favorable outcomes" (ibid.: 43). With non-university scientists and engineers reading resulted in higher quality, faster performance, and initiated broadened options (ibid.: 43). The book has a number of examples of quantitative support for the value of information.

> Engineers tend to rely much more on interpersonal and informal means of communications than other scientists. . . . Engineers tend to be self-sufficient and more direct in their approach to work. Their learning style emphasizes listening and discussion rather than observing and reading. It may be that those who enter engineering as a profession may lean toward a certain personality, way of thinking, and learning style.
>
> (Tenopir and King, 2004: 181)

At the same time the authors make it clear that engineers need quality information. A mistake by an engineer can be very serious and very obvious – building collapse, equipment fails, machinery malfunctions. In order to find the information to avoid these mistakes a wide range of resources is necessary, found in a timely manner. To keep up with advances, to become aware of errors, means it is essential to correct such errors. In addition, the field of engineering is becoming increasingly interdisciplinary, so even the engineers who are aware of the resources in their specialization will find the need to investigate engineering and science resources outside their area of expertise.

The common theme in all of these issues is the need for quality information and quality information-finding skills and that these skills tend to be lacking in the engineering profession. There is a need to address these shortcomings.

The information environment has continued to evolve from the time of these studies, yet the attitude has not changed significantly. Resources in engineering continue to proliferate. These resources may now be found in multiple formats. Electronic files have joined print and microforms. Databases range from general engineering to specialized databases in very specific areas of study. While attempts are made to simplify the search interfaces, record labels, and the means to obtain the resources for users, it is still a confusing jungle of titles, acronyms, formats, and search structures which need to be navigated by the academic, the practitioner, the researcher, the student, and the librarian.

One of the roles of guides to the literature is to provide a road-map of the information world in a particular subject. Engineering, as is the case in many subjects, has become increasingly interdisciplinary. The number of departments has grown from those in the traditional civil, mechanical, and industrial engineering to include bioengineering, nanotechnology, artificial intelligence, and smart materials. In addition to the core engineering subjects and the basic science and math which support these classes, engineers are now taught to consider issues such as ethics, environmental impact, and social implications of their technologies. The subject range continues to grow more complex at the same time that the need to navigate through the resources becomes more important.

Engineering shares a need for handbooks, manuals, journal articles, conference papers, patents, and, to some extent, monographs with the rest of the science, medical and technical areas of study. In addition, it has some resources that are more "engineering-centric." These include the heavy use of technical reports and standards. A sizable percentage of resources can be gray literature, items that are not produced by major publishers, but by societies, agencies, and institutions. While often more difficult to locate, these can be central to engineering. The tools to identify and obtain them are crucial. This, too, provides a challenge for the practitioner, student, and librarian.

In addition, information sources in engineering are increasingly being made available on the Web. While some agencies have provided free access to reports and standards, a number of databases, full text sources of handbooks, reports, standards, articles, and conference papers are available for a subscription fee. Desktop access to engineering resources is becoming a reality but adds to the complexity of knowing where and how to look. Evaluation of resources, always an issue even in the days of paper-only resources, is crucial when considering the growth of Internet resources. Engineering relies on verified, reliable information, two terms that are not necessarily synonymous with many Web resources.

To return to the metaphor of a road-map, the field of engineering has branched into a number of side roads with many intersections. These intersections lead into areas of the sciences with strong connections to physics, biology, chemistry, and agriculture. However, engineering has also recognized the strong linkage with non-science and technology fields of study. Ethics, human factors, business, social, and psychological issues must now be part of the research process, and resources in these fields must be available for consideration in the research, design, and application process. Underlying all of these is the crucial concept of evaluation as to the validity and authority of the information.

While these issues provide additional road-blocks, the modern trends in librarianship have been able to meet the challenge. Rich collections are organized and described in library catalogs; subject databases provide multiple access points to the proliferation of reports, standards, papers, and articles; and Web portals provide entry into the growing wealth of evaluated information that is available, sometimes for free and sometimes for a fee.

Using the Engineering Literature is meant to be a road-map to these access points and resources. It provides directions to find the needed information, as well as a description of the resources to help readers determine if they need to "stop" for a closer look. The guide cannot be comprehensive, as the information environment changes constantly. However, it highlights the core resources as well as specialized titles. It provides a basic guide to the resources for the newcomer either to the field of engineering or to engineering resources. In addition, it provides an exploration of more specifically focused resources for the experienced engineering professional or librarian. It will also provide an entry into a newer fields emerging in engineering, such as nanotechnology.

While the lines between engineering departments may have blurred somewhat in recent years, they still define the basic areas of study. Resources are described within these traditional descriptions with emphasis being placed on some of the newer areas of study. Unless a resource is considered a classic, all titles are from the 1990s or more recently. All formats are considered and, where there are multiple formats, they will be noted.

While the primary users will be librarians in engineering, the book will help guide the practicing engineer and the engineering student to resources that will enhance their work and studies. The value of information is almost without measure in the saving of time and resources, and in the increase in efficiency and safety; all of vital importance in engineering. If information is power, then the goal of this book is to provide the means to empower each librarian and engineer.

References

England, Mark M. (1995) "Information Literacy for Engineers: The Problem and Its Solution," *AIAA, Aerospace Sciences Meeting and Exhibit*, paper AIAA-95-0705.

Holland, Maurita Peterson and Powell, Christina Kelleher (1995) "A Longitudinal Survey of the Information Seeking and Use Habits of Some Engineers," *College and Research Libraries*, 56 (1): 7–15.

Pinelli, Thomas E. (1991) "The Information Seeking Habits and Practices of Engineers," *Science and Technology Libraries*, 11 (3): 5–25.

Tenopir, Carol and King, Donald W. (2004) *Communication Patterns of Engineers*, Piscataway, NJ: IEEE Press; Hoboken, NJ: Wiley-Interscience.

Acknowledgments

I have had the privilege of meeting and working with some of the contributors to this book. The choice of writers was made carefully, based on their experience and reputation. There was no way to have all of the exceptional librarians I have met work on this book, but those selected represent a group dedicated to providing quality service to the profession. I thank them all for the diligent work and the long hours spent to put this book together. I have learned so much from their expertise.

I thank Russell Dekker for allowing me this opportunity, and Rosemary Doherty of Marcel Dekker for her assistance and friendliness. I would also like to thank The University Libraries at Penn State for the time and encouragement to take on this challenge and Kelly Riley for her editorial assistance. Last but never least, I thank my husband Tom for his patience and encouragement as I worked through all the edits and reviews of the text.

1 Introduction

Bonnie A. Osif

Pick up a newspaper or a magazine and odds are that there will be an article about the role of information in our lives. This has been called the information age, and we are bombarded with twenty-four-hour news, electronic news sources, text messaging, instant messaging, Blogs, RSS, and so much more. Yet we also hear that many of our information resources are not used or are used inefficiently. Information overload or overkill? Possibly both. However, there may be other reasons beyond too little time, too many resources, such as too little knowledge of information sources that are available and how to use them, lack of recognition of the real value of the resources, and a perception of no real need for additional resources to add to our repertoire since we already have the Web. But what if the best information is still in print format?

Engineers have a serious need for quality, timely information. The discipline is charged with creating, building, maintaining, and testing the infrastructure of our lives. When we get up in the morning we use the products of architectural engineers. When we get into our car we are trusting the automotive, mechanical, and safety engineers. Our drive to work or our trip on public transit owes much to the transportation, civil, and petroleum engineer. As we eat (agricultural engineering), get a medical check-up (bioengineering), or relax in front of our television or computer (electrical and/or computer engineering), we are using the skills of a wide range of people. If we consider almost any aspect of our daily lives, there is an engineering specialty involved to some extent. To do their job at the highest level, engineers must have access to evaluated information that can be accessed in a timely manner and in a useful format. Increasingly that format is desktop access to full text documentation. However, not everything is available electronically at this point in time, nor does every engineer have access to these materials. Therefore, it is important to have a guide to the best resources and the ways to obtain these resources. Weichert and colleagues state: "A good engineer does not know everything from memory, but he knows, (*sic*) where and how he can find the response to his question or the solution to his problem" (Weichert *et al.*, 2001: 144). They go on to state that they must "be able to quickly find data and information, regardless of its global origin, and effectively evaluate, organize and present it" as well as to "recognize the need for continuous, career-long learning and career planning and have the knowledge and discipline to do both" (ibid.: 167). While the library is not mentioned during any of these discussions, it is still very clear that these are all needs and concerns that are library-focused, and actually very difficult without the services of a good library and librarian.

Engineering librarians have an integral role in meeting these stated needs. Librarians are trained to find solutions to problems, to find data and information that is evaluated, to find it efficiently and in a usable format, to address both immediate needs and to consider future needs, and to develop collections and services to support the work of their patrons. While many engineering librarians do not have academic training in engineering, library education provides the means to learn the language, resources, and skills to provide the necessary support. In addition, the seemingly innate curiosity of most librarians encourages continual exploration, further enhancing the services they offer.

Using the Engineering Literature is designed to address these needs and concerns. It is an evaluated guide to useful resources. Note that the book is to evaluate resources, not to provide exhaustive coverage. Any of these subject chapters could have been much longer, or even a book in itself; however, that would compound the problem of information overkill. Instead, the authors categorize major works, usually with a brief description. The goal is to guide the reader to good resources to address most needs. Additional resources may be found using the bibliographies in suggested titles, by literature searches in recommended indexes or abstracts, and by searching library catalogs or browsing (physically or virtually) library shelves using suggested classification schemes. Or, that ubiquitous Web can be searched, but instead of an unfocused search with a vast number of results of uncertain quality, evaluated and subject-focused web pages can be searched.

Intended audiences

There are three reader groups that were considered in the planning of this book.

Students of library science or new engineering librarians will find an overview of the formats, major resources, and subdisciplines in engineering. Experienced engineering librarians can use it as a reference source to aid in answering reference questions, when doing collection development, and as a refresher or aid when doing work in an area of engineering that is not familiar. Practicing engineers will see it as a guide/map to the myriad range of resources in their area of expertise, an introduction to new resources when doing research in a new specialty, or when they don't have a reference librarian to assist them.

It is suggested that Chapter 2 on general engineering be reviewed, as many of the references apply to all the subdisciplines. All literature searches should include *Compendex* and *NTIS*, as well as the more specific abstracts and indexes reviewed in the subsequent, discipline-appropriate chapter. In some cases, resources are mentioned in more than one chapter because the authors felt that these resources were important enough to be duplicated, and they highlight special features for the subject specialist. In other cases, a "see" mention may be used. In addition, engineering, as in so many other fields, has become increasingly interdisciplinary, and it can be difficult to distinguish where one area of engineering ends and another begins. So, it is important to scan both the general engineering resources as well as the appropriate engineering discipline.

The literature of engineering

Engineers use a wide variety of information formats. They share many formats with other professionals, while others are more closely associated with engineering than other subjects.

Periodical publications range from the scholarly journal to the new-oriented trade magazine. Journals provide relatively up-to-date information with detailed reports of the materials, methods, and results of research. The articles are normally refereed, or reviewed by experts in the field who evaluate the quality of the research and determine whether it should be published. Journal articles allow the engineer to keep up with the new discoveries, trends, and issues in their discipline, determine lead researchers and institutions, and provide a record to consult when additional information is needed. Journals are published by a number of commercial companies that specialize in scientific and technical publications or by learned societies. Trade journals are not refereed and serve a different purpose. They normally cover the business trends, industry information such as new products and methods, job opportunities, and news about the industry. Both serve a vital purpose, and it is important that engineers be aware of and have access to the titles most important to their discipline and job activities.

One of the major trends in journal and trade magazines is the move to electronic access. Some may be read for free on the Web, others may be subscribed to individually or through aggregators, companies that provide access to a number of titles "bundled" together. The trend toward desktop access to full text journals is extremely popular and provides fast, efficient access. Titles, length and type of access, and licensing regulations, along with cost, must be considered when evaluating this option. However, the popularity and convenience is considerable and appears to be the trend in journal access.

Conferences are an important means of communication in engineering. Most societies sponsor some conferences and they range from general to extremely specialized topics. Conference papers are usually but not always published. Some conference proceedings are easily purchased, while others are only available for a short time after the conference or even made available only at the conference. Increasingly, conference proceedings are published in electronic format. Conference papers are in some cases the only account of a research project if the findings are not published as a journal article. This increases the importance of monitoring conference literature carefully. One added benefit of attending a conference is the ability to network, meet other researchers or practitioners, and to ask questions of the presenter to gain additional insight into the work presented. Most of the subject databases described in the subsequent chapters index conference papers.

While journals and conferences are very important in most academic fields, reports, while not unique to engineering, are extremely important. The report literature covers all engineering disciplines and in the official record of research that is often done with money from the government. Reports are required of researchers under government contract, although research findings from other sources may be reported. These reports must be filed with the National Technical Information Service for inclusion in their *NTIS* database. Most reports are then available for purchase from NTIS, although the trend is for many reports to be available as full text on the Web. A number of portals have made locating these reports easier and are detailed in the appropriate chapters. Increasing the value of reports is the fact that many of these studies will not be available in other formats, but only made public as a report. Reports tend to be very detailed, providing valuable in-depth information to the researcher, and can be published very rapidly. Even with the advent of electronic publication, there is often some delay in electronic publication of journals and conference proceedings. A review of the report literature is crucial if a thorough search is required.

Many reports are one type of literature commonly referred to as gray (or grey) literature. Gray literature is defined as "information produced on all levels of government, academics, business and industry in electronic and print formats not controlled by commercial publishing i.e. where publishing is not the primary activity" (Grey Literature Network Services, 2005). This form of communication is an important, if somewhat elusive form of communication. Types include reports, preprints, white papers, working papers, research in progress, presentations, datasets, and so on. While indexed to some extent in many databases, including GrayLIT Network <http://www.osti.gov/graylit/index.html>, many requests will come for items located in the bibliographies rather than a database search. A librarian will often need to go directly to the author or sponsor of a gray document to obtain it. Chapter 2 on general engineering resources includes a guide to gray literature. Access is improving as libraries make an effort to systematically collect gray literature that is relevant to their research needs and as more is made available on the Web. However, gray literature can be both one of the most valuable and most difficult to obtain resource type.

Monographs are extremely important in many of the liberal arts and humanities, but less so in engineering. Since so much of the relevant research is published in a more timely manner in reports, journal articles, or conference papers, or even faster as a Web paper, the time needed to write an entire book and to go through the publication process limits the usefulness of monographs. However, there are a number of books that are published in engineering and they serve many useful purposes, including introductory textbooks, reviews, analysis, historical studies, and in-depth research. While a monograph is not the primary format for much of engineering communication, they are still important and should be included in a literature review.

Handbooks, encyclopedias, and dictionaries are an integral part of the engineer's library. Engineering dictionaries define words in appropriate context and detail for the engineer and the student. Technical dictionaries are essential for understanding engineering terms, since general purpose dictionaries either fail to include the many technical terms or define them in a general, relatively useless, manner. The same may be said for the benefit of using engineering encyclopedias. Encyclopedias range from the general (*AccessScience*) to the highly specific (*Dekker Encyclopedia of Nanoscience and Nanotechnology*), and they provide the same service that the encyclopedias did in our school days – they supply basic background information that is most useful when working in a new field. However, the engineering encyclopedias are often very detailed and provide an excellent introduction to the topic.

Handbooks provide fast access to charts, tables, short descriptions, formulas, and so on needed to do everyday tasks. While most engineers will have a copy of the handbooks that are most relevant to their work on their desk or on their computer desktop, they will seldom have access to the wide range of handbooks to support the increasingly interdisciplinary nature of engineering. A number of companies are now bundling specialized handbooks and making these available in electronic format.

An easily overlooked resource is the dissertation or thesis. The research done by an engineering student will often be rewritten and published as an article or a conference paper, less often as a book. In some cases it will be published in a shorter format, as a technical report since the research was funded by government money. However, the dissertation may be the only publication, or may be available significantly before it is published elsewhere. While this in itself is a good reason to search *Dissertation Abstracts*, now called *Digital Dissertations*, an even better one is the quality of many bibliographies in a dissertation. These bibliographies will often represent the results of a thorough literature search and are an excellent short cut up to the time of the dissertation. While these bibliographies should not substitute for a well-thought-out and executed literature search, they can be very useful.

Standards, while important to some other disciplines, are basic to engineering. A standard "is a document, established by consensus that provides rules, guidelines or characteristics for activities or their results" (American National Standards Institute, 2005). Standards define terms, provide common methods of testing, interchangability of parts, and work to improve quality, safety, and efficiency. Much of modern society depends on the adherence to standards so that our interconnected world can actually interconnect. General standards organizations are reviewed in Chapter 2, and more specific standards are covered in their appropriate chapters.

As defined by the US Patent Office, a patent "is the grant of a property right to the inventor" and may be for a design, a new or improved process, machine, and so on, or even for a plant. While there are very specific reasons to search patents, and to search carefully if there is interest in filing a patent, valuable information may be found in patents while doing research or trying to understand a product or process. Patent research is much easier now that there are options to search at no cost in relatively user-friendly websites.

The World Wide Web is ubiquitous, and we ignore it at our peril. Almost all of the electronic databases we search on are now on the Web. We can say almost all, because there may still be locations that have workstations with a database or handbook on a CD-ROM. Full text journal articles, conference papers, technical reports, white papers, preprints, even encyclopedias, handbooks, patents, and entire monographs are hosted on the Web and accessed from our desktop. It is difficult to name an organization that does not have a website, usually with a wealth of information a few clicks away. Most people think everything is on the Web, and a vast amount of material is.

However, one thing that is not recognized is that everything, or almost everything, is not available on the Web for free. While the trend is strong to make report literature available on the Web for free and there are a number of fantastic sites with chemical data or computer information free to all, most of the detailed, quality, and essential reference materials are still the product of a commercial publisher and are not free. Quite the contrary, these resources can be very expensive, and the researchers with desktop access may not realize that their organization, usually through a library, is making this service available after contract negotiation and the payment of fees. The ease of access and the look makes it appear to be another web product, but the difference is that this is a web product that is not free. A number of the resources in this book are available in multiple formats; the most popular, and sometimes the most costly, is desktop access to the electronic files.

Communication patterns continue to evolve, and two newer trends are Blogs (weblogs) and RSS (real simple syndication). Blogs may be written by anyone on any topic. Value varies widely but in engineering a local Blog by an information specialist can be an extremely valuable way to keep up with new resources and services. Check with your local librarian to see if they are hosting one. Two that have received attention and may be useful are *EngLib: for the scitech librarian* and *scitech library question*. RSS requires a downloadable newsreader, some of which are free. Subscribe to your favorite web services with RSS and your newsreader will gather the information you want and display it. RSS allows you to access a wide array of web services but only get the information you want in an easy, customized package. This provides an excellent way to efficiently monitor the sites of interest. Keep monitoring for new communication trends!

In addition to the format types, engineering has become increasingly interdisciplinary, not only within different disciplines of engineering, but with science and business. At different times it may be important to review the specialized databases, journals, handbooks, and so on in physics, chemistry, biology, or medicine, or to consider business models and trends. As this is a book devoted to the engineering literature, there are specialized titles that introduce these subjects and may be found by searching a catalog or asking a librarian. From this description of resource formats and increasingly blurry boundaries of disciplines, it is clear that very few libraries will have all the resources. Interlibrary loan, or cooperative borrowing among libraries is a necessity, and librarians can assist engineers in obtaining materials not held locally or available on the Web.

The value of information

"Libraries sometimes aren't used by engineers because they are unaware of important services that are provided or because they don't fully appreciate the benefits of their use" (Tenopir and King, 2004: 184). One of the oft overlooked and under-researched areas of study is the value of information. Engineers are trained to look at value, risks, and cost/benefits as they relate to their work. It can be very useful to use these same skills to document the value of libraries and library services. While librarians have relied on anecdotes to indicate the value of their resources and services, a few documented cases can provide much stronger support. Return on investment studies have been done by several and the results can be staggering (Baldwin, 2002; Metcalf, 2003; Portugal, 2000). Baldwin's study showed a cost/benefit ratio of 12:1. A study funded by the Transportation Research Board surveyed state departments of transportation and others to obtain a number of case studies where a short literature review saved significant amounts of money in indicating technologies that could be used to perform tasks better and less expensively (Dresley and Lacombe, 1998).

In addition to the case studies with an actual dollar amount, there are a number of studies which indicate that access to information can save lives, time, equipment, and improve safety (Osif, 2004). One final way to look at the value of information is that anything which can save engineers time in doing their jobs, help avoid expensive or wasteful dead-ends, save equipment, and improve safety and efficiency has real value. This translates into better work, better evaluation, and, most likely, better pay. The many ways to indicate the value of information cannot be overlooked and is a useful tool to market the value of libraries and library resources.

How to use this book

Many readers may only read the chapter for their specialty, but it is suggested that the general engineering resources be reviewed as well as the appropriate engineering discipline chapter. In addition, check the index for extra resources. Many references may be located in several chapters. In some cases, they are mentioned more than once because chapter authors felt that the titles were important enough to be duplicated, and they highlighted special features of the resource for their discipline. In other cases, "see" references are used to remind the reader of the value of a title to more than one discipline.

Each chapter represents advice from "front-line" librarians. They represent some of the best in the profession, and their collective expertise provides a guide to the disciplines of engineering and the resources to address the information questions.

Change is inevitable and to be embraced. A common expression "may you live in interesting times" is thought to be both a curse and a blessing. We do live in interesting and ever-changing times, and both engineers and engineering librarians are in the midst of these changes by the very nature of their profession. The technologies, the resources, the needs keep changing, and this provides a great challenge. But these changing technologies and resources are giving us the tools to continue to meet the challenge and actually be leaders in changing the profession of librarianship and the perception of libraries.

References

American National Standards Institute, <http://www.ansi.org/about_ansi/faqs/faqs.aspx?menuid=1> (accessed June 16, 2005).

Baldwin, Jerry (2004) "The Crisis in Special Libraries," *Sci-Tech News* 56 (2): 4–11.

Dresley, Susan and Lacombe, Annalynn (1998) *Value of Information and Information Services*, Washington, DC: Federal Highway Administration.

Grey Literature Network Services, <http://www.greynet.org/pages/1/index.htm> (accessed 16 June 2005).

Lavallee-Welch, Catherine EngLib: for the scitech librarian <http://www.englib.info> (accessed June 16, 2005).

Metcalf, John (2003) "The Business Case for Translation of Transportation Documents," *Sci-Tech News* 57 (4): 4–9.

Osif, Bonnie (2004) "The Value of Information: The Missing Piece in the Puzzle," in *Knowledge and Change: Proceedings of the 12th Nordic Conference for Information and Documentation*, Copenhagen: Royal School of Library and Information Science.

Portugal, Frank H. (2000) *Valuating Information Intangibles*, Washington, DC: Special Libraries Associations.

Reichardt, Randy, *scitech library question*, <http://stlq.info/> (accessed June 16, 2005).

Tenopir, Carol and King, Donald W. (2004) *Communication Patterns of Engineers*, Piscataway, NJ: IEEE Press; Hoboken, NJ: Wiley-Interscience.

Weichert, D., Rauhut, B., and Schmidt, R. (eds) (2001) *Educating the Engineer for the 21st Century*, Dordrecht: Kluwer Academic.

2 General engineering resources

Jean Z. Piety and John Piety

Engineering resources: a brief introduction

Engineering encompasses all of modern civilization. There seems to be a different engineering specialty for every field of study. However, all of them have commonalities; therefore, a number of basic concepts and practices are needed to support all the specializations. This chapter covers general engineering resources and provides a foundation for engineering studies, engineering ethics, standards, practices, and major concepts common to all engineering subfields. It includes dictionaries, thesauri, encyclopedias, and resources of information relevant and necessary to all engineers. Some of these items are far from current, but are older publications. They are included because they are still needed in every sense. Just as a road or bridge cannot stand without a foundation, a thorough knowledge of basic engineering has to include these classic titles.

Besides the classics, sources for understanding engineering as a profession are also mentioned. A succinct description is found in the *Occupational Outlook Handbook*, which describes what an engineer does, what working conditions may prevail, what training and qualifications are needed, and suggests sources for more information. The *Handbook* continues with descriptions of specific branches of engineering. *Masterworks of Technology: The Story of Creative Engineering, Architecture, and Design* is recommended for high school students thinking about careers in engineering. Examples of successful engineers are found in *Real People Working in Engineering*. This book not only gives vignettes, but also demonstrates some of the varied engineering fields. The websites for the National Society of Professional Engineers and the American Society for Engineering Education help in the decision-making process.

In today's global economy, a successful engineer may be employed in a company which has offices in another country. That engineer should know the industrial standards needed for implementation of the company's product, and should also understand the culture and mores of that country. Multi-language dictionaries help with words pertinent to the professional field. Using Honda's *Working in Japan* as an example shows the necessity of understanding what an engineer working in another country should learn about the social life and customs of that country.

As resources appear electronically, some traditional methods of research may be neglected. Because gaps in information retrieved may be expensive, the emphasis in this chapter is to provide resources that help toward a basic understanding of how to do research. Looking up one subject often leads to tangent ideas and encourages further research. While electronic information may be favored, often the necessary resource is still only available in print.

Searching the catalog

Whether the Dewey Decimal, the Library of Congress or another classification scheme is used for the contents in a library does not matter, since the classification scheme exists solely for shelving purposes. What is important to know is how to search the library catalog. Each library catalog uses a controlled thesaurus, a predetermined set of descriptors to describe a topic. This insures that searching the catalog will have a more focused result than searching sites with uncontrolled terminology, for example, the Web. Using the subjects in a catalog may result in a smaller, but more pertinent, quantity of material than a random web search would provide. Keyword searching produces the same result as it does in web searching, but it seems more direct due to the control exercised by library staff in assigning headings. Useful tools include the *Library of Congress Subject Headings* and *Super LCCS: Gale's Library of Congress Classification Schedules*. Specific thesauri, such as *Ei Thesaurus*, also aid in developing search strategies.

Most libraries offer online guides in their online public access catalogs (OPACs). Some may be more extensive than others with detailed instructions. Learning how to search the OPAC is parallel to knowing the nuances of a book. With a book, the contents should be checked for an index, a table of contents, a preface, and special instructions that may be entitled "How to use this source." A site map in an online resource serves a similar purpose. It should be noted that while most OPACs are available to all users, many of the online databases may be limited to a set group of subscribers, depending upon the license agreements, while others may be accessed remotely or by a broader audience. One element that is essential and common to most, if not all systems is that instructions should be understood before plunging into the contents of a book, a catalog, an index, or an encyclopedia, or doing an online search.

Books

Library of Congress Subject Headings LCSH26 (2003), (26th edn, 5 vols), Washington, DC: Library of Congress, Cataloging Distribution Service. End papers in each volume list the abbreviated aids for broader and narrower terms. Produced from records in the MARC21 subject authorities format, the headings help in constructing the right terms to use in searches. Words in bold letters are real headings, while lighter ones are cross-references. Broader and narrower terms, those that are used for another word and appropriate subdivisions are also included. All are similar to the "see and see also" references formerly found in library card catalogs. About 36 percent of the headings are followed by Library of Congress class numbers. The explanations found within a heading are helpful. For example, the heading: "Engineering – graphic methods" (TA337) starts with "Here are entered works on the use of scales, graph papers," and finishes with "as opposed to Engineering graphics" which is a broader term. Updates are found online in *Library of Congress Subject Headings Weekly Lists.* New headings and changes are at the Library of Congress Cataloging Policy and Support Office website <http://www.loc.gov/ catdir/cpso/> (accessed May 18, 2005).

Super LCCS: Gale's Library of Congress Classification Schedules, Class T (2002), Detroit, MI: Gale Group. Subtitled "combined with additions and changes through 2001," this series serves as companion volumes to the *Library of Congress Subject Headings* (LSCH). Individual volumes issued on each Library of Congress classification show the breakdown of the classification structure. The volume on Class T covers technology. The Library of Congress classification on engineering starts with TA and continues with the specific specialties. No classification scheme is perfect, so researchers must be aware of useful materials that might class elsewhere. Books on engineering design and specific materials may be in fine arts, Class N. To add to the confusion, books on design of specific buildings may class with the subject, such as hospital design, RA967. Knowing the LCSH scheme is not essential, since keyword searching is an option, but it provides a useful entry to resources specific to a subject. Later chapters in this book will note particular subject headings and classifications that are pertinent to their subject specialization.

Milstead, J.L. (1995) *Ei Thesaurus* (2nd edn), Hoboken NJ: Engineering Information Inc. Not only does this tool serve as the indexing tool for *Engineering Index* and *Ei Compendex*, it also replaces the older tool, *SHE* (*Subject Headings for Engineering*), and provides keywords that may be useful in any engineering research.

Thesaurus of Engineered Materials (2003) (4th edn), Bethesda, MD: Cambridge Scientific Abstracts, 228 pp. ISBN: 0883873788. Compiled by the editors of *Materials Information*, this tool provides vocabulary in numerous materials fields, such as polymers, ceramics, and composite materials.

Traditional research

Martha Schenk and James Webster's book, *What Every Engineer Should Know about Engineering Information Resources*, was published in 1984 and still serves as a model for research. The engineer should be aware of specific handbooks, dictionaries and encyclopedias, indexes and abstracts, standards, journals, specialized guides, and online resources. A comprehensive bibliography of the engineering literature was compiled by Charles Lord and published in 2000. Although he emphasized the literature of the 1990s, the resources are useful today. Some titles may now be e-books, as even his book now appears as one. His overview describes the historical roots to engineering, and the complexities and challenges to engineering today. The contents of both Lord's book and Malinowsky's *Reference Sources in Science, Engineering, Medicine, and Agriculture* serve the librarian in collection-building.

What the engineer will find while researching is often a lot of gray literature, resources that do not seem to fit into a standard category. Gray literature is defined as "Information produced on all levels of government, academics, business and industry in electronic and print formats not controlled by commercial publishing i.e. where publishing is not the primary activity" (Greynet website, February 25, 2005). Auger's bibliography describes the obscure literature that offers a challenge to researchers trying to identifying sources. Research on conference titles presents challenges. There are sources that help to identify conferences. *Science Citation Index*, described in the index section, is a unique tool that shows where and how often an author is cited.

Words challenge engineers and librarians. Several dictionaries are listed to help with definitions, as are encyclopedias that describe unfamiliar concepts. Both *Kirk-Othmer* and McGraw-Hill encyclopedias offer possibilities in describing new concepts and new processes, with bibliographic listings that lead to other sources for additional information. McGraw-Hill has a dictionary format with citations concluding each category and an excellent index. *Kirk-Othmer*, described in Chapter 7 on chemical engineering, is a classic, with each of the five editions not superseding each other, but each edition adding emerging technologies and newer uses of processes and materials. Contents of the lengthy chapters are reached by a detailed index. Chapters end with extensive bibliographies.

As engineers work, they may be in a situation in which they need to design, to create, or to develop a product, or to be a member of a professional committee that helps to develop standards. They need to understand engineering ethics. They need to know how to design, how to write, how to interpret industrial standards, and how to do a patent search. Separate sections in this chapter on general engineering provide some guides in these categories.

To practice in a specific state in the Union, engineers have to be registered with that state. Most states maintain online sites for registered professional engineers, but two directories are listed as examples of sources for information. While some titles were published years ago, they continue to provide useful information for current work, historical perspectives, or may still be the primary resource on the topic.

Bibliographies and guides to the literature

Auger, C.P. (1998) *Information Sources in Grey Literature* (4th edn), London and New Providence, NJ: Bowker-Saur. Unique tool for showing the more difficult and obscure types of technical literature.

European Sources of Scientific and Technical Information (1993) (10th edn), Harlow, Essex: Longman; distributed in the USA and Canada by Gale Research Co. The editors make an effort to include a source for every country by providing the name, address, library facility, consultant information, and types of publications for the prime technical body within each country of the world.

Lord, C. (2000) *Guide to Information Sources in Engineering*, Englewood, CO: Libraries Unlimited. Published as an e-book in 2004, this is an excellent, comprehensive bibliography of all engineering specialties.

Malinowsky, H.R. (1994) *Reference Sources in Science, Engineering, Medicine, and Agriculture*, Phoenix, AZ: Oryx. As a teacher, editor, and librarian in engineering and science, the author wrote a collection development tool for librarians. Several chapters cover serial price increases, impact of information technology, sources of science and technology information, and types of reference sources. Contents of the other chapters list the basic bibliographic tools in science, engineering, medicine, and agriculture.

Macleod, R.A. and Corlett, J. (2005) *Information Sources in Engineering* (4th edn), New York: Bowker-Saur. Another title in the "Guides to Information Sources" series includes experts presenting and evaluating the variety of reference sources in engineering, from conferences to standards.

The Reader's Advisor. Vol. 5: The Best in Science, Technology, and Medicine (1994) (14th edn), New Providence, NJ: Bowker. Part of a series aimed at advising the librarian and the reader about the best guides in the literature, this volume lists the classics useful for collection development in science, technology, and medicine.

Schenk, M.T. and J.K. Webster (1984) *What Every Engineer Should Know about Engineering Information*, New York: Marcel Dekker. A good introduction to basic research information.

Walker, R.D. and C.D. Hurt (1990) *Scientific and Technical Literature: An Introduction to Forms of Communication*, Chicago, IL: American Library Association. Engineers must know how to communicate their research. This book offers a philosophical approach to scientific research, but shows an engineer the vital forms, such as patents, technical reports, maps, conference proceedings, and journals.

Directories

Associations Unlimited (online) or *Encyclopedia of Associations* (print) (2005) Detroit, MI: Thomson Gale. International listing of associations by name, location or subject. Includes contact information, publications, membership, and so on.

International Directory of Engineering Societies and Related Organizations (1999/2000) (16th edn), Washington, DC: American Association of Engineering Societies. Print copy ceased with this edition. It provides a clue to engineering societies and related organizations that existed at the turn of the century.

ThomasNet <http://www.thomasnet.com> (accessed May 31, 2005). This is an electronic version of *Thomas Register*, New York: Thomas Industrial Network. The two allow searching of products and companies, CAD drawings and more. A free site that requires registration. see Chapter 14 on industrial and manufacturing engineering.

Who's Who in Engineering, (1995) (9th edn), New York: American Association of Engineering Societies. Originally published by the Engineers Joint Council, this edition seems to be the last published by the American Association of Engineering Societies. Over 15,000 engineers are included, to the exclusion of listing the numerous awards presented in the engineering fields. Even limiting to those who fit the posted criteria makes this edition almost unwieldy to publish. The Association website at <www.aaes.org> (accessed May 18, 2005) lists statistical and salary survey reports, rather than directories.

Who's Who in Science and Engineering (2005–2006) (8th edn), New Providence, NJ: Marquis Who's Who. Biographies of over 40,000 scientists, doctors, and engineers who submitted profiles to Marquis are entered in this series. Even some names from the "soft sciences" representing sociologists, economists, and psychologists have been included, plus major executives from technology and science-related businesses.

Encyclopedias and dictionaries

ASTM Committee on Technology (2000) *ASTM Dictionary of Engineering, Science, and Technology* (9th edn), West Conshohocken, PA: ASTM. Committee E02 on Terminology has sponsored the editions of the *Compilation of ASTM Standard Definitions*. The Committee changed the name of the latest edition to *Dictionary*, a more fitting title than the previous one, since the volume serves as a resource that defines engineering terms.

Brady, G.S. and H.R. Clauser (1991) *Materials Handbook* (13th edn), New York: McGraw-Hill. Subtitle provides an accurate description – "an encyclopedia for managers, technical professionals, purchasing and production managers, technicians, supervisors, and foremen."

Bucher, W. and C. Madrid (1996) *Dictionary of Building Preservation*, New York: Preservation Press. This is an example of a dictionary that sounds specialized, but which provides useful definitions of unusual engineering terms in construction, and includes illustrations and line drawings from the Historic American Buildings Survey.

Cardarelli, F. (2000) *Materials Handbook: A Concise Desktop Reference*, New York: Springer. This concise version provides quick access to physical and chemical properties of all classes of materials. Topically arranged, the usability of the chapters is reinforced with an index and a bibliography.

Elsevier's Dictionary of Engineering: in English/American, German, French, Italian, Spanish and Portuguese/Brazilian (2004) New York: Elsevier. Edited by M. Bignami, over 10,000 terms are listed in this translation source.

Ernst, R. (1985) *Comprehensive Dictionary of Engineering and Technology* (2 vols), New York: Cambridge University Press. The set serves as a useful tool for translations. Volume 1 covers French–English and Volume 2, English–French.

Ernst, R. (1989) *Dictionary of Engineering and Technology, with Extensive Treatment of the Most Modern Techniques and Processes* (5th edn), New York: Oxford University Press. This multi-volume set covers German–English and English–German and serves as a useful tool for translators.

Eshbach, O.W., B.D. Tapley and T.R. Poston (1990) *Eshbach's Handbook of Engineering Fundamentals* (4th edn), New York: Wiley. From mathematics to properties of materials, this handbook covers the formulas needed in all engineering disciplines.

Harris, C.M. (1993) *Dictionary of Architecture & Construction* (2nd edn), New York: McGraw-Hill. Good source for terms that could apply either to architecture or engineering.

Heisler, S.I. (1998) *The Wiley Engineer's Desk Reference: A Concise Guide for the Professional Engineer* (2nd edn), New York: Wiley. This volume includes the resources that the engineer needs for practical applications and daily problems, from mathematics to structures. A number of helpful tables range from a guide on writing for publication to the periodic table of elements.

Keller, H. and Erb, U. (2004) *Dictionary of Engineering Materials*, New York: Wiley. This comprehensive resource covers 25,000 materials and trademarks in the dictionary format, and includes an appendix listing databases, reference materials, and websites. Whenever possible, the trademarks list specific ingredient percentages.

Knight, E.H. (1979) *Knight's American Mechanical Dictionary: Being a Description of Tools, Instruments, Machines, Processes, and Engineering; History of Inventions; and Digest of Mechanical Appliances in Science and the Arts* (limited edn) (3 vols). Missouri: Mid-West Tool Collectors Association: Early American Industries Association. Reprint of the 1876 edition published by Hurd and Houghton, New York, this set includes one supplementary volume and is noted for its thousands of illustrations in a wide range of science and technology subjects.

McGraw-Hill Dictionary of Scientific and Technical Terms (2003) (6th edn), New York: McGraw-Hill. First published in 1974, the latest edition covers 5,000 new terms. Credit is given to numerous sources, many of them federal government publications.

McGraw-Hill Encyclopedia of Science and Technology (2002) (9th edn) (20 vols), New York: McGraw-Hill. This set is a good starting point for unfamiliar terms, since the extensive index leads to short, concise articles, ending with sources that provide additional information. A single-volume complement is the *McGraw-Hill Concise Encyclopedia of Science and Technology* (2004) (5th edn). Also available as an electronic resource (*Access Science*).

Parker, S.P. (1993) *McGraw-Hill Encyclopedia of Engineering* (2nd edn), New York: McGraw-Hill. About 700 articles were adapted from the (7th edn) (1992) of the *McGraw-Hill Encyclopedia of Science and Technology* and published in this single volume.

Thomann, A.E. (1990) *Elsevier's Dictionary of Technology: English–Spanish* (2 vols), New York: Elsevier. This two-volume set serves as an example for translating English terminology into Spanish. Its extensive contents are published in two parts.

Thomann, A.E. (1993) *Elsevier's Dictionary of Technology: Spanish–English* (2 vols), New York: Elsevier. This companion volume serves those translating technical terms from Spanish into English.

Webster, L.F. (1995) *The Contractors' Dictionary of Equipment, Tools, and Techniques*, New York: Wiley. Over 30,000 words are defined and described in the fields of civil engineering, construction, forestry, open-pit mining and public works. It includes conversion factors.

Handbooks and manuals

Berinstein, P. (1999) *The Statistical Handbook on Technology*, Phoenix, AZ: Oryx Press. Statistical information is compiled on a number of technical subjects. The book serves as a tool for watching new technologies emerge, since the numbers can serve as a baseline for flourishes or declines in technology.

Bies, J.D. (1986) *Mathematics for Mechanical Technicians and Technologists: Principles, Formulas, Problem Solving*, New York: Macmillan. Audel was a publisher of practical books for the technician. This one retains the series title "an Audel book," since it is a useful, practical guide for technical personnel. It contains the mathematical formulas for solving everyday problems, from cams and gears to economics in machining, and includes answers to the exercises.

Darton, M. and J. Clark (1994) *The Macmillan Dictionary of Measurement*, New York: Macmillan. The dictionary format includes 4,250 entries on weights and measures, values, volumes, frequencies, temperatures, and speeds. It includes a theme index.

Dorf, R.C. (2005) *The Engineering Handbook* (2nd edn), Boca Raton, FL: CRC Press.

A ready reference for practicing engineers includes key equations, mathematical formulae and tables, besides a wealth of sections covering all the engineering fields. It remains an informative, well-organized source of fundamental knowledge.

ENGnetBASE CRC Press <http://www.engnetbase.com> (accessed May 31, 2005). Electronic collection of the CRC handbooks, it provides full text searching of these valuable reference works.

Fischer, K. (1999) *Handbook of Technical Formulas*, Cincinnati, OH: Hanser Gardner. English translation of *Taschenbuch der Technischen Formeln*, originally published in 1996, includes formulas and tables, plus bibliographic references.

Ganic, E.N. and T.G. Hicks (2002) *The McGraw-Hill's Engineering Companion*, New York: McGraw-Hill. This book helps an engineer with the information needed in daily operations with handy mathematical formulas, measurement tables, engineering units, and general properties of materials. Also available as an e-book.

Gieck, K. (1997) *Engineering Formulas* (7th edn), New York: McGraw-Hill. This translation of *Technische Formelsammlung* contains all the formulas an engineer would normally use.

Glover, T.J. (2002) *Pocket Ref* (3rd edn), Littleton, CO: Sequoia Publishing. One example of a handy tool for the hip pocket is this little tome that includes tables ranging from those written by the American Society of Heating, Refrigeration and Air-conditioning Engineers' committees to *Water Well Handbook.* The tables go from sandpaper grit to glue types. Sources are included.

Grazda, E.E., M. Brenner and W.R. Minrath (1966) *Handbook of Applied Mathematics* (4th edn), Princeton, NJ: Van Nostrand. In spite of its age, this handbook remains useful in its clear and practical pages on shop mathematics, business mathematics, and engineering mathematics.

Hicks, T.G., S.D. Hicks, and J. Leto (2005) *Standard Handbook of Engineering Calculations* (4th edn), New York: McGraw-Hill. Each chapter is devoted to one of seven specialized engineering fields (civil, architectural, mechanical, electrical, chemical and process plant, water and waste-water, and environmental).

Knovel <http://www.knovel.com> (accessed May 31, 2005). Knovel is a large and growing collection of reference works in science and engineering from a number of publishers. Full text, deep searching is possible. Value-added tools allow manipulation and analysis of the results.

Kurtz, M. (1991) *Handbook of Applied Mathematics for Engineers and Scientists*, New York: McGraw-Hill. Provides a practical approach to mathematics.

Tuma, J.J. and R.A. Walsh (1998) *Engineering Mathematics Handbook* (4th edn), New York: McGraw-Hill. Newer and more comprehensive than Tuma's *Technology Mathematics Handbook*, this volume could also serve as a textbook. Useful appendices cover numerical tables, glossary, units of measurement, and sample problems.

Journals and periodicals

Few periodical titles are exclusively about general engineering. Even the classic *Engineering News-Record (ENR)* and *Engineering (London)* emphasize civil engineering, since their contents cover the construction industry more than any other field. To be knowledgable about engineering is to know current events and trends in science and technology. Reading some of the science and technology periodicals can be helpful. *Science* and *New Scientist* are examples of newsworthy titles. Those interested in design and inventions would enjoy reading *American Heritage of Invention and Technology*. Some engineering schools publish magazines, such as MIT's *Technology Review*. Some titles may be found online, an increasingly popular option. Two sources to help locate titles in engineering include *The Standard Periodical Directory* and *Ulrich's Periodicals Directory*. Once a title is found, the OCLC *WorldCat* database helps to find the closest library with holdings.

Periodical titles

American Heritage of Invention and Technology. (1985–) Quarterly. New York: American Heritage (8756–7296). American Heritage partnered with the National Inventors Hall of Fame in providing a quarterly magazine to encourage creativity and recognize inventiveness. The magazine provides a mechanism for telling the story of the cultural impact of technology and innovation.

Engineering (London) (1866–) Monthly. London: Gillard Welch Associates (0013–7782). The issues cover innovation in technology, manufacturing, and management. Subscription information is available at the website <www.engineeringnet.co.uk> (accessed May 18, 2005).

ENR (1874–) Weekly. New York: Engineering News-Record (0891–9526). Former title was *Engineering-News Record*, which is still the publisher, a subsidiary of McGraw-Hill Construction Information Group. *ENR* may be considered a construction magazine, but it also contains general business information for engineers and contractors. For more information, go to <www.enr.com> (accessed May 18, 2005).

New Scientist (1956–) Weekly. London: Reed Business (0262–4079). Latest news and hot topics found worldwide. Good source for keeping up-to-date in the technical world. The website <www.newscientist.com> (accessed May 18, 2005) provides full text articles.

Science (1880–) Weekly. Washington, DC: American Association for the Advancement of Science (0036–8075). *Science* provides news of recent international developments and research in all fields of science. A subscriber must belong to the American Association for the Advancement of Science (AAAS) in order to receive *Science* in paper, on CD-ROM, or online at <www.scienceonline.org> (accessed May 18, 2005).

Science News (1921–) Weekly. Washington, DC: Science Service (0036–8423). Its mission is to advance the understanding and appreciation of science through publications and educational programs. Full text is available from a number of sources. Its website is <www.sciencenews.org> (accessed May 18, 2005).

Technology Review (1899–) Ten times per year. Cambridge, MA: Massachusetts Institute of Technology (1099–274X). Its banner heading is "MIT's magazine of innovation." Its strength lies in reporting on emerging technologies, their impact and potential in the marketplace. For more details, go to <www.techreview.com/> (accessed May 18, 2005).

Periodical directories

The Standard Periodical Directory (2004) (27th edn), New York: Oxbridge (0085–6630). Published annually since 1964, this guide covers U.S. and

Canadian periodicals. Although there is a ten-page section on engineering, *ENR* is listed under construction, so some scouting helps among the categories. Title index leads to the subject categories. It is online at <www.mediafinder.com> (accessed May 18, 2005).

Ulrich's Periodicals Directory (2004) (42nd edn), New Providence, NJ: Bowker (0000–2100). This multi-volume set has provided international periodical information since 1932. It includes a classified list of serials, annuals, and irregular publications with an alphabetical title index. Other indexes, such as ceased titles, and International Standard Serial Number (ISSN) listings are also useful.

Electronic sources for periodicals

Fulltext Sources Online (1989–) Semiannual (16th edn) (2004), Medford, NJ: Information Today, Inc. (1040–8258). Each complete, cumulated directory provides sources for finding over 22,000 periodicals, newspapers, newswires, and TV or radio transcripts with full text online. Input comes from twenty-eight major aggregator products. Online edition is updated weekly at <www.fso-online.com> (accessed May 18, 2005).

Worldcat <http://newfirstsearch.oclc.org> (accessed May 18, 2005). One of the OCLC's FirstSearch databases, *Worldcat* is the world's most comprehensive bibliography covering over forty million bibliographic records representing 400 languages. It covers the contents of member libraries worldwide. Some libraries include serial holdings while others list only the title, but this database remains a good search tool for obscure bibliographic information.

Electronic research

Traditional resources are moving increasingly toward electronic versions. Some publishers offer a choice, paper or electronic. Some services are now available exclusively on the Internet. The National Academy of Engineering still publishes the results of research in book format, but its website provides full text to many documents with links in the references to online resources. Myriad attempts at providing resources and databases online are occurring, thus making the future look interesting and challenging. Thirty years ago Dialog provided files of databases from numerous information publishers. Competition arose and today there are numerous choices in searching online. Joint efforts range from consortiums such as OhioLink that provide online resources to subscribers, to search engines such as Google that show links to numerous sites and to books available from Amazon.com. Even *WorldCat* in OCLC sometimes cites Amazon as a link. *JStor* and *Project Muse* offer archives of older journal articles. CrossRef developed as an online consortium among a number of technical publishers. It provides linking between online publications. All are subscription-based.

What is a database? A database is a collection of citations, often with abstracts or full text, concerning a specific subject area. These citations are usually to journals in a given field and often include conference papers. Some specialized databases include *Ei Compendex**, the electronic version of *Engineering Index*, and Inspec, the resource published by the Institution of Electrical Engineers. Both were early files on Dialog. Both started print versions before the twentieth century, are starting points for in-depth research, and provide thesauri to aid in searching techniques.

How authoritative are databases? Professional societies and recognized publishers can be easily accepted as authoritative developers of databases. With the increase of information available through the Internet, guidelines for analyzing an online source are important. Who developed and maintains the site and how current it is are some of the questions to ask. An understanding of the structure of the Web helps. Tutorials assist in developing search skills and analyzing search engines. Sherman and Price's book on the invisible Web offers an even deeper approach to web-searching than just knowing standard universal resource locators (URLs). Other books show the development of the Internet, help with electronic terms and acronyms, provide an understanding of how to search the Web, and give some hints from those who have devised short cuts to interesting resources. One print and online resource, the *Gale Directory of Databases*, gives a description of all online vendors, producers, databases, and electronic search tools on the market today. With the increase in the number of web resources, it is useful to have a basic identification of Internet databases, search engines, and so on.

Books

Abbate, J. (1999) *Inventing the Internet*, Cambridge, MA: The MIT Press. This is a readable, extensive history of the Internet starting with packet-switching and the cold war in the 1960s, the U.S. Department of Defense Advanced Research Projects Agency (ARPANET) to today's global picture. An extensive bibliography leads to additional information on the subject.

Bidgoli, H. (2004) *The Internet Encyclopedia* (3 vols), Hoboken, NJ: Wiley. This three-volume work encapsulates the Internet, from active server pages (ASP) to wireless technology. In encyclopedic fashion, chapters include cross-references and references for further reading.

Calishain, T., R. Dornfest and D.J. Adams (2003) *Google Pocket Guide*, Sebastopol, CA: O'Reilly. How to understand what Google can and cannot do is the theme of this helpful little book.

Cooke, A. (1999) *Neal-Schuman Authoritative Guide to Evaluating Information on the Internet*, New York: Neal-Schuman. This book serves as an aid to evaluating information on the Internet. In spite of its date, it helps in deciphering what is a worthwhile Internet source.

Gale Directory of Databases (2004) (2 vols in 4 parts), Detroit, MI: Gale Research, Inc. This semiannual multi-volume set grew from a single volume published in 1979 and compiled by Martha Williams, who offers a beginning essay on the state of databases today in the first volume of the 2004 edition. Comprehensive in its coverage of the electronic services on the market, this set covers over 18,000 databases, over 4,000 producers, and some 3,300 online services and vendors or distributors of database products. This product is also available as an online subscription.

Hock, R. (2004) *The Extreme Searcher's Internet Handbook: A Guide for the Serious*

Searcher, Medford, NJ: CyberAge Books. With a foreword by Gary Price, this book serves as a basic tool for general information on the Web. It includes specialized directories, search engines, news resources, finding products online, and how to publish on the Internet.

Langford, D. (2000) *Internet Ethics*, New York: St. Martin's Press. Thoughts from an international team of specialists, each chapter forms an essay on a specific legal aspect of the Internet. More issues and discussion may be found in the extensive bibliography.

Moschovitis, C.J.P. (ed.) (1999) *History of the Internet: A Chronology, 1843 to the Present*, Santa Barbara, CA: ABC-CLIO. Capsule chronologies of telecommunications from Charles Babbage to the Internet's impact in 1998. Dated, but the early history is interesting.

Sherman, C. and G. Price (2001) *The Invisible Web: Uncovering Information Sources Search Engines Can't See*, Medford, NJ: Information Today. Chapters give a history of the Internet and the visible Web. The contents include all kinds of searching possibilities, numerous suggestions of web sources arranged topically, and a detailed index.

Online tools and indexes

Abstracts in New Technologies and Engineering (ANTE) (1997–) Bimonthly. East Grinstead, Sx: Bowker-Saur. The latest title to the British equivalent of *Applied Science and Technology Index* and *Engineering Index*, its predecessor title was *Current Technology Index*. The title is descriptive of the content as it covers all the engineering and technology fields. It is bimonthly with annual cumulations, and is available as an online subscription through Cambridge Scientific Abstracts.

Applied Science and Technology Index (1958–) Monthly. Bronx, NY: H.W. Wilson Co. One online equivalent with abstracts is *Applied Science and Technology Abstracts*, and is a basic tool for searching English-language periodical articles in numerous technical subjects. The paper edition is published ten times a year with cumulations and a permanent annual edition. *Applied Science and Technology Full Text* is an online combination of the abstracts and the index. Subscriptions are currently available for a CD-ROM version. For information contact WilsonWeb or WilsonDisk at <www.hwwilson.com> (accessed May 18, 2005).

Cambridge Scientific Abstracts (CSA) (2002–) <www.csa.com> (accessed May 18, 2005). Bethesda, MD: Cambridge Scientific Abstracts. Subscriptions are available in numerous combinations for major technical categories, ranging from aeronautics to zoology. Contents in the categories cover a wide range of journals. CSA is continually expanding as a producer and provider of information databases. Some classic indexes, such as *Metals Index (Metadex)*, may be accessed through this service.

CrossRef (2000–) <www.crossref.org> (accessed May 18, 2005). Lynnfield, MA: Publishers International Linking Association. Founded by several international publishers, *CrossRef* is designed as the official digital object identifier (DOI) site to link bibliographic citations, abstracts and full text together. DOI is an open standard, created to facilitate online searching among the publications from a number of scientific publishing houses. Jointly, the publishers created the Publishers International Linking Association (PILA). Additional information is available on the website.

Dialog Corporation (1973–) <www.dialog.com> (accessed May 18, 2005). Cary,

NC: Dialog Corporation. One of the oldest database vendors in the world, Dialog provides access to numerous business and technical files in which to search for information and numerous aids to help in searching. The Corporation's newsletter, *Dialog Chronolog Summary*, was first published in 1973, when Dialog was owned by Lockheed Information Systems. In 2004, *Chronolog* (ISSN 0163–3732) shows that Dialog is a subsidiary of Thomson.

Dissertation Abstracts: see *ProQuest Dissertations and Theses.*

EEVL <http://www.eevl.ac.uk/> (accessed May 18, 2005). Formerly called the *Edinburgh Engineering Virtual Library* and sometimes referred to as the *Enhanced and Evaluated Virtual Library*, *EEVL* was developed in the United Kingdom. This free source is run by a team of information specialists from Heriot-Watt University, which partnered with Cranfield University, University of Birmingham, and University of Ulster. The website is sponsored by Adept Scientific. It includes an Internet bibliography, events databases, and links to other technical sites.

Engineering Index (1884–) Monthly. Stockton, NJ: Engineering Information, Inc. (Ei). This comprehensive interdisciplinary engineering service has an electronic version called *Ei Compendex*. Records date back to 1884, depending on the agreement with the publisher. Both print and electronic versions cover the entire spectrum of engineering, with abstracts from over 5,000 international journals, conferences, proceedings, and technical reports. The online subscription is available through several vendors. Ei, now owned by Elsevier, provides a powerful electronic platform in engineering fields in its Engineering Village. For more information, access Engineering Village 2, at <www.engineeringvillage2.org> (accessed May 18, 2005).

IEEE/IEE Electronic Library/IEEEXplore (IEL) <http://ieeexplore.ieee.org> (accessed May 18, 2005). Piscataway, NJ: Institute of Electrical and Electronic Engineers (IEEE). A technical information database that includes all the journals and conference proceedings produced by IEEE, indexing the literature back to 1988 and some selections back to 1950; IEEE Xplore offers numerous subscription packages. Browsing the table of contents and abstracts of IEEE transactions, journals, conference proceedings, and standards is free. Accounts may be set up for online subscriptions.

Inspec (1968–) <www.iee.org> (accessed May 18, 2005). Stevenage, Herts: UK or Edison, NY: Institution of Electrical Engineers. Founded by the Institution of Electrical Engineers (IEE), the *Inspec* database was created in 1968 based on *Science Abstracts* (1898–) which split into three sections: *Physics Abstracts* (ISSN: 0036-8091), *Electrical & Electronic Abstracts* (ISSN: 0036-8105), and *Computer & Control Abstracts* (ISSN: 0036-8113). It is updated weekly and is available online through vendors or in CD-ROM format. At present the paper indexes covering physics, computer science, engineering, science, and technology are still published. The numerous tools that aid in refining search techniques are listed below.

Inspec Classification (2004). This publication provides the period of use of each classification entry and indicates previous codes. The codes help in refining a search. An index leads to the appropriate codes. *Inspec List of Journals* (2004). A valuable tool that lists all the serials scanned for the *Inspec* database. It is useful for locating titles in the scientific literature.

Inspec Thesaurus (2004). Because the *Inspec* database uses controlled terms, this thesaurus aids in structuring the search strategy and shows relationships between terms.

Inspec Search Aids on CD-ROM (2004). The electronic version of the *Inspec Thesaurus*, classification and list of journals may be purchased as a stand-alone or as a network product.

Journal Storage Project (JSTOR) (1995–) <www.jstor.org> (accessed May 18, 2005), Ann Arbor, MI: University of Michigan. This site is a large digital archive of scholarly journals, providing full-text access in several scientific fields. Started as a pilot project sponsored by the Mellon Foundation to provide electronic access to the back files of ten journals, it has expanded to include a large range of scholarly titles, a number of which are in engineering and technology.

National Academies <http://www.nas.edu/> (accessed May 31, 2005) includes the National Academy of Sciences, the National Academy of Engineering, and the Institute of Medicine. Private and non-profit organizations, they operate to advise the federal government on science and technology issues. The National Research Council is the "principal operating agency of both the National Academy of Sciences and the National Academy of Engineering." National Academy of Engineering <http://www.nae.edu> (accessed May 31, 2005) has the mission "to promote the technological welfare of the nation by marshaling the knowledge and insights of eminent members of the engineering profession." News links are an excellent means to monitor some engineering news. Results of their findings are published in hard copy and may be found on the website.

National Technical Information Service (NTIS) (1964–) <http://www.ntis.org> (accessed June 15, 2005). Indexes government-sponsored research in all areas of science and technology. Some international reports are also included. Database available from several services. Reports may be ordered individually or by subscription; some reports available full text on the Web.

Project MUSE (1995–) <http://muse.jhu.edu/about/index.html> (accessed May 18, 2005). Baltimore, MD: Johns Hopkins University Press. Launched in 1995 by the Johns Hopkins University Press to offer full text of scholarly journals archived with a moving wall via the Web. Tables of contents and sample articles may be viewed without a subscription to the journal titles published by the university presses at Johns Hopkins, Duke University, Carnegie Mellon, Indiana, and the University of Hawaii.

ProQuest Dissertations & Theses <http://www.umi.com/umi/dissertations/> (accessed June 15, 2005). Database of over two million dissertations and theses from over 1,000 universities.

Scirus <http://www.scirus.com/srsapp/> (accessed May 5, 2005). New York: Elsevier. The home page describes itself as "the most comprehensive science-specific engine on the Internet." It searches only for specific science web pages using academic, company, societies' websites, scientists' home pages, conferences, patents, products, and e-prints/preprints. From the advanced search page, searches may be limited to specific subjects, content sources, or information types. More sophisticated searching tools allow use of Boolean connectors, field abbreviations, and phrase searching. An international advisory board provides guidance and aid in ongoing development.

Scopus <http://info.scopus.com> (accessed May 31, 2005). A product created by Elsevier, "Scopus is a multidisciplinary navigational tool that contains records going back to the mid 1960s, offering newly-linked citations across the widest body of scientific abstracts available in one place." There are over 14,000 titles in the database. Subscription is required (from the website).

Scout Report <http://scout.wisc.edu/> (accessed May 31, 2005). Hosted at the University of Wisconsin and funded by the National Science Foundation, the *Scout Report* provides evaluated, annotated updates of online information. A must to keep up with changes in websites. Select the "Math, Engineering, and Technology" link.

Search Engine Showdown: The Users' Guide to Web Searching <http://www.searchengineshowdown.com/> (accessed May 18, 2005). A Notess.com website copyrighted by Greg R. Notess starts with search engine news. The site has several features that will help in understanding how to search the Web.

ScienceDirect (1997–) <www.sciencedirect.com> (accessed May 18, 2005). New York: Elsevier. Starting as a web database of Elsevier journals, this has become one of the largest electronic collections of science, technology, and medicine in full-text bibliographic format. Several licensing arrangements are possible.

STN <http://www.cas.org/stn.html> (accessed June 15, 2005). A collection of over 200 scientific and technical databases may be searched for a fee.

Technology Research. Cambridge Scientific. Over 5,000,000 records to articles, proceedings, reports, and patents in engineering and technology. Similar to but less extensive than *Compendex.*

US Government Agencies <http://Firstgov.gov> and <http://Science.gov> (accessed May 18, 2005). Although Firstgov.gov is the website for the federal web pages, Science.gov is the portal for searching government science agencies. It provides links to several science websites that range from agriculture to science education and allows deep searching of the sites.

University of California, Berkeley Library. *The BEST Search Engines* <http://www.lib.berkeley.edu/TeachingLib/Guides/Internet/SearchEngines.html> (accessed May 18, 2005). "Finding Information on the Internet: A tutorial" is the topic of this online tutorial found at UC Berkeley. This site compares the bigger search engines, gives recommendations, and describes how search engines work. This citation is included as one example of what universities develop to help students in their online searching.

Conferences

Conferences form main sources of information in the engineering field, as a conference, a symposium, or a meeting gives an engineer the chance to express work in progress or research accomplished. They also give peers the chance to voice opinions, and they provide useful forums to listen and learn. They enable participants to discuss developments and present new ideas. The results often appear as published proceedings of a conference or a symposium, thus providing a time frame for work in progress or aiding future researchers from reinventing the wheel. Some are regular events while others vary in dates, sponsors, and format. Some are considered gray literature. Some contents of sessions never get published in their entirety. The challenge lies in trying to find the one that is needed.

Papers presented at a session may appear as one print document, as individual papers, in electronic format, or only as abstracts. Most of the engineering conferences fall into specific fields of interest. There are resources and guidelines that help develop the search strategy for locating any conference or paper within a conference. Two titles, available in print and in online subscriptions, include the *Directory of Published Proceedings* and *Index to Scientific and Technical Proceedings (ISTP).* The *Directory* identifies the conference, the place, and the sponsor. *ISTP* includes tables of contents to the conferences, which help in identifying a paper within a given conference. The OCLC *WorldCat* database locates the library having a title of a conference.

Some societies append a number to each paper. Attempts to index by number vary by society. SAE International provides ample resources to make retrieval possible by using the assigned number. That society produces an online digital library or a CD-ROM subscription. Print indexes cover SAE literature from 1906 to 1993 and help to locate older papers. The Institution of Electrical and Electronic Engineers (IEEE) added a file number to each of its numerous conferences. Several years ago, the library staff at General Electric published an index to the American Society of Mechanical Engineers (ASME) papers that appeared in the ASME journals. For a number of years, the American Society of Civil Engineers arranged its publications by using a coding system for each section within its proceedings. Subsequently, those sections became distinct journal titles. Searching the older years can present a challenge to those unfamiliar with the system and the titles.

Searching for reports that may or may not have appeared in a conference is also challenging. Report literature tends to have filing codes with letters and numbers. Many reports written under contract to a federal government agency start with distinctive letters identifying the specific agency. Others, such as the Rand Corporation, assign obvious letters for its research publications, papers, and monographs, and add numbers at the beginning of a project. Godfrey's *Dictionary of Report Series Codes* helps locate many that were results of research from World War II and later.

A unique resource is the *Science Citation Index* with paper subscription back to 1945 and electronic to 1964. With its multiple indexes, an author may be searched for published articles and where those articles are cited in the scientific literature. The unique features enable the Institute for Scientific Information to provide a bibliometric analysis of science journals in their database to list the most frequently used journals and those journals with the highest impact. This analysis is published as *Journal Citation Reports.*

Sources

Directory of Published Proceedings: Series SEMT: Science, Engineering, Medicine, and Technology (1965–2005) Harrison, NY: InterDok) (0012–3293). Publication started in 1965 and is published ten times a year, with annual and four-year cumulations. Series SEMT covering science, engineering, medicine, and technology is one of several series published by InterDok that provides citations to published conferences. Print edition is ceasing in 2005 and series is exclusively online. Coverage is international and contents are arranged by date of conference. Issues may be searched by name of conference, sponsors, and location. Future conferences may be searched at InterDok's website for its electronic database, *MInd: the Meetings Index, at* <http://www.interdok.com/mind/> (accessed May 18, 2005).

Godfrey, L.E. and Redman, H. (1973) *Dictionary of Report Series Codes* (2nd edn),

New York: Special Libraries Association. Older but still useful, the *Dictionary* provides a listing of report initials and numbers to research conducted by numerous agencies that contracted with the U.S. government after World War II.

Index to Place of Publication of ASME Papers 1950–1977 (1981) Schenectady, NY: General Electric Company, Technology Marketing Operation. Compiled by the staff at General Electric's main library at Schenectady, the *Index* provides a cross-reference between preprint numbers and corresponding journal volume, pages, and dates. If the number is not listed, then the paper was probably not published in ASME journals from 1950 through 1977.

Index to Scientific and Technical Proceedings (1978–) Philadelphia, PA: Thomson (0149–8088). Produced by the Institute for Scientific Information, this service provides a table of contents for published conference proceedings. Several indexes cover category, author or editor, sponsor, location, corporate, and subject of conference. It is available also in electronic formats.

Journal Citation Reports (JCR), a Bibliometric Analysis of Science Journals in the ISI Database (science edn) (1989–) Philadelphia, PA: Institute for Scientific Information, Inc. (1524–5047). This annual analysis contains statistical information on nearly 6,000 of the world's leading journals. The coverage tells which are the largest journals, the most frequently used, and those with the highest impact. The data may be used for collection-building or for citation value. Microfiche accompanies the printed guide, or the resource may be purchased electronically.

Science Citation Index (1945–) Philadelphia, PA: Thomson (0036–827X). Produced by the Institute for Scientific Information six times a year, this unique series provides sources of where and how often an author is cited. The index covers the literature back to 1945, and may be searched by subject, author, journal, and/or author address. Coverage includes more than 3,700 major journals across 100 scientific disciplines. The online version, *SciSearch*, is part of *Web of Science*, which also includes *Social Sciences Citation Index* and *Arts and Humanities Citation Index*. It is updated weekly. Format and delivery options may be found on the website: <www.isinet.com/products/citation/sci/> (accessed May 18, 2005).

Scientific Meetings (1977–) San Diego, CA: Scientific Meetings Publications (0487–8965). With the slogan "The International Source for Meeting Information," this quarterly continues *Scientific Meetings* published by the Special Libraries Association. It lists future technical, medical, scientific, and management meetings. Arranged by title/sponsor, date, location, subject, it is also available on microfilm or on disk for IBM and compatible PCs in several formats.

Industrial standards

Engineers depend on industrial standards for their work. Standards also simplify our daily lives. Transportation engineers ease driving by following the *Manual on Uniform Traffic Control Devices* (also included in Chapter 20, Transportation engineering). Civil engineers surface highways following standards developed by the American Concrete Institute. Other engineers conform to building codes to erect safe structures. Electricity works safely, thanks to the National Electric Code. Clocks show time measurement through standardization. Thompson's update of Batik's *A Guide to Standards* explains the background and development of industrial standards. Another simple guide on understanding the standards process is *Standards Make the*

Pieces Fit, a pamphlet written and published by the American Society for Testing and Materials in 1985.

Standards developing organizations identify standards by test methods, guidelines, recommended practices, codes, specifications, and even manuals. Usually the names of the societies change into acronyms, so books on abbreviations are useful for identifying the myriad acronyms. Single volumes such as De Sola and multiple volumes (e.g. the set compiled by Peschke) identify the initials or acronyms. Online sources provide another approach to identification. Information Handling Service developed its Engineering Resource Center as a site that could be used to identify a specific standard or perform subject searches.

One area of confusion can be distinguishing report numbers from specification numbers. Tools, such as Godfrey and Global listing document codes and sources, help to distinguish a report from a specification. Engineers and librarians confuse letter and number combinations that are not standards but grades of metal. Indexes and handbooks help by listing alloys to clear that confusion. *ILI Metals Infobase* identifies specific numbers, gives chemical content, and includes equivalencies. *Metals & Alloys in the Unified Numbering System* correlates many international number systems.

In the United States, one starting point to learn about standards is the American National Standards Institute (ANSI) website <www.ansi.org/>. Another is the National Information Standards Organization (NISO). NISO identifies and publishes information management standards, such as Z39.50. Accessible at its website <www.niso.org> is the document, *Understanding Metadata*. This paper explains the complexity of retrieving, managing, and storing information in traditional ways (e.g. formatting catalog cards), and, in newer ways (e.g. accessing electronic resources). Not only is ANSI the body for approving domestic standards, but it also serves as the United States international representative to standards developing organizations. Over the past decade it has developed NSSN, the national resource for global standards database located at <www.nssn.org>. Originally named the National Standards Systems Network, it grew beyond "national" but retained its acronym.

ANSI serves as the U.S. representative to the International Organization for Standardization (ISO), an organization that serves as the world's largest developer of standards. From testing to trading a product, officials could not operate without the network of standardization provided by ISO. International standards provide a common technological language between suppliers and customers. The World Trade Organization (WTO) recognizes the preparation, adoption, and application of standards. By understanding international standards, engineers have the framework to succeed in international projects. Engineers working in companies that want certification or have global concerns need to be familiar with the ISO 9000 and the ISO 14000 series. There are several books that explain the implementation of those standards.

Industry standards are developed by many engineering societies, acting as standards developing organizations, each covering its own specialty. Some society members are part of technical committees developing and reviewing domestic and international standards. Other societies are members of trade associations that publish in a specific area of expertise. Published standards may range from one or two (e.g. Standards Engineering Society), to hundreds in specific disciplines (e.g. the Aerospace Materials Specifications (AMS) from SAE International).

Several federal government agencies publish safety standards. The National Institute for Occupational Safety and Health (NIOSH) addresses industrial hazards and the Occupational Safety and Health Administration (OSHA) continues with more safety standards. Other regulations come from the Environmental Protection Agency (EPA). The Federal Aviation Administration (FAA) issues safety regulations for flying. The Federal Communications Commission (FCC) regulates radio stations. Other federal branches, such as the U.S. Bureau of Mines, publish specific regulations in their area of expertise. Updates appear in the *Code of Federal Regulations*, <http://www.archives.gov/federal_register/code_of_federal_regulations/code_of_federal_regulations.html>. The OSHA website cross-references the Standard Industrial Classification (SIC) four-digit codes. Since 1997 SICs became the North American Industry Classification System (NAICS), which classifies all industry. Only a few of the federal agencies that are involved in creating standards are mentioned in this paragraph. The National Institute of Standards and Technology (NIST) provides a web source for locating federal agencies producing standards. A number of agencies, national and international, impact upon standards, but engineers must have access to appropriate standards to accomplish their jobs. Specific websites help in learning and locating standards. One example is the ISO link to other worldwide sites at the World Standards Services Network (WSSN). International Electrotechnical Commission (IEC) is also helpful in understanding about standards. Many ISOs are adopted as British Standards Institution (BSI) engineering standards. Domestic ones (e.g. American Society for Testing and Materials (ASTM)), and federal government ones (e.g. US Department of Defense and the National Institute for Standards and Technology (NIST)) provide starting points for learning more about standards and specifications. Since so many societies and organizations are involved in industrial standards, links to all their websites are helpful. Sources for linking both domestic and international sites include NSSN at <www.nssn.org>, Standards Engineering Society at <www.ses-standards.org>, the University of Kentucky at <www.uky.edu/Subject/standardsall.html>, and WSSN at <www.wssn.net> (all accessed May 18, 2005).

Books

Acronyms, Initialisms & Abbreviations Dictionary (2005) (34th edn), Detroit, MI: Thomson/Gale Research Co. Not only a guide to acronyms, abbreviations, contractions, alphabetic symbols, and condensed appellations, the set also includes a reverse edition. Almost annual, this multi-volume set is updated with supplements and is available as an online subscription.

American Society for Testing and Materials (1985) *Standards Make the Pieces Fit*, Philadelphia, PA: ASTM. Subject headings for this little gem include method engineering and industrial engineering. It is a readable, succinct description of what is an industrial standard.

Batik, A.L. (1992) *The Engineering Standard: A Most Useful Tool*, Ashland, OH: Book Master/El Rancho. This book explains why standards are often overlooked and shows the broad spectrum of the standards picture, the good, the bad, and the ugly. Batik's aim is to educate executives, engineers, and government officials on the usefulness and importance of engineering standards.

De Sola, R., D. Stahl and K. Kerchelich (1995) *Abbreviations Dictionary* (9th edn), Boca Raton, FL: CRC Press. Signs and symbols from A to Z, with appendices of special lists (i.e. airline and airport abbreviations, earthquake data, ports of the world, weather symbols, and winds of the world) that make this a source of helpful information on a variety of topics.

Directory of Engineering Document Sources (1997) Englewood, CO: Global Engineering Documents. A cross-reference guide helps to match the document initialism with the issuing organization, including many federal agencies. A third section offers a subject index to the standard developing organizations.

Erb, U. and H. Keller (1989) *Dictionary of Engineering Acronyms and Abbreviations*, New York: Neal-Schuman. An older volume, but useful for codes in the engineering field.

Fire Protection Handbook (2003) (19th edn, 2 vols), Boston, MA: National Fire Protection Association. This is a handbook for more than fire safety, as, besides being a source for the fire codes, it refers to building codes, electrical codes, and chemical substances. It is an example of a quick reference tool for engineering, published by one of the largest standards developing organizations.

Frick, J.P. and N.E. Woldman (2000) *Woldman's Engineering Alloys* (9th edn), Materials Park, OH: ASM. A classic work originally edited by Woldman and now in its 9th edition, a good source for named engineering alloys.

Index and Directory of Industry Standards (2001) (16th edn, 8 vols), Englewood, CO: Global Engineering Documents. Last in the print series, this set provides numerical and subject indexes to numerous domestic and international standards. Available only to subscribers.

Libicki, M.C. (1995) *Information Technology Standards: Quest for the Common Byte*, Boston, MA: Digital Press. API does not only stand for American Petroleum Institute, but also for Applications Programming Interface. The author shows the complexity of information standards, why machines can communicate, and the difficulties that ensue in the process. An extensive bibliography at each chapter's end leads to more information on this complex topic. In looking for the common byte, the author projects the fate of various standards in a thought-provoking timeline.

Metals and Alloys in the Unified Numbering System (2004) (10th edn), Warrendale, PA: Society of Automotive Engineers, or West Conshohocken, PA: American Society for Testing and Materials. Jointly developed by SAE International and ASTM International, this effort results in the Unified Numbering System (UNS) that correlates many international numbering systems, with cross-referencing that eases searching for a grade or a standard.

National Information Standards Organization (2004) *Understanding Metadata*,

Bethesda, MD: NISO Press. A revised expansion of *Metadata Made Simpler: A Guide for Libraries* published in 2001 and a free downloadable publication on the NISO website. Its title encompasses the content, explaining metadata and its implications.

Peschke, M. (1996) *International Encyclopedia of Abbreviations and Acronyms in Science and Technology* (17 vols), Munchen: K.G. Saur. Conceived by Peter Wennrich, this is a multi-volume set that includes reversed phrases and is updated with supplements, all translated from the German.

Ricci, P. (1992) *Standards: A Resource and Guide for Identification, Selection, and Acquisition* (2nd edn), Woodbury, MN: Pat Ricci Enterprises. A useful directory of domestic and international standards sources. Limited to dates in the 1990s, it provides a clue to help identify standards sources.

Ross, R.B. (1992) *Metallic Materials Specification Handbook* (4th edn), London and New York: Chapman and Hall. A valuable source for identifying what sounds like an industrial standard, but is really a grade of metal. Useful as a source for the chemical content of a material and for locating materials that may be similar in content.

Rothery, B. (1995) *ISO 14000 and ISO 9000*, Brookfield, VT: Gower. This book explains the difference between the two international series of standards so vital to companies that want to be recognized for quality control and European Union certification. While the ISO 9000 series provides guidelines and emphasizes the implementation of quality management systems, the ISO 14000 series extends to environmental and safety issues.

Standards Activities of Organizations in the United States (NIST Special Publication 806) (1996) Washington, DC: U.S. Government Printing Office. United States public and private sector organizations that develop, publish, and revise standards are described in detail. This is a good source for locating federal agencies that issue standards. Staff at the National Institute of Standards and Technology (NIST), formerly known as the U.S. Bureau of Standards, compiled the directory.

Thompson, D.C. (2003) *A Guide to Standards*, Miami, FL: Standards Engineering Society. An update to the 1989 edition by Albert L. Batik, this is an excellent overview to the industrial standards process from initial committee work of the standards developing organizations to litigation.

Tibor, T. and I. Feldman (1997) *Implementing ISO 14000: A Practical, Comprehensive Guide to the ISO 14000 Environmental Management Standards*, Chicago, IL: Irwin Professional Publications. Written for companies that practice environmental management standards, this guide also helps those who need to understand the complexities of the ISO series.

Electronic sources for standards

Nearly all standards-developing organizations have websites. This list shows the links to organizations and agencies that provide guides.

American National Standards Institute (ANSI) <www.ansi.org> (accessed May 18, 2005). The ANSI is a private, non-profit organization that administers and coordinates the U.S. voluntary standardization system. Although headquartered in Washington, DC, its New York City office is the point of contact for most activities. ANSI serves as the US representative to numerous international standards organizations. Its membership includes over 1,000 agencies, organizations, and governmental bodies. Its web source provides a starting point for locating domestic and global standards. It developed the National Standards Systems Network (NSSN) with ten of the largest standards-developing organizations. The service now contains information from more than 600 national, regional and international bodies. NSSN may be found at <www.nssn.org> (accessed May 18, 2005).

American Society for Testing and Materials (ASTM) <www.astm.org/standard-source.htm> (accessed May 18, 2005). Known today as ASTM International, this society is one of the oldest and largest standards-developing organizations. Founded in 1898, it now produces more than seventy-seven volumes of standards, whose contents contain standards arranged in subject groups. Its members come from over 100 countries and form technical committees that develop, review, and revise standards. The entire annual set, single volumes, or subsets are available on subscriptions, and individual standards may be purchased online.

Document Engineering Company (DECO) (1958–) <http://www.doceng. com> (accessed May 18, 2005). This vendor has been a source since 1958 for current and historical standards from US government, industry, and international standards organizations. Standards may be ordered individually or through a subscription.

ILI <http://www.ili-info.com/> (accessed May 18, 2005). Started as London Information in 1949, ILI is now a producer of nine databases and a supplier of standards. Two of the databases are *Standards Infobase* and *Metals Infobase.* Most of the current standards may be purchased through ILI. The *Metals Infobase* contains 70,000 world metal grades, their properties, related standards on which they are based, and the suppliers.

Information Handling Services <www.ihs.com> (all accessed May 18, 2005). Founded in 1959, Information Handling Services provides the world's largest collection of technical standards. The Engineering Resource Center, <www.ihserc.com>, available to subscribers, is a source for technical, energy, regulatory, and business information. Its Worldwide Standards Service includes engineering and military standards, and specifications that may be searched with time-saving tools. Items may be purchased through Global Engineering Documents <www.global.ihs.com>.

International Organization for Standardization (ISO) <www.iso.org> (accessed May 18, 2005). International standards began with the International Electrotechnical Commission (IEC) in 1906. In 1946 the International Organization for Standardization (ISO) was created to fill the need for international coordination of industrial standards. In 2004, the ISO website announced the joint ISO/IEC Information Centre <www.standardsinfo.net> (May 18, 2005) in order to facilitate world trade. ISO serves as a network of the national standards bodies of 146 countries, with a Central Secretariat in Geneva, Switzerland.

Manual on Uniform Traffic Control Devices (MUTCD) <http://mutcd. fhwa.dot.gov/> (accessed May 18, 2005). Growing from a slim document from a single society in 1927, *MUTCD* is now an accepted standard developed by several societies and the United States Federal Highway Administration, and is published as a thick loose-leaf binder. Available also online with amendments through 2003 and the comment period extended to February 1, 2005, this document provides the signs and signals used for traffic control in all fifty states. It is an example of a federal government agency and several societies, the American Traffic Safety Services

Association, the Institute of Transportation Engineers, and the American Association of State Highway and Transportation Officials, working together for traffic safety.

National Information Standards Organization (NISO) <www.niso.org> (accessed May 18, 2005). The National Information Standards Organization (NISO), founded in 1939, assumed its current name in 1984. As the name implies, NISO provides the information standards such as Z39.50 that manages information retrieval, or Z39.9 that covers the International Standard Serial Number. NISO is designated by ANSI to represent US interests on the ISO Technical Committee 46 on Information and Documentation.

National Institute for Standards and Technology (NIST) <www.nist.gov> (all accessed May 18, 2005). Formerly the National Bureau of Standards, this federal government agency provides numerous services to the industrial community. A part of the United States Commerce Department's Technology Administration, the NIST laboratories perform research and develop standards. NIST promotes and recognizes organizational performance excellence through the Baldrige National Quality Program and partners with business and the private sector to develop quality services. Within NIST is the National Center for Standards and Certification Information, a source for U.S. and international standards. It may be accessed at <http://ts.nist. gov/ts/htdocs/210/ncsci/sources.htm> (accessed May 9, 2005). To aid in locating information about the use of standards in government, NIST developed a direct portal to standards <http://standards.gov> (accessed May 10, 2005).

NSSN <http://www.nssn.org> (accessed May 5, 2005). Developed by the American National Standards Institute as a national standards systems network, NSSN has expanded into a shopping center for standards information from many sources. It serves as a gateway connecting suppliers with seekers.

Standards Engineering Society <www.ses-standards.org> (accessed May 18, 2005). Established in 1947, the Standards Engineering Society (SES) promotes the use of standards and standardization. SES members range from information specialists to standards-developing organizations. Some societies issue hundreds of standards, but currently, SES produces and publishes one entitled *Recommended Practice for Designation and Organization of Standards (SES1)*. The SES Board of Directors has proposed a second one with the tentative title of *Model Standards Development Procedure (SES2)*. The website provides links to domestic and international standards sources.

Techstreet <http://www.techstreet.com> (accessed May 18, 2005). Thomson Scientific owns this service that is an index to more than 300,000 international industry codes and standards. Brief descriptions are available for some records. Standards may be ordered individually or on a subscription basis.

U.S. government <http://firstgov.org> (accessed May 18, 2005). Firstgov provides the first stop for websites of federal government agencies. A click or two leads to the appropriate agency. Important to those needing military standards and specifications is the online site for current editions. The Department of Defense Quicksearch is <http://assist.daps.dla.mil> (accessed May 18, 2005). NIST developed <http://standards.gov> (accessed May 18, 2005) to aid in understanding the use of standards in government.

University of Kentucky <www.uky.edu/Subject/standardsall.html> (accessed May 18, 2005). The World Wide Web Subject Catalog at the University of Kentucky contains the Standards Database maintained in the Shaver Engineering Library.

The site provides links to domestic and international standards produced by standards-developing organizations and government agencies.

World Standards Services Network <www.wssn.net> (accessed May 18, 2005). This site links to a network of standards organizations around the world through alphabetical and geographical lists. Prominently displayed are the links to ISO, IEC, and International Telecommunication Union (ITU). The objective is to simplify access to international, regional, and national standards through the Web.

Patent searching

The United States Patent and Trademark Office (USPTO) has been automating for the past two decades. Brigid Quinn, Deputy Director of the USPTO's Office of Public Affairs, announced in August 2004 that patent applications are now accessible online eighteen months after being filed. The patent site <www.uspto.gov> contains a wealth of information for anyone searching patents and trademarks. Links at that site provide connections to additional resources.

There are some guides that help in basic understanding of domestic and international patents. With the changes in the Patent Office, print materials are rapidly dated, but some books are useful for their description of the patent process by showing the steps for searching prior literature and developing the patent application. The books also help to explain the U.S. design and plant patents, U.S. trademarks, and touch on the international process.

Books

Auger, C.P. (1992) *Information Sources in Patents*, New York: Bowker. Part of the series "Guides to Information Sources" this book focuses on two types of engineers, the individual inventor and the team-project one in research and development. With a decidedly British focus, Auger devotes one chapter to the United States. In preparation is the second edition by S.R. Adams (2005) as an e-book.

Gordon, T.T. and Cookfair, A.S. (2000) *Patent Fundamentals for Scientists and Engineers* (2nd edn), Boca Raton, FL: Lewis Publishers, Inc. This volume provides an overview of international patent systems for non-lawyers and describes patenting processes and principles.

Hitchcock, D. (2001) *Patent Searching Made Easy* (2nd edn), Berkeley, CA: Nolo. This book explains how to do patent searches on the Internet and in the library. Websites may need updating, but the steps outlined here will help a novice.

Jaffe, A.B. and Trajtenberg, M. (2002) *Patents, Citations, and Innovations: A Window on the Knowledge Economy*, Cambridge, MA: MIT Press. An interesting book on technology change; the authors measure the importance, generality, and originality of patented innovations.

Sharpe, C.C. (2000) *Patent, Trademark, and Copyright Searching on the Internet*, Jefferson, NC: McFarland. This book was researched and written in anticipation of the U.S. Patent and Trademark Office offering searchable databases for the general public. It serves as a guide for the lay person on what is a patent, what can be patented, and shows the different types of patents.

Walker, R.D. (1995) *Patents as Scientific and Technical Literature*, Metuchen, NJ: Scarecrow Press. For those unfamiliar with patents, the visuals help. The figures include descriptions of U.S. patents, U.S. design patents, and U.S. plant patents. Other interesting visuals show the 1601 Proclamation by the Queen, a British patent application, a page from the 5th edition of the International Patent Classification, and how the subclasses appear in the *Index of Classification.*

Wherry, T.L. (1995) *Patent Searching for Librarians and Inventors*, Chicago, IL: American Library Association. Written by a librarian who taught patent searching for thirty years, the methods described still apply.

Electronic sources

Canadian Patents Database <http://patents1.ic.gc.ca/> (accessed May 10, 2005). Developed by the Canadian Intellectual Property Office, this site covers patent data from 1920 to the present. For patents that are either laid-open documents or were granted since August 15, 1978, the database includes bibliographic data, titles, abstracts, claims, and image data. Prior patents include bibliographic data, text of titles, and images.

Community of Science, Inc. <http://patents.cos.com> (accessed May 18, 2005). The Community of Science, Inc. (COS) is a consortium of corporations and government agencies worldwide that created an Internet site for research and development. COS reference service includes access to the U.S. Patent database. The subscription makes available the first page of more than 2.6 million patents from 1971 to date.

Delphion Intellectual Property Network <http://www.delphion.com/> (accessed May 14, 2005). Owned by Thomson Scientific, this subscription service provides full text access to U.S. patents from 1971 to date and international patents from several countries.

Derwent World Patents Index <http://thomsonderwent.com> (accessed May 18, 2005). One of the oldest online patent services, Derwent, now owned by Thomson Scientific, provides abstracts for U.S. and international patents from 1963 to the present. Included in the site is a descriptive page on "information about patents."

European Patent Office Esp@cenet <http://ep.espacenet.com> (accessed May 10, 2005). Developed by the European Patent Office (EPO), this site provides access to over forty-five million patents from EPO member states and over seventy countries and regions worldwide through several databases. Included are current patent applications published by EPO or WIPO (see below) in the past twenty-four months, Japanese applications, and abstracts of non-examined Japanese applications since October 1976.

Franklin Pierce Law Center Intellectual Property Mall <http://ipmall.org/web_resources/index.php> (accessed May 18, 2005). Created by Pierce Law's Professor Jon Cavicchi and Sitesurfer Publishing LLC's Bill Shaw, the IP Mall links to worldwide online resources, giving almost everything that is needed to know about patents and intellectual property issues.

Intellectual Property Office of Singapore (IPOS) <http://ipos.gov.sg/main/index-page/index.html> (accessed May 10, 2005). Through the intellectual property portal SurfIP, access is available to patent databases from Singapore, Japan, United Kingdom, United States, Canada, WIPO, EPO, and Taipei.

Japan Patent Office <http://www.jpo.go.jp/> (accessed May 10, 2005). Besides the home page to Japan Patent Office, this site links to its Industrial Property Digital Library (IPDL) in the National Center for Industrial Property Information and Training (NCIPI) that offers public access to Japanese patents abstracts and trademarks, and information about NCIPI activities.

Lexis-Nexis Academic Universe <http://web.lexis-nexis.com/universe> (accessed May 18, 2005). Access to U.S. utility patents from 1790 to date, selective coverage from 1790 to 1974, and full text to U.S. patents from 1971 to date. Additional fees for images. Select "Legal Research" from the Lexis-Nexis Academic Universe menu.

National Inventors Hall of Fame <www.invent.org> (accessed May 18, 2005). Invent Now is a non-profit organization that fosters the spirit and practice of invention. The National Inventors Hall of Fame in Akron, Ohio showcases inventors' achievements and their impact on our lives. Included in the exhibits are hands-on activities, displays depicting famous inventors, and special educational exhibits that challenge creativity.

PatentCafe Intellectual Property Management Enterprise Solution <www.patentcafe.com> (accessed May 18, 2005). A smorgasbord of patent information for all levels of interest appears in this online intellectual property magazine. The audience ranges from the novice inventor to the experienced inventor and researcher looking for marketing information.

State Trademark Resources <http://statetm.tripod.com> (accessed May 18, 2005). This site identifies the state agencies responsible for registering trademarks; last revision was 2002 and copyright belongs to Michael White.

SurfIP <http://www.surfip.gov.sg/sip/site/sip_home.htm> (accessed May 18, 2005). This intellectual property (IP) portal is designed for searching information relating to biotechnology, information technology, telecommunications, and manufacturing industries.

United Kingdom Patent Office <http://www.patent.gov.UK/patent/dbase/espace.htm> (accessed May 10, 2005). The Home page describes this free service as a worldwide patent database with over thirty million documents plus databases from Japan, the World Intellectual Property Organization, the European Patent Office, and nineteen European countries. A current awareness capability provides searching across a limited range of Great Britain patent applications only. Further descriptions on searching combinations by dates, numbers, and International Patent Classifications (IPC) are listed on the Home page.

United States Patent and Trademark Office <http://www.uspto.gov/> (accessed May 18, 2005). Official website of the United States Patent and Trademark Office. First screen is the Home page welcome with the top patent news and links to the fee structure, products and services, and the search screens. Also available online is the United States Manual of Classification <http://www.uspto.gov/go/classification> (accessed May 18, 2005) listing class schedules and linked classification definitions, updated regularly.

World Intellectual Property Organization (WIPO) <http://www.wipo.int/portal/index.html.en> (accessed May 10, 2005). The WIPO links to the Intellectual Property Digital Library (IPDL) website providing access to intellectual property data collections hosted by the WIPO. The WIPO collects and publishes annual statistics on industrial property, by country. The site includes the Patent Cooperation Treaty (PCT) gazette containing bibliographic data, drawings, abstracts, and images of PCT applications issued since January 1997.

Gray literature

Gray (or grey) literature has been defined as "information produced on all levels of government, academics, business and industry in electronic and print formats not controlled by commercial publishing i.e. where publishing is not the primary activity" (GreyNet, accessed April 5, 2005). Auger's book *Introduction to Grey Literature* helps to explain the complexity of locating gray literature sources that may be preprints, conference papers, newsletters, white papers, working papers, reports, and other materials. The information found in this literature can be very useful and may not be published in other formats. However, these sources can be hard to identify and even harder to obtain. References to gray literature may be found in bibliographies or recommended by individuals who have used or heard of the resource. Because these resources are often difficult to obtain through routine channels, a number of portals have appeared to provide links to gray literature. One online source is GreyNet <www.greynet.org> (accessed May 18, 2005). GreyNet hosts conference and publishes proceedings on various aspects of gray literature.

Search portals

DOE's *Information Bridge* <http://www.osti.gov/bridge/> (accessed May 10, 2005). The *Information Bridge* provides the open source to full-text and bibliographic research reports from the U.S. Department of Energy (DOE) laboratories since 1995. Plans exist to add to the site as reports become available electronically.

GrayLIT Network. DOE's Office of Scientific and Technical Information (OSTI) and the Government Printing Office (GOP) <http://www.osti.gov/graylit/index.html> (accessed May 10, 2005). Described as the world's most comprehensive portal to federal gray literature, this site is available by DOE through GPO access in partnership with the Government Printing Office. Several federal agencies participate in this project.

U.S. Army Corps of Engineers (USACE) <www.usace.army.mil.inet/> (accessed May 10, 2005). The mission, the history, the array of publications, the digitized collections and the responsibilities of the largest public engineering, design, and construction management federal agency. Descriptions of the technical libraries and listings of their regulations and materials published by the Army Corp of Engineers are found through this site.

Virtual Technical Reports Center <www.lib.umd.edu/ENGIN/TechReports/Virtual-TechReports.html> (accessed May 14, 2005). Maintained by the Technical Reports Librarian at the University of Maryland, this site provides links to technical reports, preprints, reprints, dissertations, theses, and research reports of all kinds generated by domestic and international agencies.

Internet resources

In addition to Internet sites mentioned in specific categories, there a number of sites that should be browsed periodically. These include:

American Association for the Advancement of Science <http:www.aaas.org> (accessed May 31, 2005). Stated to provide "timely, objective information to Congress on current science and technology issues and assists the science and engineering community in understanding and working with Congress" (from website).

American Association of Engineering Societies <http://www.aaes.org/inside_aaes/index.asp> (accessed May 31, 2005). "A multidisciplinary organization of engineering societies dedicated to advancing the knowledge, understanding, and practice of engineering" (from website).

National Research Council Canada <http://www.nrc-cnrc.gc.ca> (accessed May 31, 2005). Canada's national organization for research and development.

Society of Women Engineers <http://www.swe.org/stellent/idcplg?IdcService=SS_GET_PAGE&nodeId=5> (accessed June 15, 2005). "SWE empowers women to succeed and advance in those aspirations and be recognized for their life-changing contributions and achievements as engineers and leaders" (from website).

Women in Engineering Programs and Advocates Network <http://www.wepan.org/> (accessed June 15, 2005). The "mission is to be a catalyst for change to enhance the success of women in the engineering profession" (from website).

World Wide Virtual Library <http://vlib.org/> (accessed June 15, 2005). A listing of virtual library sites. The engineering and sciences sites are worth a periodic look.

Yahoo <http://dir.yahoo.com/Science/Engineering> (accessed June 15, 2005). A directory to dozens of engineering disciplines, providing links to a number of websites.

Engineering design, management and ethics

Whether engineers are employed in a large corporation that maintains a research and development department or are individual inventors, a basic understanding of design, ethics, and a commitment to avoid failures is vital. Creativity helps engineers develop ideas, while management looks at the cost factors in research and development. Knowing appropriate industrial standards helps transform ideas into products. Knowing ethical standards helps avoid litigation. An understanding of engineering ethics helps in making decisions. Most of the resources on quality control and production classify in industrial engineering. Some of the newer disciplines, such as failure analysis and risk management, grew from engineering problems. Revisions to industrial standards occur after disasters. How to balance human gains of development against environmental damage becomes an ethical issue. Environmental problems provide numerous examples in ethics and values. Vesilind uses that specialty as the subject for a textbook. Many sources on ethical issues cover specific types of research (i.e. bioengineering, genetic engineering, animal research). The National Research Council (U.S.) publishes results in many of these fields. The sources listed in this chapter are offered as a starting point for the creativity found in most engineers. Some engineers who are designers can write. Those who are both authors and engineers leave a legacy of failures and successes. What they wrote will guide others in developing the creative process and in the weighing of ethical issues.

Websites provide a good starting point for finding information that will aid in an understanding of engineering ethics. Besides e-books available on the Web, succinct information is compiled on sites such as the Online Ethics Center for Engineering and Science. The National Society of Professional Engineers includes documents that help in understanding ethics and avoiding liability. The site map for the National Institute for Engineering Ethics provides direction on codes, tests, a help-line, and links to other ethics sources. The following list of books and websites are presented to encourage creativity, yet avoid failures and aid in basic understanding of engineering ethics.

Books

Babcock, D.L. (1996) *Managing Engineering and Technology: an Introduction to Management for Engineers* (2nd edn), Upper Saddle River, NJ: Prentice Hall. Part of the publisher's series on industrial and systems engineering.

Bronikowski, R.J. (1985) *Managing the Engineering Design Function*, New York: Van Nostrand Reinhold. An older volume contains useful information on basic managerial skills.

Macaulay, D. (1998) *The New Way Things Work*, Boston, MA: Houghton Mifflin. This prolific author writes in a clear, concise fashion that makes his works suitable for school libraries and understandable by everyone. The text, with numerous illustrations, explains the scientific principles and workings of hundreds of machines. This title updates his 1988 *The Way Things Work* to include new material on digital technology.

Petroski, H. (1992) *To Engineer is Human: the Role of Failure in Successful Design*, New York: Vintage Books. Originally published by St. Martin's Press in 1985, some of the content had appeared previously in *Technology Review* and other publications. As an engineer with a flair for writing, the author shares his ideas on systems failures in an understandable fashion. He is the author of several engineering books. A collection of his essays from *American Scientist* is published as *Pushing the Limits: More Adventures in Engineering* (2004) by Knopf.

Tenner, E. (2003) *Our Own Devices: The Past and Future of Body Technology*, New York: Alfred A. Knopf. "An exploration not only of inventive genius, but also of user ingenuity," this is an exploratory work for those who are curious about simple things.

Design

Hoke, J.R. Jr. (2000) *Ramsey/Sleeper Architectural Graphic Standards* (10th edn), New York: Wiley. This is a classic quick reference tool that describes all types of building design and construction detail.

McGowan, M. and K. Kruse (2004) *Interior Graphic Standards*, Hoboken, NJ: Wiley. The first student edition of architectural graphic standards, it serves as a companion for all aspects of design education, from ergonomics to specifications to construction.

Watson, D. (2001) *Architectural Details: Classic Pages from Architectural Graphic Standards*, New York: Wiley. Watson compiled selections from the 1940 to 1980 editions of *Architectural Graphic Standards*, classic works by Charles Ramsey and Harold Sleeper. This volume may be used to complement a basic textbook.

Watson, D., J.H. Callender and M. Crosbie (1997) *Time-saver Standards for Architectural Design Data* (7th edn), New York: McGraw-Hill. Subtitled "The reference of architectural design data," this volume serves as one-stop shopping for architectural and building design. Contents are compiled from numerous sources, including government agencies, trade associations, manufacturers, and professionals. The good visuals provide a balance of representation and explanation of design integration.

Ethics

Cook, R.L. (2003) *Code of Silence: Ethics of Disasters*, Jefferson City, MO: Trojan Publishing Company. The author is a professional engineer who taught engineering and technology, and in this volume asks, "How safe are we?" He covers the moral and ethical aspects of man-made disasters, from the Titanic to the World Trade Center.

Harris, C.E. Jr, Pritchard, M.S. and Rabins, M.J. (2004) *Engineering Ethics: Concepts and Cases* (3rd edn), Belmont, CA: Wadsworth Publishing. This textbook covers theory and practice by providing cases, methodology, and analysis of what is involved in practicing ethics.

Martin, M. and Schinzinger, R. (2004) *Ethics in Engineering* (4th edn), New York: McGraw-Hill. Key issues in engineering ethics are discussed. The appendix includes resources on the subject and codes of ethics from several societies.

Pinkus, R.L.B. (1997) *Engineering Ethics: Balancing Cost, Schedule, and Risk – Lessons Learned from the Space Shuttle*, New York: Cambridge University Press. By using the space shuttle program as the framework, Pinkus and others examine the role of ethical decision-making in the practice of engineering. In-depth case studies show engineers at work as they balance budgets, deadlines, and risks.

Seebauer, E.G. and Barry, R.L. (2000) *Fundamentals of Ethics for Scientists and Engineers*, New York: Oxford University Press. This is a textbook intended for ethics courses in engineering or science. It tries to emphasize ethic reasoning by including cases for graphic visualization.

Vesilind, P.A. (2003) *Introduction to Environmental Engineering* (2nd edn), Scarborough, ON: Nelson Thomson Learning (Canada). The author emphasizes materials balance and environmental ethics by incorporating ethical decision-making in technical problems. The author also wrote a sixty-page guide, *Doing the Right Thing: An Ethics Guide for Engineering Students*, published by Lakeshore Press (2004).

Online sources

National Institute for Engineering Ethics (NIEE) <http://www.niee.org/site_map.htm> (accessed May 18, 2005). The site map offers contact information about the NIEE and numerous directions to other sites covering an understanding of ethics.

National Society of Professional Engineers <www.nspe.org> (accessed May 18, 2005). The site includes documents that help in professional liability and risk management, including downloadable ones such as "Guidelines for NSPE State Societies in Addressing Unlicensed Practice of Engineering."

Online Ethics Center for Engineering and Science at Case Western Reserve University <http://www.onlineethics.org/> (accessed May 18, 2005), Cleveland, OH:

Case Western Reserve University. According to the mission statement, this site provides engineers, scientists, and science and engineering students with resources for understanding and addressing ethically significant problems that arise in their work. It also serves those who are promoting learning and advancing the understanding of responsible research and practice in science and engineering.

Career information

Basta, N. (2003) *Opportunities in Engineering Careers* (rev. edn), Chicago, IL: VGM Career Books. The author talks about the career opportunities in engineering professionals, including some of the specialties, working profiles, and everyday impact. Salary information and tips about colleges, licensing, and professional registration help to guide the student.

Camenson, B. (1998) *Real People Working in Engineering*, Lincolnwood, IL: VGM Career Horizons. Part of the series "On the Job," this book describes real people who choose engineering as a career.

Davis, M. (1998) *Thinking Like an Engineer*, New York: Oxford University Press. Excellent chapter on the history of engineering gives the educational background and how the engineering fields evolved. Responsibilities, code of ethics, the social questions are woven into the decision-making process. Examples of wrongdoing are described.

Garner, G.O. (2002) *Great Jobs for Engineering Majors* (2nd edn), Chicago, IL: VGM Career Books. Also available as an e-book. Aimed at engineering majors, the book is divided into two parts, from job-searching to career paths. Resumés and graduate school choices form the basis of the first section. The second section divides career paths into the industry, consulting, government, education, Internet, and non-technical areas.

Honda, H. (2000) *Working in Japan: An Insider's Guide for Engineers and Scientists*, New York: ASME Press. The theme of this book is the importance of understanding the social life and customs of the country in which the engineer works.

Lewis, E.E. (2004) *Masterworks of Technology: The Story of Creative Engineering, Architecture, and Design*, Amherst, NY: Prometheus Books. Written to inspire students to pursue a career in engineering, the author interweaves personal stories to illustrate creating in engineering. Chapter bibliographies aid in further reading to help high school students and technical program students to decide on specific careers in engineering.

National Council of Examiners for Engineers and Surveying <http://www.ncees.org> (accessed May 18, 2005). Site provides links to exam and license information: www.ncees.org/licensure/licensing_boards/ (accessed December 3, 2005). All the states and the District of Columbia require engineers and surveyors to register with the state in which they want to offer their services. This website provides a quick source to locate state licensing boards. The link leads to the address, telephone numbers, contacts, and website address. For example, in Ohio, it is www.ohiopeps.org (accessed May 18, 2005).

National Society of Professional Engineers <http://www.nspe.org> (accessed May 18, 2005). Links include employment, licenses, scholarships, and so on.

U.S. Department of Labor, Bureau of Labor (2002–2003) *Occupational Outlook Handbook, Statistics Bulletin 2540*, Washington, DC: Government Printing Office. An annual publication, also available online at <www.bls.gov/oco/cg/home.htm> (accessed May 18, 2005), this resource provides the starting point for necessary information about engineering and its specialties. Several sources of additional information include web addresses, such as the American Society for Engineering Education <http://www.asee.org> (accessed May 18, 2005). Site provides several links to information on careers, fellowships, and international opportunities.

Guides for writing

Once those research techniques are mastered, the student or engineer must learn how to write up that information. Whether it is writing a memo, outlining a project, or composing a report, some skills are necessary. Publishers usually provide specific guidelines, but some general resources are provided as aids in writing. The following titles will assist in writing the research paper, the report, or the memo.

Books

Alexander, I.F. (2002) *Writing Better Requirements*, Boston, MA: Addison-Wesley. Emphasis is on writing software for the computer. Useful in systems engineering, it also offers guidelines that may be used in a broader context.

Alred, G.J., Brusaw, C.T. and Oliu, W.E. (2003) *Handbook of Technical Writing* (7th edn), New York: St. Martin's Press. Arranged alphabetically with a comprehensive index, contents range from formulating abbreviations to web design, and offers guidelines that may be used in a broader context than technical writing.

Beer, D.F. (2005) *A Guide to Writing as an Engineer* (2nd edn), Hoboken, NJ: John Wiley. Subject-specific guide to improving written and spoken communication.

Blicq, R. and Moretto, L. (1999) *Technically-Write!* (5th edn), Upper Saddle River, NJ: Prentice-Hall. Covers technical communication from e-mail to formal reports.

Davis, M. (2005) *Scientific Papers and Presentations*, Burlington, MA: Academic Press. Covers written reports and presentations to groups.

Finkelstein, L. (2004) *Pocket Book of Technical Writing for Engineers and Scientists*, New York: McGraw-Hill. Step-by-step, the author shows students the skill of technical writing. He uses practical outlines in explaining the options. Part of McGraw-Hill's BEST (Basic Engineering Series and Tools) series of modularized textbooks for introductory courses.

Hart, H. (2004) *Introduction to Engineering Communication*, Upper Saddle River, NJ: Prentice-Hall. A textbook aimed at introductory courses in engineering or computer science shows students how to communicate their ideas, whether writing documents or giving oral presentations.

Matthews, C. (2000) *A Guide to Presenting Technical Information: Effective Graphic Communication*, London, Professional Engineering Publications. This guide shows how to communicate technical information by providing methods and guidelines in a clear and effective way.

McGuire, M. (2002/2003) *The Internet Handbook for Writers, Researchers, and Journalists*, New York: Guilford Press. This handbook will help writers locate and cite computer network resources.

Michaelson, H.B. (1990) *How to Write and Publish Engineering Papers and Reports* (3rd edn), Phoenix, AZ: Oryx Press. This book gives basic information on technical writing.

Microsoft Manual of Style for Technical Publications (1998) (2nd edn), Redmond, WA: Microsoft Press. Specifically written for Microsoft Windows 98 or NT, this book serves as a guide to technical writing in the computer age.

White, J. (1997) *From Research to Printout: Creating Effective Technical Documents*, New York: ASME Press. Excellent guidelines show how to write effectively, especially technical reports. References may be specific to mechanical engineers, but the principles hold true for any engineers.

Current awareness

Many electronic journals and databases support e-mail services in which the table of contents and/or bibliographic information are sent to users on a periodic basis. For engineers and researches this is a convenient way to keep up to date in specific areas. Depending upon the provider, there may be charges for these services. Many providers do not charge for current awareness services, but do charge for document delivery. If an institution or an individual subscribes to a journal or a database, these electronic services are generally free of charge.

Electronic services

Amedeo.com (12 to 24 months) AmedeoGroup.com <http://www. amedeo.com> (accessed May 18, 2005). A free e-mail alert service to more than 1,380 core medical journals. Coverage is a window of the past twelve to twenty-four months. Subscribers receive a weekly Amedeo newsletter to new articles published within user-defined subject areas and a weekly update of abstracts from user-defined journals.

BiblioAlerts (2002–) Bethseda, MD: Cambridge Scientific Abstracts <http://www.biblioalerts.com> (accessed May 18, 2005). Customized electronic reports, updated monthly, compiled by CSA staff on technical topics, including engineering, biotechnology, genetics, and neuroscience. Reports include summaries and bibliographic information from journals, conference proceedings, reports, patents, and websites. Coverage begins when a report is generated and monthly updates proceed from that date.

Current Contents Connect (Monthly) Philadelphia, PA: Thomson Scientific <http://www.isinet.com/cap> (accessed May 18, 2005). Individual products that provide table of contents, cover-to-cover indexing, abstracts and bibliographic information to journals and books. International coverage to important journals in specific fields, clinical medicine, life sciences, engineering, computing and technology.

Dialog (Monthly) Philadelphia, PA: Thomson Scientific <http://www. dialog.com> (accessed May 18, 2005). Access to more than 100 publishers and information aggregators and coverage to more than 11,000 specialized journals. Alert and table of contents services available.

Infotrieve (Monthly) Los Angeles, CA: Infotrieve <http://www4.infotrieve.com/products_services/current_awareness/default.asp> (accessed May 18, 2005). A fee-based information service that includes database searching, document delivery and current awareness options to scientific, medical, and technical publications.

Ingenta Select (1998–) Oxford: Ingenta <http://www.ingentaselect.com/newaler.htm> (accessed May 18, 2005). Subscribed users can set up to five journals for table of contents and alert services from more than 6,000 academic and professional journals. Also offers standard index searching and document delivery options.

Kluwer Alert Services/Springer Link Alert (Monthly) New York: Springer <http://www.kluweralert.nl> (accessed May 18, 2005). A free e-mail table of contents service to Kluwer and Springer journals, as well as announcements on new books.

Scholarly Articles Research Alerting (SARA) (Monthly) Oxfordshire: Taylor & Francis Group Ltd. <http://www.tandf.co.uk/sara> (accessed May 18, 2005). A free e-mail table of contents service to more than 950 academic journals covering chemistry, biological sciences, biomedical education, engineering and technology, health and medical sciences, genetics, and biotechnology.

ScienceDirect (Monthly) Amsterdam, the Netherlands: Elsevier <http://www.sciencedirect.com> (accessed May 18, 2005). ScienceDirect provides access to more than 1,800 scientific, technical, and medical journals. Alert and table of contents services are available to subscribers (section by Linda Martinez).

Conclusion

This chapter offers suggestions for doing research in general engineering. It includes career information to show the numerous engineering disciplines. Regardless of the specialty there are basic resources and traditional methods that should be understood by any engineering student or librarian. How books classify and sit on the shelf will help the beginner and the specialist. Frustrations in searching for papers that appeared at a general engineering conference also apply to those sponsored by specific engineering societies, and often held in a foreign country. Acronym dictionaries define societies' names, but the same acronym can apply to more than one society. For example, ASSE means American Society of Safety Engineers or American Society of Sanitary Engineers. Drop an "E" from IEEE (Institute of Electrical and Electronic Engineers) and it becomes IEE, the Institution of Electrical Engineers. Both publish resources indexed in *IEEE Xplore*. The trend toward the international results in many societies changing name. The Web helps to locate society names and author locations. Gray literature abounds in every discipline. Tools listing report numbers or resources on how to search for documents may help in locating specific ones.

Titles of reference tools belie their content. They sound specific, but contain information valuable as a starting point. When Paul Thrush compiled the mining dictionary for the Bureau of Mines, he complained about the limitations of subject headings to describe its content, since it covers more than mining and serves as a basic dictionary in many engineering fields. *Architectural Graphic Standards* is more than architecture. It serves as a first choice in quick reference, as it mentions sources that lead to specialties (i.e. the *Americans with Disabilities Act of 1990*, Underwriters Laboratory or a specific society). An overview of industrial standards guides an engineer in design and development. Ethical standards apply in all aspects of engineering. An understanding of the patent process is helpful in design engineering. The desire to have full-text resources downloaded from the Internet is paramount and basic searching techniques ensure success. Skills learned by using resources in this chapter apply to searches in specific engineering fields.

3 Aeronautical and aerospace engineering

Thomas W. Conkling

Introduction, history, and scope of discipline

The idea of flight has always intrigued mankind, but it was only a little over 100 years ago that powered manned flight was achieved. The first successful demonstration of a flying machine is usually credited to the Montgolfier brothers' hot air balloon that flew in June of 1783 in Annonay, France. The first manned flight in a balloon followed just four months later. Sir George Cayley investigated the aerodynamics of fixed wing flight in the early 1800s, and designed the first man-carrying glider. Samuel Henson and John Stringfellow later attempted to design and construct a steam-powered aircraft based on Cayley's work. Otto Lilienthal's pioneering work with manned glider flight between 1891 and 1896 advanced the field, and Samuel Langley successfully tested a number of model aircraft powered by small steam-engines in 1896. The Wright brothers built and flew a number of gliders before they designed the aircraft that became the first to demonstrate powered manned flight in 1903.

The theoretical foundations for the discipline of aeronautical engineering began to take form with some of the early work on fluid mechanics by Newton, Bernoulli, Euler, and d'Alembert in the seventeenth and eighteenth centuries. This work was expanded during the nineteenth century by scientists such as Navier, Stokes, Rankine, Helmholz, Kirchhoff, and Rayleigh (Anderson 1997). A number of the experimenters mentioned earlier, such as Cayley and Lilienthal, also compiled data files on the airfoils that they had been working with. The field received a boost in the early 1900s, when governments realized the potential military applications of this new technology and began supporting research into the principles of flight and propulsion. Two of the earliest government-sponsored research bodies established were the Advisory Committee for Aeronautics in Great Britain (1909) and the National Advisory Committee for Aeronautics in the U.S. (1915). The earliest research concentrated on aerodynamics, with much experimental work conducted in wind tunnels.

The period before World War I saw many advances in aircraft design, particularly from Germany, France, and Russia. These countries invested considerable effort in aircraft development and pushed Europe ahead of the

United States in this area. Developments in the U.S. during this period were slowed somewhat by patent disputes between various individuals. At the start of the war, aircraft were used primarily for reconnaissance and artillery spotting, but their role quickly expanded with the development of effective machine-guns, better engines, and improved design. The developmental cycles for aircraft happened rapidly as countries tried continuously to counter improvements in enemy aircraft.

Aviation was in the public eye after the war. Air races and other forms of competition for flying records made headlines regularly. Passenger service and airmail routes were started. Great advances were made in aircraft design and piloting. Radial engines were developed, followed by liquid-cooled in-line piston engines. Fixed-pitch propellers were increasingly replaced by the more efficient controllable-pitch propellers. Metal replaced wood as the key structural element, and almost all new aircraft had a single wing. A number of exceptional aircraft were designed, including the Douglas DC-3. This airplane formed the backbone of both commercial air service and military transport, and examples of this craft are still flying on a daily basis in several countries.

Aviation technology advanced rapidly during World War II, and before it had ended, examples of jet-powered aircraft had been developed by several countries. The sound barrier was broken in 1947 by the rocket-powered Bell X-1, and the X-15 experimental aircraft set many altitude and speed records in the 1960s. Swept-wings and jet engines are now the standard for fighters and bombers around the world. Propeller-driven civilian transport aircraft such as the Douglas DC-6 and DC-7 and the Lockheed Constellation began to give way in the 1950s to faster jet-powered craft such as the Boeing 707. A substantial amount of research and development has been done in recent years on stealth technology and pilotless aircraft for military applications. Many of the innovations in the civilian transport fleet in the past decade have centered on advanced materials and electronics, and improvements in engine technology.

Space exploration has been made possible by the rocket engine. The first rockets were probably built in China almost a thousand years ago, powered by black powder. By the 1500s, rockets were widely used for fireworks and in elementary weapons. Their use as weapons became more widespread by the late 1700s and into the 1800s. The founding fathers of modern rocketry are considered to be Konstantin Tsiolkovsky, Robert Goddard, and Hermann Oberth (Von Braun and Ordway 1966). Tsiolkovsky did important theoretical work on rockets in the latter part of the nineteenth century. He is credited with developing the basic theories of rocket propulsion and using liquid oxygen and hydrogen as fuels, and the idea of multiple-stage rockets. He also did theoretical work on the problem of escaping the Earth's gravity with a rocket. Robert Goddard did extensive experimental work with rockets during the first half of the twentieth century. He developed a series of liquid-propelled rockets and was awarded a number of patents on a variety of rocket components. Oberth was a contemporary of Goddard – he published two influential books on space travel and worked tirelessly to generate interest in space flight.

Rocket technology improved during World War I, but the pace of innovation and development accelerated during the 1920s and 1930s. The military became increasingly interested in rockets during this period. Progress was made in both solid and liquid propellants, and in many aspects of rocket design and control. Germany had a very advanced rocket program, as did Russia. Many varieties of military rockets were developed during World War II for use as surface-to-surface, surface-to-air, air-to-surface, and air-to-air weapons. Germany developed the V-2 missile during this period, a vehicle that heavily influenced post-war launch vehicle design.

The space age began in October 1957 with the Russian launch of the Sputnik satellite. The success of this mission created a sense of urgency in the United States to develop comparable missile capabilities. The U.S. launched the Vanguard satellite in January 1958. Military and civilian missile technology advanced rapidly, particularly in the U.S. and Russia, as each country attempted to build more powerful launch systems for weapons and for manned spacecraft. The Space Shuttle, first flown in 1981, is probably the most complex spacecraft and launch system developed to date. This reusable craft is launched using a liquid-fueled main engine, assisted by solid propellant booster rockets.

Aeronautical engineering as an academic discipline began in the early years of the twentieth century. In 1913, the first formal course in aeronautics was taught at the University of Michigan, followed the same year by a similar offering at MIT. The first Master of Science degree in aeronautical engineering and the first doctorate in the field were awarded at MIT in 1915 and 1916, respectively. The University of Michigan created the country's first official Aeronautics Department in 1916. Other early programs in aeronautical engineering were established at the University of Washington, State University of New York, Stanford, Caltech, Georgia Tech, and the University of Dayton (McCormick *et al.* 2004). An additional group of schools began offering programs in the early 1940s, again influenced by the war. There are currently about sixty aeronautical and aerospace engineering programs in the United States.

Searching the library catalog

While traditional card catalogs may have offered some unique benefits to the library user, the capabilities of web-based catalogs are impressive. Keyword searching can help the user retrieve material in aerospace or aeronautical engineering, but also allows them to discover materials in other disciplines that may have direct application to their information needs. If a keyword search returns excessive items, searching with Library of Congress subject headings is one way to refocus a search. The following is a listing of some important subject headings for this field:

Aerodynamics
Aerofoils
Aeronautics
Airframes
Airplanes
Artificial Satellites
Astronautics
Ballistic Missiles
Boundary Layer
Drag (Aerodynamic)
Flight
Gas Turbines – Aerodynamics
Helicopters
Jet Planes
Jet Propulsion
Lift (Aerodynamic)
Reynolds Number
Rocket Engines
Rockets (Aeronautics)
Rotors (Helicopters)
Space Vehicles
Stability of Airplanes
Turbines – Aerodynamics
Turbomachines
Wings

The Library of Congress classification schedule places most of the materials in the field between TL500 and TL4000. However, when searching the catalog or browsing the shelves, it will be apparent that many of the items needed by students, engineers, and researchers will lie outside of these areas. Subjects such as turbomachinery, combustion, fluid flow, computational methods, electronics, computers, and structural materials are examples. Some call numbers of interest are as follows:

TL500–589	Aeronautics
TL600–688	Aircraft
TL690–697	Electrical and communication systems
TL698	Materials
TL701–704	Aircraft engines
TL709	Jet propulsion
TL710–713	Flight
TL780	Rocket propulsion
TL787	Astronautics
TL873	Manned space flight
TL1050	Astrodynamics
TL3000–3285	Astrionics and electrical equipment
TL4000–4050	Ground support systems

Indexes and abstracts

Indexes and abstracts were originally produced to allow researchers to keep track of the new technical information being issued in their specialties. Some indexes concentrated on articles and conference papers, while others covered only technical reports. The technical report has a special place in the communication of research results in aeronautical and aerospace engineering. Government agencies in several industrialized nations recognized the military potential of aircraft early on and began supporting research into this new technology. Research results were often issued as technical reports, since this format offered rapid publication time and the ability to contain large amounts of technical data. Their distribution could also be controlled fairly easily if they contained sensitive information. Technical reports started appearing in the early 1900s, and they are still produced in quantity today, with the web being the key distribution mechanism.

Print indexes have been superseded to a large degree by their database counterparts on the Web, but they still play an important role in access to the older materials not yet covered online. The records for articles, papers, and technical reports are usually arranged in some type of classified subject scheme, and in a few instances they are alphabetically arranged by subject heading. Print indexes almost always contain multiple indexes in each issue, with annual cumulative indexes to facilitate use.

Engineering Index (1884–) Hoboken, New Jersey: Elsevier Engineering Information. The most comprehensive index to the world's engineering literature; provides very good coverage of the journal and conference literature in aeronautical and aerospace engineering. Available online as the *Compendex* database (see Chapter 12, this volume).

Government Reports Announcement and Index (1975–1996) Springfield, VA: National Technical Information Service. Although no longer published, this publication and its predecessors provide an in-depth guide to the technical report literature that was produced in the U.S. from 1946 to 1996. Some reports from other countries are also included. It provided very good coverage of the aeronautical and astronautical sciences. This index was produced by NTIS, which acts as the national clearinghouse for the technical report literature in the U.S. Issues were published twice a month with annual indexes. Entries are arranged by subject category and provide bibliographic information and an abstract. This publication was preceded by the *Bibliography of Scientific and Industrial Reports* (1946–June 1949), *Bibliography of Technical Reports* (July 1949–June 1954), *U.S. Government Research Reports* (July 1954–1964), *U.S. Government Research and Development Reports* (1965–1971), and *Government Reports Announcement* (1971–1975). Technical report information may now be found on the *NTIS* database; recently issued technical reports may be searched at the NTIS website.

Index of NACA Technical Publications (1915–1958) Washington, DC: National Advisory Committee for Aeronautics. The definitive index to all unclassified and unlimited publications from the National Advisory Committee for Aeronautics (NACA), NASA's predecessor. Entries are arranged by subject category and contain concise information: author, title, report number, and date. Subject and author indexes are included. The period 1915 to 1949 is covered in one volume, and several volumes cover 1950 through 1958. The NACA Technical Report Server now provides online access to the digitized version of many NACA reports.

International Aerospace Abstracts (1961–) Bethesda, MD: Cambridge Scientific Abstracts. An excellent printed index to the world's published literature in aeronautics, astronautics, and the space sciences. Provides in-depth coverage of journals and conference papers. Issue contents are arranged by subject, and each issue has subject, author, meeting paper, and report number indexes. Through 2000, this index was produced by the American Institute of Aeronautics and Astronautics (AIAA) – AIAA now produces it in cooperation with Cambridge Scientific Abstracts. Published monthly with annual cumulative indexes. Also available as a web-based database, *Aerospace and High Technology Database.*

Scientific and Technical Aerospace Reports (STAR) (1963–1995), print version (1996–) web version. Lithicum Heights, MD: NASA Center for Aerospace Information. Originally printed but now published online in a PDF version, this index provides comprehensive coverage of the technical report literature of aeronautics, astronautics, and the space sciences. Covers technical reports from NASA and its contractors, as well as reports from other agencies, companies, and labs in the U.S. and elsewhere. Issue contents are arranged by subject with multiple indexes at the end of each issue. *STAR* was preceded by the *Index of NASA Technical Publications and Technical Publication Announcements* from 1958 through 1962. When used in conjunction with *International Aerospace Abstracts*, comprehensive coverage of all publications in the field is provided. The NASA Technical Report Server provides web-based access to the more recent NASA-related reports in *STAR* <www.sti.nasa.gov/pubs/star/Star.html>.

Databases

Bibliographic databases began revolutionizing access to published information in engineering and the sciences in the mid-1970s. Before that time, it was necessary for researchers to conduct manual searches through the printed indexes and abstracts to locate publications of interest. While this could be done effectively, it was time-consuming. Once the information in the indexes became available electronically, the dynamics of searching changed. Vast quantities of journal articles, conference papers, and technical reports could be reviewed in minutes, and search strategies could be changed on the fly to improve results. Today, the web-based databases deliver an enormous amount of access to the engineer's desktop.

The information needs of aerospace and aeronautical engineers are well served by bibliographic databases. The databases covered in this section deal directly with publications in the field, but there are additional databases mentioned in this book that would also be useful, particularly in materials, electronics, and computer science.

Aerospace and High Technology Database (1962–) Bethesda, Maryland: Cambridge Scientific Abstracts. This is the best commercially available database for aeronautical and astronautical engineering. The database is compiled by the American Institute of Aeronautics and Astronautics. Provides excellent coverage of the journal, conference, and technical report literature, and is the online equivalent of *International Aerospace Abstracts* and *STAR*. Over 1,000 publications are scanned for information, and 6,000 new records are added monthly; more than 2,300,000 records are on the database.

Compendex (1884–) Hoboken, New Jersey: Elsevier Engineering Information, Inc. The electronic version of *Engineering Index*, this database provides very good coverage of the journal and conference literature in aeronautics and astronautics (see Chapter 2, this volume).

NACA Technical Report Server (NACATRS) <http://naca.larc.nasa.gov> (accessed May 18, 2005). Lithicum Heights, MD: NASA Center for Aerospace Information. Database provides coverage of the technical reports produced by the National Advisory Committee for Aeronautics (NACA) from 1917 through 1958. Many of these reports are still used in aeronautical research. Most reports have been scanned and are available full text as PDF files.

NASA Technical Report Server (NTRS) <http://ntrs.nasa.gov> (accessed May 18, 2005). Lithicum Heights, Maryland: NASA Center for Aerospace Information. NTRS provides free public access to NASA-generated technical reports and other publications from NASA authors. Some publications from non-NASA sources are included as well. The database indexes NASA materials back to the beginning of the agency in 1958, and also covers NACA reports back to 1917. All unclassified and unlimited NASA reports are indexed in NTRS. Many full-text versions of NASA reports are on the database, particularly from those published in 2004 and later. The "advanced search" option allows one to limit searches to specific NASA research center publications.

National Technical Information Service (NTIS) database (1964–) Springfield, Virginia: National Technical Information Service. Large database (2,500,000 records) covering technical reports produced by government agencies and their contractors, including NASA, U.S. Department of Defense, and the U.S. Department of Energy. The entire database is available from commercial vendors, and a segment covering back to 1990 is available to the public at the NTIS website. Provides very good coverage of the technical report literature in aeronautics and astronautics (see Chapter 1, this volume).

Bibliographies and guides to the literature

There are several guides to the literature that are useful in their coverage of aeronautical and aerospace engineering. Most of the guides are not written exclusively for these areas, but they do contain sections discussing these subjects. These broader works have certain advantages in that they review resources for some of the other fields that are also of interest to aeronautical and aerospace engineers.

Anthony, L.J. (ed.) (1985) *Information Sources in Engineering* (2nd edn), London: Butterworths. Primary and secondary information sources are discussed, and these are followed by chapters on specific engineering fields. The "Aerospace Engineering" chapter has an excellent discussion of the technical information programs in a number of countries, key periodicals, abstracts, and a short list of monographs.

Auger, C.P. (1994) *Information Sources in Grey Literature* (3rd edn), London: Bowker Saur. This book concentrates on technical reports and other gray literature. It has a very strong chapter on aerospace technical reports and the organizations which have produced them over the years.

DePetro, T.G. and Naylor, T.E. (eds) (1997) *Selective Guide to Literature on Aerospace Engineering*, Washington, DC: American Society for Engineering Education. This work is one of the few recent guides to the aeronautical and aerospace engineering literature. The guide covers key indexes and abstracts, databases, reference works, handbooks, directories, technical report series, and journals.

Lord, C.R. (2000) *Guide to Information Sources in Engineering*, Englewood, CO: Libraries Unlimited. This is one of the newer guides to the literature. The work is divided up by format (handbooks, gray literature, journals, Internet resources) and then subdivided by engineering specialty. Aeronautical and aerospace entries have their own section under most formats.

Macleod, R.A. and Corlett, J. (eds) (2005) *Information Sources in Engineering* (4th edn), New York: Bowker Saur. A collection of chapters describing primary and secondary sources in all fields of engineering, and then listing the specific literature in over twenty fields. This edition provides a chapter on aerospace engineering.

Mildren, K.W. (1976) *Use of Engineering Literature*, London: Butterworths. One of the classic guides to the engineering literature, the book discusses the important information formats for engineering (journals, technical reports, standards), and abstracts and indexes. Individual chapters are devoted to specific disciplines. The "Aeronautics and Astronautics" chapter provides excellent background information on many of the subdisciplines such as propulsion and gas dynamics, as well as a lengthy discussion of classic texts in all of these areas.

Directories

Aeronautical and aerospace engineers work in a large industry and their efforts result in the production of complex products. The nature of this business lends itself to the production of directories to help individuals and companies keep track of new aircraft, space, and launch vehicles, and to find vendors of related products and services. Some of the major directories in the field are presented here, but there are more including a number produced by Jane's Information Group Ltd. With a powerful search engine such as Google, the Web is also a rich source of similar information.

Baker, D. (ed.) *Jane's Space Directory*, Coulsdon, U.K.: Jane's Information Group. An annual publication that provides extensive coverage of the space industry. Over 3,000 entries, arranged by sections including the space industry, space centers, launch vehicles, propulsion systems, satellites (military and civilian), and contractors. Full of illustrations, technical specifications, launch histories, and photos.

International Satellite Directory (2 vols), Sonoma, CA: Satnews Publishers. An annual directory of the space industry, oriented toward satellite owners and operators, programmers, broadcasters, and related companies. Entries provide basic company information, chief personnel, contacts, and a description of services. Includes a listing of operating satellites with launch dates, design life, geographical coverage, and orbit specifications.

Isakowitz, S.J., Hopkins, Jr, J.P., and Hopkins, J.B. (eds) (2004) *International Reference Guide to Space Launch Systems* (4th edn), Reston, VA: American Institute for

Aeronautics and Astronautics. International directory of launch vehicles and their related systems. Describes the vehicles (technical specifications, design, payload, and performance), flight histories, and launch pad operations. Includes appendices with abbreviations and acronyms, and a bibliography of technical references for each launch vehicle covered.

Jane's All the Worlds Aircraft, Coulsdon, U.K.: Jane's Information Group Ltd. This annual is the premier directory of the world's military and civilian aircraft. Entries are arranged alphabetically by country and then by manufacturer. Aircraft descriptions include development histories, technical and performance specifications, configurations, instrumentation, and armaments. Most entries contain photographs and diagrams. There are sections on air-launched missiles, engines, world flight records, a glossary, and a listing of first flights made during the year. Jane's produces a range of similar publications covering helicopters, air-launched weapons, avionics, and unmanned aerial vehicles.

World Aviation Directory and Aerospace Database (2 vols), New York: McGraw-Hill. Issued twice yearly, this two-volume set provides a comprehensive guide to the aviation industry. Entries in the "Directory" volume give brief information on companies and organizations, their products and services, and contact information. The "Buyer's Guide" serves as a guide to components and services vital to aviation, and covers avionics, parts, and equipment. Publication is available in print, on CD, or on the Web.

Encyclopedias and dictionaries

Encyclopedias fill a role in engineering by offering summaries of a subject that may be used either as an introduction to a new area for a reader, or as a review of a familiar one. Dictionaries complement encyclopedias by providing very concise meanings for thousands of words, phrases, and acronyms that would be too specific or narrow to cover as encyclopedia entries. The encyclopedias and dictionaries mentioned here are representative of what is available in aeronautical and aerospace engineering. A few multilingual dictionaries are listed, but others covering additional languages have been published.

Angelo, J.A. Jr (ed.) (2004) *Facts on File Dictionary of Space Technology*, New York: Facts on File. Comprehensive coverage of the basic concepts of space technology, space flight, and their underlying principles. Over 1,500 entries, with useful photos and illustrations.

Beck, S. and Aslezova, S. (eds) (2002) *Elsevier's Dictionary of Civil Aviation*, Amsterdam, the Netherlands: Elsevier. An English/Russian dictionary covering all aspects of civil aviation including air traffic control, navigation, flight, meteorology, communications, and airports. The English–Russian section has 19,000 entries and the Russian–English section has 21,000.

Bristow, G.V. (ed.) (2003) *Encyclopedia of Technical Aviation*, New York: McGraw-Hill. An encyclopedia geared to the needs of pilots. Coverage is very broad – includes aircraft systems and engines, aerodynamics, flight, air traffic control, aviation-related operations, and selected regulations and rules. Many diagrams and illustrations.

Cheremisinoff, N.P. (ed.) (1986) *Encyclopedia of Fluid Mechanics* (13 vols), Houston,

TX: Gulf Publishing Company. This thirteen-volume set brings together theoretical and practical engineering information on all types of flow phenomena. Contains contributions from hundreds of engineers and scientists. Individual volumes cover particular areas (e.g., gas-liquid flow, flow phenomena and measurement), with chapters exploring the topics in detail. All chapters contain numerous illustrations, tables, formulas, and bibliographic references.

Crocker, D. (ed.) (1999) *Dictionary of Aeronautical English*, London: Fitzroy Dearborn Publishers. English is the standard language of international civil aviation, and this volume brings together the vocabulary needed by pilots, crew members, and maintenance and ground staff. Covers terms that are used in flight, ground, and maintenance operations, and includes terms from air traffic control communications and weather.

Gunston, B. (ed.) (2004) *Cambridge Aerospace Dictionary*, Cambridge: Cambridge University Press. Comprehensive dictionary of aerospace and aeronautical terms. Entries are concisely written, and those for acronyms, which are very well represented, often indicate which country or organization developed the term. Appendices cover such diverse topics as electromagnetic frequency bands, phonetic alphabets, U.S. military aircraft and missile designations, and a guide to civilian aircraft registration numbering schemes for countries around the world.

Mark, H. (ed.) (2003) *Encyclopedia of Space Science and Technology* (2 vols), Hoboken, NJ: John Wiley & Sons. A two-volume set made up of eighty articles covering major topics in the space sciences. Articles treat the subject matter in depth, and are intended for a technically literate audience. Numerous photos, illustrations, diagrams, and a bibliography accompany the articles.

Multilingual Aeronautical Dictionary (1980) Neuilly sur Seine, France: Advisory Group for Aerospace Research and Development. A multilingual dictionary of major aeronautics and aerospace terminology. Ten languages are represented – English, French, Dutch, German, Greek, Italian, Portuguese, Turkish, Spanish, and Russian. The core is comprised of English terms and their definitions; alphabetic indexes of terms in other languages refer back to the English section.

Tomsic, J.L. and Eastlake, C.N. (eds) (1998) *SAE Dictionary of Aerospace Engineering* (2nd edn), Warrendale, PA: Society of Automotive Engineers. Includes 20,000 terms from aeronautics, aerospace, astronomy, geophysics, and computing. Intended for anyone interested in aerospace engineering as a student, educator, engineer, scientist, or technician. Some entries reference SAE-related standards.

Verger, F., Sourbes-Verger, I., and Ghirardi, R. (eds) (2003) *Cambridge Encyclopedia of Space: Missions, Applications, and Exploration*, Cambridge: Cambridge University Press. Arranged by chapters comprising in-depth articles. Covers all aspects of space science. Extensive graphics, tables, charts, illustrations, and photos. Sections include the space environment, orbits, satellites, launch vehicles, earth observations, telecommunications, and navigation.

Walker, P.M.B. (ed.) (1990) *Cambridge Air and Space Dictionary*, New York: Cambridge University Press. A subset of the *Cambridge Dictionary of Science and Technology*, containing 6,000 definitions from aeronautics, astronomy, meteorology, and space science. Special articles are interspersed among the entries, presenting fuller treatments of topics. Entries occasionally include formulas, illustrations, graphs, and tables.

Williamson, M. (2001) *Cambridge Dictionary of Space Technology*, Cambridge: Cambridge University Press. Contains 2,300 entries on all aspects of space science and technology. Concisely written with some formulas, illustrations, and photos. Many acronyms and abbreviations are included.

Handbooks and manuals

Engineering is a discipline known for an abundance of handbooks and manuals. There are several reasons for this – engineers often work across multiple fields in their assignments, and handbooks allow them to gather information that may be useful in such situations. Engineers involved in design projects often need numerical data and formulas to complete their work. Thus, handbooks in aeronautical and aerospace engineering may contain lengthy listings of materials property data, conversion factors, vehicle components, fluid mechanics, mathematical functions, formulas, and equations. Handbooks and manuals are usually excellent sources for background information on topics in the field.

Abbot, J.H. and van Doenhoff, A.E. (1958) *Theory of Wing Sections Including a Summary of Airfoil Data*, New York: Dover. Classic work on wing sections. Includes detailed theoretical and experimental data on most NACA airfoils.

Aerospace Structural Metals Handbook (2004) (6 vols), West Lafayette, IN: Purdue University. A six-volume set sponsored by the U.S. Department of Defense and prepared by CINDAS at Purdue University. The set provides comprehensive property data on alloys of interest to the aerospace industry. A wide variety of data is presented for each alloy, and references are provided to the original sources of the data.

AIAA Aerospace Design Engineers Guide (1998) (4th edn), Reston, VA: American Institute of Aeronautics and Astronautics. Intended to help design engineers develop aerospace products. Concise collection of commonly used aeronautical, mechanical, and electrical engineering reference data. Covers mathematics, geometric section properties, conversion factors, structural elements, mechanical and electrical components, and aircraft and spacecraft design factors.

Avallone, E.A. and Baumeister III, T. (eds) (1996) *Mark's Standard Handbook for Mechanical Engineers* (10th edn), New York: McGraw-Hill. Classic reference tool for engineering. Useful materials for aerospace engineers and students. Chapters cover a broad range of topics – thermodynamics, strength of materials, fuels, materials properties, mechanics, instrumentation, and the design of machine elements. Sections provide bibliographic references.

Damage Tolerant Design Handbook: A Compilation of Fracture and Crack Growth Data for High Strength Alloys (1994) (5 vols), West Lafayette, IN: Purdue University. A five-volume handbook sponsored by the U.S. Department of Defense and prepared by CINDAS at Purdue University. The set provides extensive property data on damage-tolerant materials of interest to aeronautical and aerospace engineers. Tables of contents in each volume facilitate access to the alloy data.

Chase, M.W. Jr (1998) *NIST-JANAF Thermochemical Tables* (4th edn) (2 vols), New York: American Institute of Physics. Provides critically evaluated physical and chemical property data, primarily of interest to those involved in rocket propulsion and other areas of combustion. The volumes contain detailed thermochemical properties of hundreds of chemicals and compounds, and access is provided by chemical name and chemical formula indexes. Entries give various properties, enthalpy of formation, heat capacity and entropy, phase and decomposition data, and bibliographic references to the original sources of the data.

Davies, M. (ed.) (2002) *The Standard Handbook for Aeronautical and Astronautical Engineers*, New York: McGraw-Hill. Contributions written by over sixty experts. Broad coverage of all areas of aeronautical and astronautical engineering including propulsion, structures, aerodynamics, stability and control, avionics and astrionics, aircraft systems, design, astrodynamics, spacecraft, space environment, aircraft safety and maintenance, human factors, and review materials on mathematics, fluid mechanics, electronics, and computers. Many diagrams, tables, graphs, charts, and illustrations, with an extensive index.

Johnson, R.W. (ed.) (1998) *Handbook of Fluid Dynamics*, Boca Raton, FL: CRC Press. Intended to help professionals new to the field as well as experts. Materials are arranged into six parts – basics, classic fluid dynamics, high-Reynolds number theories, numerical solutions, experimental methods, and applications. Each of these sections is made up of articles contributed by experts, providing comprehensive coverage of the field. Appendices cover mathematics, a table of dimensionless numbers, and properties of gases and vapors. An index is included.

Matthews, C. (ed.) (2002) *Aeronautical Engineer's Data Book*, Oxford: Butterworth Heinemann. Written for practicing engineers and students. Brings together a broad range of information on aeronautical engineering and aviation in a compact package. Chapters cover aerodynamics, flight dynamics, aircraft design and performance, fluid mechanics, and airport design.

Schetz, J.A. and Fuhs, A.E. (eds) (1996) *Handbook of Fluid Dynamics and Fluid Machinery* (3 vols), New York: Wiley. A three-volume work for the practicing engineer and researcher. The volumes are titled "Fundamentals of Fluid Dynamics," "Experimental and Computational Fluid Dynamics," and "Applications of Fluid Dynamics." Chapters provide in-depth information on specific areas within these fields.

Streeter, V.L. (ed.) (1961) *Handbook of Fluid Dynamics*, New York: McGraw-Hill. A classic handbook on fluid dynamics with contributions from distinguished experts. Written for engineers and scientists in the field. Deals with both fundamental concepts and applications. Covers fluid flow (one-dimensional, ideal, laminar, compressible, two-phase, open channel, stratified), turbulence, boundary layers, sedimentation, turbomachinery, fluid transients, and magnetohydrodynamics. Includes many formulas, equations, tables, graphs, and illustrations. Each chapter has a bibliography, and the volume has subject and author indexes.

Yang, W.J. (2001) *Handbook of Flow Visualization* (2nd edn), New York: Taylor & Francis. Covers techniques used to visualize flow in liquids and gasses. Includes both underlying theory and experimental applications. Techniques presented include Schlieren, shadowgraph, speckle, interferometry, light sheet, and plasma fluorescence. Numerous applications are given in subsequent chapters – medical, aerospace, wind tunnels, turbines, and indoor airflow.

Monographs and textbooks

There is no single printed or online resource to consult for comprehensive listings of monographs and textbooks in aeronautical and aerospace engineering. However, collection development in this area is facilitated by the activity of a relatively robust group of commercial and society publishers.

Some of the most active publishers for aero materials include Wiley, McGraw-Hill, Cambridge University Press, Oxford University Press, Kluwer, Springer, CRC, Taylor & Francis, and Academic Press. New offerings may be found at their websites or in printed catalogs and flyers. The AIAA publishes many high-quality items every year. AIAA issues texts aimed at the student audience as well as advanced monographs for practicing engineers and researchers. A number of publishers specialize in niches in the industry. For example, the Iowa State University Press publishes a series of texts and handbooks for commercial and private pilots.

A tool such as *Books in Print* is very useful for discovering what is available. Amazon.com and similar sites are also good resources for books in the field. For an excellent listing and discussion of some of the older classics in aeronautical and aerospace engineering, consult the chapter devoted to these subjects in *Use of Engineering Literature* (Mildren 1976).

Journals

Scholarly journals form the cornerstone of technical information exchange in aerospace and aeronautical engineering. Journals serve both as an announcement vehicle for new advances and as an archival medium. The conference literature and the technical report literature are also important pathways for announcing research results in the field. Trade journals are common and they serve as more of a current awareness service for industry news, events, new products, and similar topics of interest to professionals. This section includes representative scholarly and trade journals, from both engineering society and commercial publishers, with an emphasis on English language titles.

Acta Astronautics (1974–) New York: Pergamon Press (0094–5765).
Advances in Space Research (1981–) New York: Pergamon Press (0273–1177).
Aeronautical Journal (1897–) London: Royal Aeronautical Society (0001–9240).
Aerospace America (1963–) Reston, VA: American Institute of Aeronautics and Astronautics (0740–722X).
Aerospace International (1974–) London: Royal Aeronautical Society (1467–5072).
AIAA Journal (1963–) Reston, VA: American Institute of Aeronautics and Astronautics (0001–1452).
Atomization and Sprays (1991–) New York: Taylor & Francis (1044–5110).
Automatica (1963–) New York: Pergamon Press (0005–1098).
Aviation Week and Space Technology (1916–) New York: McGraw-Hill (0005–2175).
Canadian Aeronautics and Space Journal (1955–) Ottawa, Canada: Canadian Aeronautics and Space Institute (0008–2821).
Combustion and Flame (1957–) New York: Elsevier (0010–2180).
Combustion Science and Technology (1969–) London: Taylor & Francis (0010–2202).
Flight International (1909–) Sutton, U.K.: Reed Business Information (0015–3710).
Flow, Turbulence, and Combustion (1998–) Boston, MA: Kluwer (1386–6184).
IEEE Transactions on Aerospace and Electronic Systems (1965–) New York: Institute of Electrical and Electronic Engineers (0018–9251).
International Journal of Chemical Kinetics (1969–) New York: Wiley (0538–8066).

International Journal of Heat and Fluid Flow (1979–) New York: Elsevier (0142–727X).

International Journal of Heat and Mass Transfer (1960–) New York: Pergamon Press (0017–9310).

International Journal of Robust and Nonlinear Control (1991–) Chichester, U.K.: Wiley (1049–8923).

Journal of Aerospace Computing, Information, and Communication (2004–) Reston, VA: American Institute of Aeronautics and Astronautics (1542–9423).

Journal of Aircraft (1964–) Reston, VA: American Institute of Aeronautics and Astronautics (0021–8669).

Journal of the British Interplanetary Society (JBIS) (1934–) London: British Interplanetary Society (0007–084X).

Journal of Chemical Physics (1933–) New York: American Institute of Physics (0021–9606).

Journal of Computational Physics (1966–) New York: Academic Press (0021–9991).

Journal of Fluid Mechanics (1956–) London: Taylor & Francis (0022–1120).

Journal of Fluids Engineering (1973–) New York: American Society of Mechanical Engineers (0098–2202).

Journal of Guidance, Control, and Dynamics (1978–) Reston, VA: American Institute of Aeronautics and Astronautics (0731–5090).

Journal of Heat Transfer (1959–) New York: American Society of Mechanical Engineers (0022–1481).

Journal of Intelligent Material Systems and Structures (1990–) Lancaster, PA: Technomic Publishing (1045–389X).

Journal of Physical Chemistry (1896–) Washington, DC: American Chemical Society (0022–3654).

Journal of Propulsion and Power (1985–) Reston, VA: American Institute of Aeronautics and Astronautics (0748–4658).

Journal of Quantitative Spectroscopy and Radiative Transfer (1961–) New York: Pergamon Press (0022–4073).

Journal of Spacecraft and Rockets (1964–) Reston, VA: American Institute of Aeronautics and Astronautics (0022–4650).

Journal of the American Helicopter Society (1956–) Washington, DC: American Helicopter Society (0002–8711).

Journal of the Astronautical Sciences (1954–) New York: American Astronautical Society (0021–9142).

Journal of Thermophysics and Heat Transfer (1987–) Reston, VA: American Institute of Aeronautics and Astronautics (0887–8722).

Journal of Turbomachinery (1986–) New York: American Society of Mechanical Engineers (0889–504X).

Physics of Fluids (1958–) New York: American Institute of Physics (1070–6631).

Proceedings of the Institution of Mechanical Engineers, Part G: Journal of Aerospace Engineering (1989–) London: Mechanical Engineering Publications Ltd. (0954–4100).

Progress in Aerospace Sciences (1961–) New York: Pergamon Press (0376–0421).

Propellants, Explosives, Pyrotechnics (1982–) Weinheim, Germany: Verlag Chemie (0721–3115).

Smart Materials and Structures (1992–) New York: Institute of Physics Publishing (0964–1726).

Spaceflight (1956–) London: British Interplanetary Society (0038–6340).

Vertiflite (1963–) New York: American Helicopter Society (0042–4455).

Patents and standards

Designing a modern aircraft or space vehicle is an extremely complicated and expensive process. A commercial airliner may contain hundreds of thousands of components, manufactured by thousands of companies around the world. Engineering standards help insure the quality, fit, and performance of the materials and components used in these complex products. Aerospace engineers may need access to a variety of standards from both industry and military sources. There are numerous online resources for finding and ordering industry and military standards. Patents also serve as useful resources in aeronautical and aerospace engineering. As is the case with standards, the Web has made it easy to access these documents from the U.S. and other industrialized nations.

Acquisitions Streamlining and Standardization Information System (ASSIST) – Quick Search <http://assist.daps.dla.mil/quicksearch/> (accessed May 17, 2005). Washington, DC: U.S. Department of Defense. Much of the work in the aeronautical and aerospace industries is done under U.S. Department of Defense (DOD) contracts. Military standards are important to engineers working under these agreements. ASSIST was developed by the DOD to facilitate access to DOD and other government agency standards. The system provides for title and document number searching. Many of the retrieved active standards are available as PDF downloads on ASSIST.

American National Standards Institute (ANSI) <http://www.ansi.org> (accessed May 17, 2005). Washington, DC: American National Standards Institute. The American National Standards Institute serves as the main administrative and coordinating organization for the development and production of industry standards in the U.S. Almost one thousand companies and organizations are members of ANSI, and many of these groups participated in the development of the 10,000 ANSI standards now in existence. These standards cover all areas of engineering practice, materials, safety, and procedures. The website offers keyword searching and the ability to purchase electronic or print copies of documents.

NSSN: A National Resource for Global Standards <http://www.nssn.org> (accessed May 17, 2005). Washington, DC: American National Standards Institute. NSSN serves the engineering community as a one-stop location for identifying and ordering standards. The site is a cooperative effort between ANSI (American National Standards Institute), standards organizations in the U.S and other countries, and government agencies. Its search engine is powerful and can find standards by keyword or document number. The retrieved standards records give complete identifying and ordering information for print or downloadable copies. NSSN is a free service.

Society of Automotive Engineers (SAE) <http://www.sae.org> (accessed May 17, 2005). The SAE is an active developer of standards for aerospace engineering – it has 250 technical committees, subcommittees, and workgroups in this area. Standards are issued in these series: AMS (Aerospace Material Specifications), AS (Aerospace Standards), ARP (Aerospace Recommended Practice), and AIR (Aerospace Information Report).

Search engines and websites

The aerospace industry is a complex enterprise with research and production activities taking place in the commercial, military, and government sectors. The web is well suited for facilitating the communication of information for engineers in the field. There are search engines and numerous sites available for retrieving information on technical reports, parts and components, vendors, standards, patents, images, and companies. Several of these sites are covered in other sections of this chapter. The sites discussed here are representative of the types of additional resources that may be found on the Web.

Embry-Riddle Aeronautical University <http://amelia.db.erau.edu/gen/ref/genirt.html> (accessed May 17, 2005). The "Internet Research Tools" page at the Embry-Riddle website presents an excellent selection of evaluated web resources to the user. They are grouped into main categories (aerospace, aviation, business), and then divided into sub-topic. The site covers associations, U.S. and international agencies, job resources, industry news sources, technical reports, and other information.

The European Space Agency <http://www.esa.int/esaCP/index.html> (accessed May 17, 2005). The European Space Agency, a consortium of seventeen countries, provides extensive resources to space-related information and activities at their website.

Google <http://www.google.com> (accessed May 17, 2005). This powerful search engine is almost always a good starting point for finding information. *The Google Directory – Aerospace* <http://directory.google.com/Top/Science/Technology/Aerospace/> offers a subject-based listing of hundreds of sites that may be useful to aerospace engineers.

Mission and Spacecraft Library <http://msl.jpl.nasa.gov/home.html> (accessed May 17, 2005). A resource developed by the Jet Propulsion Laboratory and intended for the public – covers spacecraft from all nations. Provides basic information on spacecraft, launches, and orbits. Keyword searchable, or browsable by spacecraft, program name, or mission type.

National Aeronautics and Space Administration <http://www.nasa.gov> (accessed May 17, 2005). The main NASA website serves as a gateway to the whole family of NASA sites for their laboratories (Langley, Goddard, Ames), educational and research resources, mission news, and their technical report servers.

Yahoo Index of Space Sciences <http://dir.yahoo.com/Science/Space/> (accessed May 17, 2005). This Yahoo page indexes hundreds of aerospace and aviation websites – a good place to seek information if keyword searching fails to return appropriate sites.

Associations, societies, and organizations

Societies serve a number of important functions in engineering. Their overall role is to keep the membership current and informed about developments and trends in the discipline. Societies often promote short courses and other programs to help members grow professionally. They convene conferences that bring engineers together to exchange new ideas and expand networks of contacts. Their publication programs offer an outlet for research, and publishing in society journals is usually considered to be prestigious. Committees can advance the profession through the development of standards, and provide avenues for professional cooperation. Larger societies work to influence legislation and public policy at the highest levels. Most societies maintain websites that are full of information and are sometimes divided into public and member-only segments. The sites present news of the organization, career and employment information, conference calendars, and listings of publications. The site may include a searchable bibliographic database of recent society papers or other publications. In addition to the selected societies described here, there are other engineering societies important to members of the aerospace community. They include the American Society of Mechanical Engineers <http://www.asme.org>, Institute of Electrical and Electronics Engineers <http://www.ieee.org)\>, Acoustical Society of America <http://asa.aip.org>, and the Society of Automotive Engineers <http://www.sae.org>. These will be covered in detail in other chapters of this book.

American Astronautical Society (AAS) <http://www.astronautical.org> (accessed May 17, 2005). Professional society dedicated to advancing space sciences, education, and exploration. AAS sponsors professional, scientific, and engineering meetings, and has an active publishing program. Key objectives of the society include assessing public and private space programs, providing guidance to space-planning efforts, and promoting research in the various sciences required for the exploration and utilization of space.

American Helicopter Society International (AHSI) <http://www.vtol.org> (accessed May 17, 2005). Founded in 1944, the society has dedicated itself to the advancement of vertical flight, and has exerted an ongoing influence on the rotorcraft industry, both military and civilian. The AHSI conducts a variety of professional and technical meetings, and publishes journals and proceedings. There are 6,000 individual members and 100 corporate members in AHSI.

American Institute of Aeronautics and Astronautics (AIAA) <http://www.aiaa.org> (accessed May 17, 2005). The world's largest professional society devoted to advancing engineering and science in aviation, space, and defense. AIAA has an extensive array of committees that concentrate on both technical and member support activities. The AIAA publications program is extensive – they publish seven scholarly journals and hundreds of technical papers annually.

International Astronautical Federation (IAF) <http://www.iafastro.com/index.htm> (accessed May 17, 2005). The IAF is a professional group whose membership comprises of corporations, educational institutions, space agencies, research institutes, and other organizations. Its role is to advance the knowledge of space and the development of space assets for the benefit of society. It regularly convenes technical and professional meetings and publishes papers presented at the various IAF congresses.

International Scientific and Technical Gliding Organization (OSTIV) <http://www.isd.uni-stuttgart.de/OSTIV> (accessed May 17, 2005). OSTIV is a society created to encourage the "science and technology of soaring and development and use of the sailplane in pure and applied research" (OSTIV web page).

The group holds a congress at each World Gliding Championship and publishes the papers presented at these meetings.

Royal Aeronautical Society (RAeS) <http://www.raes.org.uk/> (accessed May 17, 2005). A well-known society with members in almost 100 countries. Membership is open to everyone from aviation enthusiasts and students through aerospace professionals. Corporate memberships are also available. The society's goal is the advancement of the global aerospace community. The RAeS publishes a number of journals and conference proceedings.

Conclusion

The aeronautical and aerospace industries continue to expand in many countries around the world, driven by commercial and defense-related demand for aircraft, missiles, and satellites, and for their support services and systems. The literature of the field is broad and centered around journal articles, technical reports, and conference papers. Most of the information is publicly available, but a portion falls under various distribution controls due to its militarily sensitive or proprietary nature. The distribution mechanisms and the databases covering aerospace and aeronautical information are well established and provide engineers working in these fields with ready access to the data they require.

References

Anderson, J.D. Jr (1997) *A History of Aerodynamics and Its Impact on Flying Machines*, Cambridge, UK: Cambridge University Press.

McCormick, B., Newberry, C., and Jumper, E. (2004) *Aerospace Engineering Education During the First Century of Flight*, Reston, VA: American Institute of Aeronautics and Astronautics.

Mildren, K.W. (1976) *Use of Engineering Literature*, London: Butterworths.

Von Braun, W. and Ordway, F.I. (1966) *History of Rocketry and Space Travel*, New York: Thomas Y. Crowell Company.

4 Agricultural and food engineering

Kathy Fescemyer and Helen Smith

Introduction

Agricultural engineering is a multi-disciplinary profession that relies on expertise in both the engineering and agricultural fields. The roots of agricultural engineering go back to the earliest civilizations with the origin of the hoe, early irrigation systems, and other early farming methods. Over time, advances in agricultural machinery resulted in more efficient crop production; irrigation and drainage developments produced more usable land resources while conserving natural resources; advances in agricultural buildings resulted in a healthier environment for livestock and increased efficiencies in crop storage and production; and the use of electricity on the farm made possible the automation of many farm processes (Isaacs, 2003). Food engineers were also developing advances in food processing to increase the safety of foods with cost-efficient technology. Research in the heating, refrigeration, drying, chemical preservation, and packaging of food has advanced the development of these areas (Farkas, 2003).

Prior to 1907, engineering for agriculture was done by mechanical, architectural, electrical, and civil engineers. As a profession, formalized agricultural engineering began with the American Society of Agricultural Engineers (ASAE) in 1907. The scope of agricultural engineering has grown over time. The ASAE now calls itself the Society for Engineering in Agricultural, Food and Biological Systems, and defines the profession as follows: "Agricultural, Food and Biological Engineers develop efficient and environmentally sensitive methods of producing food, fiber, timber, and renewable energy sources for an ever-increasing world population" (ASAE, 2004). The ASAE has eight specialty areas which show the depth of their interest. These areas include biological engineering; food and process engineering; information and electrical technologies; power and machinery; soil and water; structures and environment; ergonomics, safety and health; and emerging areas.

Many basic engineering information resources such as *Compendex* and *Perry's Chemical Engineers' Handbook* are essential to the study of agricultural engineering. These basic resources are described in Chapter 2. The chapters in this book on mechanical, electrical, civil, and architectural engineering should also be consulted for important information resources because these engineering disciplines still provide basic concepts and research that are applied to agricultural engineering. Biological engineering is a new engineering discipline that has its roots in agricultural engineering. One recent definition of biological engineering is "the discipline of engineering that integrates biology, physics, chemistry and mathematics with engineering principles in the design of biologically based products and processes from the biomolecular to organisms to the ecosystem" (Dooley, 2003). Because of its importance, a separate chapter on bioengineering is included in this work. This chapter encompasses the traditional areas of agricultural engineering and food engineering and will include some broader biosystems and biotechnology resources that are relevant to agricultural engineering. The focus is on the information resources that will be useful to agricultural engineering librarians, upper-level undergraduates, graduate students, and professionals.

Searching the library catalog

The advent of keyword searching has greatly enhanced the retrieval of materials via library catalogs. However, title word searching should not be relied on to the neglect of controlled vocabulary. Controlled vocabulary searches provide a precise means of finding a specific subject in online catalogs. The following list of Library of Congress Subject Headings contains most of the specific agricultural and food engineering terms.

Agricultural engineering
Agricultural implements
Agricultural machinery
Agricultural mechanics
Agricultural processing
Agriculture safety measures
Aquacultural engineering
Dairy engineering
Dairy processing
Drainage
Electricity in agriculture
Farm buildings
Farm engines
Farm equipment
Feed processing
Food industry and trade
Food packaging
Food processing industry
Forestry engineering
Horticultural machinery
Irrigation engineering
Livestock housing

Seeds processing
Tractors

In addition to these terms, combining individual types or groups of plants or livestock with subheadings such as: Housing; Equipment and supplies; Processing; or Postharvest technology (e.g. "Swine – equipment and supplies" or "Corn – processing" or "Vegetables – postharvest technology") will locate relevant publications.

The following Library of Congress call number ranges cover most of the specifically agricultural engineering material, but again, materials in other basic engineering or agricultural areas should not be ignored.

S671–S790 (agricultural engineering, equipment, machinery, buildings)
TJ1480–TJ1500 (agricultural machinery)
TK4018 (electricity in agriculture)
TP370–465 (food technology, manufacturing and processing)

Abstracts and indexes

The general engineering indexes included in other chapters should always be consulted when searching for agricultural and food engineering information. In addition, depending on the subject of the search in question, indexes from other chapters should also be consulted. The following resources focus specifically on agricultural and food engineering, or are basic agriculture or biological resources, and should not be ignored. All these resources are available in electronic format and more information is provided at the websites listed.

AGRICOLA (1970–present) Beltsville, MD: National Agricultural Library <http://agricola.nal.usda.gov/> (accessed October 7, 2004). This is a core agriculture database from the National Agricultural Library that indexes journal articles as well as books, book chapters, USDA, State Experiment Station, and State Extension service publications. Agriculture is defined broadly and includes animal and plant science, entomology, agronomy, horticulture, rural sociology, agricultural economics, family living, food and nutrition, and agricultural and biosystems engineering. This database is available from many vendors as well as directly from the National Agricultural Library via a public web access version (see web address above).

Agricultural Engineering Abstracts/CAB Abstracts (1976–present) Wallingford, Oxon, UK: CAB International <http://www.cabi-publishing.org> (accessed October 8, 2004). The *Agricultural Engineering Abstracts* database is available as a stand-alone product, or is included in the multi-disciplinary *CAB Abstracts* product. It indexes all aspects of internationally published research on agricultural engineering. *CAB Abstracts* is a premier index to the agricultural literature, covering everything from production agriculture to nutrition and economics. *Biofuels Abstracts* (biofuels and bioenergy research) and *Irrigation and Drainage Abstracts* (all aspects of water resource management, soil water, crop irrigation and resulting environmental aspects) are also available as stand-alone products, or within *CAB Abstracts*. Some overlap between the subsets does occur. *CAB Abstracts* is available from many vendors or directly from CABI Publishing. *Agricultural Engineering Abstracts, Biofuels Abstracts* and *Irrigation and Drainage Abstracts* are available directly from CABI Publishing.

Agricultural Engineering Index (1999) <http://bae.engineering.ucdavis.edu/AgIndex/aeindex.html> (accessed October 7, 2004). William J. Chancellor, Compiler. This is a freely downloadable database, available courtesy of UC Davis that includes all technical articles appearing in ASAE periodicals from 1950 to 1999 (plus some articles appearing in publications from other international societies). It is no longer being updated. It is also searchable via a telnet session (sweetpea.engr.ucdavis.edu).

AGRIS and CARIS Homepage (1975–present) Rome: Food and Agricultural Organization of the United Nations. <http://www.fao.org/AGRIS/> (accessed October 7, 2004). AGRIS is the international information system for the agricultural sciences and technology. Participating countries input references to the literature that are produced within their boundaries. The system identifies literature dealing with all aspects of agriculture, and covers agricultural engineering as a part of that. CARIS is the Current Agricultural Research Information System, created by FAO in 1975 to identify and facilitate the exchange of information about current agricultural research projecting being carried out in, or on behalf of, developing countries. The system identifies projects dealing with all aspects of agriculture. The basic unit in CARIS is a set of data describing a single project, giving such details as project title, objectives, inception and termination dates; names of researchers and their specializations, contact addresses of individuals and institutes carrying out the research. Coverage is from 1975 to the present. All searching and viewing is free.

ASAE Technical Library (2001–present) St. Joseph, MI: American Society of Agricultural Engineers <http://asae.frymulti.com/> (accessed October 7, 2004). This is the online publication site of the American Society of Agricultural Engineers and is a core resource for agricultural and biosystems engineering information. It includes the full text of all ASAE documents (journals, conference proceedings, monographs, standards) published since 2001 with plans to add older information. The site is searchable, or you can browse the table of contents of the publications. Searching and viewing the abstracts is free. Access to the full text is via ASAE membership or site license.

Biological and Agricultural Index (1916–present) New York: H.W. Wilson Co. <http://www.hwwilson.com/> (accessed October 7, 2004). This index cites articles from more than 250 English-language periodicals published in the United States and elsewhere. Periodical coverage includes a wide range of scientific journals, from popular to professional, that pertain to biology and agriculture. Although not suitable for extensive research in any area, this database does cover the core journals in all of agriculture and can be used as a starting point. Online coverage is from 1985 to the present and is available via many vendors.

BIOSIS Previews (1969–present) Philadelphia, PA: Thompson Scientific <http://www.biosis.org/> (accessed October 7, 2004). BIOSIS is the premier database for the biological sciences and includes information on biochemistry, microbiology, human biology, physiology, botany, and zoology. Biosystems and biological engineering are covered as part of the biological sciences. The subset of *Biological Abstracts* includes journal article information, and *Biological*

Abstracts/RRM (Reports, Reviews, Meetings) includes conference proceedings and meeting reports, book contents, and patents. Online coverage is 1985 to the present and it is available via many vendors.

Food Science and Technology Abstracts (FSTA) (1969–present) Reading, UK: International Food Information Service <http://www.ifis.org/> (accessed October 7, 2004). This database contains comprehensive coverage of all aspects of food science and technology research, including: raw materials and ingredients; manufacturing and distribution; food safety; and product development and consumer issues. Food engineering is covered extensively from harvesting to processing and packaging technology. Online coverage is 1969 to the present and it is available via many vendors.

National Ag Safety Database (1994–) Atlanta: United States Centers for Disease Control and Prevention <http://www.cdc.gov/nasd/> (accessed October 7, 2004). This is a full-text consumer-oriented database that seeks to provide a national resource for the dissemination of information; to educate workers and managers about occupational hazards associated with agriculture-related injuries, deaths and illnesses; to provide prevention information; to promote the consideration of safety and health issues in agricultural operations; and to provide a convenient way for members of the agricultural safety and health community to share educational and research materials with their colleagues. Safety professionals and organizations from across the nation have contributed the information contained in NASD. Only current information is included. There is no charge for searching or viewing.

Bibliographies and guides to the literature

Included in this section are publications that provide information on the literature of the disciplines of agricultural engineering and food science. Many are useful for providing an historical overview to the literature of agricultural engineering, or for verifying incomplete or inaccurate references obtained from other sources.

American Society of Agricultural Engineers (1907/60–1986/90) *Agricultural Engineering Index*, St. Joseph, MI: American Society of Agricultural Engineers. This five-volume set provides a complete bibliographic overview of the field of agricultural engineering. The entries are arranged alphabetically by subject.

American Society of Agricultural Engineers (1985–1997) *Comprehensive Index of Publications*, St. Joseph, MI: American Society of Agricultural Engineers. This publication, and its predecessors *Comprehensive Index of ASAE Publications* (1979–1984) and *Comprehensive Keyword Index of ASAE Publications* (1971–1978), provides keyword and author indexes to ASAE publications and a select group of other agricultural engineering publications.

Brogdon, J. and Olsen, W.C. (1995) *The Contemporary and Historical Literature of Food Science and Human Nutrition*, Ithaca, NY, and London: Cornell University Press. Brogdon and Olsen provide quality historical background in the area of food science. In addition, there are lists of core monographs, primary journals, databases, and primary historical literature on this topic.

Cloud, G.S. (1985) *Selective Guide to Literature on Agricultural Engineering*, College Station, TX: American Society for Engineering Education, Engineering Libraries

Division. This is a selective annotated list of agricultural engineering sources for researchers.

Green, S. (1985) *Keyguide to Information Sources in Food Science and Technology*, London and New York: Mansell Publishing Ltd. Green provides a complete overview in the area of food science and technology.

Hall, C.W. (1976) *Bibliography of Agricultural Engineering Books*, St. Joseph, MI: American Society of Agricultural Engineers. This book provides a complete listing of books focusing on agricultural engineering and published prior to 1976.

Hall, C.W. (1976) *Bibliography of Bibliographies of Agricultural Engineering and Related Subjects*, St. Joseph, MI: American Society of Agricultural Engineers. This is a list of the bibliographies of agricultural engineering published prior to 1976.

Hall, C.W. and Olsen, W.C. (eds) (1992) *The Literature of Agricultural Engineering*, Ithaca, NY: Cornell University Press. This volume surveys the traditional subjects such as power and machinery, soil and water, structures and environment, and electric power and processing. Subject specialists provide essays describing the literature of agricultural, food, forest, and aquacultural engineering. The volume also supplies core lists of monographs and primary journals and historical literature that made important contributions to the subject.

Hutchinson, B.S. and Greider, A.P. (eds) (2002) *Using the Agricultural, Environmental, and Food Literature*, Dekker. The most recently published information guide in agriculture. Chapters on agricultural engineering and food science review the most important sources in the discipline.

Morgan, B. (1985) *Keyguide to Information Sources in Agricultural Engineering*, London and New York: Mansell. Despite being twenty years old, this volume provides a complete overview of the literature of agricultural engineering except for Internet resources. Included is information on general searching of the literature, language problems, library classification, journals, directories, handbooks, monographs, and information about organizations that focus on agricultural engineering.

Directories

Many directories are now available on the Internet, and as such are often more up-to-date than printed resources. Institutions or organizations may often be found via the Internet; however, some societies limit access to their membership directories to members only. The educational and equipment directories listed here may facilitate searching for addresses or contact information.

ASAE Membership Roster <http://www.asae.org> (accessed October 8, 2004). Available only to ASAE members. From the site: "Conduct in-depth searches of the entire ASAE membership by name, address, city, state, zip code, country, member grade, professional registration and years of membership. Links are then provided to individual member information."

Directory of Universities and Approved Programs <http://www.ift.org/cms/?pid=1000426> (accessed October 8, 2004). The universities on this list offer undergraduate programs that "meet the "IFT Undergraduate Education Standards for Degrees in Food Science." Includes United States, Canadian and Mexican institutions.

Educational Programs in Agricultural Engineering and Related Fields <http://www.asae.org/membership/students/edprogrm.html> (accessed October 8, 2004). The ASAE provides this list of United States and Canadian educational programs and includes contact details for further information.

Food Master <http://www.foodmaster.com/> (accessed October 8, 2004). This is an online database listing equipment, ingredients, supplies, and services for food and beverage manufacturers.

Food Processing Machinery <http://www.processfood.com/productLocator/index.cfm> (accessed October 8, 2004). An online directory of food processing equipment.

Graduate Program Directory <http://www.ift.org/cms/?pid=1000624> (accessed October 8, 2004). The Institute of Food Technologists listing of United States and Canadian graduate programs in food science.

Guide to Consultants <http://www.asae.org/resource/guideforms.html> (accessed October 8, 2004). The ASAE provides this guide as a service to businesses and consumers interested in agricultural or biological engineering assistance.

International Academic Programs Agricultural, Food, or Biological Engineering Departments <http://www.asae.org/membership/students/intlacademic.html> (accessed October 8, 2004). The ASAE provides this list of links to international (including the United States) educational programs and departments.

International Directory of Agricultural Engineering Institutions <http://www.fao.org/intdir/> (accessed October 8, 2004). This directory includes more than 640 institutions from approximately 100 countries. It contains names and addresses of the institutions, as well as a description of their activities. It is searchable by keywords or geographic areas, names, activities, or all fields combined.

International List of Agricultural Engineering Societies <http://www.asae.org/membership/international.html> (accessed October 8, 2004). The ASAE provides this directory of other agricultural engineering societies. It includes related professional engineering and agricultural science societies as well.

Worldwide Agricultural Machinery and Equipment Directory <http://www.agmachine.com/> (accessed October 8, 2004). Agmachine.com is a large specialized directory of agricultural machinery and farm equipment manufacturers on the internet. Categories are browsable, but there is no search feature.

WorldFoodNet <http://www.worldfoodnet.com/> (accessed October 8, 2004). This site contains an online food supplier's directory and buyers' guide. You will also find the latest product launches, links to upcoming events, and much more.

Encyclopedias and dictionaries

Encyclopedias and dictionaries provide excellent basic definitions and descriptions of terms in agricultural and food engineering. The following dictionaries and encyclopedias are good sources, but none are comprehensive. One may need to consult several dictionaries to locate the definition for a specific term because all define a very different set of terms. The encyclopedias are also varied and provide different types of information. Although not listed here, older resources are also valuable in providing historical terminology and descriptions.

Arntzen, C.J. and Ritter, E.M. (eds) (1994) *Encyclopedia of Agricultural Science* (4 vols), San Diego, CA: Academic Press. This encyclopedia provides articles in all areas of agriculture. Articles contain an outline, a glossary, cross-references to other articles in the same encyclopedia, and a bibliography.

Bains, W. (2004) *Biotechnology from A to Z* (3rd edn), Oxford: Oxford University Press. The entries in this dictionary provide a quick description of the concept, describe related terms, and give an indication of achievements accomplished with the technology.

Bender, D.A. and Bender, A.E. (1999) *Benders' Dictionary of Nutrition and Food Technology*, Boca Raton: CRC Press. A resource of over 5,000 terms, this dictionary provides definitions for foods and food-related terms.

Caballero, B., Trugo, L., and Finglas, P. (2003) *Encyclopedia of Food Sciences and Nutrition* (2nd edn) (10 vols), Amsterdam: Academic Press. A comprehensive ten-volume encyclopedia that covers all aspects of food and food technology; each entry is extensive, illustrated, and contains a broad list of references for further reading.

Considine, D.M. and Considine, G.D. (1982) *Foods and Food Production Encyclopedia* (2 vols), New York: Van Nostrand Reinhold. Although older, this encyclopedia contains information on the cultivation of food plants and livestock and the processing of the resulting food materials into refined products.

Farrall, A.W. and Basselman, J.A. (1979) *Dictionary of Agricultural and Food Engineering* (2nd edn), Danville, IL: Interstate. An older dictionary that provides definitions for terms in agriculture engineering and food engineering.

Flickinger, M.C. and Drew, S.W. (1999) *Encyclopedia of Bioprocess Technology: Fermentation, Biocatalysis and Bioseparation* (5 vols), New York: Wiley. This five-volume set presents the applications and established theories in biotechnology, focusing on industrial applications of fermentation, biocatalysis and bioseparation. It is part of the Wiley biotechnology encyclopedias.

Francis, F.J. (2000) *Encyclopedia of Food Science and Technology* (2nd edn) (4 vols), New York: Wiley. This 4 volume set provides articles in all areas of food science. Each entry gives a detailed description of the topic with graphs and illustrations, and a lengthy bibliography.

Hall, C.W., Farrall, A.W., and Rippen, A.L. (1986) *Encyclopedia of Food Engineering* (2nd edn), Westport, CT: AVI Publishing. Although dated, this encyclopedia provides detailed articles with emphasis on the equipment and facilities used in food handling, manufacture, and transportation.

Heldman, D.R. (ed.) (2003) *Encyclopedia of Agricultural, Food and Biological Engineering*, New York: Marcel Dekker. This encyclopedia focuses on the processes that produce raw agricultural materials and convert them into products for distribution. Each entry contains a detailed description of the topic, several illustrations, and relevant references.

Lewis, R.A. (2002) *CRC Dictionary of Agricultural Sciences*, Boca Raton, FL: CRC Press. This is a specialized agricultural dictionary that defines many agricultural terms.

Philippsborn, H.E. (2002) *Elsevier's Dictionary of Nutrition and Food Processing in English, German, French and Portuguese*, Amsterdam: Elsevier. This dictionary provides translations from English into three other languages and with indexes of German, French, and Portuguese words translated back into English.

Spier, R.E. (2000) *Encyclopedia of Cell Technology* (2 vols), New York: Wiley. Part of the Wiley biotechnology encyclopedias. This encyclopedia gives an overview of both animal and plant cell technology. It provides in-depth articles on bioreactors, and other topics of interests to biotechnologists.

Stewart, B.A. and Howell, T.A. (2003) *Encyclopedia of Water Science*, New York: Marcel Dekker. This one-volume encyclopedia focuses on agricultural water management. Topics include aquifers, drainage, erosion, evaporation, ground water, irrigation, precipitation, soil water, surface water, and many others.

Tosheva, T., Djarova, M., and Deliiska, B. (2000) *Elsevier's Dictionary of Agriculture in English, German, French, Russian and Latin*, Amsterdam; New York: Elsevier. This dictionary contains 9,389 terms and over 4,000 cross-references with indexes in each language.

Troeh, F.R. and Donahue, R.L. (2003) *Dictionary of Agricultural and Environmental Science*, Ames, IA: Iowa State Press. This dictionary emphasizes the terminology of the ecological aspects of agriculture.

Van der Leeden, F., Troise, F.L., and Todd, D.K. (1990) *The Water Encyclopedia* (2nd edn), Chelsea, MI: Lewis Publishers. This book provides over 600 tables of information about the hydrologic environments. Tables include information on pollution, contamination of surface and ground water, use of pesticides and fertilizers, waste disposal, water treatment, and other topics.

Handbooks and manuals

As with other formats, the handbooks and manuals used by agricultural engineers are often the same as those engineers in the supporting engineering disciplines. The handbooks listed below are those that focus specifically on the information used by agricultural and food engineers.

Brown, R.H. (1988) *CRC Handbook of Engineering in Agriculture*, Boca Raton, FL.: CRC Press, vol. 1. Crop production engineering; vol. 2. Soil and water engineering; vol. 3. Environmental systems engineering. Although somewhat outdated, this handbook includes basic information on most agricultural engineering processes.

CIGR, the International Commission of Agricultural Engineering (1999) *CIGR Handbook of Agricultural Engineering*, St. Joseph, MI: American Society of Agricultural Engineers, vol. 1. Land and water engineering; vol. 2. Animal production and aquacultural engineering; vol. 3. Plant production engineering; vol. 4. Agro-processing engineering; vol. 5. Energy and biomass engineering. This handbook covers all the major fields of agricultural engineering and, although lacking in some topics, it is still quite comprehensive and valuable.

Ibarz, A. and Barbosa-Cánovas, G.V. (2003) *Unit Operations in Food Engineering*, Boca Raton, FL: CRC Press. This handbook presents the basic information required to design food processes.

Nebraska Tractor Tests <http://tractortestlab.unl.edu/> (accessed October 8, 2004). The University of Nebraska Tractor Test Laboratory is the officially designated tractor-testing station for the United States, and tests tractors according to the codes of the Organization for Economic Co-operation and Development (OECD). Specifications and performance of specific tractor models are tested. Tractor test reports from 1999 to the present are available online; others may be purchased from the website.

Rao, M.A. and Rizvi, S.S.H. (1995) *Engineering Properties of Foods* (2nd edn), New York: Marcel Dekker. This handbook is a classic in describing the physical properties of foods.

Valentas, K.J., Rotstein, E., and Singh, R.P. (1997) *Handbook of Food Engineering Practice*, Boca Raton, FL: CRC Press. An overall guide on food engineering that includes specific food-processing systems (freezing, drying), as well as some general information on areas such as shelf life, packaging, and food properties.

Monographs and textbooks

Agricultural engineering students will be using information from a wide spectrum of subjects. They take courses in engineering mechanics, thermodynamics, fluid mechanics, as well as basic mathematics and physics courses. The following are a few recent general textbooks that focus on agricultural and food engineering.

Roth, L.O. and Field, H. (1999) *An Introduction to Agricultural Engineering: A Problem Solving Approach*, Gaithersburg, MD: Aspen Publishers.

Singh, R.P. and Heldman, D.R. (2001) *Introduction to Food Engineering*, London: San Diego, CA: Academic Press.

Tollner, E.W. (2002) *Natural Resources Engineering*, Ames: Iowa State Press.

Other materials are available from traditional publishers such as Elsevier, Marcel Dekker, and CRC Press. The American Society of Agricultural Engineers also publishes excellent basic and advanced books. Two organizations that publish practical information are the Natural Resource, Agriculture, and Engineering Service at <http://www.nraes.org/> (accessed October 8, 2004) and the Midwest Plan Service at <http://www.mwpshq.org/> (accessed October 8, 2004). Searching library union catalogs such as WorldCat, or library catalogs at agricultural universities, will often retrieve these items.

Journals

Scholarly journals provide the most important form of primary literature in the sciences. This list includes important research-oriented titles focused specifically on agricultural and food engineering. A selection of journals in the area of biotechnology is included due to the interdisciplinary research performed in biosystems engineering. In addition, basic engineering and agricultural journals, though not listed here, include articles in agricultural and food engineering. Relevant articles may also appear in specialized engineering journals, especially in the areas of biological, mechanical, and civil engineering.

Acta Biotechnologica (1981–2003) Weinheim: Wiley-VCH Verlag GmbH. Online www3.interscience.wiley.com/cgi-bin/home. Wiley Interscience (0138–4988). In 2004 this journal merged with *Engineering in Life Sciences* which focuses on biotechnology in the life sciences.

Agricultural Engineering International: The CIGR Journal of Scientific Research and Development (1999–). International Commission of Agricultural Engineering

(1682–1130). Online: <http://cigr-ejournal.tamu.edu/>. This electronic journal is produced by the International Commission of Agricultural Engineering (CIGR) and is freely accessible to all. The journal emphasizes efficiency in agricultural production, the responsible use of natural resources, sustainability, value-added processing and other agricultural engineering subjects.

Agricultural Mechanization in Asia, Africa, and Latin America (AMA) (1981–) (0084–5841). Tokyo: Farm Machinery Industrial Research Corp. Continues *Agricultural Mechanization in Southeast Asia* and is produced in Japan.

Agricultural Systems (1976–) Amsterdam: Elsevier (0308–521X). Online: ScienceDirect. Topics included are the development and application of systems methodology, including system modeling, simulation and optimization; ecoregional analysis of agriculture and land use; studies on natural resource issues related to agriculture; impact and scenario analyses related to topics such as GMOs, multifunctional land use, and global change; and the development and application of decision and discussion support systems; approaches to analyzing and improving farming systems; technology transfer in tropical and temperate agriculture; and the relationship between agricultural development issues and policy.

Agricultural Water Management (1976–) Amsterdam: Elsevier (0378–3774). Online: ScienceDirect. Articles cover the subjects of irrigation and drainage of cultivated areas, collection and storage of precipitation water in relation to soil properties and vegetation cover; the role of ground and surface water in nutrient cycling, water balance problems, exploitation and protection of water resources, control of flooding, erosion and desert creep, water quality and pollution both by, and of, agricultural water, effects of land uses on water resources, water for recreation in rural areas, and economic and legal aspects of water use.

Agriculture, Ecosystems & Environment (1983–) Amsterdam: Elsevier (0167–8809). Online: ScienceDirect. Scope includes topics such as the influence of agricultural production methods on the environment, including soil, water, and air quality, the use of energy and non-renewable resources; agroecosystem management, including agro-biodiversity and response of multi-species ecosystems to environmental stress; the effect of pollutants on agriculture; agro-landscape values and changes, landscape indicators, and sustainable land use; farming system changes and dynamics; integrated pest management and crop protection; and problems of agroecosystems from a biological, physical, economic, and socio-cultural standpoint.

Applied and Environmental Microbiology (1976–) Washington, DC: American Society for Microbiology (0099–2240). Online: PubMed Central; Highwire. The articles in this journal focus on biotechnology, microbial ecology, food microbiology, and industrial microbiology, and highlight research in the development of new processes or products.

Applied Engineering in Agriculture (1985–) St. Joseph, MI: American Society of Agricultural Engineers (0883–8542). Online: American Society of Agricultural Engineers. This journal publishes practical applications of current research in all areas of agricultural engineering. Articles focus on the latest techniques and approaches for field equipment, food processing, farmstead systems, energy, irrigation, drainage, farm structures, storage and handling, electronics, natural resources, and other related topics.

Applied Microbiology and Biotechnology (1984–) Heidelberg: Springer Verlag Heidelberg (0175–7598). Online: SpringerLink. Continues *European Journal of Applied Microbiology and Biotechnology*.

Appropriate Technology (1974–) London: Research Information Ltd. (0305–0920). The articles in this journal are international in focus and emphasize developing and using sustainable techniques in agriculture.

Aquacultural Engineering (1982–) Amsterdam: Elsevier (0144–8609). Online: ScienceDirect. The focus is on the design and development of effective aquacultural systems for marine and freshwater facilities. Articles emphasize the engineering and design of aquaculture facilities, construction experience and techniques, materials selection and their uses, and quantification of biological data and constraints for developing facilities.

Biochemical Engineering Journal (1998–) Amsterdam: Elsevier (1369–703X). Online: ScienceDirect. This journal promotes progress in the chemical engineering aspects of the development of biological processes associated with everything from raw materials preparation to product recovery, and is relevant to many industries in the fields of medical/healthcare, food, and environmental protection.

Biodegradation (1990–) Dordrecht, the Netherlands; Boston: Kluwer (0923–9820). Online: Kluwer. *Biodegradation* focuses on research pertaining to the detoxification, recycling, amelioration, or treatment of waste materials and pollutants by naturally occurring microbial strains or associations of recombinant organisms

Biomass & Bioenergy (1991–) Amsterdam: Elsevier (0961–9534). Online: ScienceDirect. This journal concentrates on biomass, biological residues, bioenergy processes, bioenergy utilization, and biomass and the environment. Articles will be found on topics such as energy crop production processes, genetic improvement, wastes from agricultural production and forestry, processing industries, and municipal sources, fermentations, thermochemical conversions, liquid and gaseous fuels, and petrochemical substitutes, direct combustion, gasification, electricity production, chemical processes, by-product remediation, the net energy efficiency of bioenergy systems, assessment of sustainability, and biodiversity issues.

Bioprocess and Biosystems Engineering (1986–) Heidelberg: Springer-Verlag Heidelberg (1615–7591). Online: SpringerLink. Focuses on multi-disciplinary approaches for integrative bioprocess design. From the website: "Of special interest are the rational manipulation of biosystems through metabolic engineering techniques to provide new biocatalysts as well as the model based design of bioprocesses (up-stream processing, bioreactor operation and downstream processing) that will lead to new and sustainable production processes."

Bioresource Technology (1991–) Amsterdam: Elsevier (0960–8524). Online: ScienceDirect. Topics included are biomass, biological waste treatment, bioenergy, biotransformations and bioresource systems analysis, and technologies associated with conversion or production. Topics include productivity enhancement and bioenergy resources, agricultural and food-processing residues, energy crops, bioremediation, thermochemical conversion technologies, and biochemical conversion technologies.

Bioscience, Biotechnology, and Biochemistry (1992–) Tokyo, Japan: Japan Society for Bioscience, Biotechnology and Agrochemistry (0916–8451). Online: Japan Society for Bioscience, Biotechnology and Agrochemistry. This journal continues *Agricultural and Biological Chemistry* and focuses on research into the lives of plants, animals, and micro-organisms, and the chemical structures and functions of their bio-products.

Biosensors & Bioelectronics (1985–) Amsterdam: Elsevier (0956–5663). Online: ScienceDirect. *Biosensors & Bioelectronics* studies the research, design, development, and application of biosensors and bioelectronics Articles focus on the exploitation of biological materials and designs in diagnostic and electronic devices including DNA chips, electronic noses and micro-total analysis systems, and other aspects of bioelectronics.

Biosystems Engineering (2002–) Amsterdam: Elsevier (1537–5110). Online: ScienceDirect. *Biosystems Engineering* continues the *Journal of Agricultural Engineering Research* and promotes research in the physical sciences and engineering to understand, model, process, or enhance biological systems for sustainable developments in agriculture, food, land use, and the environment. Papers are arranged into eight categories: automation and emerging technologies, information technology and the human interface, precision agriculture, power and machinery, post-harvest technology, structures and environment, animal production technology, and soil and water.

Biotechnology and Bioengineering (1959–) Hoboken, NJ: Wiley (0006–3592). Online: WileyInterscience. This journal focuses on all aspects of applied biotechnology including cellular physiology, metabolism and energetics of cells, enzyme systems, animal cell biotechnology, bioseparation, environmental biotechnology, applied genetics and metabolic engineering, plant cell biotechnology, biochemical engineering, biosensors, thermodynamic aspects of cellular systems, mineral biotechnology, biological aspects of biomass and renewable resources engineering, and food biotechnology.

Canadian Agricultural Engineering (1959–2000) Ottawa, Canada: Canadian Society of Agricultural Engineering (0045–432X). Continued by *Canadian Biosystems Engineering.*

Canadian Biosystems Engineering: The Journal of the Canadian Society for Engineering in Agriculture, Food, and Biological Systems (2001–) Saskatoon, SK: Canadian Society for Engineering in Agriculture, Food, and Biological Systems (1492–9058). Online: Canadian Society for Engineering in Agriculture, Food, and Biological Systems

This journal continues *Canadian Agricultural Engineering* and publishes agricultural engineering research in seven areas. These areas are soil and water systems engineering, machinery systems engineering, bioprocessing systems engineering, biological systems engineering, building systems engineering, waste management engineering, and information systems engineering.

Computers and Electronics in Agriculture (1985–) Amsterdam: Elsevier (0168–1699). Online: ScienceDirect. *Computers and Electronics in Agriculture* provides research articles on the advances in the application of computer hardware, software and electronic instrumentation and control systems to agriculture, forestry, and related industries. Topics include: computerized decision-support aids, electronic monitoring or control of any aspect of livestock/crop production, and post-harvest operations. Research includes articles using artificial intelligence, sensors, machine vision, robotics, and simulation modeling.

Critical Reviews in Biotechnology (1983–) Philadelphia, PA: Taylor & Francis (0738–8551). Online: Taylor & Francis; Ingenta. This journal presents articles on biotechnological techniques from fermentation to genetic manipulation which have become increasingly relevant to the food and beverage, fuel production, chemical and pharmaceutical, and waste management industries.

Critical Reviews in Food Science and Nutrition (1970–) Philadelphia, PA: Taylor & Francis (1040–8398). Taylor & Francis. Online: Taylor and Francis Online Journals. This source presents information on food science and technology and human nutrition. Articles include information on nutrition, functional foods, food safety, diet and disease, antioxidants, allergenicity, flavor chemistry, food colors pesticides, regulation, risk assessment, food processing, government regulation and policy, effects of processing on nutrition, food labeling, and functional/bioactive foods.

Current Opinion in Biotechnology (1990–) Amsterdam: Elsevier (0958–1669). Online: ScienceDirect. This journal arranges research articles into the following topics: analytical biotechnology, plant biotechnology, food biotechnology, environmental biotechnology, systems biology, protein technologies and commercial enzymes, biochemical engineering, tissue and cell engineering, chemical biotechnology, and pharmaceutical biotechnology.

Engineering in Life Sciences (2000–) Weinheim: Wiley-VCH Verlag GmbH (1618–0240). Merged with *Acta Biotechnologica.* Online: Wiley Interscience. Title concentrates on technology rather than biological fundamentals, and provides articles on engineering applications in microbiology, genetics, biochemistry, and chemistry.

Food and Bioproducts Processing (1970–) Official Journal of the European Federation of Chemical Engineering: Part C. Rugby, UK: Institution of Chemical Engineers (0960–3085). Online: Ingenta. This journal focuses on research for safe processing of biological products. Topics that will be found in this journal are biocatalysis and biotransformations, bioprocess modeling, bioseparation, fermentation and bioreactor design, biocompatible materials and scaffolds, bioreactor design and control, scale-up and preservation technology, food and drink process engineering, engineering for food safety, environmental issues in food manufacture, minimal processing techniques, processing and microstructure interactions, and hygienic manufacture

Food Research International (1992–) Amsterdam: Published on behalf of the Canadian Institute of Food Science and Technology by Elsevier (0963–9969). Online: ScienceDirect. *Food Research International* continues the *Canadian Institute of Food Science and Technology Journal.* Topics covered by the journal include physical properties of food, microbiology, chemistry and analysis, process science, food safety, food quality sensory studies, and nutritional properties.

Food Science and Technology International (2001–) Thousand Oaks, CA: Sage (1082–0132). Online: <http://online.sagepub.com/> Focuses on the many aspects of food science and technology such as food-processing engineering, composition, food safety, nutritional quality, biotechnology, quality, physical properties, microstructure, microbiology, packaging, sensory analysis, bioprocessing, and post-harvest technology.

Food Service Technology (2001–) Oxford: Blackwell Publishing (1471–5732). Online: Blackwell Synergy. This journal publishes research on the technical aspects of food service as related to practical, commercial and public health issues. It covers market trends in the usage of food service technology, and essential health and safety information. Other subjects included are market research; consumer science; psychology; economics; modeling and simulation; food technology; food preservation; safety and hazard analysis; equipment design and engineering.

International Journal of Food Science and Technology (1987–) Oxford: Blackwell Publishing (1365–2621). Online: Blackwell Synergy. This journal publishes in a wide range of subjects, ranging from pure research associated with food to practical experiments designed to improve technical processes. Subjects covered range from raw material composition to consumer acceptance, from physical properties to food engineering practices, and from quality assurance and safety to storage, distribution, marketing, and use.

International Journal of Forest Engineering (1989–) Fredericton, NB: Faculty of Forestry and Environmental Management, University of New Brunswick (1494–2119). Online: <http://www.lib.unb.ca/Texts/JFE/Index.htm>. This journal provides information on many aspects of forest operation. The scope includes tree harvesting, processing and transportation; stand establishment, protection and tending; operations planning and control; machine design, management and evaluation; forest access planning and construction; human factors engineering; and education and training.

Irrigation and Drainage (2001–) Hoboken, NJ: Wiley (1531–0353). Online: Wiley-Interscience. As the official journal of the International Commission on Irrigation and Drainage this journal provides research on irrigation, drainage, and flood control.

Irrigation and Drainage Systems (1986–) Norwell, MA: Kluwer (0168–6291). Online: KluwerOnline. Topics included in this journal are performance assessment of irrigation and drainage systems, the interrelationship between irrigation management and system design, design criteria of drainage systems for effective control of waterlogging and salinity, maintenance such as research on sedimentation, weed control and so on, the interrelationship between scheme management and water users' organizations, the adaptation of irrigation/drainage to avoid water-related diseases such as malaria, planning and construction methods for canals and related structures, and the interaction between irrigation/drainage and the environment.

Irrigation Science (1978–) Heidelberg: Springer-Verlag Heidelberg (0342–7188). Online: SpringerLink. *Irrigation Science* publishes articles on all aspects of irrigation including research from the plant, soil, and atmospheric sciences and also irrigation water management modeling. Topics include research on the problems involved in maintaining the long-term productivity of irrigated lands, increasing the efficiency of agricultural water use, physiology of plant growth and yield response to water status, physical and chemical aspects of water status and movement in the plant-soil-atmosphere system, salinity and alkalinity control by soil and water management, agricultural drainage; water requirements in irrigation practice; irrigation scheduling and ecological aspects of irrigated agriculture.

Journal of Agricultural Engineering Research (1956–2001) (0021–8634). Predecessor to *Biosystems Engineering.* (1537–5110).

Journal of Agricultural Safety and Health (1995–) St. Joseph, MI: American Society of Agricultural Engineers (1074–7583). Online: American Society of Agricultural Engineers. This journal focuses on such areas as engineering, occupational safety, social psychology, public policy, education, industrial hygiene, and public health. Subjects included are treatment and prevention of trauma and illness, engineering design and application, safety and health intervention strategies, health standards, legislation and regulation, and the development of agricultural safety.

Journal of Agromedicine (1994–) Binghamton, NY: Haworth Medical Press (1059–924X). This journal provides articles on agricultural health and safety affecting producers, consumers, and the environmental health of communities impacted by agricultural practices. Articles emphasize occupational health and safety issues in agriculture, zoonotic and emerging diseases, food safety, health education strategies, and public health.

Journal of Applied Irrigation Science (1966–) Frankfurt, Germany: DLG Verlag

(0049–8602). Partially online at <http://www.sakia.org/cms/index.php?id=155>. *The Journal of Applied Irrigation Science/Zeitschrift für Bewässerungswirtschaft* is a bilingual journal with English and German articles that provide abstracts in both languages. The research focus is on practical application and sustainable irrigation practices worldwide.

Journal of Biotechnology (1984–) Amsterdam: Elsevier (0168–1656). Online: ScienceDirect. Articles on many aspects of biotechnology are published in this journal. Topics range from nucleic acids, molecular biology, physiology, biochemistry, biochemical engineering, bioprocess engineering, industrial processes, new products, and medical biotechnology.

Journal of Chemical Technology and Biotechnology (1986–) Hoboken, NJ: Published for the Society of Chemical Industry by Wiley (0268–2575). Online: WileyInterScience. This journal continues the *Journal of Applied Chemistry and Biotechnology* and publishes research articles in chemical and biological technology that aim at economically sustainable industrial protection or are necessary for environmental protection. Some areas covered are water and air pollution reduction, waste treatment/management, remediation, economic and life cycle analysis, water and off-gas treatment/management, alternative fuels/energy sources, treatment of hazardous wastes, reactor and equipment design, fermentation and industrial biotechnology, downstream processing, industrial catalysis and biocatalysis, industrial genomics and proteomics, nanotechnology and fuel cells.

Journal of Fermentation and Bioengineering (1999–) Osaka, Japan; Amsterdam: Society of Fermentation Technology Japan (0922–338X). Distributed outside Japan by Elsevier. Continues *Journal of Fermentation Technology.* Online: ScienceDirect. The journal publishes research on fermentation technology, biochemical engineering, food technology and microbiology.

Journal of Food Engineering (1982–) Amsterdam: Elsevier (0260–8774). Online: ScienceDirect. *Journal of Food Engineering* publishes research articles in the area of engineering properties of foods, food physics and physical chemistry processing, measurement, control, packaging, storage and distribution; engineering aspects of the design and production of novel foods and of food service and catering; design and operation of food processes, plants and equipment; and economics of food engineering, including the economics of alternative processes.

Journal of Food Process Engineering (1977–) Oxford, UK: Blackwell Publishing (0145–8876). Online: Blackwell Synergy. This journal focuses on the engineering aspects of post-production handling, storage, processing, packaging, and distribution of food. Papers focus on processes that change the physical properties, or changes to the food product that result in preservation of food, extending to transportation, product shelf-life, or improvements in the product quality attributes.

Journal of Food Processing and Preservation (1977–) Oxford: Blackwell Publishing (0145–8892). Online: Blackwell Synergy. This journal focuses on both fundamental and applied research relating to food processing and preservation. It features important discussions of current economic and regulatory policies and their effects on the safe and quality processing and preservation of a wide array of foods.

Journal of Hydrologic Engineering (1996–) New York: American Society of Civil Engineers (1084–0699). Online: American Society of Civil Engineers. *The Journal of Hydrologic Engineering* publishes research on the development of new hydrologic methods, theories, and applications to current engineering problems, and includes articles on analytical, numerical, and experimental methods for the investigation and modeling of hydrological processes.

Journal of Hydrology (1963–) Amsterdam: Elsevier (0022–1694). Online: ScienceDirect. The scope of *The Journal of Hydrology* is all areas of the hydrological sciences including the physical, chemical, biogeochemical, stochastic and systems aspects of surface and ground water hydrology, hydrometeorology, and hydrogeology. Papers are also published on related topics such as climatology, water resource systems, hydraulics, agrohydrology, geomorphology, soil science, instrumentation and remote sensing, civil and environmental engineering.

Journal of Industrial Ecology (1997–) Cambridge, MA: MIT Press (1088–1980). Online: Ingenta. As the official journal of the International Society for Industrial Ecology, the *Journal of Industrial Ecology* was created for the emerging field of industrial ecology. The journal publishes research on material and energy flows studies ("industrial metabolism"), dematerialization and decarbonization, life cycle planning, design and assessment, design for the environment, extended producer responsibility ("product stewardship"), eco-industrial parks ("industrial symbiosis"), product-oriented environmental policy, and eco-efficiency.

Journal of Irrigation and Drainage Engineering (1983–) New York: American Society of Civil Engineers (0733–9437). Online: American Society of Civil Engineers. This journal publishes articles on all aspects of irrigation, drainage, engineering hydrology, and related water management subjects, such as watershed management, weather modification, water quality, ground water, and surface water.

Journal of Soil and Water Conservation (1946–) Ankeny, IA: Soil and Water Conservation Society (0022–4561). Focuses on research in the conservation of soil, water, and related natural resources. Topics included are in the areas of agronomy, conservation education, conservation planning, ecosystem management, environmental quality, erosion and sediment control, geology, floodplain management, farmland preservation, forage management, forestry, GIS, GPS, irrigation, mined land reclamation, nonpoint source pollution, rangeland management, soil science, sustainable agriculture, watershed management, wetland restoration, and wildlife management.

Landwards (1996–) Silsoe, UK: Institution of Agricultural Engineers (1363–8300). *Landwards* continues *Agricultural Engineer* and provides articles on developments in engineering and technology for practical application in agriculture, horticulture, forestry, and environment.

Resource: Engineering & Technology for a Sustainable World (1994–) St. Joseph, MI: American Society of Agricultural Engineers (1076–3333). Online: American Society of Agricultural Engineers.

Resource continues *Agricultural Engineering* and provides information trends on employment and other related information to the membership of the American Society of Agricultural Engineers.

Soil & Tillage Research (1980–) Amsterdam: Elsevier (0167–1987). Online: ScienceDirect. The articles in this journal study the changes in the physical, chemical, and biological parameters of the soil environment brought about by soil tillage and field traffic, their effects on both below- and above-ground environmental quality, crop establishment, root development and plant growth, and the interactions between these various effects. In 1998 this journal incorporated *Soil Technology*. (0933–3630).

Soil Use and Management (1985–) Wallingford, UK: CABI Publishing (0266–0032). Online: Ingenta. This journal publishes research on applying scientific principles to soil problems and how they affect crop production and environmental issues. Topics included are environmental protection, soil–crop interactions, soil erosion and conservation, pollution control, restoration and reclamation of land, evaluation of soil surveys, and the development of methodology.

Transactions of the ASAE (1958–) St. Joseph, MI: American Society of Agricultural Engineers (0001–2351). Online: American Society of Agricultural Engineers. *Transactions of the ASAE* provides research on a broad range of topics including agricultural machinery, drainage, irrigation, electronics, biological engineering, forestry, food engineering, agricultural structures, crop production, natural resources, soils, and other related subjects.

Trends in Biotechnology (1983–) Amsterdam: Elsevier (0167–7799). Online: ScienceDirect. This journal publishes research and reviews on many aspects of the applied biosciences and includes molecular biotechnology, gene transfer and expression, applied microbiology, environmental biotechnology, fermentation and bioprocessing, rDNA therapeutics and vaccines, plant biotechnology, and patenting and regulatory issues.

Trends in Food Science and Technology (1990–) Amsterdam: Elsevier (0924–2244). Online: ScienceDirect. Publishes reviews on the science and technology of food analysis, development, manufacture, storage and marketing, from the molecular and microstructural level, through raw material processing to food engineering, novel processing methods, automation, quality control and assurance, microbiological safety issues, advances in preservation, and packaging technologies and sensory analysis.

World Journal of Microbiology & Biotechnology (1990–) Heidelberg: Springer Verlag Heidelberg (0959–3993). Online: SpringerLink. This journal continues *MIRCEN Journal of Applied Microbiology and Biotechnology* and publishes research on all aspects of applied microbiology and biotechnology, including management of culture collections, foodstuffs, and biological control agents. The journal focuses on microbiological and biotechnological solutions to global problems, such as agriculture and food supplies and environmental issues including pollution and waste management, and emphasizes biotechnological advances in the developing world.

Standards

Engineering standards included in other chapters should be consulted when searching for standards in agricultural or food engineering information that would be based on those areas. In addition, the following resource focuses specifically on agricultural engineering.

ASAE standards (1985–) (Annual) St. Joseph, MI: The Society. Standards, engineering practices, and data adopted by the American Society of Agricultural Engineers. Also available online in the *ASAE Technical Library* (see information in the Abstracts and indexes section of this chapter).

Search engines, web guides, and discussion lists

The advent of the Internet has fostered great improvements in communication and information discovery. The other sections of this chapter cover relevant information in agricultural and food engineering, whether or not it is accessed through the Internet. This section describes search engines, web guides, and discussion lists not previously listed.

Search engines

These searchable indexes to agricultural information include many commercial enterprises and are useful for finding products. They also provide access to information at some colleges, universities, and government agencies, and may provide more focused results than general search engines such as Google.

Agrisurf <http://www.agrisurf.com> (accessed October 14, 2004). This claims to be the world's largest searchable agricultural www index.

WebAgri <http://www.web-agri.com/> (accessed October 14, 2004). This search engine is useful for finding agricultural information in universities.

Web guides

Guides to information on the Internet provide selected resources in their given subject areas.

Agriculture Network Information Center (AGNIC) <http://www.agnic.org> (accessed October 14, 2004). AGNIC is a voluntary partnership retrieving selected quality agricultural information. The database is searchable, or may be browsed through the member sites.

Agriculture Virtual Library <http://cipm.ncsu.edu/agVL/> (accessed October 8, 2004). Along with the engineering virtual library there will be relevant information in the agricultural section of the virtual library. Information is accessed by browsing through the categories.

Irrigation Virtual Library <http://www.vl-irrigation.org/> (accessed October 14, 2004). Although currently in a developmental stage, dedicated and structured information related to the wider area of irrigation and hydrology are planned. This site also provides access to Irrigation-L listserv and some articles in the *Journal of Applied Irrigation Science.*

Discussion lists

E-mail communication via discussion lists has become the norm for many disciplines. While agricultural engineering does not have an overall discussion list, the following may prove useful.

CIGR-FAO Global Network. There are seven e-mail discussion lists jointly sponsored by CIGR and FAO that are used as communication tools for the respective fields and special interest groups. Descriptions of the lists and subscribing instructions may be found at <http://www.ucd.ie/cigr/globnet.htm> (accessed October 14, 2004).

IRRIGATION-L is the discussion list in the field of irrigation theory and practice.

See additional information and subscribing instructions at <http://www.irriga-tion-l.org> (accessed October 14, 2004).

DRAINAGE-L provides a forum for drainage professionals and interested parties around the world to discuss issues related to the role of drainage in providing a suitable agronomic environment in humid, semi-arid, and arid areas. To subscribe, send an e-mail message to: LISTSERV@UNL.EDU. In the body of the message (not in the subject) type: SUBSCRIBE DRAINAGE-L.

TRICKLE-L focuses on drip irrigation technology in its widest sense and also covers micro-irrigation issues. To subscribe, send an e-mail message to: LISTSERV@crcvms.unl.edu. In the body of the message (not in the subject) type: SUBSCRIBE TRICKLE-L.

Associations, organizations, societies, and conferences

Societies provide information in many forms and often have well-developed websites. Some types of information provided are publications, meeting schedules, history of the discipline and society, and opportunities for networking. Listed below are the major societies in agricultural and food engineering. For other societies consult the *International List of Agricultural Engineering Societies*; <http://www.asae.org/membership/international.html> (accessed October 8, 2004) which provides information on agricultural engineering societies throughout the world.

American Society for Agricultural Engineering (ASAE) <http://www.asae.org/> (accessed September 21, 2004). ASAE was renamed the Society for Engineering in Agriculture Food and Biological Systems. According to the website The American Society of Agricultural Engineers "is an educational and scientific organization dedicated to the advancement of engineering applicable to agricultural, food, and biological systems. Founded in 1907 and headquartered in St Joseph, Michigan, ASAE comprises 9,000 members in more than 100 countries." The ASAE sponsors annual meetings and specialty conferences.

Canadian Agricultural Safety Association (CASA) <http://www.casa-acsa.ca/> (accessed September 21, 2004). CASA was established in 1993 in response to an identified need for a national farm safety networking and coordinating agency to address problems of illness, injuries and accidental death in farmers, their families, and agricultural workers. Since then, its mission has been to improve the health and safety conditions of those who live, and or work on Canadian farms.

Canadian Institute of Food Science and Technology (CIFST) <http://www.cifst.ca/> (accessed September 21, 2004). "Founded in 1951, CIFST is the national association for food industry professionals. Its membership of more than 1500 is comprised of scientists and technologists in industry, government and academia who are committed to advancing food science and technology" (from website).

Canadian Society for Engineering in Agricultural, Food and Biological Systems (CSAE) <http://www.csae-scgr.ca/> (accessed September 21, 2004). The website provides information about publications, meetings, and news for students and members.

Council on Forest Engineering (COFE) <http://www.cofe.org/> (accessed October 11, 2004). This is an international professional organization interested in matters relating to the field of forest engineering.

Institute of Biological Engineering (IBE). <http://www.ibeweb.org/> (accessed October

14, 2004). This institute was established to encourage inquiry and interest in biological engineering in the broadest and most liberal manner and to promote the professional development of its members.

Institute of Food Science and Technology <http://www.ifst.org/> (accessed September 21, 2004). This institute is the independent incorporated professional qualifying body for food scientists and technologists.

Institute of Food Technologists <http://www.ift.org/cms/> (accessed September 21, 2004). The Institute of Food Technologists is an international not-for-profit scientific society that was founded in 1939. Its website provides information on publications, continuing education and professional development, awards, and government relations and policy activities.

Institution of Agricultural Engineers <http://www.iagre.org/> (accessed September 21, 2004). According to the website, "The Institution of Agricultural Engineers (IAgrE) is the professional body for engineers, scientists, technologists and managers in agricultural and allied land based industries, including forestry, food engineering and technology, amenity, renewable energy, horticulture and the environment."

International Commission of Agricultural Engineering (Commission Internationale du Génie Rural (CIGR)) <http://www.ucd.ie/cigr/> (accessed September 21, 2004). The commission was set up during the first International Congress of Agricultural Engineering, in Liege, Belgium in 1930. It is an international, non-governmental, non-profit organization with a primary focus on networking. Every five years and from 1994 every four years, CIGR convenes a *World Congress of Agricultural Engineering*.

International Union of Food Science & Technology <http://www.iufost.org/> (accessed October 8, 2004). IUFoST, a country-membership organization, is the sole global food science and technology organization. It is a voluntary, non-profit association of national food science organizations linking the world's best food scientists and technologists. This association supports the *World Congress of Food Science and Technology* every year, and the *International Congress on Engineering and Food* every four years.

National Institute for Farm Safety <http://www.ag.ohio-state.edu/~agsafety/NIFS/nifs.htm> (accessed September 21, 2004). According to the website, the NIFS "is an organization dedicated to the professional development of agricultural safety and health professionals, providing national and international leadership in preventing agricultural injuries."

Summary

In conclusion, because of the multi-disciplinary nature of agricultural and food engineering, knowledge of the basic engineering information resources mentioned in other chapters of this work is extremely important. Within the strict bounds of agricultural and food engineering, the traditional journal literature is of high importance, as well as the ASAE publications and conferences. There are trends toward more specializations within the traditional realms of agricultural engineering, and increasing focus on the microtechnology and cellular engineering aspects. Astute students of the subject would do well to keep abreast of such changes.

References

ASAE (Society for Engineering in Agricultural, Food and Biological Systems) <http://www.asae.org/> (cited October 10, 2004).

Dooley, J.H. (2003) "Biological engineering definition," in: Heldman, D.R. (ed.) *Encyclopedia of Agricultural Food and Biological Engineering*, New York: Marcel Dekker, pp. 60–63.

Farkas, D.F. (2003) "Food engineering history," in: Heldman, D.R. (ed.) *Encyclopedia of Agricultural Food and Biological Engineering*, New York: Marcel Dekker, pp. 346–349.

Isaacs, G.W. (2003) "Agricultural engineering history," in: Heldman, D.R. (ed.) *Encyclopedia of Agricultural Food and Biological Engineering*, New York: Marcel Dekker, pp. 14–17.

Ochs, M.A. and Patterson, M.E. (2002) "Agricultural and biosystems engineering," in: Hutchinson, B.S. and Greider, A.P. (eds) *Using the Agricultural, Environmental and Food Literature*, New York: Marcel Dekker. pp. 53–74.

5 Architectural engineering

Barbara Opar

Introduction

> Architectural Engineering is the discipline concerned with the planning, design, construction, and operation of engineered systems for commercial, industrial, and institutional facilities. Engineered systems include electric power, communications and control; lighting; heating, ventilation, and air conditioning; and structural systems. An Architectural Engineer works closely with those in all areas of the building process to design and possibly to construct the engineered systems that make buildings come to life for their inhabitants.
>
> (Architectural Engineering Institute web page)

To the layman, architectural engineering is less clearly defined as a field than electrical engineering, where the parameters of the profession are more firmly set. Architectural engineering is a highly complex field involving advanced knowledge of both architecture and engineering. Although licensed as engineers rather than as architects, architectural engineers collaborate with architects, and therefore must fully understand design concepts and the architect's vision. They are responsible for building system integration and for ensuring that the mechanical and electrical systems operate properly. Their understanding of building system integration is what differentiates them from civil engineers – although some architectural engineers do specialize in certain areas, such as structures.

Architectural engineering may be seen as an outgrowth of civil engineering, but very little has been written on the history of the profession. As architectural projects have become more sophisticated and have required more technological expertise, architectural engineers, by bridging this gap with their understanding of the design process alongside engineering issues, have emerged as part of a distinct profession.

While the American Society of Civil Engineers is 150 years old, the Architectural Engineering Institute, the major professional organization aimed at architectural engineers, was only recently created on October 1, 1998, as the result of a merger between the National Society of Architectural Engineers (NSAE) and the American Society of Civil Engineers Architectural Engineering

Division (AED). Sixteen universities currently offer accredited programs in architectural engineering (as opposed to 113 for architecture, and 207 for civil engineering). These are California Polytechnic State University-San Luis Obispo, Drexel University, Kansas State University, Milwaukee School of Engineering, North Carolina A&T, Tennessee State University, the University of Colorado at Boulder, the University of Kansas, the University of Miami, the University of Texas at Austin, the University of Wyoming, Oklahoma State University, the Pennsylvania State University, the Illinois Institute of Technology, the University of Nebraska, and the University of Missouri at Rolla.

Architectural engineering is likely to grow in importance as new building systems evolve, as technology becomes more complicated, and as we search for ways to conserve our built environment.

The literature of architectural engineering

The monographic literature devoted to architectural engineering is somewhat limited and rather dated. Few book titles appear annually, and those that do are often cached under headings other than architectural engineering, such as building systems or structural design, or under elements of a particular topic, such as sustainable building materials. In addition, it can prove difficult to obtain a general overview of new developments in the field from the engineering or architecture journals, as most articles focus on highly technical and specific subjects, for example, the design of concrete cladding or issues related to indoor air quality. The engineering periodical literature provides important studies on various technical aspects of architectural engineering, but the student or practitioner should also scan new developments in the architecture literature by browsing the major architectural design magazines, including *Architecture, Architectural Record, and Architectural Review*. In addition, titles such as *Architecture Today* often bridge the gap between building technology and design practice.

Searching the library catalog

Architectural engineering is not a Library of Congress (LC) subject heading. Users are instructed to see the following LC headings: Building; Building, iron and steel; Strains and stresses; Strength of materials; and Structural analysis *(Engineering)*. Most academic library catalogs employ the same terminology; moreover, they also allow natural language or keyword searching, thereby permitting the use of the term "architectural engineering." Since most of the current information on architectural engineering comes from periodicals, researchers should consult specific periodical indexes for in-depth information (see list of indexes below under Abstracts and indexes). The *Avery Index*, the foremost architectural periodical index, uses "architectural engineering" in its subject word search. The *Civil Engineering Database* includes "architectural engineering" as a subject heading, but other appropriate search terms include "building design" and "construction industry." The *Applied Science & Technology*

Index denotes structural engineering as a way of searching for this type of information. Researchers should consider the focus of their topic prior to selecting a heading in a catalog or database, and may wish to consider searching in more than one way. Other broad headings include Architecture and technology, Structural engineering, and Structures. In addition to the above list, a few of the narrower headings include Building materials and/or the specific material (e.g., Steel, Concrete); Specifications; Building envelope; Exterior walls; Air-conditioning; Heating, lighting, and ventilation; and Seismic design. Library of Congress call numbers commonly associated with architectural engineering (other call number areas may apply) include:

NA2540–2545 Architecture in relation to special subjects
NA2640–2645 Specifications
NA2750 Architectural design
NA2800 Architectural acoustics
NA2835–3300 Architectural details
NA4050–9000 Special classes of buildings
TA151 Engineering handbooks and manuals
TA160 Engineering research
TA170 Environmental engineering
TA178 Specifications
TA401–492 Materials of engineering and construction
TA625–695 Structural engineering
TH (entire class) Building construction
TH7005–7699 Heating and ventilation: Air-conditioning

Abstracts and indexes

Largely because the monographic literature of architectural engineering is somewhat dated, students and professionals are advised to consult the following indexes to supplement their readings and to maintain currency in the field.

Applied Science and Technology Abstracts (1958–) (electronic resource (1983–) http://www.hwwilson.com (accessed November 3, 2004) New York: H.W. Wilson (see Chapter 2, this volume).

Architectural Index (1950–) Boulder, CO. Electronic database <http://www.archindex.com/free/copyright.htm> free for years 1982–1988. Also in paper format. Jerry T. Moore, AIA, editor and publisher; Ervin J. Bell, AIA, editor and publisher, emeritus. Indexes the following major architectural journals: *Architecture, Architectural Record, Architectural Review, Builder, Building Design & Construction, Interior Design, Journal of Architectural Education, Landscape Architecture, Residential Architect*, and *The Construction Specifier*. Emphasis is on aesthetics rather than engineering.

ASCE Civil Engineering Database (1970–) Reston, VA (see Chapter 1).

Avery Index to Architectural Periodicals. New York: Avery Architectural and Fine Arts Library, Columbia University. Electronic database (1930s–), with selective coverage dating back to the 1860s. Indexes more than 2,000 periodicals published worldwide on archaeology, city planning, interior design, and historic preservation, as well as architecture. "Avery indexes not only the international scholarly and popular periodical literature, but also the publications of professional associations, US state and regional periodicals, and the major serial publications on architecture and design of Europe, Asia, Latin America, and Australia. Expanded coverage includes obituary citations providing an excellent source of biographical data – often the only information available for less-published architects" (updated weekly, Avery web page).

Engineering Village 2 <http://www.engineeringvillage2.org> (see Chapter 2, this volume).

Bibliographies and guides to the literature

Bibliographies and guides to the literature can provide short cuts, and most often include books and periodical articles that have been selected by subject specialists. Publications in the area of architectural engineering are limited, especially ones of a broad or generalized scope. Researchers should consult bibliographies from appropriate monographs on specific topics (e.g., exterior wall design).

Gafford, W.R. (1951) *Source Materials for Architectural Engineering* (MS thesis in Arch. E.), Austin, Texas: University of Texas at Austin. While dated, Gafford's thesis remains important for its early work on identifying architectural engineering sources.

Godel, J.B. (1977) *Sources of Construction Information, Volume one: Books*, Metuchen, New Jersey: Scarecrow Press. An annotated guide to reports, books, periodicals, standards, and codes. This reference work is divided into eleven sections, including the following topics: mechanical and electrical design, building materials, and the construction industry.

Directories

Directories of relevance to the topic of architectural engineering are becoming more widely available via the World Wide Web. Again, the literature is fragmented; for example, there are no directories of architectural engineers. However, directories do exist for architects and engineers which often include lists of consultants where specialties are noted (e.g., curtain wall design). A number of directories exist for product information, including paper and electronic versions.

ARCAT Specs <http://www.arcat.com/> (accessed May 17, 2005). "*ARCAT Specs* (SpecDisk®) are complete, accurate, and in the CSI 3-part format. They are available free for viewing and downloading in Word Perfect, Microsoft Word, and ASCII for both PC's and Macintosh computers" (ARCAT web page).

Ballast, D.K. (1998) *The Encyclopedia of Association and Information Sources for Architects, Designers, and Engineers*, Armonk, NY: Sharpe Professional. In addition to listings of professional associations and organizations, this source includes information about journals, online databases, CD-ROMs, and federal government publications in the fields of architecture, design, and engineering.

Cramer, J.P, and Yankopolus, J.E. (2005) *Almanac of Architecture and Design* (6th edn), Washington, DC: Greenway Communications. Includes information on awards, achievements, design education, and appropriate organizations.

ENR Directory of Construction Information Resources, (1993–), New York: ENR Special Projects. A useful source for locating contact information for construction-related organizations.

GreenSpec: Product Directory with Guideline Specifications (2003) (4th edn) Brattleboro, VT: BuildingGreen. A comprehensive source for sustainable products.

ProFile (Construction Market Data (Firm)) *ProFile: The Sourcebook of U.S. Architectural Design Firms. Atlanta: Construction Market Data* (1996–) <http://www.First-SourceONL.com> (accessed May 18, 2005). Notes firm specialization, principal staff, as well as standard contact information.

Sweet's Catalog File (various editions: Architects, Engineers and Contractors; Residential; Facilities and Owners; Structural and Civil Products source book) (2001–) New York: Sweets. The most comprehensive source for standard product information. Uses CSI format.

Ultimate Civil Engineering Directory <http://www.tenlinks.com/engineering/civil/index.htm> (see Chapter 8).

Who's Who in Engineering (1977–) New York: American Association of Engineering Studies. A source for locating leaders in the field (see Chapter 2).

Encyclopedias and dictionaries

Bianchina, P. (1993) *Illustrated Dictionary of Building Materials and Techniques*, New York: John Wiley & Sons. Excellent line drawings are included to illustrate specific terms.

Brooks, H. (1976) *Illustrated Encyclopedic Dictionary of Building and Construction Terms*, Englewood Cliffs, NJ: Prentice-Hall. Arranged by building function, this work includes line drawings or photographs for selected terms.

Cowan, H.J. (1988) *Encyclopedia of Building Technology*, Englewood Cliffs, NJ: Prentice-Hall. Arranged alphabetically by topic. This source provides a good overview of various technical topics (e.g., heat engines and heat pipes, pneumatic structures, and sound absorption), often including definitions, histories, and applications.

Cowan, H.J. and Smith, P.R. (2004) *Dictionary of Architectural and Building Technology* (4th edn), London; New York: Spon Press. Substantially updated with 1,750 new or revised entries, this source continues to be an important reference for the building industry.

Guedes, P. (ed.) (1979) *Encyclopedia of Architectural Technology*, New York: McGraw-Hill. Divided into six sections, this source, while somewhat dated, does provide a solid introduction to the history of topics such as building types, building services (mechanical and electrical systems), and building materials.

Keller, H. and Erb, U. (2004) *Dictionary of Engineering Materials*, Hoboken, NJ: John Wiley. An up-to-date source that defines key materials of the trade.

Putnam, R. (1984) *Builder's Comprehensive Dictionary*, Reston, VA: Reston Publishing Company. Easy to read with numerous photographs and illustrations.

Schwartz, M. (ed.) (2002) *Encyclopedia of Smart Materials*, New York: John Wiley. Schwartz's encyclopedia outlines responsive, or smart, materials. Smart materials are those products which can change their properties, structure, or function in response to environmental stimulae.

Scott, J.S. (1984) *Dictionary of Building* (3rd edn), New York: Halsted Press. A straightforward approach with clear definitions.

Stein, J.S. (1993) *Construction Glossary: An Encyclopedic Reference and Manual*, New York: Wiley. Organized by building component following CSI (e.g., masonry). Very detailed and in-depth.

Wilkes, J.A. (ed.) (1988) *Encyclopedia of Architecture: Design, Engineering & Construction*, New York: John Wiley. In five volumes, it includes historical background information on architects, building types, processes, and standard materials. Useful as an introduction.

Handbooks and manuals

ACI Manual of Concrete Practice, Detroit: American Concrete Institute. Issued annually in parts. Provides detailed information on all aspects of concrete preparation, curing, and fatigue. Includes reports, evaluations, formulas, and recommendations for identifying and addressing problems such as cracking and creep.

Allen, E. (1985) *The Professional Handbook of Building Construction*, New York: John Wiley. Well-conceived drawings help elucidate the construction process.

American Institute of Steel Construction (2001) *Load and Resistance Factor Design: Manual of Steel Construction* (3rd edn), Chicago, IL: American Institute of Steel Construction. The standard reference for steel construction.

American Institute of Timber Construction (2005) *Timber Construction Manual*, Hoboken, NJ: John Wiley. Internet version also available. The standard reference for timber construction.

ASM Handbook. Prepared under the direction of the ASM International Handbook Committee (multi-volume and multi-year). Materials Park, OH: ASM International. Emphasis on materials; technical in approach.

Buettner, D.R. (2001) *PCI Design Handbook: Precast and Prestressed Concrete* (5th edn), Chicago, IL: Precast/Prestressed Concrete Institute. The source for concrete design.

Butler, R.B. (2002) *Architectural Engineering Design* (includes CD-ROM), New York: McGraw-Hill. An important resource for architectural engineering students.

Cowan, H.J. (ed.) (1991) *Handbook of Architectural Technology*, New York: Van Nostrand. Well written and easy to read. Chapters cover materials (e.g., metal, ceramics, timber, plastics) as well as building service design topics such as durability, loads, and wind effects. Includes references and suggestions for further reading at the end of each chapter.

Daniels, K. (2003) *Advanced Building Systems: A Technical Guide for Architects and Engineers*, Basel; Boston, MA: Birkhauser. A significant resource for designers of building systems.

Grimm, N.R. and Rosaler, R.C. (1990) *Handbook of HVAC Design*, New York: McGraw-Hill. Chapters range from conceptual and preliminary design to specific HVAC considerations, such as cooling towers.

Haines, R.W. (2003) *HVAC Systems Design Handbook* (4th edn), New York: McGraw-Hill. An important and well-conceived resource for mechanical design.

Hart, R.D. (1994) *Quality Handbook for the Architectural, Engineering, and Construction Community*, Milwaukee: ASQC Quality Press. Covers quality control aspects.

Hornbostel, C. (1991) *Construction Materials: Types, Uses and Applications* (2nd edn), New York: John Wiley & Sons. A clear, well-illustrated source which discusses the history, manufacture, types, and uses of specific materials.

Meier, H.W. (1989) *Construction Specifications Handbook* (4th edn), Englewood Cliffs, NJ: Prentice-Hall. Discusses the writing of specifications, and includes draft specifications.

Pennsylvania State University. Department of Architectural Engineering (1966) *Emerging Techniques of Architectural Practice*, ed. C.H. Wheeler *et al.*, Washington, DC: American Institute of Architects. Somewhat dated, but still of historic interest.

Perry, R. and Perry, J.H. (1967) *Engineering Manual; A Practical Reference of Data and Methods in Architectural, Chemical, Civil, Electrical, Mechanical, and Nuclear Engineering* (2nd edn), New York: McGraw-Hill. A handbook referencing technical data and calculations.

Rea, M.S. (ed.) (2000) *The IESNA Lighting Handbook: Reference and Application* (9th edn), New York: Illuminating Engineering Society of North America. The standard lighting engineering manual.

Ricketts, J.T., Loftin, M.K., and Merritt, F.S. (eds) (2004) *Standard Handbook for Civil Engineers* (5th edn), New York: McGraw-Hill. A major handbook in the field.

Rush, R.D. (contributor) (1986) *The Building Systems Integration Handbook*, New York: Wiley. Includes bibliographies and index. An important source for students in the field.

Simmons, H.L. (2001) *Construction: Principles, Materials, and Methods* (7th edn), New York: John Wiley. A solid introduction to materials and technology.

Spengler, J.D., McCarthy, J.F., and Samet, J.M. (eds) (2001) *Indoor Air Quality Handbook*, New York: McGraw-Hill. A comprehensive source on an important issue.

Standard Handbook of Architectural Engineering (1999: CD-ROM) New York: McGraw-Hill. Electronic version of the print reference tool. Includes a collection of structural, mechanical, electrical, lighting and acoustical data, universal design scenarios, design tools, and interactive formulas for sizing architectural components.

Stein, B. and Reynolds, J.S. (2000) *Mechanical and Electrical Equipment for Buildings* (9th edn), New York: John Wiley & Sons. The standard HVAC textbook.

Time-Saver Standards (1995–2002) New York: McGraw-Hill. Separate editions for building types, interior design and space planning, landscape architecture, site construction details manual, housing and residential development, building materials and systems, and adding on and remodeling. Some available in electronic format. Essential tool of the trade.

Monographs and textbooks

Allen, E. (1980) *How Buildings Work: The Natural Order of Architecture*, New York: Oxford University Press. A clear and engaging explanation of building systems design.

Bachman, L.R. (2003) *Integrated Buildings: The Systems Basis of Architecture*, New York: John Wiley & Sons. Stresses integration issues between architecture and technology.

Baird, G. (2001) *The Architectural Expression of Environmental Control Systems*, London; New York: Spon Press. Connects architectural design to the need for well-conceived mechanical systems.

Bovill, C. (1991) *Architectural Design: Integration of Structural and Environmental Systems*, New York: Van Nostrand Reinhold. Clearly explains the importance of structures and HVAC in the design process.

Building Arts Forum (1990) *Bridging the Gap: Rethinking the Relationship of Architect and Engineer: The Proceedings of the Building Arts Forum*, New York Symposium Held in April of 1989 at the Guggenheim Museum. New York: Van Nostrand Reinhold. An interesting overview of practices and collaborative possibilities.

Fischer, R.E. (1980) *Engineering for Architecture*, New York: McGraw-Hill. An Architectural Record publication, this work illustrates the relationship between architecture and engineering.

Freitag, J.K. (1985) *Architectural Engineering: With Special Reference to High Building Construction, Including Many Examples of Chicago Office Buildings*, New York: John Wiley & Sons. This early title presents a solid picture of architectural engineering at the beginning of the twentieth century.

Holgate, A. (1986) *The Art in Structural Design: An Introduction and Sourcebook*, Oxford: Clarendon Press. Shows the design potential of architectural engineering.

Hordeski, Michael F. (2003) *New Technologies for Energy Efficiencies*, Lilburn, GA: Fairmount Press, New York: Dekker. Wind and solar power are discussed, as well as fuel cells.

Kim, D-H. (1995) *Composite Structures for Civil and Architectural Engineering*, London; New York: E & FN Spon. Discusses the integration of different materials and technology.

Komendant, A.E. (1987) *Practical Structural Analysis for Architectural Engineering*, Englewood Cliffs, NJ: Prentice-Hall. Komendant worked with Louis I. Kahn on many projects.

Larsen, O.P. (2003) *Conceptual Structural Design: Bridging the Gap Between Architects and Engineers*, London: Thomas Telford. Discusses the necessary connection between architecture and engineering.

Levy, M. (2002) *Why Buildings Fall Down: How Structures Fail*, New York: W.W. Norton. An important resource for students, this book discusses sources of building failure.

Liu, M. and Parfitt, K. (eds) (2003) *Architectural Engineering Conference. Building Integration Solutions: Proceedings of the Architectural Engineering 2003 Conference, September 17–20, 2003, Austin, Texas*, Sponsored by Architectural Engineering Institute of American Society of Civil Engineers, Reston, VA: American Society of Civil Engineers.

Lyall, S. (2002) *Remarkable Structures: Engineering Today's Innovative Buildings*, New York: Princeton Architectural Press. This conference covered important contemporary issues related to architectural engineering.

Mainstone, R.J. (1998) *Developments in Structural Form*, Oxford; Boston, MA: Architectural Press. An important work on structural design.

Salvadori, M.G. (1982) *Why Buildings Stand Up: The Strength of Architecture*, New York: McGraw-Hill. Aimed at students, this book presents a clear overview of the topic.

Salvadori, M.G. and Levy, M. (1981) *Structural Design in Architecture* (2nd edn), Englewood Cliffs, NJ: Prentice-Hall. Includes examples and problem solutions. Recommended for students.

Schittich, C. (2003) *Solar Architecture: Strategies, Vision, Concepts*, Basel; Boston, MA: Birkhauser. Discussion of various architectural designs for more energy-efficient buildings.

Smith, P.F. (2003) *Sustainability at the Cutting Edge: Emerging Technologies for Low Energy Buildings*, Oxford: Architectural Press. A look at architecture and solar, geothermal, wind, biogas, and more.

Szokolay, S.V. (2004) *Introduction to Architectural Science: The Basis of Sustainable Design*, Amsterdam; Boston, MA: Elsevier, Architectural Press. Discusses technology solutions toward sustainability.

West, H.H. (1989) *Analysis of Structures: An Integration of Classical and Modern Methods* (2nd edn), New York: John Wiley & Sons. A solid study useful for students of the field.

Journals

Architectural & Engineering News (1958–1970) New York: Hagan Publishing. Indexed and of use for mid-twentieth-century developments.

Architectural Design (1987) "Engineering and Architecture" (whole issue), 57 (11/12) (0003–8504). An interesting issue of an important architecture periodical.

Architectural Science Review (1958–) Sydney: Academic Press (0003–8628). Articles on specific technical issues.

Architecture Today (1989–) London: Architecture Today (0958–6407). Each issue includes the latest product developments in a particular area.

Builder (1978–) Washington, DC: Hanley-Wood, Inc., for the National Association of Home Builders (0744–1193). Trade association journal.

Construction Review (1997) Washington, DC: U.S. Department of Commerce, Domestic and International Business Administration, Bureau of Domestic Commerce: for sale by Supt. of Docs., U.S. Gov. Print. Off. Begun in 1955, frequency varies (0010–6917).

The Construction Specifier (1949–) Washington, DC: The Construction Specifications Institute (1093–846X). A professional organization journal which presents solid information on materials specifications.

Detail: Zeitschrift fur Architektur & Baudetail & Einrichtung (1961–) Munchen: Verlag Architektur + Baudetail GmbH (0011–9571). An important architectural design magazine that includes fine line drawings. Now issued in English.

ENR (1987–) Continues *Engineering News Record*, New York: McGraw-Hill (0891–9526). A long-running journal that covers current (largely civil) engineering projects and topics (see Chapter 2, this volume).

Journal of Architectural Engineering (1995–) New York: American Society of Civil Engineers (1076–0431). Electronic version (http://www.pubs.asce.org/journals/ae.html) free with print subscription. "*The Journal of Architectural Engineering* provides a multidisciplinary forum for dissemination of practice-based information on the engineering and technical issues concerning all aspects of building design. Peer-reviewed papers and case studies address issues and topics related to buildings such as planning and financing, analysis and design, construction and maintenance, codes applications and interpretations, conversion and renovation, and preservation" (ASCE web page).

Patents and standards

Patent and standard information may, in many instances, be searched online. A patent search may provide insight into new technology and processes.

Standards may be voluntary or required, and they provide information on materials, testing, specifications, and practices.

ASHRAE Handbook (1970–) New York: The Society. Standards handbook for heating, ventilation, and air-conditioning. Issued in inch-pound and metric editions. Each standard (Applications, Fundamentals, Refrigeration, Systems and Equipment) is updated on a five-year cycle, replacing the previous edition.

American Society for Testing and Materials http://www.astm.org/ (see Chapter 2).

Architectural Graphic Standards Online (2003–) <http://graphicstandards.wiley.com/> (accessed May 17, 2005), New York: John Wiley & Sons. "Graphic Standards Online, from John Wiley & Sons, is the premier source for building design and construction information. Graphic Standards Online puts the power of the complete Architectural Graphic Standards – and much more – at your fingertips, whenever you need it" (Graphic Standards web page).

International Code Council (2003–) *International Building Code*, Falls Church, VA: The Council. Comprehensive in scope, and the result of cooperation within regulatory organizations. This code establishes minimum standards for building systems performance.

International Organization for Standardization (ISO) <http://www.iso.org/> (accessed May 17, 2005). A network of the national standards institutes of 146 countries, with a coordinating Central Secretariat in Geneva, Switzerland. "ISO is a non-governmental organization: its members are not, as is the case in the United Nations system, delegations of national governments. Nevertheless, ISO occupies a special position between the public and private sectors. This is because, on the one hand, many of its member institutes are part of the governmental structure of their countries, or are mandated by their government. On the other hand, other members have their roots uniquely in the private sector, having been set up by national partnerships of industry associations" (ISO web page).

Life Safety Code (2003) Quincy, MA: National Fire Protection Association. Includes standards for egress for specific types of buildings.

National Electrical Code (2002) Quincy, MA: National Fire Protection Association. NFPA 70.

National Information Standards Organization <http://www.niso.org/> (accessed May 18, 2005) (see Chapter 2, this volume).

United States Patent and Trademark Office <http://www.uspto.gov> (see Chapter 2, this volume).

Search engines and important websites

Internet search engines grow in size and importance with each passing day. Google is currently beta testing a new service aimed at searching scholarly literature <http://scholar.google.com> (accessed May 18, 2005). While much professional literature continues to require paid subscriptions, most organizations do have free websites that provide introductions to their aims, goals, and services. The metasites listed below serve as portals to locating technical information.

iCivilEngineer <http://www.icivilengineer.com/> (see Chapter 8, this volume).

Construction WebLinks <http://www.constructionweblinks.com/> (accessed

May 17, 2005). "Thelen Reid & Priest LLP has served the construction industry for more than 75 years. Its lawyers used the knowledge and experience they gained in representing all facets of the construction industry in building this site" (Construction WebLinks website).

Ultimate Civil Engineering Directory <http://www.tenlinks.com/engineering/civil/index.htm>.

The Virtual Library: Engineering <http://vlib.org/Engineering.html> (accessed May 17, 2005). Provides links to architectural engineering (and many other types of engineering) Internet resources.

Associations, societies and organizations

The following list includes professional organizations useful to the student or practitioner of architectural engineering.

American Institute of Architects (AIA) <http://www.aia.org/> (accessed May 17, 2005). "Through a culture of innovation, The American Institute of Architects empowers its members and inspires creation of a better built environment. . . . The American Institute of Architects provides guidance, service, and standards to architects around the world. The AIA continues to strive for quality, consistency, and safety in the built environment and to serve as the voice of the architecture" (AIA web page).

American Society of Heating, Refrigerating, and Air-conditioning Engineers (ASHRAE), Inc. <http://www.ashrae.org> (accessed May 17, 2005). ASHRAE, according to its 1993 vision statement, "seeks to be the global leader in the arts and sciences of HVAC&R (heating, ventilation, air-conditioning and refrigeration); the foremost responsive, authoritative, and timely source of technical and educational information, standards and guidelines; and, the primary provider of opportunities for professional growth, recognizing and adapting to changing demographics and embracing diversity" (ASHRAE web page).

American Solar Energy Society (ASES) <http://www.ases.org/homepage.htm> (accessed June 2, 2005). Advances solar use in America. News, conferences, and publications available.

Architectural Engineering Institute (AEI) <http://www.aeinstitute.org/> (accessed May 17, 2005). "AEI is the home for all professionals in the building industry. The Architectural Engineering Institute was created through a merger of the National Society of Architectural Engineers (NSAE) and the American Society of Civil Engineers Architectural Engineering Division (AED) on October 1, 1998. AEI strives to be the premier organization providing a multi-disciplinary forum for building industry professionals to examine technical, educational, scientific and professional issues of common interest" (AEI web page).

Illuminating Engineering Society of North America (IESNA) <http://www.iesna.org/> (accessed May 17, 2005). "The IESNA is the recognized technical authority on illumination. For over ninety years its objective has been to communicate information on all aspects of good lighting practice to its members, to the lighting community, and to consumers through a variety of programs, publications, and services. The strength of the IESNA is its diversified membership: engineers, architects, designers, educators, students, contractors, distributors, utility personnel, manufacturers, and scientists, all contributing to the mission of the Society:

to advance knowledge and disseminate information for the improvement of the lighted environment to the benefit of society" (IESNA web page).

National Society of Architectural Engineers (NSAE) <http://www.aldea.com/guides/ag/b609ud.html> (accessed May 17, 2005). "The National Society of Architectural Engineers is a professional organization promoting the advancement of architectural engineering. NSAE incorporated in 1984 as the outgrowth of a student organization at the University of Kansas in cooperation with the other existing ABET-accredited architectural engineering programs" (NSAE website).

National Society of Professional Engineers <http://www.nspe.org/aboutnspe/ab-home.asp> (see Chapter 2, this volume).

Structural Engineering Institute (SEI) <http://www.seinstitute.org/> (accessed May 17, 2005). "SEI is a vibrant, 20,000 plus community of structural engineers within the American Society of Civil Engineers. SEI started on October 1, 1996 in order to serve the unique needs of the structural engineering community more effectively while also being their voice on broader issues that shape the entire civil engineering community.... SEI advances our members' careers, stimulates technological advancement, and improves professional practice" (SEI web page).

Conclusion

Research on architectural engineering will likely continue to require a multi-faceted approach. Students can gain a basic understanding through general works and textbooks on architectural engineering and integrated building systems; they should also be encouraged to consult more specific works, such as those on mechanical and electrical systems. For the professional, the literature comes largely from periodicals and is normally related to more specific and specialized topics, such as seismic design or green building products. It is also important for professionals to remain current with architectural trends, and to review contemporary design periodicals.

Acknowledgments

Architectural engineering is a highly technical, broad-based profession, and architecture students are becoming increasingly cognizant of the many ways an architectural engineer can contribute to a project. Architecture schools now offer courses on advanced building systems in which students analyze building components, and study the integration of its various systems. At Syracuse University, such a capstone course was developed and taught by Joel Bostick, Associate Professor of Architecture (1947–2004). Joel helped numerous students at Syracuse understand architectural technology and the necessity for well-designed, functional building systems. This chapter is dedicated to the memory of Joel, who helped me understand better the importance of building technology.

6 Bioengineering

Mary D. Steiner and Linda Martinez

Introduction

In 2000, readers of the journal *Mechanical Engineering* voted on the top ten engineering achievements of the twentieth century. Bioengineering was listed as number nine. Rastegar (2000: 76) states:

> The 20th century was known as the Century of Physics. The 21st century is believed to be the Century of Biology. The definition of bioengineering can be broad, but the key characteristics are the application of engineering principles to the study of biological sciences and biological systems. In addition, it includes various traditional fields of science: physics, chemistry, mathematics, and computing. Bioengineering is a field that connects physical sciences to biological sciences, making it a bridge between the two centuries and, as such, one of the major fields of engineering of the 21st century.

A 2004 article in *ASEE Prism* states that "bioengineering is among the most popular and fastest growing undergraduate majors at top research universities nationwide." In 2004, "the Whitaker Foundation is supporting 130 biomedical engineering programs, up from 42 programs in the early 1990s" (Loftus, 2004: 39).

Scope

The Institute of Biological Engineering (IBE) states that the rapid growth of bioengineering programs may have made it difficult for the field itself to develop into a true engineering discipline:

> There are two basic categories of an engineering discipline: 1) applications-based and 2) science-based. The first category defines the discipline by those served, e.g., agricultural engineering, petroleum engineering and mining engineering. Mechanical engineering and chemical engineering are examples in the second category, science-based disciplines based upon mechanics, physics and chemistry.
>
> (Johnson, 2004)

According to Johnson, it is not clear at this point to which category bioengineering belongs – biomedical engineering is definitely applications-based while biological engineering is science-based with its roots in biology. Practitioners in biomedical engineering are engineers first – either electrical or mechanical – while biological engineers are first grounded in the biological sciences. Bioengineering, as defined by the National Institutes of Health:

> integrates physical, chemical or mathematical sciences and engineering principles for the study of biology, medicine, behavior, or health. It advances fundamental concepts, creates knowledge for the molecular to the organ systems levels, and develops innovative biologics, materials, processes, implants, devices, and informatics approaches for the prevention, diagnosis, and treatment of disease, for patient rehabilitation, and for improving health.
>
> (National Institutes of Health)

Since bioengineering is an extremely interdisciplinary field, it is challenging to find a clear definition/term that encompasses its entire scope. The terms bioengineering, biological engineering, and biomedical engineering are often used interchangeably to describe the same fields of study. It may be easier to define bioengineering by describing what bioengineers do. Bioengineers may be involved in designing prosthetic limbs or orthopedic implants; developing nano-devices for drug delivery; testing materials for use in vascular stents; improving biomedical imaging systems; performing genetic engineering and analysis; investigating methods of tissue repair and interventions for rehabilitation; and much, much more.

Universities offering bioengineering programs likewise vary in their use of the terms bioengineering, biological engineering, and biomedical engineering in describing their programs. Generally, bioengineering describes a broader field than biomedical engineering. Historically, though not exclusively so, bioengineering programs grew out of institutions with strong agricultural schools, while biomedical engineering programs were located in universities with strong engineering and medical schools.

For the purposes of this chapter, we do not cover resources for the agricultural aspects of bioengineering (see Chapter 4), but focus on those resources that intersect the life sciences, medicine, and the engineering disciplines.

History

In the broadest context, bioengineering developed through the practices of medical and engineering techniques as applied to specific problems. The history of bioengineering may be traced back to the ancient peoples of Europe and Asia (Russian River Ecosystem Restoration Study and the Russian River Watershed Council). For example, a 3,000-year-old mummy from Thebes was unearthed with a wooden prosthesis attached to its foot (Whitaker Foundation). Sanitation and sewer systems were features of

Egyptian and Summerian cities. Fermentation techniques to produce beer and wine are at least 10,000 years old (Craig, 2003).

Prior to World War I, research in the areas of medicine, physics, biology, and engineering was occurring in Europe and the United States, but there was very little formal communication between these research groups. In 1913, the Rockefeller Foundation supported work in the areas of physics and engineering in biological and medical research. In Germany, Friedrich Dessauer founded the Oswalt Institute for Physics in 1921 to investigate the biological effects of ionizing radiation. During the 1920s in the U.S., the Johnson Foundation for Medical Physics at the University of Pennsylvania and the Biophysics Department of the Cleveland Clinic were established.

After World War II, advances in technology, electronics, materials, and medicine fostered the growth of professional societies in engineering, medicine, and biology. In 1948, the first U.S. conference of engineering in medicine and biology, sponsored by the Institute of Radio Engineers, the American Institute for Electrical Engineering, and the Instrument Society of America, was held. Ten years later, more than 300 researchers attended a U.S. conference on computers in medicine and biology (Nebeker, 2002).

In the 1960s, the National Institutes of Health (NIH) began supporting biomedical engineering by creating a program project committee under the General Medical Sciences Institute to evaluate program project applications in biophysics and biomedical engineering areas. It set up a biomedical engineering training study section to evaluate training grant applications, and it established two biophysics study sections. The 1960s and 1970s marked the beginning of the rise of academic programs in bioengineering and biomedical engineering. Each decade has yielded significant scientific breakthroughs that have impacted upon the fields of bioengineering. The 1950s saw the development of cardiac pacemakers, heart lung machines, x-ray imaging, the scanning electron microscope. In the 1960s, computers began to gain importance in engineering and medical research areas. In the 1970s, new techniques of medical imaging emerged – computerized tomography, and nuclear magnetic resonance. The 1980s saw the development of endoscopy, lasers, ultrasound imaging, and the increasing importance of computers in medical applications. In the 1990s, the human genome project, robotics, implantable devices, and new biomaterials yielded new research areas. Future research areas may involve cellular and tissue engineering, bionanotechnology, computational bioengineering, and genetic engineering (Nebeker, 2002).

> The field has evolved from its early years, when engineering principles and technologies were applied, adapted, or devised for biological systems, to modern bioengineering, which involves the integration of engineering principles and technologies with cellular and molecular biology.
>
> (Rastegar, 2000: 76)

The literature of bioengineering

The literature of bioengineering is dispersed far and wide, particularly when including all of biomedical engineering, biomolecular engineering, biological engineering, bioinformatics, and biotechnology. For example, in a library arranged by Library of Congress (LC) call numbers, materials are located throughout the areas of science (Q), medicine (R), and technology (T). In larger institutions, the quest for print material becomes more challenging as works may be located in numerous individual libraries housing biology, chemistry, computer science, engineering, medicine, and physics collections. Bioethics issues are likely to be of interest to the bioengineer as well. As with other scientists and engineers, bioengineers rely heavily on journal literature, much of which is increasingly available online. Conference literature is also important. Technical reports are not as relevant to bioengineers as they may be in other engineering disciplines, although NASA or Department of Transportation biomedical research reports may be of interest.

Keyword searching in library catalogs and indexes/abstracts is an important way to discover books, journal articles, and conference literature. Free-form keywords and/or standard subject headings may be used. Bioengineers should be aware of two distinct subject headings: Library of Congress Subject Headings (LCSH) and Medical Subject Headings (MeSH). The former system is used in the library catalogs of the Library of Congress and many academic institutions; the latter, developed by the National Library of Medicine, is used by many biomedical institutions and hospital libraries.

Engineers have long browsed through a library's collection by call number to foster the serendipitous discovery of materials. It is also fruitful to "virtually browse" through call number ranges in a library's online catalog, because hot topics and newer books may be checked out or stored in different physical locations (and therefore not on the shelf for discovery). Most academic and research institutions use the Library of Congress call number classification system.

Library of Congress Subject Headings (LCSH)

Although not exhaustive, this collection of terms may prove helpful in searching for books and other literature in an LCSH library catalog: Biochips; Biocompatibility; Bioengineering; Bioethics; Bioinformatics; Biological transport; Biomechanics; Biomedical engineering; Biomedical materials; Biosensors; Biotechnology; Colloids in medicine; DNA microarrays; Human mechanics; Image analysis; Image processing; Imaging systems in medicine; Medical informatics; Medical instruments and apparatus; Nanotechnology; Polymeric drugs; Polymers in medicine; Prosthesis; Recombinant molecules; Tissue engineering

Medical Subject Headings (MeSH)

Similarly, this collection of terms may prove helpful in searching a MeSH library catalog: Biocompatible materials; Biomechanics; Bioethics; Biomedical and dental materials; Biomedical engineering; Biotechnology–instrumentation; Implants, artificial; Image processing, Computer-assisted methods; Medical informatics; Nanotechnology; Polymers.

Keywords

Other potential keywords, which along with the aforementioned terms could be used in indexes/abstracts as well as library catalogs, might include: Biomaterials; Biomedicine; Clinical engineering; Drug delivery; Drug delivery system(s); Nanobiotechnology; Synthetic biology. Of course, new terminology will continue to emerge in this field.

Library of Congress call numbers

Relevant Library of Congress call numbers are presented here in alphanumeric order, i.e., sequentially as they would appear on library shelves (or in an online catalog):

Q300–342	Cybernetics
QA76.87	Neural computers; Neural networks (in computing)
QH505	Biophysics
QH507	Information theory in biology
QH508	Biological control systems
QH509	Biological transport
QH509.5	Bioelectronics
QH513	Biomechanics
QH513.5	Fluid dynamics (in biology)
QH332	Bioethics
QH324.2	Bioinformatics
QH441.2	Bioinformatics
QP303	Mechanics; Kinesiology
QP363.3	Neural circuitry; Neural networks (in physiology)
R724–726.2	Medical ethics
R856–857	Biomedical engineering; Electronics; Instrumentation (this call number range includes many special topics, such as biosensors, biomaterials, imaging systems, nanotechnology)
R858–859	Computer applications to medicine; Medical informatics; Medical information technology
RD130	Prosthesis; Artificial organs
RD132	Artificial implants and implant materials

RS199.5–RS201	Pharmaceutical technology; Drug delivery systems
RS210	Drug delivery devices
TA164	Bioengineering
TP248.13–TP248.65	Biotechnology (including biochemical engineering)

Bibliographies and guides to the literature

In contrast to some of the long-standing engineering disciplines such as civil engineering or mechanical engineering, relatively few literature guides or bibliographies exclusive to bioengineering/biomedical engineering exist. A number of these came out with the dawn of the field (Pate, 1966; Force, 1972; Kramer, 1973), but they are somewhat outdated now. Since that time, some works specific to biotechnology or biosciences have been created. Some worth noting include the following.

Crafts-Lighty, A. (1986) *Information Sources in Biotechnology* (2nd edn), New York: Stockton Press. Although this volume is about twenty years old, it is very comprehensive and worth taking a look at. Coverage of biotechnology science (even some business aspects as well) is extensive and detailed.

MacLeod, R.A. and Corlett, J. (2005) *Information Sources in Engineering* (4th edn), London; New Providence, NJ: Bowker-Saur. This work is unique in being organized by primary sources, then secondary sources, and finally information sources for specialized subject fields. Some of these specialized subject areas are broad (e.g., chemical engineering), and some narrow (e.g., marine technology). Biomedical engineering is one of these, with twenty pages written by an academic in biomedical engineering. This chapter is definitely worth taking a look at – in particular the history of the field and the various societies, and the journals section.

Schmidt, D., Davis, E.B., and Jacobs, P.F. (2002) *Using the Biological Literature: A Practical Guide* (3rd edn), revised and expanded, New York: Marcel Dekker. Sections on biochemistry, biophysics, molecular and cellular biology, genetics, and biotechnology complement other volumes in the publisher's Books in Library and Information Series.

Abstracts and indexes

To locate the most recent, cutting-edge research information, engineers, scientists, and researchers refer to journal articles, conference proceedings, and technical reports. Abstracts and indexes are tools to locate these types of documents. Given the sheer amount of information in the engineering and scientific disciplines, abstracts and indexes specialize in their coverage and scope. Abstracts and indexes provide citation information: title of an article or conference paper, authors, title of the periodical or conference, volume issue, date, and pages in which the article or conference paper appears. In addition, abstracts provide a brief description of the article or conference paper's content. Increasingly, digital abstracts and indexes not only provide citation information, but link to the full text of the article or conference paper as well.

Applied Science and Technology Abstracts (1983–) Bronx, NY: H.W. Wilson. International coverage of scientific/technical journals, trade/industrial journals, and conference proceedings. Topics include: engineering, biomedical, the environment, and biotechnology (see Chapter 2, this volume).

BioCommerce Abstracts and Directory (1981–) Berkshire, UK: BioCommerce Data Ltd. International coverage of the commercial aspects of biotechnology. Topics include research and development projects, product news, industry news from reports, and newsletters. Includes abstracts from journals and organizational profiles.

BioEngineering Abstracts (1993–) Bethesda, MD: Cambridge Scientific Abstracts. International coverage of 257 primary sources and 2,500 secondary sources (conference proceedings, journals, directories) in all aspects of the field of bioengineering. Topics include: biomedical engineering, biotechnology, biomaterials, biomechanics and human engineering, genetic engineering, and rehabilitation engineering.

Biosis Previews (1969–) Philadelphia, PA: Thomson Scientific. International coverage of journals covering the life sciences, international, interdisciplinary, biochemistry, bioengineering, biology, biophysics. Covers articles, meetings, patents, book chapters, technical reports. Also referred to as *Biosis*; print version known as *Biological Abstracts*, which goes back to 1926.

Biotechnology and Bioengineering Abstracts (1982–) Bethesda, MA: Cambridge Scientific Abstracts. International coverage of 597 primary sources and 5,500 secondary sources covering biotechnology and bioengineering. Topics include biomedical engineering, biomaterials, human rehabilitation engineering, medical applications, and drug delivery systems.

Chemical Abstracts (1907–) Columbus, OH: Chemical Abstracts Service. International coverage of 80,000 serials publications covering scientific and technical literature relevant to all aspects of chemistry, biochemistry, biotechnology, and biomolecular engineering. Formats include journal articles, patents, technical reports, conference proceedings, and dissertations. *SciFinder* is the software search interface to *Chemical Abstracts*.

Compendex (1884–) Hoboken, NJ: Elsevier Engineering Information, Inc. International coverage of 5,000 journals, technical reports, conference proceedings, dissertations to more than 175 engineering disciplines, including biology, biological materials, biomechanics, biomedical engineering, and biotechnology. Print version referred to as *Engineering Index*.

Derwent Biotechnology Resource (1982–) London: Thomson Derwent. International coverage of journals, conference proceedings, and patents focused on all aspects of biotechnology and associated sciences. Topics include bioinformatics, genomics, biotechnological processes, and microbiology.

EMBASE (1974–) New York: Elsevier Science. International coverage of more than 4,000 journals focusing on human medicine and areas of biological sciences related to human medicine. Topics include: bioengineering, biochemistry, genetics, biophysics, and medical instrumentation. Print version referred to as *Excerpta Medica*.

Inspec (1898–) London: Institution of Electrical Engineers. A collection of abstracts: *Physics Abstracts, Electrical and Electronics Abstracts*, and *Computer and Control Abstracts*. International coverage of 4,000 scientific/technical journals and 2,000 conferences, in addition to books, technical reports, and dissertations. Topics include physics, electronics and electrical engineering, computers and control, robotics, biochemistry, bioengineering, bio-optics, biocomputing, biomedical engineering, biomedical communications, and biomedical equipment.

Medline (1960–) *PubMed* (1950–) Bethesda, MD: U.S. National Library of Medicine. International coverage of 4,600 biomedical journals and genetics databases focusing on clinical medicine, medical research, and health care; covers allied health, biological and life sciences, neuroscience, neurobiology, and information science.

Directories

Directories are particularly useful for finding information about people, institutions, equipment/companies, grants, and some statistics. For example, some societies maintain print and/or online directories of their membership, and several commercial publishers provide subject directories to university programs and government-funded research. The listing given below is not comprehensive (society memberships should be investigated at each society's website), but provide a sampling of sources of potential interest to the bioengineer.

Biomedical Engineering Curriculum Database <www.whitaker.org/academic/database> (accessed May 17, 2005). This website, hosted by the Whitaker Foundation, allows colleges and universities to enter details about their academic programs, including undergraduate, Masters, and doctoral programs. Some schools list areas of concentration, as well as required courses. Browsing and searching features are available.

Computer Retrieval of Information on Scientific Projects (CRISP) <http://crisp.cit.nih.gov/> (accessed May 17, 2005). CRISP is a searchable database of federally funded biomedical research projects conducted at universities, hospitals, and other research institutions. The site covers awards from 1972 to the present.

Directory of Biomedical and Health Care Grants. Phoenix, AZ: Oryx Press. This annual directory provides grant titles, descriptions, requirements, amounts, application deadlines, contact information, Internet access, sponsor names and addresses, and samples of awarded grants (when available).

Health Devices Sourcebook. Plymouth Meeting, PA: ECRI. Annual print directory of medical product manufacturers in the United States and Canada. Has sections for product listings, trade names, manufacturer names, and equipment services.

Medical Device Register. Los Angeles, CA: Canon Communications. Available in a print two-volume set, on CD, or via the Web, this annual directory lists medical devices and manufacturers in the U.S., Canada, and Mexico.

Nature Biotechnology Directory. London: Nature Publishing Group. The annual print version is issued in February, and, according to the publisher's website at <http://guide.nature.com/> (accessed May 17, 2005), the directory "is the world's leading reference to the global biotechnology industry with profiles of more than 8,000 companies, institutions, associations and regulatory bodies. Key features include: Buyer's Guide to over 16,000 products, equipment and services; profiles of more than 10,000 commercial companies in 82 countries; translation guide to common biotech terms in French, German, and English; contact details for government and regulatory agencies; trade and industry associations, grants and grant funding bodies." Some basic information may be searched on the website itself, though extended information requires a fee.

Dictionaries

Dictionaries are alphabetical lists providing definitions of terms, concepts, and principles within a specific discipline or field. Coverage may vary from a few sentences, to lengthy, in-depth explanations of a term or concept.

Biomedical Acronym/Definition Database <http://invention.swmed.edu/argh/> (accessed May 17, 2005). Free online dictionary to acronyms, more than 206,000, used in the medical fields. Dictionary interfaces with PubMed to present the searched term within context.

Biotech's Life Science Dictionary <http://biotech.icmb.utexas.edu/search/dict-search.html> (accessed May 17, 2005). Free online dictionary to more than 8,300 terms dealing with biochemistry, biotechnology, medicine, and genetics.

Juo, P.S. (2001) *Concise Dictionary of Biomedicine and Molecular Biology*, Boca Raton, FL: CRC Press. Definitions of more than 30,000 terms used in biotechnology, molecular biology, and biomedicine, understandable to researchers and students.

Myers, R.A. (ed.) (1995) *Molecular Biology and Biotechnology: A Comprehensive Desk Reference*, New York: John Wiley & Sons. Covers theories and techniques involved in molecular basis of life and applications in genetics, medicine and agriculture; 250 articles are understandable both to scientists and the layman.

Nill, K.R. (2002) *Glossary of Biotechnology Terms* <http://biotechterms.org> (accessed May 17, 2005) (3rd edn), Boca Raton, FL: CRC Press. Detailed definitions of biotechnical, biological and chemical terms appropriate for the non-expert.

Parker, S.P. (ed.) (2003) *Dictionary of Bioscience* <http://books.mcgraw-hill.com> (accessed November 3, 2004), New York: McGraw-Hill Professional Publishing. Definitions of more than 22,000 terms from the fields of embryology, genetics, physiology, and molecular biology.

Stedman, T.L. (2000) *Stedman's Medical Dictionary* <http://www.stedmans.com> (accessed November 3, 2004) (27th edn), Philadelphia, PA: Lippincott Williams & Wilkins. The authority on medical terminology, this edition includes more than 12,000 new entries.

Steinberg, M.L. and Cosloy, S.D. (2001) *Facts on File Dictionary of Biotechnology and Genetic Engineering*, New York: Checkmark Books. Covers the basic and more technical terminology of modern biotechnology and genetic engineering.

Szycher, M. (1992) *Szycher's Dictionary of Biomaterials and Medical Devices*, Lancaster, PA: Technomic Publishing Company, Inc. Definitions of terms from the disciplines of polymer chemistry, biochemistry, metallurgy, and organic chemistry as they relate to biomaterials used in medical devices.

William, D.F. (ed.) (1999) *Williams Dictionary of Biomaterials*, Liverpool, UK: Liverpool University Press. Defines more than 6,000 words and phrases associated with biomaterials and related disciplines. Written for clinical scientists involved in engineering matters.

Encyclopedias

Encyclopedias are excellent resources for obtaining an overview of a particular topic that can be understood by practitioners as well as non-experts. The encyclopedia's authoritative articles may provide both historic and the most up-to-date information on a particular topic and usually include extensive bibliographies for further research.

Atkinson, B. and Mavituna, F. (1991) *Biochemical Engineering and Biotechnology Handbook* (2nd edn), Basingstoke: Stockton Press. Topics include general classification of micro-organisms, morphology of micro-organisms, and the life cycles of micro-organisms.

Bronzino, J.D. (ed.) (2000) *Biomedical Engineering Handbook* (2nd edn) (2 vols), Boca Raton, FL: CRC Press. Serves as a reference review of fundamental physiology, and to the latest developments in various aspects of biomedical engineering. Topics include: transport phenomena, tissue engineering, prostheses and artificial organs, human performance engineering, physiological modeling, clinical engineering, medical informatics, artificial intelligence, and ethical issues.

Finn, W.E. and LoPresti, P.G. (2003) *Handbook of Neuroprosthetic Methods*, Boca Raton, FL: CRC Press. Comprehensive resource for the techniques and methodologies to design and undertake experiments within the fields of neuroprosthetics. Topics include: microelectronics, biomolecular electronics, hearing, vision, and motor prostheses.

Flickinger, M.C. and Drew, S.W. (1999) *Encyclopedia of Bioprocess Technology: Fermentation, Biocatalysis, and Bioseparation* (5 vols), New York: Wiley. Covers the applications and theories in biotechnology, with a focus on industrial applications of fermentation, biocatalysis, and bioseparation. Topics include cell separation, biotransformations, bioremediation, and bioreactors. Also available through Knovel.

Kirk, R. and Othmer, D.F. (2004) *Encyclopedia of Chemical Technology* <http://www3.interscience.wiley.com> (accessed November 3, 2004) (5th edn), New York: Wiley. This well-known twenty-seven-volume reference work covers the properties, manufacturing, and uses of chemicals and materials, processes and engineering principles.

McAinsh, T.F. (ed.) (1986) *Physics in Medicine and Biology Encyclopedia* (2 vols), New York: Pergamon Press. Provides a review of topics of current interest in medical physics, bioengineering, and biophysics; geared to hospital physicists, bioengineers, and medical technologists, but also a reference tool for non-experts and students.

Myers, Robert A. (ed.) (2005) *Encyclopedia of Molecular Cell Biology and Molecular Medicine* (2nd edn) (16 vols), Hove, Sussex: Wiley Europe, Authoritative source to disciplines of molecular biology, cell biology, and molecular biology; includes 425 articles, and a glossary of basic terms aimed at the general reader.

Nalwa, H.S. (ed.) (2004) *Encyclopedia of Nanoscience and Nanotechnology* (10 vols), Stevenson Ranch, CA: American Scientific Publishers. Provides an introduction and overview of emerging fields of nanoscience and nanotechnology. Topics include nanostructured materials, nanobiotechnology, nanobiology, nanomedicines, drug delivery, and biomedical applications.

O'Neil, M.J. (ed.) (2001) *Merck Index: An Encyclopedia of Chemicals, Drugs, and Biologicals* (13th edn), New York: Wiley. More than 10,000 entries on chemicals, drugs, and natural substances; entries provide concise descriptions of a single substance or a small group of closely related compounds.

Payne, P.A. (ed.) (1991) *Concise Encyclopedia of Biological and Biomedical Measurement Systems*, New York: Pergamon Press. An essential reference work for biomedical engineers, coverage includes chemical and biochemical analysis *vis-à-vis* measurements of living organisms and systems.

Rehm, H.J., Reed, G., Puhler, A., and Stadler, P. (eds) (1996) *Biotechnology: A Multi-volume Comprehensive Treatise* (2nd edn), New York: Wiley. Covers all aspects of biotechnology research and development from biological and genetic fundamentals to genomics, bioinformatics, special processes, metabolism and legal, economic and ethical dimensions. Each volume covers a specific topic in depth.

Schwarz, J.A., Contescu, C.I., and Putyera, K. (2004) *Dekker Encyclopedia of Nanoscience and Nanotechnology* (5 vols), New York: Marcel Dekker. Covers the emerging fields of nanoscience and nanotechnology; topics include nanoparticles, nanomaterials, gene delivery, and nanotubes.

Skalak, R. and Chien, S. (eds) (1987) *Handbook of Bioengineering*, New York: McGraw-Hill. Covers the main areas of research and technology in bioengineering. Topics include orthopedics, biomechanics, cardiovascular mechanics, respiratory mechanics, and properties of hard and soft tissues.

Webster, J.G. (ed.) (1999) *Wiley Encyclopedia of Electrical and Electronics Engineering* (24 vols), New York: John Wiley. In the areas of biomedical engineering the topics include bioelectric phenomena, biomedical sensors, biomedical telemetry, clinical engineering, medical information systems, and neurotechnology.

Wikipedia – Biology <http://en.wikipedia.org/wiki/Biology> (accessed May 17, 2005), Wikimedia Foundation Inc. Wikipedia is a free content encyclopedia written collaboratively by contributors from all around the world. The site is a wiki, which means that anyone can edit articles. *Wikipedia – Biology* covers diverse fields of biology, including genetics, biomechanics, biophysics, bioinformatics, and biotechnology.

Wise, D. (ed.) (1995) *Encyclopedic Handbook of Biomaterials and Bioengineering* (4 vols), New York: Marcel Dekker. Covers the materials used in or on the human body. Topics include polymer chemistry, biochemistry, orthopedics, pharmaceutics, biomaterials, and biocompatibility.

Wnek, G.E. and Bowlin, G.L. (2004) *Encyclopedia of Biomaterials and Biomedical Engineering* (2 vols), New York: Marcel Dekker. Comprehensive coverage of more than 400 topics; geared to clinicians, engineers, scientists, and researchers.

Handbooks, data, and manuals

There are many different types of handbooks for bioengineering – in fact, some works with the word "handbook" in the title are actually more like an encyclopedia or dictionary. Typically, handbooks provide quick look-up information, often tabular or graphical data, in a conveniently indexed form. They can also be a condensed treatment of a subject arranged in a systematic way; therefore some works which have the word "handbook" in the title may be found in the Encyclopedias section of this chapter. Material property and selection information is an important component of many handbooks.

Datasets are described as organized collections of related information; a resource that is a collection of pieces of data; a homogenous collection of data records managed as a single entity, or a collection of data that originated from the same source. Datasets are of particular importance to scientific, medical, and engineering disciplines.

Manuals offer guidelines on how to carry out procedures, particularly in a laboratory setting. Some of the descriptions that follow are taken from the work itself or from its website.

Atkinson, B, and Mavituna, F, (1991) *Biochemical Engineering and Biotechnology Handbook* (2nd edn), New York: Stockton Press. This is a substantial volume at over 1,200 pages and contains a wealth of information on topics including properties of micro-organisms, lab techniques, cell cultures, rheology, instrumentation, costs of bioprocesses, and much more.

Biodynamics Database <www.biodyn.wpafb.af.mil> (accessed May 17, 2005). The Biodynamics Data Bank (BDB) was established in 1984 by a team of researchers at the U.S. Air Force Research Laboratory (AFRL). The primary objective was to provide a national repository of biodynamics test data accessible to the entire research community. The contents of the BDB include data from approximately 7,000 acceleration impact tests conducted at the Biodynamics and Acceleration Branch of AFRL, as well as information on the associated test facilities, test programs, and test subjects. The test data include measured forces, accelerations, and motions of both human and manikin subjects. Users must register to use the site.

BioMolecular Engineering Research Center (BMERC) <http://bmerc-www.bu.edu> (accessed May 17, 2005). From Boston University, BMERC provides research support in the areas of biology and bioinformatics, molecular evolution, and gene regulation. Links to several statistical, genome datasets are provided.

Black, J. and Hastings, G. (1998) *Handbook of Biomaterial Properties*, London; New York: Chapman & Hall. This work is organized in three parts: composition and properties of biological tissues; different biomaterials in use; and applications issues such as biocompatibility, systems interactions and responses.

Bronzino, J.D. (ed.) (2000) *The Biomedical Engineering Handbook* (2nd edn), Boca Raton, VA: CRC Press, IEEE Press. This substantive work, in two volumes, covers all the major areas of the discipline. Individual contributions provide explanations, property information, processes, devices, and more.

Current Protocols <http://www3.interscience.wiley.com/aboutus/currentProtocols. html> (accessed May 17, 2005). Available as looseleaf binders which can be updated, or online, the numerous *Current Protocols* manuals specify methodologies for various laboratory procedures. According to the publisher's website, "these extensively annotated, step-by-step manuals are composed by globally recognized experts and carefully screened for maximum accuracy. To ensure easy duplication in any lab, the protocols are thoroughly field tested and scrutinized by our editorial boards prior to inclusion." A number of these should be of interest to the bioengineer, including *Current Protocols in Bioinformatics, Current Protocols in Protein Science*, among others.

Davis, J.R. (ed.) (2003) *Handbook of Materials for Medical Devices*, Materials Park, OH: ASM International. This work provides a review of the properties, processing, and selection of materials used in the environment of the human body. Among the application areas described are orthopedics (hips, knees, and spinal and fracture fixation), cardiology (stents, heart valves, pacemakers), surgical instruments, and restorative dentistry. Materials discussed include metals and alloys, ceramics, glasses, and glass-ceramics, polymeric materials, composites, coatings, and adhesives and cements.

Dyro, J. (2004) *The Clinical Engineering Handbook*, London; Burlington, MA: Elsevier Academic Press. As the title suggests, this volume concentrates on clinical engineering, where the focus is health care and medical device technologies. Chapters cover the discipline in scientific, managerial, and societal aspects; clinical engineering in a number of countries worldwide; safety and regulatory issues; and more.

Kress-Rogers, E. (ed.) (1997) *Handbook of Biosensors and Electronic Noses: Medicine, Food, and the Environment*, Boca Raton, VA: CRC Press. Descriptions of different types of biosensors are noted in this work.

Kutz, M. (ed.) (2003) *Standard Handbook of Biomedical Engineering and Design*, New York: McGraw-Hill. This one-volume work focuses on human biomechanics, biomaterials, and bioelectricity, then segues into design principles for medical devices, imaging systems, and prostheses.

Molecules to Go <http://molbio.info.nih.gov/cgi-bin/pdb> <http://www.ncbi.nlm.nih.gov> (accessed May 17, 2005). This National Institutes of Health site provides access to the molecular data provided by the Brookhaven Protein Data Bank.

Moore, J. and Zouridakis, G. (ed.) (2004) *Biomedical Technology and Devices Handbook*, Boca Raton, VA: CRC Press. A good reference for anyone wanting details about medical devices and technologies, processes for measurement, imaging techniques, biological assays, tissue engineering, interventional and rehabilitation treatments.

National Center for Biotechnology Information <http://www.ncbi.nlm.nih.gov> (accessed May 17, 2005). From the National Library of Medicine, this site is a clearinghouse for resources in molecular biology. Links to molecular databases, genomic biology resources, and data-mining tools.

Protein Data Bank <http://www.rcsb.org/pdb> (accessed May 17, 2005). The Research Collaboratory for Structural Bioinformatics (RCSB)'s international repository for the processing and distribution of 3-D structure data of large molecules of proteins and nucleic acids.

Ratner, B.D. (ed.) (2004) *Biomaterials Science: An Introduction to Materials in Medicine* (2nd edn), Amsterdam; Boston, MA: Elsevier Academic Press. A textbook with a handbook feel; for those with particular interests in biomaterials.

Visible Human Project <http://www.nlm.nih.gov/research/visible/visible_human.html> (accessed May 17, 2005). The National Library of Medicine's database of digital images of "complete, anatomically detailed, three dimensional representations of the normal male and female human bodies." Of interest to practitioners of clinical medicine and biomedical researchers.

von Recum, A.F. (ed.) (1999) *Handbook of Biomaterials Evaluation: Scientific, Technical, and Clinical Testing of Implant Materials* (2nd edn), Philadelphia, PA: Taylor & Francis. This volume addresses the needs of those who are involved in inventing, developing, and testing implants and are concerned about the interactions between biomaterial and body tissue. Topics include bulk and surface characterization of various materials, biocompatibility, implants, soft and hard tissue histology, regulations and clinical trials.

Whole Brain Atlas <http://www.med.harvard.edu/AANLIB/home.html> (accessed May 17, 2005). From Harvard Medical School, a collection of clinical data and magnetic resonance images of the brain. In addition to the "normal" brain, various disease states of the brain are illustrated.

Wiley Database of Polymer Properties. According to its website, the *Wiley Database of*

Polymer Properties is the single most comprehensive source of physical property data for polymers commercially available, with experimentally determined and selected data for over 2,500 polymers. The initial content is derived from the *Polymer Handbook*, an essential print reference for polymer physical data.

Wise, D.L. (ed.) (2000) *Biomaterials and Bioengineering Handbook*, New York: Marcel Dekker. Chapters cover "hard" (ceramic and metallic) and "soft" (collagen, polymer) biomaterials and bioactive materials for use in drug delivery systems, orthopedics, and skin applications.

Monographs and textbooks

The dissemination of scientific information begins with research results/data, which migrates into conference proceedings, technical reports, and peer-reviewed journals. This process can take from one to three years to occur. *Annual Reviews* and monographs generally deal with a specific topic in-depth in a systematic manner. These materials may appear between four to five years after the initial research (Garvey and Belver, 1971). Monographs are frequently titled "Principles of...," "Treatise on...," or "Series on...." For example, the CRC Press Biomedical Engineering Series includes titles such as *Handbook of Neuroprosthetic Methods, Medical Image Registration*, and *Noninvasive Instrumentation and Measurement in Medical Diagnosis*; or Springer's Biological and Medical Physics, Biomedical Engineering Series with titles such as *Molecular and Cellular Signaling, Medical Applications of Nuclear Physics*, and *Biological Imaging and Sensing*.

Textbooks are designed for instruction and the audience would be those requiring an overview of basic information moving into more advanced topic treatment (Malinowsky, 1994). Textbooks may appear as much as ten to thirteen years after the initial research. Textbooks frequently have titles such as "Introduction to..." or "Fundamentals of...." For example, the following textbooks were assigned to bioengineering and biomedical engineering courses from various universities: *Introduction to Bioengineering; Essential Physics of Medical Imaging; Introduction to Biomedical Engineering; Methods of Tissue Engineering;* and *Biomaterials Science: An Introduction to Materials in Medicine.*

Journals and other periodicals

Journal literature is the most important scientific communication tool for bioengineering, crossing the fields of biology, engineering, chemistry, physics, materials science, and medical literature. The listing given below focuses on biomedical engineering, biomaterials, and biomechanics. Titles from established societies, with high-impact factor rankings, and newer titles with prospects for success, are noted below.

The Institute for Scientific Information's annual *Journal Citation Reports* may be used to determine rankings of journals, including the ubiquitous "impact factor." Broad categories such as "Engineering, Biomedical,"

"Materials Science, Biomaterials," "Biophysics," "Medical Informatics," "Biotechnology & Applied Microbiology," among others, may be of interest.

The National Cancer Institute at Frederick, MD, hosts the *Biological Journals and Abbreviations Periodicals Index* website at http://home.ncifcrf.gov/research/bja/ (accessed May 17, 2005). Admittedly not complete, the site contains the abbreviations, full titles, and links to some www pages for a large variety of biological and medical journals.

Scientific journal publishing as a whole is undergoing much scrutiny concerning how it is being produced and paid for – with biomedical literature at the forefront of this dialogue. Consequently, researchers in this area should become familiar with "open access" and "self-archiving" discussions, initiatives, and mandates. Worldwide, governments and funding agencies (including the NIH) are examining the processes by which scientific literature can be most effectively disseminated, with some recommending or even mandating that the peer-reviewed content in some form or another be made quickly and freely accessible on the Internet, through institutional, discipline-specific, or personal websites.

Journals

Advances in Bioengineering (0360–9960) American Society of Mechanical Engineers Bioengineering Division (actually an annual conference proceedings).

Annals of Biomedical Engineering (0090–6964) Biomedical Engineering Society, Kluwer Academic.

Annual Review of Biomedical Engineering (1523–9829) Annual Reviews.

Biomaterials (0142–9612) Elsevier.

Biomedical Chromatography (0269–3879) Wiley.

Biophysical Journal (0006–3495) Biophysical Society, Highwire Press.

Biorheology (0006–355X) IOS Press.

Clinical Oral Implants Research (0905–7161) European Association for Osseointegration, Blackwell.

Critical Reviews in Biomedical Engineering (not supplied) Begell House.

IEEE Transactions on Biomedical Engineering (0018–9294) Institute of Electrical and Electronics Engineers, Engineering in Medicine and Biology Society.

IEEE Transactions on Information Technology in Biomedicine (1089–7771) Institute of Electrical and Electronics Engineers, Engineering in Medicine and Biology Society.

IEEE Transactions on Medical Imaging (0278–0062) Institute of Electrical and Electronics Engineers.

IEEE Transactions on Nanobioscience (not supplied) Institute of Electrical and Electronics Engineers, Engineering in Medicine and Biology Society.

IEEE Transactions on Neural Systems and Rehabilitation Engineering (1534–4320) Institute of Electrical and Electronics Engineers, Engineering in Medicine and Biology Society.

Journal of Biomaterials Science, Polymer Edition (0920–5063) VSP International Science Publishers.

Journal of Biomechanical Engineering (0148–0731) American Society of Mechanical Engineers.

Journal of Biomechanics (0021–9290) American Society of Biomechanics *et al.*, Elsevier.

Journal of Biomedical Materials Research, Parts A and B. Part A (1549–3296), Part B (1552–4973) Society for Biomaterials, Wiley.

Journal of Materials Science: Materials in Medicine (0957–4530) European Society for Biomaterials, Chapman Hall.

Journal of Orthopaedic Research (0736–0266) Orthopaedic Research Society, Elsevier.

Medical and Biological Engineering and Computing (0140–0118) International Federation for Medical and Biological Engineering, Institution of Electrical Engineers.

Medical Engineering and Physics (1350–4533) Institute of Physics and Engineering in Medicine, Elsevier.

Medical Image Analysis (1361–8415) Elsevier.

Nature (0028–0836) Nature Publishing Group.

Physics in Medicine and Biology (0031–9155) Institute of Physics and Engineering in Medicine, Institute of Physics.

Physiological Measurement (0967–3334) Institute of Physics and Engineering in Medicine, Institute of Physics.

PLoS Biology (1544–9173) Public Library of Science.

PLoS Computational Biology (1553–734X) Public Library of Science.

PLoS Genetics (1553–7390) Public Library of Science.

PLoS Medicine (1549–1277) Public Library of Science.

Proceedings of the Institution of Mechanical Engineers. Part H: Journal of Engineering in Medicine (0954–4119) Institution of Mechanical Engineers.

Science (0036–8075) American Association for the Advancement of Science.

Spine (0362–2436) Lippincott Williams & Wilkins.

Tissue Engineering, (1076–3279) Tissue Engineering Society International *et al.*, Mary Ann Liebert.

News and Trade Magazines

Biomaterials Forum (1527–6031) Society for Biomaterials.

BioMechanics: The Magazine of Body Movement and Medicine (1075–9662) CMP Media

Biomedical Instrumentation and Technology (0899–8205) Association for the Advancement of Medical Instrumentation.

BMES Bulletin (not supplied) Biomedical Engineering Society.

IEEE Engineering in Medicine and Biology Magazine (0739–5175) IEEE Engineering in Medicine and Biology Society.

IFMBE News (1359–0944) International Federation for Medical and Biological Engineering.

Medical Design and Technology (1096–1801) Reed Business Information.

SCOPE (not supplied) Institute of Physics and Engineering in Medicine.

Electronic journal aggregators/repositories

BioMed Central <http://www.biomedcentral.com/> (accessed May 17, 2005). BioMed Central is an independent publishing house committed to providing immediate free access to peer-reviewed biomedical research, offering over 100 open access journal titles covering all areas of biology and medicine, including *BioMedical Engineering OnLine.*

Directory of Open Access Journals <http://www.doaj.org/> (accessed May 17, 2005). Maintained by Lund University Libraries, the site includes a growing number of freely available, full-text, quality controlled scientific and scholarly journals. Titles in "Biology and Life Sciences," "Health Sciences," and "Technology and Engineering" categories may be of interest.

IEEE Biomedical Engineering Library <http://biomed.ieee.org> (accessed May 17, 2005). This online collection, available by subscription, is a subset of the IEEE/IEE Electronic Library (or IEEE Xplore). According to IEEE, it offers researchers access to full-text biomedical and biotechnical documents drawn from journals, magazines, conference proceedings, and standards published by the IEEE since 1988, plus journal and conference papers from the Institution of Electrical Engineers (IEE).

PubMed Central (PMC) <http://www.pubmedcentral.nih.gov/> (accessed May 17, 2005). *PubMed Central* is a freely available digital archive of biomedical and life sciences journal literature, developed and managed by the National Center for Biotechnology Information (NCBI), a division of the National Library of Medicine (NLM) at the U.S. National Institutes of Health (NIH).

Synthesis: The Digital Library of Engineering and Computer Science <http://www.morganclaypool.com/> (accessed May 17, 2005). Neither a journal nor a monograph, rather these are "Lectures." According to the publisher, Morgan & Claypool, the basic component of the library is a 50–100-page self-contained electronic document that synthesizes an important research or development topic authored by an expert contributor to the field. These lectures offer value to the reader by providing more synthesis, analysis, and depth than the typical research journal article. "Lectures in Biomedical Engineering" is under development.

Patents

Patents are an important source for unique, detailed, technical information for scientists and engineers. It has been stated that approximately 70 percent of the information found in patents is unique and not published in any of the standard technical literature, such as journal articles or conference papers (Trepane, 1978: 273).

Patent overviews

Grubb, P. (2005) *Patents for Chemicals, Pharmaceuticals and Biotechnology: Fundamentals of Global Law, Practice and Strategy* (4th edn), Cary, NC: Oxford University Press. Covers international patent processes for patent lawyers as well as scientists, researchers, and engineers.

Standards

"A standard is something that is accepted as an authority or an acknowledged measure of comparison for quantitative or qualitative values. Standards become accepted through the reputation of those who create the standard." (Malinowsky, 1994: 22). Similar to patents, standards are detailed documents that can provide the engineer with a wealth of unique technical information which is critical to design engineers.

Association for the Advancement of Medical Instrumentation (AAMI) <http://www.aami.org> (accessed May 17, 2005). Composed of more than 100 technical committees responsible for standards, recommended practices, and technical reports for medical devices.

Institute of Electrical and Electronics Engineers (IEEE) <http://standards.ieee.org> (accessed May 17, 2005). Global standards in technical areas, including biomedical areas, healthcare, information technology, and nanotechnology.

International Electrotechnical Commission (IEC) <http://www.iec.ch> (accessed May 17, 2005). An organization of sixty nations providing standards for all electrical, electronic, and related technologies (medical instrumentation).

International Organization for Standardization (ISO) <http://www.iso.org> (accessed May 17, 2005). An organization of 100 nations whose goal it is to develop standards for the international exchange of goods and services, including medical devices.

Internet search engines, portals, and discussion forums

The Internet has changed the landscape of scholarly communication, and this section highlights web-based information resources and communication tools for bioengineers. Many of the descriptions are taken from the respective websites.

Search engines

National Library of Medicine (NLM) Gateway <http://gateway.nlm.nih.gov/> (accessed May 17, 2005). The NLM Gateway allows users to search in multiple retrieval systems at the U.S. National Library of Medicine (NLM). The current Gateway searches MEDLINE/PubMed, TOXLINE Special, NLM Catalog, MedlinePlus, ClinicalTrials.gov, DIRLINE, Genetics Home Reference, Meeting Abstracts, HSRProj, OMIM, and HSDB. Definitions of all these databases are at http://gateway.nlm.nih.gov/gw/Cmd?GMBasicSearch&loc=lhc (accessed December 21, 2004).

National Science Digital Library (NSDL) <http://nsdl.org/> (accessed May 17, 2005). NSDL is the National Science Foundation's online library of resources for science, technology, engineering, and mathematics education. Searches for "biomedical engineering" and "biotechnology" are straightforward, but for some reason index terms "bio engineering" and "bio sensors" are used (i.e., with spaces).

Science.gov: First.gov for Science <http://www.science.gov/> (accessed May 17, 2005). The "Applied Science and Technologies" and "Health and Medicine" areas may be of particular interest to the bioengineer.

Portals

Biomaterials Network <http://www.biomat.net> (accessed May 17, 2005). Primarily sponsored by the Instituto de Engenharia Biomédica (INEB), of the University of Porto, the site acts as a resource center to disclose resources, organizations, research activity, educational initiatives, scientific events, journals, books, articles, funding opportunities, industrial developments, market analyzes, jobs, and every other initiative related to biomaterials science and associated fields.

Biomedical Engineering Network (BMEnet) <http://www.bmenet.org/BMEnet> (accessed May 17, 2005). Developed and maintained by Purdue University's College of Engineering, the Biomedical Engineering Network may be browsed or searched for links to news, jobs, societies, upcoming conferences, funding opportunities, journals, and other information portals.

National Center for Biotechnology Information (NCBI) <http://www.ncbi.nih.gov/> (accessed May 17, 2005). Established in 1988 as a national resource for molecular biology information, NCBI creates public databases, conducts research in computational biology, develops software tools for analyzing genome data, and disseminates biomedical information – all for the better understanding of molecular processes affecting human health and disease. This is a very data-rich site, with links to literature indexes such as PubMed, genetic and molecular data banks, software tools to manipulate and analyze data, educational resources, and more.

National Institute of Biomedical Imaging and Bioengineering (NIBIB) *see* National Institutes of Health Bioengineering Consortium (BECON)

National Institutes of Health Bioengineering Consortium (BECON) <http://www.becon.nih.gov/becon.htm> (accessed May 17, 2005). BECON is the focus of bioengineering activities at the NIH. The Consortium consists of senior-level representatives from all of the NIH institutes, centers, and divisions plus representatives of other federal agencies concerned with biomedical research and development. BECON is administered by the National Institute of Biomedical Imaging and Bioengineering (NIBIB). The website contains information about the structure and operation of BECON, consortium activities, BECON news, events, and symposia, a bioengineering calendar, funding opportunities coordinated by the BECON and other agencies, and general information about the field. Links to information on NIH policies (technology transfer, intellectual property, commercialization, and sharing of biomedical research resources), and NIH programs associated with biomedical imaging and bioengineering are included.

National Science Foundation. Various funding opportunities, jobs, activities, news, and so on of particular interest to the bioengineer may be found at the Directorate for Biological Sciences <http://www.nsf.gov/bio/> (accessed May 17, 2005) and the Division of Bioengineering and Environmental Systems <http://www.eng.nsf.gov/bes/> (accessed November 2, 2004).

Discussion forums

BIOMAT-L Electronic Biomaterials Forum <http://www.biomaterials.org.au/biomat-L.htm> (accessed May 17, 2005). This electronic discussion list is intended for members of the Australian, Canadian, European, Japanese, US and other Societies for Biomaterials, and for anyone with an interest in biomaterials. The scope of the list includes the application of all types of materials in medicine and biology. Discussion of all aspects of biomaterials (from materials production and testing to issues of biocompatibility and tissue interactions and specific clinical applications) is encouraged. Also welcome are announcements of meetings and conferences, and discussions of particular technical problems associated with biomaterials research.

BIOMCH-L Biomechanics Electronic Discussion List <http://isb.ri.ccf.org/biomch-l/> (accessed May 17, 2005). The Biomch-L electronic discussion list was started in 1988 by Herman Woltring, and focuses on biomechanics and human/animal movement science.

Clinical Gait Analysis <http://guardian.curtin.edu.au/cga/> (accessed May 17, 2005). This is the companion website to the e-mail list CGA@info.curtin.edu.au, providing a forum for the discussion of clinical gait analysis: technical aspects, clinical cases, and new developments. Links to other biomechanics and gait analysis websites, including a data/software/videos section.

Nanodot <http://nanodot.org/> (accessed May 17, 2005). "News and discussion of Coming Technologies," (i.e., nanotechnologies). This weblog comes from the Foresight Institute (described in the Associations, organizations, and societies section of this chapter, below).

Associations, organizations, and societies

The 1960s and 1970s gave birth to a number of the bioengineering-related scholarly societies, sometimes growing out of a larger parent society. A number of these groups offer regular conferences and collaborate to jointly sponsor congresses, meetings, and symposia. Some of the explanatory text for each organization is provided directly from respective websites.

American Institute for Medical and Biological Engineering (AIMBE) <http://www.aimbe.org> (accessed May 17, 2005). AIMBE was established in 1993 with the following purposes: to establish a clear and comprehensive identity for the field of medical and biological engineering; to promote public awareness of medical and biological engineering; to establish liaisons with government agencies and other professional groups; to improve intersociety relations and cooperation within the field of medical and biological engineering; to serve and promote the national interest in science, engineering, and education; to recognize individual and group achievements and contributions to the field of medical and biological engineering. Its membership includes individual fellows, professional societies, industrial organizations, and academic institutions.

American Society of Biomechanics (ASB) <http://asb-biomech.org> (accessed May 17, 2005). ASB was founded in October 1977. The purpose of the Society is to provide a forum for the exchange of information and ideas among researchers in biomechanics. The term biomechanics is used here to mean the study of the structure and function of biological systems using the methods of mechanics.

American Society for Engineering Education, Biomedical Engineering Division (ASEE BME) <http://www.asee.org/members/organizations/divisions> (accessed May 17, 2005). Founded in 1893, the American Society for Engineering Education is a non-profit organization of individuals and institutions committed to furthering education in engineering and engineering technology. The Biomedical Engineering Division provides a forum for those interested in biomedical engineering education through workshops, paper sessions, and panel discussions of current topics in the area.

American Society of Mechanical Engineers, Bioengineering Division <http://www.asme.org/divisions/bed> (accessed May 17, 2005). Founded in 1973, the Bioengineering Division of ASME is focused on the application of mechanical engineering knowledge, skills and principles from conception to the design, development, analysis, and operation of biomechanical systems.

Association for the Advancement of Medical Instrumentation (AAMI) <http://www.aami.org> (accessed May 17, 2005). AAMI was founded in 1967 and is an alliance of members united by the common goal of increasing the understanding and beneficial use of medical instrumentation.

Biomedical Engineering Society (BMES) <http://www.bmes.org> (accessed May 17, 2005). In response to a manifest need to provide a society that gave equal status to representatives of both biomedical and engineering interests, the Biomedical Engineering Society was incorporated in Illinois on February 1, 1968. The Articles of Incorporation posted on its web page state that the purpose of the Society is: "To promote the increase of biomedical engineering knowledge and its utilization."

European Society for Biomaterials (ESB) <http://www.esbiomech.org> (accessed May 17, 2005). The ESB was set up in Italy in 1975, and its objectives are: to encourage, foster, promote and develop research, progress and information concerning the science of biomaterials, as well as to promote, initiate, sustain and bring to a satisfactory conclusion research with others and programs of development and information in this particular field.

European Society of Biomechanics (ESB) <http://www.utc.fr/esb> (accessed May 17, 2005). ESB was founded at a meeting of twenty scientists from eleven countries in Brussels on May 21, 1976. The participants decided to create a new society, the European Society of Biomechanics. The primary goal of the ESB was formulated as "To encourage, foster, promote and develop research, progress and information concerning the science of Biomechanics."

Foresight Institute <http://www.foresight.org> (accessed May 17, 2005). Founded by controversial nanotechnology pioneer K. Eric Drexler, the Institute is a nonprofit educational organization formed to help prepare society for anticipated advanced technologies. Their primary focus is molecular nanotechnology: the coming ability to build materials and products with atomic precision, the development of which they claim has broad implications for the future of our civilization.

Gordon Research Conferences <http://www.grc.org> (accessed May 17, 2005). As an organization, the Gordon Research Conferences provide an international forum for the presentation and discussion of frontier research in the biological, chemical, and physical sciences, and their related technologies. These conferences place a premium on the "off-the-record" presentation of previously unpublished scientific results and on the consequent ad hoc peer discussion.

IEEE Engineering in Medicine and Biology Society (EMBS) <http://www.eng.unsw.edu.au/embs/index.html> (accessed May 17, 2005). One of the earliest groups, and part of the Institute of Electrical and Electronics Engineers (IEEE), the EMBS established itself in 1952 (with a number of name changes over the years). The society advances the application of engineering sciences and technology to medicine and biology, promotes the profession, and provides global leadership for the benefit of its members and humanity by disseminating knowledge, setting standards, fostering professional development, and recognizing excellence. *Note:* EMBS has its own conferences, but other IEEE groups (e.g., IEEE Computer Society, IEEE Neural Networks Society) sponsor the annual "IEEE Symposium on Bioinformatics and Bioengineering."

Institute of Biological Engineering (IBE) <http://www.ibeweb.org> (accessed May 17, 2005). IBE was established to encourage inquiry and interest in biological engineering in the broadest and most liberal manner, and to promote the professional development of its members. It supports scholarship in education, research, and service; professional standards for engineering practices; professional and technical development of biological engineering; interactions among academia, industry, and government; public understanding and responsible uses of biological engineering products.

Institute of Physics and Engineering in Medicine (IPEM) <http://www.ipem.org.uk> (accessed May 17, 2005). The Institute of Physics and Engineering in Medicine is a United Kingdom-based registered charity which promotes, for the public benefit, the advancement of physics and engineering applied to medicine and biology; to advance public education in the field; and to represent the needs and interests of engineering and physical sciences in the provision or advancement of health care.

International Federation for Medical and Biological Engineering (IFMBE) <http://www.ifmbe.org> (accessed May 17, 2005). In 1959, a group of medical engineers, physicists, and physicians met at the 2nd International Conference of Medical and Biological Engineering, in the UNESCO Building, Paris, France to create an organization entitled International Federation for Medical Electronics and Biological Engineering. At that time there were few national biomedical engineering societies and workers in the discipline joined as Associates of the Federation. Later, as national societies formed, these societies became affiliates of the Federation. In the mid-1960s, the name was shortened to International Federation for Medical and Biological Engineering. The objectives of the IFMBE are scientific, technological, literary, and educational. Within the field of medical, biological, and clinical engineering, IFMBE's aims are to encourage research and the application of knowledge, and to disseminate information and promote collaboration.

International Society of Biomechanics (ISB) <http://www.isbweb.org> (accessed May 17, 2005). The ISB, founded on August 30, 1973 at Pennsylvania State University, promotes the study of the biomechanics of movement with special emphasis on human beings, encouraging international contacts among scientists in this field, promoting knowledge of biomechanics on an international level, and cooperating with related organizations.

International Society for Prosthetics and Orthotics (ISPO) <http://www.ispo.ws> (accessed May 17, 2005). ISPO is a multi-disciplinary organization composed of persons who have a professional interest in the clinical, educational, and research aspects of prosthetics, orthotics, rehabilitation engineering, and related topics. The organization was founded in Copenhagen, Denmark in 1970 by a group of surgeons, prosthetist-orthotists, physiotherapists, occupational therapists, and engineers to promote improvements in the care of all persons with neuromuscular and skeletal impairments.

International Union for Physical and Engineering Sciences in Medicine (IUPESM) <http://www.iupesm.org> (accessed May 17, 2005). IUPESM represents the combined efforts of more than 40,000 medical physicists and biomedical engineers working on the physical and engineering science of medicine. Its principal objective is to contribute to the advancement of physical and engineering sciences in medicine for the benefit and well-being of humanity.

Materials Research Society (MRS) <http://www.mrs.org> (accessed May 17, 2005). Founded in 1973, MRS is a non-profit organization which brings together scientists, engineers, and research managers from industry, government, academia, and research laboratories to share findings in the research and development of new materials of technological importance. MRS sponsors two major annual meetings with numerous topical symposia, some of which may be of particular interest to those involved in biomaterials research.

Public Library of Science (PLoS) <http://www.plos.org> (accessed May 17, 2005). Founded in October 2000, PLoS comprised a coalition of research scientists and physicians dedicated to making the world's scientific and medical literature a public resource. Their first action was to encourage scientific publishers to make the archival scientific research literature available for distribution through free online public libraries of science. This initiative prompted some significant and welcome steps by many scientific publishers towards freer access to published research, but in general the publishers' responses fell short of the policies advocated. In the summer of 2001, the group concluded that the only way forward was to develop plans for launching their own PLoS journals. In October 2003, *PLoS Biology* was launched, followed by *PLoS Medicine* in October 2004 – both open access journals and relevant to the bioengineer.

Society for Biomaterials <http://www.biomaterials.org> (accessed May 17, 2005). The Society for Biomaterials is a professional society which promotes advances in all phases of materials research and development by encouragement of cooperative educational programs, clinical applications, and professional standards in the biomaterials field.

Society for Industrial Microbiology (SIM) <http://www.simhq.org/> (accessed May 17, 2005). Organized in 1949, SIM is a professional association dedicated to the advancement of microbiological sciences, as applied specifically to industrial materials, processes, products, and their associated problems. Its members constitute scientists employed in industry, government, and university laboratories.

Whitaker Foundation <http://www.whitaker.org> (accessed May 17, 2005). The Whitaker Foundation supports research and education in biomedical engineering, and was created and funded after the death of Uncas A. Whitaker in 1975. His wife, Helen, who shared in his philanthropy during his lifetime, bequeathed a significant portion of her estate to the foundation when she died in 1982. Since its inception, the foundation has primarily supported interdisciplinary medical research, with the principal focus being on biomedical engineering. It has contributed more than $700 million to universities and medical schools to support faculty research, graduate students, program development, and the construction of facilities. The foundation sponsored two Biomedical Engineering Educational Summits, the first in 2000 and the second in 2005. *Note:* According to its website, the Whitaker Foundation is set to close in 2006, after phasing out its grant programs.

Conclusion

Bioengineering is clearly a discipline – or agglomeration of interdisciplines – on the rise, reflected by the growth of academic programs and funding for research. As a "twenty-first-century subject" at the locus of life sciences, physical sciences, and engineering, the prospects for its expansion and further development are vast. Although this compilation of resources represents only a snapshot in time, it is hoped that it will provide a useful entry point to those engaged in this exciting and evolving field.

Acknowledgments

Many thanks to Dr. Beth A. Winkelstein and Dr. Matthias Steiner for careful review of the text and helpful suggestions, and Professors Morton Friedman and Roger Barr of Duke for their "definitions of bioengineering."

References

Craig, C. (2003) *Biological Engineering: From Blue Roses to Space Suits*, <http://www.lib.uchicago.edu/cinf/225nm/presentations/225nm008.pdf> (accessed May 19, 2005).

Force, R. (1972) *Guide to Literature on Biomedical Engineering*, Washington, DC: American Society for Engineering Education, Engineering School Libraries Division.

Garvey, W.D. and Belver, C.G. (1971) "Scientific Communication: Its Role in the Conduct of Research and Creation of Knowledge," *American Psychologist* 26: 349–362.

Johnson, A.T. *Defining the Body of Knowledge for the Discipline* http://www.ibeweb.org/engineering/defining.cgi (accessed December 17, 2004).

Kramer, J. (1973) *Biophysics and Biomedical Engineering: A Bibliography*, Sacramento: California State University Library.

Loftus, M. (2004) "BioBoom," *ASEE Prism* 14 (3): 39–41.

Malinowsky, H.R. (1994) *References Sources in Science, Engineering, Medicine, and Agriculture*, Phoenix, AZ: Oryx Press.

National Institutes of Health <http://www.becon2.nih.gov/bioengineering_definition.htm> (accessed 3 December 2004).

Nebeker, F. (2002) "Golden Accomplishments in Biomedical Engineering," *IEEE Engineering in Medicine and Biology Magazine* 21 (3): 17–47.

Pate, A.G. (1966) *Bio-engineering: A Guide to Sources of Information in a New Interdisciplinary Activity*, Manchester, UK: Manchester College of Science and Technology.

Rastegar, S. (2000) "Life Force: Bioengineering Gets a Burst of Energy When the Century of Physics Meets the Century of Biology," *Mechanical Engineering* 122 (3): 74–79.

Russian River Ecosystem Restoration Study and the Russian River Watershed Council. *History of Bioengineering* <http://www.spn.usace.army.mil/russian/biohistory.html> (accessed December 3, 2004).

Trepane, J.F. (1978) "A Unique Source of Information," *Chemtech* 8 (5): 272–274.

Whitaker Foundation <http://www.whitaker.org/glance/history.html> (accessed December 3, 2004).

7 Chemical engineering

Dana Roth

Introduction, history and scope

The first "applied chemists" date from the Bronze Age cultures of Egypt and Mesopotamia that developed techniques for extraction of metals from naturally occurring ores, for the brewing of beer, and for the production of pottery and glass. The beginnings of chemistry came later with the Greeks and their conceptions of elements and compounds. The melding of these ancient techniques and chemical concepts formed the basis for alchemy, which reigned until the eighteenth century. Chemistry then evolved with the ability to synthesize an increasing variety of chemical compounds.

Chemical engineering, as we know it today, developed in the nineteenth century to meet the need for industrial-scale production of chemicals and materials. Since then, its main focus has undergone a series of transitions.

The development of a "Unit Operations" paradigm in the 1920s was based on the observation that many industrial chemical processes shared common basic operations (crystallization, distillation, evaporation, extraction) and these operations should be the focus rather than the production of specific products. The next transition, in the 1950s, was the development of expertise in "Transport Phenomena" (fluid flow, heat and mass transfer, microscopic level process rates). With these developments as a foundation, the chemical engineering profession is now coming full circle to focus again on the development of specific products (e.g. biologicals, catalysts, optical fibers, polymers, semiconductors) (Ritter, 2001).

Examples of the current diversity in chemical engineering research are exemplified by the two dozen topics listed on the CACHE Teaching Resource Center web page, and the MIT OpenCourseWare website. These include bioengineering, biomechanics, nuclear technology, and sustainable energy. Additional information on specific topics is widely available on individual professors' websites and in the open literature. For example:

> *Catalysis:* "Catalytic selective oxidation of alkanes is one of the primary tools of the current chemical industry." For example, "the oxidation of n-butane to form maleic acid) . . . over a vanadium phosphorus oxide (VPO) catalyst is a well established commercial process" (Mark Davis).

Polymers: "One of the fundamental goals of polymer science is to understand how the characteristics of the polymer molecule affect the way in which the bulk material behaves during its processing . . . in other words, the relationship between molecular architecture and processing properties" (Hadlington, 2004: 13–14).

Chemical engineering education is currently undergoing a transformation as computing experiences now go far beyond numerical analysis (Octave) to include simulation and molecular modeling. MATLAB is the de facto standard for dynamic simulation and an extensive listing of molecular modeling and viewing programs is available on the WWW (Molecular Modeling). Equally important is the need for chemical engineering graduates to "Know how to use a modern technical library to search for information located in electronic databases, and how to access electronic information services through the World Wide Web" (Edgar, 2000).

Because of the wide diversity of research areas and industrial applications, chemical engineering enrollments rise and fall with changes in national economies. Chemical engineering departments are fairly quick to recognize and incorporate new curricula, which explains the current emphasis on biology and nano-based engineering programs (Halford, 2004).

A layman's explanation of "what do chemical engineers do" includes: "construct the synthetic fibers that make our clothes more comfortable and water resistant; . . . develop methods to mass-produce drugs, making them more affordable; . . . create safer, more efficient methods of refining petroleum products, making energy and chemical sources more productive and cost effective." (Career Choices)

Additional reading:

These resources will provide a better understanding of the nature of chemical engineering both from a historical perspective and its current practice.

History of Chemical Engineering Web Page <http://www.pafko.com/history/> (accessed May 18, 2005). A light-hearted albeit serious look at the growth of chemical engineering as a profession, using the petroleum industry as a case study.

AccessScience. McGraw-Hill <http://www.accessscience.com/> (subscription required) (see Chapter 2, this volume). McGraw-Hill's *Access Science* is the continuously updated online version of both *McGraw-Hill Encyclopedia of Science and Technology* (9th edn) and *McGraw-Hill Dictionary of Scientific and Technical Terms* (6th edn). Included, for example, are chapters on Chemical engineering, Biomedical chemical engineering, Biochemical engineering, and Chemical process industry.

Felder, Richard. "Resources in Science and Engineering Education" <http://www.ncsu.edu/felder-public/RMF.html> (accessed May 18, 2005). A comprehensive discussion of effective instructional techniques that have been validated for both individuals and groups.

Library catalogs

Given recent trends toward adding book chapter titles in library catalog records, and the inappropriateness of many subject headings, keyword-searching is often more effective than the use of Library of Congress Subject Headings. This is exemplified by the full-text keyword-searching option in Amazon.com's "Search inside this Book" program (Price, 2003), which also provides cited and citing references for books.

LibWeb <http://lists.webjunction.org/libweb/> (accessed 25 May 2005) is a growing list of over 7,200 web pages from 125 countries worldwide, providing access to academic, public, national, and state libraries. It provides an excellent tool for searching catalogs for chemical titles. Amazon.com may also be considered as a "library" catalog, since it lists both in-print and out-of-print books, as well as an excellent substitute for "Books-in-Print." Online library catalogs for top-ranked graduate chemical engineering programs are freely available for searching. List of schools include Massachusetts Institute of Technology (http://library.mit.edu/); University of California–Berkeley (http://melvyl.cdlib.org/); University of Minnesota–Twin Cities (http://www.lib.umn.edu/). All of these will list books in the various areas of chemical engineering for both research and collection development.

Since most academic libraries arrange their resources using the Library of Congress classification system, the "Index to Some Library of Congress Classification Numbers Relevant to Chemistry" (ChemInfo) <http://www.indiana.edu/~cheminfo/01-03.html> (accessed May 18, 2005) should also prove useful for chemical engineering. For example, the major divisions are as follows:

Analytical Chemistry QD 71–145
Biochemistry QD415–431.7, QH345, QP501–801
Chemical Technology TP 1–1185
Chemical Engineering TP 155–159
- Manufacturing and use of chemicals TP 200–248
- Fuels TP 316–360
- Food processing and manufacturing TP 368–456
- Fermentation industry TP 500–660
- Petroleum refining and products TP 690–692.4
- Polymers, plastics and their manufacture TP 1080–1185

Chemistry (General) QD 1–69
Crystallography QD 911–999
Fluid Mechanics QA 901
Genetic Engineering QH442
Heat and Mass Transfer QC 320, TJ 260
Inorganic Chemistry QD 146–196
Kinetics QD 501–505, TP 149–157
Macromolecules QD 380–388

Molecular Biology QH506
Organic Chemistry QD 241–449
Organometallic Compounds QD 410–412
Physical and Theoretical Chemistry QD 450–731

Article abstracts and indexes

Chemists and chemical engineers are extremely well served by the *Chemical Abstracts* (in print) and its online equivalent *SciFinder/SciFinder Scholar*. These complementary products offer very broad subject coverage, extending to the chemical aspects of astronomy, biology, education, engineering, economics, geology, history, mathematics, medicine, and physics. In addition, the extensive format coverage includes articles from journals, preprints (2,000+) and regularly published conference proceedings (73 percent), articles from one-time or first-time conference proceedings (7 percent), dissertations (2 percent), technical reports (1 percent), patents (16 percent) and edited research monograph chapters (1 percent). In addition, *SciFinder Scholar* offers cited reference searching beginning with articles indexed in 1997.

Chemical Abstracts Service. *SciFinder Scholar* <http://www.cas.org/SCIFINDER/SCHOLAR/index.html> An integrated family of databases (CAplus, CA Patent Index, CA Reg File, CASreact, ChemCats, and ChemList). The CAplus file (which includes the CA Patent Index) contains references to articles, patents, theses, and book chapters, while the CA Reg File contains chemical structures, names, and physical property data, and is linked to CASreact. These companion files are searchable by chemical structures/substructures, substance names, reactions, CAS Registry Numbers (sequentially assigned compound numbers), keywords, author names, patent numbers, company/organization names, and so on. CAplus has 13,000 records containing the term "chemical engineering" and over 42,000 records on this general topic.

ChemCats (Chemical Catalogs Online) is a catalog database containing information on nearly 8,000,000 commercially available chemicals and their worldwide suppliers.

ChemList (Regulated Chemicals Listing) is a collection of more than 232,000 regulated chemical substances.

Bibliographies and guides to the literature

The Web provides a wealth of library and literature resource guides for chemical engineering. These generally follow similar patterns with listings of library materials and URLs for additional information.

Chemical Engineer's Resource Page. Online Chemical Engineering Information <http://www.cheresources.com/chelinks/index.shtml> (accessed May 18, 2005). This site includes a number of useful links. In addition to expected topics, ethics, writing guides, and an introduction for beginners to the field add to the value of this resource.

The University of Alberta Library Resource Guide for Chemical Engineering <http://www.library.ualberta.ca/subject/chemicaleng/chemguide/index.cfm> (accessed May 18, 2005). Topics include: Finding Books, Finding Journal Articles, Guides to the Literature of Engineering, Handbooks and Manuals, Chemical Properties and Data, Dictionaries and Encyclopedias, Chemical Engineering Design, Planning and Estimation, Pressure Vessel and Boiler Resources, Chemical and Biotechnology Guides, Directories and Catalogues, Chemical Industry, Prices and Production, Associations and Organizations, Standards and Patents, Selected Web Sites and Databases.

A number of other universities and organizations maintain excellent online sites to the literature of chemical engineering. A few are noted here with a brief description.

University of Buffalo. Chemical Engineering. Selected Information Resources <http://ublib.buffalo.edu/libraries/asl/guides/engineering/chemical.html> (accessed May 18, 2005). This site provides an extensive listing of handbooks and encyclopedias, bibliographic and properties databases, electronic journals, patents, and internet sites.

University of Florida. Chemical Engineering <http://www.che.ufl.edu/WWW-CHE/index.html> (accessed May 18, 2005). This site provides an interesting subject listing of links to chemical and process engineering resources, organizations, and relevant information resources.

There are two historical references published by the American Chemical Society that are essential reading for chemical engineers and engineering librarians.

American Chemical Society (1954) *Literature Resources for Chemical Process Industries: A Collection of Papers*, Advances in Chemistry Series, 10, Washington, DC: American Chemical Society. While dated, this has some timeless articles including: Sound and Unsound Short Cuts in Searching the Literature; Foreign Alphabetization Practices; Pitfalls of Transliteration in Indexing and Searching.

American Chemical Society (1968) *Literature of Chemical Technology*, Advances in Chemistry Series, 78, Washington, DC: American Chemical Society. While again dated, this provides brief introductions to virtually all fields dealing with chemical technology; for example, Photographic Chemistry, Cosmetics Industry, Synthetic Dyes, Leather and Adhesives, Rubber Industry, Explosives.

Directories

Directories are an important resource for locating individuals, institutions, companies, or products. The listings below obviate the need to maintain the print equivalents.

American Chemical Society. *Directory of Graduate Research (DGR)*, Washington, DC: The Society <http://pubs.acs.org/dgrweb> (accessed May 18, 2005). Published biannually, the DGR provides a listing of faculty members, publications and doctoral and Master's theses in chemistry, chemical engineering, biochemistry, medicinal/pharmaceutical chemistry, polymers and materials science, marine science, toxicology, and environmental science at universities in the U.S. and Canada. The DGR is searchable by personal and/or research topic and university name and/or specialization.

Chemical Engineering Buyer's Guide. New York: McGraw-Hill <http://www.cebuyers-guide.com/public/> (Personal registration required). Provides information on companies, products, and trade names including contact information. Browsable by category (e.g. Instrumentation and controls, Process chemicals).

University of Karlsruhe, Germany. *International Directory of Chemical Engineering URLs* <http://www.ciw.uni-karlsruhe.de/chem-eng.html> (accessed May 18, 2005). Worldwide listings of academic chemical engineering websites, and companies.

University of Texas at Austin, Department of Chemical Engineering, Chemical Engineering Faculty Directory <http://www.che.utexas.edu/che-faculty/> (accessed May 18, 2005). Records comprise brief biographical information (provided by the faculty member). Listings of chemical engineering departments may be displayed by institution name or geographic location (either U.S. or Worldwide).

Dictionaries

Since there is extensive overlap between the terminologies of chemistry and chemical engineering, there are only a few dictionaries specifically for chemical engineering. The *SciFinder/Scifinder Scholar* Registry File provides extensive synonym listings, especially for commercially available chemical compounds.

Comyns, A.E. (1999) *Encyclopedic Dictionary of Named Processes in Chemical Technology* (2nd edn), Boca Raton, VA: CRC Press. Concise descriptions of over 2,600 chemical processes known by special names (e.g. inventors, companies, institutions, places, acronyms) with an index that relates product names to processes.

Kent, J.A. (2003) *Riegel's Handbook of Industrial Chemistry* (10th edn), New York: Kluwer Academic/Plenum. Reviews the economic, pollution, and safety aspects of the chemical industry and specific processes used to create fertilizers, petroleum, adhesives, dyes, and so on. The tenth edition includes a chapter on industrial cell culture.

Lewis, R.J. (2002) *Hawley's Condensed Chemical Dictionary* (14th edn), New York: Wiley Interscience. Primarily a dictionary of chemical substances providing physical properties, source of occurrence, CAS number, chemical formula, potential hazards, derivations, synonyms, and applications.

Encyclopedias

Chemical engineers and chemical engineering students are fortunate that there are two standard multi-volume encyclopedias, *Kirk-Othmer* and *Ullmann's*, whose electronic versions promise to stay relatively current with the latest developments. Both also have regularly published print sets. In addition, there is a multi-volume encyclopedia on chemical processing that describes the design of key unit operations, and a multi-volume encyclopedia on separation science. The primary benefits of encyclopedia articles are their length and depth of coverage.

Kirk-Othmer Encyclopedia of Chemical Technology (2004–) (5th edn, 27 vols), Chichester: Wiley; and *Kirk-Othmer Encyclopedia of Chemical Technology* [online] <http://www3.interscience.wiley.com/cgi-bin/mrwhome/104554789/HOME> (accessed May 18, 2005). (Advanced search – freely available to non-subscribers) <http://www.mrw.interscience.wiley.com/kirk/kirk_search_fs.html> (accessed May 18, 2005). The *Kirk-Othmer Encyclopedia of Chemical Technology* has been a mainstay for chemists, biochemists, and engineers at academic, industrial, and government institutions since publication of the first edition in 1949. Since each edition focuses on currently relevant topics, research libraries will hold all print editions. This is especially important as lawyers and consultants, for example, use it for "state-of-the-art" in litigation and/or patent support. New and revised articles are posted monthly; older versions of such articles may still be accessed in an archived form.

The electronic version allows browsing by either article title (pull-down menu . . . ablative materials to zone refining) or article subject (pull down menu . . . analytical techniques to special topics). "Search in this title" searches "all text" or can be limited to article title; section title; author; author affiliation; keywords; DOI; tables; figures; Chemical Abstract Registry Numbers (CASRN); company, brand, and trade names. Boolean searching (and, or, not) is available in advanced product search. Complementary abstracts for all articles are also provided.

Ullmann's Encyclopedia of Industrial Chemistry, (2003) (6th edn, 40 vols), Weinheim; Cambridge: Wiley-VCH; and *Ullmann's Encyclopedia of Industrial Chemistry* [online] (2005–) (7th edn), Wiley-VCH <http://www3.interscience.wiley.com/cgi-bin/mrwhome/104554801/HOME> (accessed May 18, 2005) (advanced search) <http://www.mrw.interscience.wiley.com/ueic/ueic_search_fs.html> (accessed May 18, 2005). *Ullmann's Encyclopedia of Industrial Chemistry* is the European counterpart to *Kirk-Othmer* and, while providing similar coverage, better reflects European and Asian sources. First published in 1914 by Professor Fritz Ullmann in Berlin, the *Enzyklopädie der Technischen Chemie* (the 1st to 4th edns, 1914–1972, were published in German) quickly became the standard reference work in industrial chemistry. Beginning with the fifth edition, 1985 *Ullmann's* is published in English.

The electronic version allows browsing by article title, article subject or author. "Search in this title" searches "all text" or can be limited to article title; section title; author; DOI; tables; figures; Chemical Abstract Registry Numbers (CASRN). Boolean searching (and, or, not) is available in advanced product search. Additional limits include subject category and date range. An extensive listing of abbreviations is given on the home page.

These two encyclopedic works are complementary. *Kirk-Othmer* has extensive references to both journal articles and the worldwide patent literature, while *Ullmann's* primarily references the journal literature and its previous editions. *Ullmann's* also has a broader selection of figures while *Kirk-Othmer* provides more tables.

Both *Kirk-Othmer* and *Ullmann's* are subscription databases but are freely searchable. This, in effect, provides an electronic index to the print volumes.

Encyclopedia of Chemical Processing and Design (1976–1999) (69 vols), New York: Marcel Dekker <http://www.dekker.com/servlet/product/productid/2620-0/sub?n=e> (accessed May 18, 2005).

Encyclopedia of Chemical Processing and Design (2nd edn, 5 vols), New York: Marcel Dekker <http://www.dekker.com/servlet/product/productid/E-ECHP> (accessed May 18, 2005). The *Encyclopedia of Chemical Processing* provides detailed descriptions of both chemical processes and unit operations such as reactors and separation systems, process system peripherals, pilot plant design, and scale-up criteria. Research and production phases of the chemical industry are highlighted with entries on design principles as well as engineering fundamentals. Emerging areas such as nanotechnology, microreactors and microreactor engineering, plant metabolic engineering are also covered. Sixty-nine volumes were published between 1976 and 2003, and a five-volume Supplement (2nd edn) is planned. The URLs are linked to the contents listing of individual volumes.

Encyclopedia of Separation Science (2000) (10 vols) San Diego: Academic Press <http://www.sciencedirect.com/science/referenceworks/0122267702> (accessed May 18, 2005). Articles in the *Encyclopedia of Separation Science* fall into three categories (levels) as follows: Level 1 provides expert overviews of a separation technique (e.g., flotation, distillation, crystallization). These articles provide an introduction to the subject and to the Level 2 articles. For example, the Level 1 article on chromatography provides the background for the Level 2 article that describes the theory, development, instrumentation and practice of gas, liquid and supercritical fluid chromatography. Level 3 articles then describe the various uses of the methods described in Levels 1 and 2 (e.g., sample preparation of pesticides or drugs and chromatographic techniques used for their analysis). The articles are all extensively cross-referenced and indexed, providing easy access to relevant information. Each article contains a selected bibliography of key books, review articles, and important research papers. The *ScienceDirect* web page provides access to the forward, preface, introduction, and all color plates as well as the contents of each level, and an author index.

Handbooks and compendia of property data

Handbooks provide fast access to information and are an important information resource in chemical engineering. The number of handbooks in the field is staggering and only a few will be listed here. The content of most is evident from the title. Many of the handbooks are available in electronic format as a growing number of publishers are developing electronic libraries for their reference materials. To find additional titles, search a library catalog or an Internet bookseller's list.

Property data are the *sine qua non* for both chemical engineers and students. An example would be vapor pressure, a measure of a liquid's volatility at a given temperature (e.g., the vapor pressure of a liquid is equal to atmospheric pressure at boiling-point). The design of storage tanks, for example, requires both hazard analysis and vent system technology, which in turn are based on the vapor pressure of the liquid being stored. Vapor-liquid operations, such as distillation, also require knowledge of vapor pressures as a function of temperature.

Handbooks

Branan, C. (2005) *Rules of Thumb for Chemical Engineers: A Manual of Quick, Accurate Solutions to Everyday Process Engineering Problems* (4th edn), Boston, MA: Elsevier.

CHEMnetBASE <http://www.chemnetbase.com/> (accessed June 16, 2005). CHEMnetBASE is a collection of handbooks and other reference materials from CRC and Chapman Hall. Searching the database is free, but to view the full text of the resources requires a subscription.

Chopey, N.P. (2004) *Handbook of Chemical Engineering Calculations* (3rd edn), New York: McGraw-Hill.

Green, D.W., Maloney, J.O., and Perry, R.H. (1999) *Perry's Chemical Engineers Handbook* (7th edn), New York: McGraw-Hill <http://books.mcgraw-hill.com/getbook.php?isbn=0071355405> (accessed May 18, 2005). *Perry's* has sections on both chemical and physical property data, and chemical engineering fundamentals; processes operations (heat transfer, distillation, kinetics); construction materials, process machinery, waste management, and safety. Also available in electronic format.

Griskey, R.G. (2000) *Chemical Engineers' Portable Handbook*, New York: McGraw-Hill.

Knovel <http://www.knovel.com/knovel2/Overview.jsp> (accessed May 18, 2005). Knovel's "Chemistry and Chemical Engineering" section lists books on the following topics: analytical chemistry, dispersion and aggregation, electrochemistry, environmental chemistry, equipment, general references, industrial chemistry and chemicals, industrial safety, physical chemistry, polymer chemistry, separation, and transport processes.

Although the book selection is both eclectic and heterogeneous, it allows display of both a book's contents and provides keyword searching results that display chapter title references. In addition, Knovel offers "free" access to several physical property databases (with personal registration): Knovel Critical Tables, International Critical Tables of Numeric Data, Smithsonian Physical Tables (9th rev. edn). For example, searching Knovel <http://www.knovel.com/knovel2/default.jsp> for "diethyl ether" and "vapor pressure," and limiting to chemistry and chemical engineering, quickly identifies the following titles (which will be easily found in academic libraries or those serving chemical engineers in industry): *Chemical Properties Handbook* (vapor pressure; organic compounds – Live Eqns), *Perry's Chemical Engineers Handbook* (vapor pressures of organic compounds, up to 1 atm), *Industrial Solvents Handbook* (vapor pressure of various ethers), and *Knovel Critical Tables* (basic physical properties of common solvents).

Lide, D.R. (ed.) (2004) *CRC Handbook of Chemistry and Physics* (85th edn), Boca Raton, VA: CRC Press <http://www.chemnetbase.com/tours/hbcp/index.html> (an online tour) (accessed May 18, 2005). The *CRC Handbook* is an excellent starting point for physical property data searches. Data tables listing both specific chemical compound data (e.g., bp, mp, mw) and physical properties (e.g., azeotropic data for binary mixtures, critical solution temperatures of polymer solutions) are provided along with literature references for additional data.

Riegel, E.R. and Kent, J.A. (eds) (2003) *Riegel's Handbook of Industrial Chemistry* (10th edn), New York: Kluwer Academic.

Property data

ChemFinder.com <http://chemfinder.cambridgesoft.com/> (accessed May 18, 2005). ChemFinder.com is a compound database providing chemical structures (plus molecular formulas, CAS RNs, chemical names and synonyms), physical properties (e.g., boiling-point, density, evaporation rate, flashpoint, refractive index, vapor pressure), and hyperlinks to over 400 freely available websites with additional information. ChemFinder also provides direct links to chemical vendors on ChemACX.Com and to MSDS sheets.

Daubert, T.E. and Danner, R.P. (1989–) *Physical and Thermodynamic Properties of Pure Chemicals: Data Compilation*, Washington, DC: Taylor & Francis. This title is an encyclopedic guide to pure chemical properties. Sponsored by the American Institute of Chemical Engineers' Design Institute for Physical Property Data (DIPPR), it contains physical and thermodynamic property data for 1,708 compounds. Tables for each compound include both physical (e.g., melting point, dipole moment, refractive index) and temperature-dependent properties (e.g., vapor pressure, heat capacity, viscosity). An excellent source of uniform data for either engineers or students. Fundamental SI units are used.

DIPPR Data Compilation of Pure Compound Properties <http://dippr.byu. edu/public/chemsearch.asp> (subscription required). The *DIPPR Chemical Database* consists of both experimental data and temperature-dependent properties for over 1,600 pure chemicals. Data have been evaluated, correlated, and checked for thermodynamic consistency. Datasets, DIPPR-approved property constants, and regressed correlation coefficients for temperature-dependent properties are included. Calculation of temperature-dependent properties is possible in a variety of units (Standard, cgs or British). Compounds are searched with either chemical names (or name fragments) or Chemical Abstracts Registry Numbers. Individual compound records also include both CAS and IUPAC names as well as trivial names.

Landolt-Börnstein: Substance and Physical Property Indexes <http://springerlink.com/media/public/profiles/springerlink/0284/indexes/0_start.pdf> (accessed May 18, 2005). Inorganic substances and physical properties indexes are linked from page 3. The LB Index of Inorganic Substances and their Properties (p. 10) is a substance/property/index for elements, binary and higher compounds, and alloys. This online index corresponds to the "print" index edited by Madelung (1993) that has been updated (1996/1997). Typically, an element system Au-Cu is followed by a substance Cu3Au and one or many physical properties (e.g., electrical resistivity), and the respective volume citation (e.g., III/15c).

There are also additional substance/subvolume/page indexes for a single subvolume, or a group of subvolumes (p. 3). For example: Low and high frequency properties of dielectric materials (p. 6). Index to Group III/vol. 29a, b, & vol. 30a, b Semiconductors (p. 9). Substance/property index for Group III/vol. 41 A1, A2, B, C, D, E Index of Organic Compounds and their Physical Properties

<http://springerlink.com/media/public/profiles/springerlink/0284/indexes/organic_index/organic_index.pdf> (accessed May 18, 2005).

The *Landolt-Börnstein "Index of Organic Compounds"* lists organic compounds that have data (e.g. molecular constants, density, surface tension, dielectric constants, refractive index, vapor pressure) in the New Series volumes. In total, nearly 17,000 compounds are listed with nearly 34,000 references. The default display is in molecular formula order and provides a chemical structure diagram, the molecular formula, chemical names, and references to New Series volumes. Links at the top of the page allow searching with specific CAS Registry Numbers or chemical names.

Searching the *Index of Organic Compounds* by structure, substructure, name, formula, CAS RN, and so on is also available: Landolt-Börnstein Search <http://lb.chemie.uni-hamburg.de/> (accessed May 18, 2005).

NIST/ASME Steam Properties Database: Version 2.21 <http://www.nist.gov/srd/nist10.htm> (purchase required). Thermophysical properties include: in the *STEAM Database*: temperature, Helmholtz energy, thermodynamic derivatives, pressure, Gibbs energy, density, fugacity, thermal conductivity, volume, isothermal compressibility, viscosity, dielectric constant, enthalpy, volume expansivity, dielectric derivatives, internal energy, speed of sound, Debye-Hückel slopes, entropy, Joule-Thomson coefficient, refractive index, heat capacity, surface tension. The STEAM database generates tables and plots of property values. Vapor-liquid-solid saturation calculations with either temperature or pressure specified are available.

NIST Chemistry WebBook <http://webbook.nist.gov/chemistry/> (accessed May 18, 2005). The *NIST Chemistry WebBook* provides Internet access to chemical and physical property data for nearly 50,000 chemical species (e.g., compounds, ions, radicals). The data are derived from collections maintained by both the NIST Standard Reference Data Program and outside contributors. The available data include: thermodynamic data, gas phase, IR spectrum, condensed phase, mass spectrum, phase change, UV/Vis spectrum, reaction, vibrational and electronic energy levels, ion energetics, constants of diatomic molecules, ion cluster, and Henry's Law.

The *NIST Chemistry WebBook* supports a variety of searches for chemical species. Each search type has its own associated web page. The following search types are currently available: General searches, Physical property-based searches, Formula, Ion energetics searches, Chemical name, Ionization energy, CAS registry number, Electron affinity, Reaction, Proton affinity, Author, Acidity, Structure/substructure, Appearance energy product, Vibrational and electronic energy level searches, Vibrational energy, Electronic energy level, and Molecular weight search.

Each search can be limited by data types (see above). For example, a Formula search with both the "Gas phase thermochemistry data" and "Mass spectrum" checked will only retrieve species whose records include either of these data types. Searches involving physical quantities may also be restricted by chemical formula. Searching for Electron affinity = 0–10), Formula = C H and clicking "Allow more atoms of elements in formula than specified" will list all the hydrocarbons in the database with electron affinity data.

Truncation is also allowed. C5H? will find all species with five carbon atoms and one or more hydrogen atoms. C5C*H? will search for species with five or more carbon atoms and one or more hydrogen atoms. Either SI or calorie-based unit systems may be specified.

Numerical Data and Functional Relationships in Science and Technology. (*Zahlenwerte und Funktionen aus Naturwissenschaften und Technik*) New Series (1961–) Berlin; New York: Springer-Verlag <http://www.landolt-boernstein.com> (accessed May 18, 2005)

Group I Nuclear and Particle Physics
Group II Atomic and Molecular Physics
Group III Crystal and Solid State Physics
Group IV Macroscopic and Technical Properties of Matter
Group V Geophysics and Space Research
Group VI Astronomy, Astrophysics, and Space Research
Group VII Biophysics
Group VIII Advanced Materials and Technologies

There is an additional volume on "Units and Fundamental Constants in Physics and Chemistry" and a web index, which are freely available.

Subvolume a: *Units in Physics and Chemistry*, 1991 <http://springerlink.com/media/public/profiles/springerlink/0284/indexes/units/t000_units_a.pdf> (accessed May 18, 2005).

Subvolume b: *Fundamental Constants in Physics and Chemistry*, 1992 http://springerlink.com/media/public/profiles/springerlink/0284/indexes/units/t000_units_b.pdf (accessed May 18, 2005).

Parry, W.T. (2000) *ASME International Steam Tables for Industrial Use*, New York: ASME (American Society of Mechanical Engineers). A standard set of thermodynamic and transport properties for water and steam. Based on the International Association for the Properties of Water and Steam's Formulation 1997. This is a complementary reference to the NIST/ASME STEAM PROPERTIES DATABASE. Based upon the new International Association for the Properties of Water and Steam (IAPWS) 1995 formulation for general and scientific use for the thermodynamic properties of water, this updated version provides water properties from the international standards over a wide range of conditions.

Thermodex: An Index of Selected Thermodynamic and Physical Property Resources (1997) Austin, TX: Mallet Chemistry Library, University of Texas-Austin, 1997 <http://thermodex.lib.utexas.edu/> (accessed May 18, 2005). *ThermoDex* is an index to printed and web-based compilations of thermochemical and thermophysical data and is the next choice after searching standard handbooks which are not indexed (e.g., *CRC Handbook*, *Perry's*, *Critical Tables*). Proprietary databases such as *SciFinder Scholar* and *DIPPR* are also not indexed. Searches for compounds, or preferably compound types, linked to a specific physical property or properties, display a list of handbooks that *may* contain data of interest. Actual data values are not displayed. Thermodex records include the book's title, a brief abstract defining the scope and arrangement, properties and types of compounds included. Web datasets include URL links. While based primarily on the holdings of UT-Austin's Mallet Chemistry Library, most of these resources will be available in both university and major industrial libraries.

Yaws, C. (2001) *Matheson Gas Data Book* (7th edn), Parsippany, NJ: Matheson Tri-Gas; New York: McGraw-Hill. The *Matheson Gas Data Book* contains individual sections with property data on over 150 industrial gaseous elements and compounds (from Acetylene to Xenon). Data include, thermodynamic properties, IR spectra, vapor pressure-temperature curves, Henry's Law constants, explosion limits, and viscosity.

Zahlenwerte und Funktionen aus Physik, Chemie, Astronomie, Geophysik und Technik (Landolt-Bornstein), (1950–1981) (6th edn), Berlin: Springer. Landolt-Bornstein was intended to be a compilation of all critically evaluated physical property data in the fields of physics, chemistry, astronomy, geophysics, and technology. The sixth edition (in German) was published in four volumes (Band), with multiple parts (Teile), between 1950 and 1981: (1) *Atomic and Molecular Physics*, Parts 1–5; (2) *Properties of Matter in its Aggregated State*, Parts 1–10; (3) *Astronomy and Geophysics*, Part 1; (4) *Technology*, Parts 1–4.

Because of the difficulty in adhering to the rigid overall plan of the sixth edition, a New Series was begun in 1961 and is arranged to accommodate future information, without new editions. The preface, table of contents, and introductory chapter in each volume are given in both English and German.

Textbooks and monographs

Textbooks

Some important textbooks include:

Bird, R,B. (ed.) (2002) *Transport Phenomena* (2nd edn), New York: Wiley. *Transport Phenomena* is a fundamental engineering concept with a wide variety of applications. This text and the first edition (1966) are generally considered classics.

Deen, W.M. (1998) *Analysis of Transport Phenomena*, New York: Oxford University Press. A graduate-level text covering both the concepts and techniques applicable to momentum, heat and mass transfer, and presenting unique approximation and scaling techniques.

Felder, R.M. (2000) *Elementary Principles of Chemical Processes* (3rd edn), New York; Chichester: John Wiley. A text that introduces thermodynamics, unit operations, kinetics, and process dynamics. The third edition is revised to reflect curriculum changes that include biotechnology, environmental engineering, and microelectronics.

Hinds, W.C. (1999) *Aerosol Technology: Properties, Behavior, and Measurement of Airborne Particles* (2nd edn), New York: Wiley. An upper-division/graduate-level text covering, for example, bioaerosols, Brownian motion and diffusion, respiratory deposition models, measurement and sampling.

Kyle, B.G. (1999) *Chemical and Process Thermodynamics* (3rd edn), Upper Saddle River, NJ: Prentice Hall PTR. A text describing techniques, applications, and mathematical analysis, with problems and examples. The CD-ROM includes spreadsheets and programs for both numerical analysis and graphics. This edition has a chapter on modeling thermodynamic systems.

Lauffenburger, D.A. and Linderman, J.J. (1993) *Receptors, Models for Binding, Trafficking and Signalling*, New York: Oxford University Press. A classic text describing the principles of modern systems biology for chemical engineers.

Shul, R.J. (2000) *Handbook of Advanced Plasma Processing Techniques*, Berlin; New York: Springer. A multi-authored volume covering the fundamental physics of plasmas, diagnostics, modeling, and microelectronic applications for development of transistors, sensors, and so on.

Amazon.com has an interesting feature called Listmania, which provides an individual's selection of favorite texts and handbooks. For example (listed under the Amazon entry for *Perry's Chemical Engineers Handbook* Platinum edn (1999), a reader will find several lists of books recommended by users of the website. This feature should prove useful for a wider sampling of textbooks, since chemical engineering curricula are heavily dependent on faculty research interests.

Research monographs and symposia publications

Research monographs and symposia publications are generally focused on very specific topics. Several publishers have active publication programs related to chemical engineering including Taylor & Francis, CRC Press <http://www.crcpress.co>, AIChE <http://www.aiche.org/pubcations>, DECHEMA, and ICE and the American Chemical Society <http://www.pubs.acs.org>. Some representative examples include:

Advances in Industrial Heat Transfer: A Two Day Conference (1996) Rugby: Institution of Chemical Engineers.

Ahmad, M. and Benson, R. (1999) *Benchmarking in the Process Industries*, Rugby, U.K.: Institution of Chemical Engineers.

AIChE (1970–) *Ammonia Plant Safety and Related Facilities*, New York: American Institute of Chemical Engineers.

Baum, E.J. (1998) *Chemical Property Estimation: Theory and Application*, Boca Raton, VA: Lewis Publishers.

Choy, B. and Reible, D.D. (2000) *Diffusion Models of Environmental Transport*, Boca Raton, VA.: Lewis Publishers: CRC Press.

Danner, R.P. and High, M.S. (1993) *Handbook of Polymer Solution Thermodynamics*, New York: Design Institute for Physical Property Data, American Institute of Chemical Engineers.

Datta, A.K. (2002) *Biological and Bioenvironmental Heat and Mass Transfer*, New York: Marcel Dekker.

DECHEMA Chemistry Data Series (2nd edn), Frankfurt/Main: Dechema; Port Washington, NY: Distributed by Scholium International (volumes include *Electrolyte Data Collection* and *Vapor-liquid Equilibrium Data Collection*).

Felgner, P.L. (1996) *Artificial Self-assembling Systems for Gene Delivery*, Conference Proceedings Series, Washington, DC: American Chemical Society.

Korugic-Karasz, L.S. (ed.) (2005) *New Polymeric Materials*, ACS Symposium Series 916, Washington, DC: American Chemical Society.

Nowack, B. and VanBriesen, J.M. (eds) (2005) *Biogeochemistry of Chelating Agents*, ACS Symposium Series 910, Washington, DC: American Chemical Society.

Seolover, T.B. and Chen, C. (1994) *Thermophysical Properties for Industrial Process Design*, AIChE Symposium Series 298, New York: American Institute of Chemical Engineers.

Smallwood, I.M. (2002) *Solvent Recovery Handbook* (2nd edn), Boca Raton, VA: Blackwell Science; CRC Press.

Spurny, K.R. (ed.) (2000) *Aerosol Chemical Processes in the Environment*, Boca Raton, VA: Lewis Publishers.

Suthersan, S.S. (2002) *Natural and Enhanced Remediation Systems*, Boca Raton, VA Lewis Publishers.

Wang, Y. and Holladay, J. (2005) *Microreactor Technology and Process Intensification*, ACS Symposium Series 914, Washington, DC: American Chemical Society.

Chemical safety

Every research facility that handles chemicals should have a safety manual. A useful example is *Caltech's Chemical Safety Manual* (2004 revision) <http://www.cce.caltech.edu/resources/Safety1.pdf> (accessed May 18, 2005). Additional published resources include the following.

Hazardous Substances Data Bank (HSDB). HSDB is a peer-reviewed toxicology data file on the National Library of Medicine's (NLM) Toxicology Data Network (TOXNET) <http://toxnet.nlm.nih.gov/cgi-bin/sis/htmlgen?TOXLINE> (accessed 16 June 2005). The primary focus is on the toxicology of about 5,000 potentially hazardous chemicals. It also includes information on chemical and physical properties, emergency handling procedures, environmental fate, regulatory requirements, and related areas. Data are extensively referenced from a wide variety of books, government reports, and the journal literature.

Material safety data sheets

Material safety data sheets (MSDS) provide the proper procedures for handling or working with specific chemicals. In addition, MSDS include physical data, toxicity, health effects, first aid, reactivity, storage, disposal, protective equipment, and spill/leak procedures. As a general rule, an MSDS should be available for every on-site chemical. Aldrich-Sigma is an example of the many chemical suppliers that freely provide MSDS on their website <http://www.sigmaaldrich.com/> (accessed June 16, 2005) with links from specific chemicals. Additional information on MSDS is available at:

Material Safety Data Sheets, Cornell University, Department of Environmental Health and Safety <http://msds.ehs.cornell.edu/msdssrch.asp> (accessed June 3, 2005). A database of over 250,000 MSDS.

Material Safety Data Sheets on the Internet, Oklahoma State University, Department of Health and Safety <http://www.pp.okstae.edu/ehs/LINKS/Msds.htm> (accessed June 3, 2005). Information about and links to MSDS as well as other chemistry sites.

National Research Council, Committee on Hazardous Substances in the Laboratory (1981) *Prudent Practices for Handling Hazardous Chemicals in Laboratories*, Washington, DC: National Academy Press.

Journals

The journal article, in any science or engineering discipline, is its primary literature. Articles in referred journals (e.g., those with editorial boards) have undergone critical review by both editors and reviewers. A journal's quality may be assessed by its ISI impact factor which is calculated by dividing the number of a given year's citations by the number of source items published in that journal during the previous two years (e.g., citations in 2003/source items in 2001 and 2002).

In addition to the journals specific to chemical engineering, many journals identified with other fields are found in chemical engineering libraries. Since each institution's selection is generally very idiosyncratic and dependent on current faculty interests, the list below gives the top ten chemical engineering research journals indexed by ISI in 2003. One way of determining the journals important to a given community is to analyze the references in articles they have recently authored or requested.

AIChE Journal (American Institute of Chemical Engineers) (0001–1541)
Catalysis Today (0920–5861)
Chemical Engineering Science (0009–2509)
Chemical and Engineering News (ACS) (0009–2347)
Chemical Engineering (CE) (0009–2460)
Chemical Engineering Progress (AIChE) (0360–7275)
Chemical Market Reporter (CMR) (1092–0110)
Chemical Week (CW) (0009–272X)
Chemistry and Industry (SCI) (0009–3068)
Combustion and Flame (0010–2180)
Energy and Fuels (0887–0624)
Industrial and Engineering Chemistry Research (0888–5885)
Journal of Aerosol Science (0021–8502)
Journal of Catalysis (0021–9517)
Journal of Membrane Science (0376–7388)
Journal of Supercritical Fluids (0896–8446)

Chemical engineering reviews

Advances in Biochemical Engineering/Biotechnology (Springer) (0724–6145)
Advances in Polymer Technology (Wiley) (0730–6679)
Catalysis Reviews: Science and Engineering (Taylor & Francis) (0161–4940)
Chemistry and Physics of Carbon (Dekker) 0069–3138
Critical Reviews in Biotechnology (Taylor & Francis) (0738–8551)
Progress in Energy and Combustion Science (Elsevier) (0360–1285)
Reviews in Chemical Engineering (Freund) (0167–8299)

Chemical engineering research journals

Adsorption (Springer) (0929–5607)
AIChE Journal (AIChE/Wiley) (0001–1541)

Applied Catalysis (Elsevier) (0166–9834)
Biochemical Engineering Journal (Elsevier) (1369–703X)
Bioprocess and Biosystems Engineering (Springer) (1615–7591)
Canadian Journal of Chemical Engineering (NRC) (0008–4034)
Catalysis Letters (Springer) (1011–372X)
Catalysis Today (Elsevier) (0920–5861)
Chemical Engineering and Processing (Elsevier) (0255–2701)
Chemical Engineering and Technology (Wiley-VCH) (0930–7516)
Chemical Engineering Communications (Taylor & Francis) (0098–6445)
Chemical Engineering Education (CEE) (0009–2479)
Chemical Engineering Journal (Elsevier) (1385–8947)
Chemical Engineering Research and Design (ICE) (0263–8762)
Chemical Engineering Science (Elsevier) (0009–2509)
Chemie-Ingenieur-Technik (Wiley) (0009–286X)
Combustion and Flame (Elsevier) (0010–2180)
Combustion Science and Technology (Taylor & Francis) (0010–2202)
Computers and Chemical Engineering (Elsevier) (0098–1354)
Energy and Fuels (ACS) (0887–0624)
Environmental Progress (AIChE/Wiley) (0278–4491)
Fluid Phase Equilibria (Elsevier) (0378–3812)
Fuel (Elsevier) (0016–2361)
Fuel Processing Technology (Elsevier) (0378–3820)
Heat and Mass Transfer (Springer) (0947–7411)
Heat Transfer Engineering (Taylor & Francis) (0145–7632)
Industrial and Engineering Chemistry Research (ACS) (0888–5885)
International Communications in Heat and Mass Transfer (Elsevier) (0735–1933)
International Journal of Chemical Reactor Engineering (Berkeley Electronic Press) (1542–6580)
International Journal of Heat and Mass Transfer (Elsevier) (0017–9310)
Journal of Applied Polymer Science (Wiley) (0021–8995)
Journal of Biotechnology (Elsevier) (0168–1656)
Journal of Catalysis (Elsevier) (0021–9517)
Journal of Chemical Engineering Data (ACS) (0021–9568)
Journal of Chemical Engineering of Japan (SCEJ) (0021–9592)
Journal of Chemical Technology and Biotechnology (SCI/Wiley) (0268–2575)
Journal of Loss Prevention in the Process Industries (Elsevier) (0950–4230)
Journal of Membrane Science (Elsevier) (0376–7388)
Journal of Process Control (Elsevier) (0959–1524)
Journal of Separation Science (Wiley) (1615–9306)
Journal of Supercritical Fluids (Elsevier) (0896–8446)
Macromolecular Materials and Engineering (Wiley) (1438–7492)
Microporous and Mesoporous Materials (Elsevier) (1387–1811)
Organic Process Research and Development (ACS) (1083–6160)
Polymer Engineering and Science (SPE/Wiley) (0032–3888)
Polymer Reaction Engineering (Taylor & Francis) (1054–3414)
Powder Technology (Elsevier) (0032–5910)
Process Biochemistry (Elsevier) (1359–5113)
Process Safety and Environmental Protection (ICE) (0957–5820)
Process Safety Progress (AIChE/Wiley) (1066–8527)
Separation and Purification Reviews (Taylor & Francis) (1542–2119)

Separation and Purification Technology (Elsevier) (1383–5866)
Separation Science and Technology (Taylor & Francis) (0149–6395)
Theoretical Foundations of Chemical Engineering (Springer) (0040–5795)
Transport in Porous Media (Springer) (0040–5795)

Patents

Patents are an excellent source of information on particular inventions or chemical processes. A book that would be useful to all chemical engineers is *What Every Chemist Should Know About Patents* (2001) Washington, DC: American Chemical Society http://www.chemistry.org/portal/resources/ACS/ACSContent/government/publications/Chem_patent2001.pdf (accessed May 18, 2005).

Patents may be searched by patent number, inventor, assignee, keywords, classification numbers, and so on. *SciFinder Scholar* also allows searching of chemical patents with CAS Registry Numbers. Both *Web of Science* (1945+) and *SFS* (1997+) provide patent citation searching.

The Chemical Abstracts Service routinely abstracts the first publicly available document. This is often a Japanese, German, European, or PCT (WO) patent application, which may be the first in what becomes a family of related patent documents. Chemical Abstracts' definitions follow:

"A patent family is a collection of patent documents concerned with a particular invention. A family member is considered equivalent to the first-abstracted document if the family member and the abstracted document contain only one priority number and that number is common to both documents." Additional information on "related or non-priority patents and definitions of the terms 'Division', 'Continuation-in-part', 'addition', 'continuation', etc." is available (4).

Thus, simply ordering a patent document (especially a patent application), without first checking the family history and other details, may overlook additional information; an English language equivalent; changes in the patent's legal status; and significant cost (e.g., WO2000058473 to CuraGen has 5,509 pages and US3400371 to IBM was issued in three volumes and 27,861 pages).

Both SciFinder Scholar and the fee-based World Patent Index (STN/Dialog) provide extensive information on patent families.

SciFinder Scholar (Chemical Abstracts) includes information on both chemical patents and patent family members from forty-four countries. Since 1907, Chemical Abstracts (CA) has abstracted and indexed over three million patents worldwide in all areas of chemistry and chemical engineering. CA is currently abstracting and indexing over 100,000 patents each year. Patent titles may be altered by the Chemical Abstracts Service to make them more descriptive.

SciFinder Scholar is an excellent resource for article, paper and patent searching, since (1) both patent numbers <http://www.cas.org/STNEWS/JULAUG01/patent.html> (accessed May 18, 2005) and inventor/assignee names are searchable from 1907 to date; (2) bibliographic and abstract (machine translations for JP and DE patents) information for patents from major industrial countries is available within two days of publication, and full indexing is completed within thirty days, and (3) patent family information is available from 1957 to date

<http://www.cas.org/ONLINE/STN/STNOTES/stnote19.html> (accessed May 18, 2005). Several journals routinely provide listings of recent patents. These include *Applied Catalysis B: Environmental, Chemistry and Industry, Drug Discovery Today, Environment International, Microporous and Mesoporous Materials, and Organic Process Research and Development.*

Others works on patents that might be useful include: *Kirk-Othmer Encyclopedia of Chemical Technology*'s section on "Patents and Trade Secrets," a basic, step-by-step approach to the practice and management of patents and trade secrets, and "Patents, Literature," a basic guide to sources, databases, searching, and more. J.M. Cogen's "Technical Disclosures; Advanced Tactics" (in *Chemical Innovation* (2001) 31 (7): 33–38) should also be read http://pubs.acs.org/isubscribe/journals/cinnov/31/i07/html/07cogen.html (accessed May 18, 2005).

Describes the use of technical disclosures to establish prior-art and preclude patentability and/or avoid costs of patenting.

Associations, organizations, societies and conferences

American Chemical Society, Division of Biochemical Technology <http://membership.acs.org/b/biochem/> (accessed May 18, 2005. "The mission of this division is to promote the exchange of information among academic and industrial researchers regarding technology utilizing life-based systems to produce useful products and services"(from website).

Division of Fuel Chemistry <http://www.anl.gov/PCS/acsfuel/> (accessed May 18, 2005). "The ACS Division of Fuel Chemistry provides a forum for documentation and communication to the international community of research and development results, in order to promote efficient and environmentally acceptable fuel production and use" (from website).

Division of Industrial and Engineering Chemistry <http://membership.acs.org/I/IEC/> (accessed May 18, 2005). "I&EC is a multi-disciplinary Division helping individuals convert science into commercially relevant products and processes" (from website).

Division of Petroleum Chemistry <http://membership.acs.org/P/PETR/> (accessed May 18, 2005). "A vibrant professional network of scientists and engineers interested in the chemistry of petroleum exploration, production, refining, and utilization" (from website).

Division of Polymeric Materials: Science and Engineering <http://membership.acs.org/P/PMSE/> (accessed May 18, 2005). "Providing a forum for the exchange of technical information on the chemistry of polymeric materials including plastics, paints, adhesives, composites and biomaterials" (from website).

American Institute of Chemical Engineers (AIChE) <http://www.aiche.org/> (accessed May 18, 2005) <http://www.aiche.org/conferences/> (accessed May 18, 2005). AIChE is an international association of professions in all fields of chemical engineering.

American Society for Engineering Education, Chemical Engineering Division <http://www.asee-ched.org/> (accessed May 18, 2005) (includes link to conferences). "The Chemical Engineering Division of the ASEE is dedicated to the promotion and improvement of Chemical Engineering Education. The purpose of the Chemical Engineering Division site is to provide information about ASEE ChED events, awards, meetings, and more" (from website).

DECHEMA <http://www.dechema.de/About_the_DECHEMA-lang-en.html> (accessed May 18, 2005). DECHEMA (Gesellschaft fur Chemische Technik und Biotechnologie e.V.) was founded in 1926, and currently has over 5,000 personal and institutional members. Their focus is the promotion of research and technological advances in chemical engineering, biotechnology, and environmental protection. DECHEMA is responsible for the organization of the ACHEMA, ACHEMASIA, and ACHEMAMERICA exhibition congresses, as well as a wide variety of local and international workshops, congresses, and colloquia through its subject divisions (Zeolite, Catalysis, Membrane technology, Biotechnology, Adhesive technology, Safety technology, and Reaction engineering).

DECHEMA is also a database publisher (CEABA-VtB – a chemical technology and biotechnology bibliographic database, CHEMSAFE – for explosion and fire protection safety parameters for flammable gases, liquids and dusts, and DETHERM – a compilation of thermophysical properties of pure substances and mixtures).

Style guide

Dodd, J.S. (ed.) (1997) *The ACS Style Guide; A Manual for Authors and Editors* (2nd edn), *The ACS Style Guide* provides chapters on: Writing a scientific paper, Communicating in other formats: posters. . ., Grammar, punctuation and spelling, Editorial style, Numbers, mathematics, and units. . ., References (with journal abbreviations for the 1000+ most commonly cited journals), Names and numbers for chemical compounds, Conventions in chemistry, Illustrations and tables, Peer review (short descriptions of peer review authored by about forty prominent chemists/chemical engineers), Copyright and permissions, Making effective oral presentations, ACS publications, ACS divisions, Ethical guidelines. . ., Chemist's code of conduct, Proofreaders' marks. A new edition is scheduled for 2005/2006 that will update new ACS policies on copyright and permissions.

Websites and listservs

Cheme-l <http://www.lsoft.se/scripts/wl.exe?SL1=CHEME-L&H=LISTSERV.LOUISVILLE.EDU>. "Informal discussion forum on chemical engineering and related topics."

Chemcyclopedia <http://www.mediabrains.com/client/chemcyclop/BG1/search.asp> (accessed June 16, 2005). Subtitled the "Worldwide Guide to Chemicals and Services" the site lists websites for products and services in the chemical industries.

CHMINF-L <https://listserv.indiana.edu/archives/chminf-l.html>. The Chemical Information Sources Discussion List has been in existence since May 1991 and is hosted at the University of Indiana. It is the primary listserv for members of both the ACS Division of Chemical Information and the Special Libraries Association Chemistry Division.

Connecting Industry: Process and Control <http://www.connectingindustry.com/processandcontrol/> (accessed May 18, 2005). Process and Control today and Connecting Industry: Process and Control are complimentary websites that provide news, jobs, suppliers, products, and so on.

EDA Incorporated, Chemical Engineering <http://www.edasolutions.com/Groups/ChemicalEngineering.htm>. A very extensive listing of websites related to industrial chemistry.

George Porter's Chemical Engineering Web Page <http://library.caltech.edu/collections/engineering/chemical.htm#journal> (accessed May 18, 2005). Associations, societies, and groups, crystallography resources, databases, chemicals and equipment, journals and newsletters, patents.

Google Directory <http://directory.google.com/Top/Science/Technology/Chemical_Engineering/> (accessed May 18, 2005). An extensive listing of subcategories, and web pages in Google Rank Order.

International Directory of Chemical Engineering URLs <http://www.ciw.uni-karlsruhe.de/chem-eng.html> (accessed May 18, 2005). A comprehensive listing of databases, newsgroups, publishers, organizations, companies, and so on.

Process and Control Today <http://www.pandct.com/> (accessed May 18, 2005).

University of Delaware, Internet Resources for Chemical Engineering <http://www2.lib.udel.edu/subj/chee/internet.htm> (accessed June 16, 2005). Very good portal to information on organizations, reference data, and more.

University of Santa Barbara, InfoSurf: Information Resources for Chemical Engineering <http://www.library.ucsb.edu/subjects/chemeng/chemeng.html> (accessed June 16, 2005). Another university site that offers good access to Internet information on chemical engineering.

Yahoo! Chemical Engineering <http://dir.yahoo.com/Science/engineering/chemical_engineering> (accessed June 16, 2005). Site includes employment opportunities as well as links to organizations, discussion groups, and other Internet sites.

Conclusion

Information searching is a talent that must be developed and maintained. Imagination, persistence and, for librarians, an ethic of service to others is absolutely essential.

While a wide variety of resources are available electronically, there are a significant number of print resources available in both research and technical libraries. This highlights one of the ever-present dangers of overusing electronic resources, namely the inability to make serendipitous discoveries in books on the shelf near the one you are looking for.

As with all engineering disciplines the chemical engineer needs to rely not only on the primary resources of his own profession, but to refer to the reference materials of chemistry. This chapter is a survey of the major works in chemical engineering. The research in both engineering and chemistry is important, supporting the significance of the role of the librarian in assisting the practitioners to stay current in their knowledge.

References

American Institute of Chemical Engineers (2000) *Career Choices for Chemical Engineers* <http://www.aiche.org/careers/>.

CACHE Teaching Resource Center <http://www.che.utexas.edu/cache/trc.html>.

Edgar, T.F. (2000) "Chemical Engineering Education and the Three C's: Computing, Communication, and Collaboration," *CACHE Newsletter* (fall issue) <http://www.che.utexas.edu/cache/newsletters/fall2000_chemengedu.pdf>.

Hadlington, S. (2004) "Polymer Science Unlocked," *Chemistry and Industry* (London) 14: 13–14.

Halford, B. (2004) "Chemical Engineering Education in Flux," *Chemical Engineering News* 82 (10): 34.

Madelung, O. (1993) *Substance Index 1993*, Heidelberg: Springer-Verlag.

Mark Davis Research Group [website] Caltech ChemE <http://www.che.caltech.edu/groups/med/>.

MATLAB <http://www.mathworks.com/products/matlab/>.

MIT OpenCourseWare website <http://ocw.mit.edu/OcwWeb/Chemical-Engineering/index.htm>.

Molecular Modeling, Viewing and Drawing <http://ep.llnl.gov/msds/orgchem/molmodl.html>.

Octave <http://www.octave.org/>.

Price, G. (2003) "Amazon Debuts New Book Search Tool," *Publishers Weekly*, October 27, 2003, <http://searchenginewatch.com/searchday/article.php/3098831> (accessed May 18, 2005).

Ritter, S.K. (2001) "The Changing Face of Chemical Engineering," *Chemical Engineering News* 79 (23): 63–66, http://pubs.acs.org/cen/education/7923/7923education.html. A review of the current expansion of chemical engineering into materials and biomedical product development.

8 Civil engineering

Carol Reese and Michael Chrimes

Introduction

While people have been building structures since ancient times, the profession of civil engineering was not defined until 1828 when the Institution of Civil Engineers in London, England received its charter. The charter stated that civil engineering was:

> the art of directing the great sources of power in nature for the use and convenience of man, as the means of production and of traffic in states, both for external and internal trade, as applied in the construction of roads, bridges, aqueducts, canals, river navigation and docks for internal intercourse and exchange, and in the construction of ports, harbours, moles, breakwaters and lighthouses, and in the art of navigation by artificial power for the purposes of commerce, and in the construction and application of machinery, and in the drainage of cities and towns.
>
> (Merdinger, 1953)

In 1961, the American Society of Civil Engineers expanded on this definition by stating that it was:

> a profession in which a knowledge of the mathematical and physical sciences gained by study, experience, and practice is applied with judgment to develop ways to utilize, economically, the materials and forces of nature for the progressive well-being of humanity in creating, improving and protecting the environment, in providing facilities for community living, industry and transportation, and in providing structures for the use of humanity.
>
> (ASCE, 2004: 118)

Based on these definitions, this chapter will cover the areas of general civil engineering, construction management, geotechnical engineering, maritime engineering, municipal engineering, structural engineering, and water resources engineering. Due to the size of the fields, energy, environmental, and transportation engineering have their own chapters, although there is much overlap between these fields of engineering and the fields discussed in this chapter.

Literature of the civil engineering field

The literature of civil engineering is scattered throughout the various specialized subject areas of the field. While engineering is as old as human history, its technical literature is a recent development. For this reason, information is mainly found in the journal literature, handbooks and manuals, standards, and conference proceedings. Professional societies play an important part in the development of the literature both for civil engineering in general and for the specialized areas. Since its literature is still in the developmental state and is scattered throughout the specialized fields, there are no single guides to the literature or encyclopedias to the entire field.

General civil engineering

Searching the library catalog

Searching for general information on civil engineering is simple. There is one subject heading to look under – Civil engineering. Under this heading, one will find all the subheadings such as:

Civil engineering – biographies
Civil engineering – dictionaries
Civil engineering – handbooks and manuals
The Library of Congress classification code for civil engineering is the same for general engineering: TA 1–2040.

Abstracts and indexes

American Society of Civil Engineers' *Civil Engineering Database* <http://www.asce.org/> (accessed May 18, 2005). *The Civil Engineering Database* (CEDB) is an index to all ASCE publications. This includes all journals, proceedings, books, standards, manuals, and *CE Magazine*. Journal papers with abstracts are covered back to 1970 while those without abstracts are covered back to the 1920s. Book records are complete dating back to the 1900s. The database is constantly being expanded.

CSA Civil Engineering Abstracts <http://www.csa.com/> (accessed May 18, 2005). CSA *Civil Engineering Abstracts* provides indexing to the serial literature of civil engineering and its complementary fields such as forensic engineering and management. It provides international coverage by monitoring over 3,000 serial titles and numerous non-serial titles.

Emerald Abstracts – *International Civil Engineering Abstracts* <http://www.emeraldinsight.com/> (accessed May 18, 2005). *International Civil Engineering Abstracts* (ICEA) covers the major civil engineering subjects. ICEA is selective in the journals it covers so that searchers do not waste time searching inappropriate materials.

ICONDA – the *International CONstruction Database* <http://www.irbdirekt.de/iconda/> (accessed May 18, 2005). *ICONDA* provides worldwide coverage of the technical literature on civil engineering, urban planning, architecture, and construction. It indexes periodicals, books, research reports, conference proceedings, business reports, theses, and non-conventional literature from over twenty different countries. It provides coverage back to 1976.

Bibliography

The following is the only bibliography available on civil engineering.

Stapleton, D.H. and R.L. Shumaker (1986) *History of Civil Engineering Since 1600: An Annotated Bibliography*, New York: Garland Press. This bibliography lists works according to broad categories such as general works, the Renaissance, the Industrial Revolution, and the Modern Age. Within these categories specific topics in civil engineering are covered.

Dictionaries

American Society of Civil Engineers (1972 and 1991) *Biographical Dictionary of American Civil Engineers* (2 vols), New York: American Society of Civil Engineers. These volumes provide biographical sketches that emphasize the engineering accomplishments of individual engineers born before 1900.

Chrimes, M., Sir Alec Skempton, R.W. Rennison, R.C. Cox, T. Ruddock, and P. Cross-Rudkin (eds) (2002) *Biographical Dictionary of Civil Engineers in Great Britain and Ireland – Volume 1: 1500–1830*, London: Thomas Telford Ltd. A reference work on the lives, works, and careers of individuals engaged in the practice of civil engineering. It provides the background, training, and achievements of engineers who began their careers before 1830.

Scott, J.S. (1984) *Penguin Dictionary of Civil Engineering*, New York: Penguin Group. This paperback is a concise dictionary of engineering terms. Unfortunately, it is out of print, but may be found in libraries.

Webster, L. (1997) *Wiley Dictionary of Civil Engineering and Construction*, New York: John Wiley & Sons. This dictionary provides a broad coverage of technical disciplines such as architecture, engineering, building, construction, forestry, and mining. Entries include terms, concepts, names, abbreviations, techniques, and tools. All entries are thoroughly cross-referenced.

Handbooks and manuals

Blake, L.S. (ed.) (1989) *Civil Engineer's Reference Book* (4th edn), London: Butterworth-Heinemann. This handbook provides a survey of the fundamentals, theory, and current practice in the different branches of civil engineering. It gives practical guidance for both student and practicing civil engineers. In addition, it incorporates chapters on construction, site practice, and contract management.

Chen, W-F. and J.Y.R. Liew (2003) *Civil Engineering Handbook* (2nd edn), Boca Raton, FL: CRC Press. A comprehensive handbook covering nearly every aspect of civil engineering including design techniques, construction methods, instrumentation, material properties, and calculations. The volume includes bibliographic references and an index.

Hicks, T.G. (ed.) (1999) *Handbook of Civil Engineering Calculations*, New York: McGraw-Hill. Over 500 key calculations covering the entire field of civil engineering including, but not limited to, structural engineering, timber engineering, soil mechanics, fluid mechanics, and waste-water treatment. Examples of each calculation procedure are included.

Merritt, F.S., M.K. Loftin, and J. Ricketts (eds) (1996) *Standard Handbook for Civil Engineers* (4th edn), New York: McGraw-Hill. This handbook is a guide to the principles and techniques for effective civil engineering practice. It covers all areas of the field such as highway construction, design, and geotechnical engineering. It also contains information on the various codes and standards, and over 700 tables, formulas, and drawings to clarify every explanation and procedure.

Journals

Professional societies are usually the publishers of the serials that provide general coverage of the field. Unlike the more technical, specialized journals, these publications provide the non-specialist with a good understanding of the variety of training needed by civil engineers in order to produce the many different projects required of them. These publications are usually designated as official publications of the societies. Examples would be ASCE's *Civil Engineering Magazine*, the Canadian Society for Civil Engineering's *Canadian Journal of Civil Engineering*, and South African Society of Civil Engineering's *Civil Engineering*.

Websites

There are several portals on civil engineering that can provide guidance to important Web resources.

EEVL <http://www.eevl.ac.uk/> (accessed May 18, 2005). Started in 1996 as an engineering gateway, the section on civil engineering includes links to sites concerned with general civil engineering, construction and building, engineering geology, hydraulics and waterworks, structural and transportation engineering, and water, sewage and waste treatment. They have a Z39.50 target for their Internet Resource Catalog of quality engineering resources that, in theory, could be searched from intranets.

iCivilEngineer.com <http://www.icivilengineer.com> (accessed May 18, 2005). This portal was designed for civil engineering professionals and students. It has sections on civil engineering news, career sites, and web resources in civil engineering arranged by subject.

TenLinks.com: Ultimate Civil Engineering Directory <http://www.tenlinks.com/engineering/civil/> (accessed May 18, 2005). This site is aimed at helping professionals to find technical information faster. TenLinks uses technical professionals to comb through the Internet catalogs to locate the most relevant sites.

Associations, organizations, and societies

Most countries have their local civil engineering society. To determine if a particular country has such a society, check the online ports such as iCivilEngineer.com to find complete lists. Some of the major societies are the following.

American Society of Civil Engineers <http://www.asce.org> (accessed May 18, 2005). Founded in 1852, it is the oldest national engineering society in the United States. It supports conferences, continuing education courses, professional development of the field, and publishes journals, monographs, and conference proceedings.

Asian Civil Engineering Coordinating Council (ACECC) <http://www.ksce.or.kr/acecc/index.html> (accessed May 18, 2005). The ACECC works to promote and advance the science and practice of civil engineering for sustainable development in the Asian region. To achieve this aim, ACECC sponsors the Civil Engineering Conference in the Asian Region (CECAR) every three years.

Canadian Society for Civil Engineering <http://www.csce.ca> (accessed May 18, 2005). Founded in 1887, the society works to develop and maintain high standards of civil engineering practice in Canada. It does so by supporting the work of academics, private institutions, and different organizations that deal with civil engineering.

European Council of Civil Engineers (ECCE) <http://www.eccenet.org> (accessed May 18, 2005). ECCE was formed in 1985 to encourage civil engineers to work together across Europe in order to advance its built environment and to protect its natural environment. The ECCE formulates guidelines to maintain and raise standards of civil engineering education, professional competence, and to assist in achieving compatibility of codes and standards.

Institution of Civil Engineers (ICE) <http://www.ice.org.uk> (accessed May 18, 2005). Established in 1818, ICE seeks to advance the knowledge, practice, and business of civil engineering. It aims to promote the value of the civil engineer's global contribution to sustainable, economic growth through supporting research, conferences, and publications.

Japan Society of Civil Engineers (JSCE) <http://www.jsce.or.jp/e/index.html> (accessed May 18, 2005). The most important civil engineering society outside the English-speaking world, JSCE is responsible for a large number of publications, increasingly in English, including important seismic codes.

Geotechnical engineering

Geotechnical engineering, originally called soil mechanics, is the study of the physical properties and utilization of soils, especially soils used in the planning of foundations for structures and subgrades for highways. Soil is a natural aggregate that has three phases: solid, liquid, and gaseous. Geotechnical engineers study the stresses put upon soils by the weight of different structures (*Encyclopedia Britannica*, 2004). Today, geotechnical engineering embraces subjects such as rock mechanics, engineering geology, and tunneling.

Searching the library catalog

Geotechnical engineering covers a wide area of subject areas. The following Library of Congress Subject Headings are a sample of just some of the major subject headings for the field:

Earthquake engineering
Earthwork
Embankments
Excavation
Fills (earthwork)
Foundations
Geology
Hydrogeology
Retaining walls
Sediments (Geology)
Soil dynamics
Soil mechanics
Soils – testing
Tunnels
Underground construction

Materials are also scattered throughout various classification codes.

GB3–5030 Physical geography; including Slopes GB448, Hydrogeology GB 1001–1199.8
QE1–996.5 Geology; including Mineralogy QE420, Dynamic and structural geology QE521–545, Volcanoes and earthquakes QE601–613.5
TA 703–712 Engineering geology
TA 715–787 Foundations, earthwork
TA 800–820 Tunnels

Abstracts and indexes

Earthquake Engineering Abstracts (EEA) Available through CSA <http://www.csa.com> (accessed May 18, 2005). EEA provides coverage of such topics as geotechnical earthquake engineering, engineering seismology, and soil dynamics. It is produced by the National Information Service for Earthquake Engineering, University of California, Berkeley.

GeoRef Information Services Produced by the American Geological Institute <http://www.agiweb.org> (accessed May 18, 2005), the *GeoRef* database provides access to the geoscience literature of the world including journal articles, books, maps, conference papers, reports, and theses.

Geotechnical Abstracts Produced twice a year by Research Resources, Inc., Geotechnical Abstracts <http://www.geotechnical-abstracts.com> (accessed May 18, 2005) is available in print or on CD-ROM. It covers such topics as soil mechanics, foundation engineering, rock mechanics, and engineering geology.

Quakeline Database Produced by the Multidisciplinary Center for Earthquake

Engineering Research <http://mceer.buffalo.edu> (accessed May 18, 2005), it covers earthquakes, earthquake engineering, and related topics.

Bibliography

Terzaghi, K. (1960) *From Theory to Practice in Soil Mechanics*, New York: John Wiley. Karl Terzaghi is considered to have started modern soil mechanics with his theories of consolidation, lateral earth pressures, bearing capacity, and stability. He pioneered a great range of methods and procedures for investigation, analysis, testing, and practice that defined much of the geotechnical engineering field (Goodman, 2002). These seminal papers and reports are important to anyone who wishes to understand the field.

Dictionaries

Barker, J.A. (1981) *Dictionary of Soil Mechanics and Foundation Engineering*, New York: Construction Press. This volume presents definitions of a wide variety of terms dealing with soil mechanics and foundations that range from a single sentence to an entire paragraph. It also provides cross-references within the publication.

Kurtz, J-P. (ed.) (2004) *Dictionary of Civil Engineering*, New York: KluwerAcademic. This English version of the French publication (*Dictionnaire du Génie Civil*, published in 1997) is a valuable reference tool for civil engineers. There are over 12,000 definitions accompanied by more than 1,300 charts, tables, and graphs. This dictionary is a comprehensive compilation of definitions, examples, and descriptions – from the study of soils, and the various materials and equipment used, including the most common architectural terms as they relate to civil engineering. This compendium will be an invaluable tool not only for civil engineers, but also for lawyers, contractors, architects, and so on, and all trade associations involved with this discipline.

Somerville, S.H. and M.A. Paul (1983) *Dictionary of Geotechnics*, London: Butterworths. Definitions in the areas of soil and rock mechanics, hydrology, ground stabilization, and excavation are included. This volume also provides a series of tables dealing with a variety of aspects of geotechnical engineering at the end of the book.

Van der Tuin, J.D. (ed.) (1989) *Elsevier's Dictionary of Soil Mechanics and Geotechnical Engineering*, New York: Elsevier Science. This dictionary provides definitions of terms in multiple languages covering such topics as soils, rocks, sediments, geological survey methods, geophysics, and the application of geotextiles.

Handbooks and manuals

Brown, R.W. (2000) *Practical Foundation Engineering Handbook* (2nd edn), New York: McGraw-Hill Professional Publishing. From site assessment through design and construction to remediation of failed foundations, this handbook provides design alternatives for substandard soil and challenging site conditions with example problems for different types of structures. It also has illustrations, charts, tables, and case study examples.

Chen, W-F. and C. Scawthorn (eds) (2003) *Earthquake Engineering Handbook*, Boca Raton, FL: CRC Press. This is a comprehensive resource that covers the spectrum of topics relevant to designing for and mitigating earthquakes. The handbook presents engineering practices, research, and developments in North America, Europe, and the Pacific Rim countries. Included are formulas, tables, and illustrations to answers to practical questions.

Day, R.W. (2001) *Soil Testing Manual: Procedures, Classification Data and Sampling Practices*, New York: McGraw-Hill Professional Publishing. For engineers, geologists, contractors, and on-site construction managers, this handbook simplifies each step of the process of soil testing from selecting the appropriate method to analyzing the results. It has handy tables, charts, diagrams, and formulas.

Day, R.W. (2002) *Geotechnical Earthquake Engineering Handbook*, New York: McGraw-Hill Publishing. This handbook features field and laboratory testing methods and procedures, current seismic codes, site improvement methods, in-depth analysis of soils, and problems with solutions to illustrate these analyses.

Fang, H-Y. (1991) *Foundation Engineering Handbook*, New York: Van Nostrand Reinhold. This handbook provides coverage throughout the field of applied geotechnics including explanations, methods, and examples of more efficient analysis, design, and construction of foundations.

Macnab, A. (2002) *Earth Retention Systems Handbook*, New York: McGraw-Hill Professional. This publication discusses temporary excavation shoring and earth retention systems used to construct permanent structures inside them. Each chapter presents a different shoring system and describes how it is constructed and the equipment needed.

Paz, M. (ed.) (1994) *International Handbook of Earthquake Engineering: Code, Programs, and Examples*, New York: Chapman & Hall. This handbook stresses the international approach to earthquake engineering. It presents national seismic codes from over thirty countries and demonstrates their application with quantitative examples.

Persson, P-A., R. Holmberg, and J. Lee (1994) *Rock Blasting and Explosives Engineering*, Boca Raton, FL: CRC Press. This handbook covers the practical engineering aspects of the different kinds of rock blasting. It covers the fundamental sciences of rock mass and material strength, the thermal decomposition, burning, and detonation behavior of explosives. Based on practical industrial experience, the handbook provides both students and practitioners with a source for understanding the constructive use of explosives.

Smoltczyk, U. (ed.) (2003) *Soil Construction and Geotechnics*, vols 1–3, New York: Wiley. Volume 1 of this handbook covers the basics in foundation engineering including laboratory and field tests. Volume 2 covers the geotechnical procedures used in manufacturing anchors and piles. The third volume deals with the basic designs of different types of foundations.

U.S. Bureau of Reclamation (2004) *Earth Manual*, Washington, DC: Department of the Interior <http://www.usbr.gov/pmts/writing/earth/index.html> (accessed May 18, 2005). Available for downloading from the Bureau of Reclamation website, this manual covers the engineering of earthen structures. It also provides references and cross-references of hundreds of terms.

Monographs

Craig, R.F. (2004) *Soil Mechanics* (7th edn), London: Spon Press. This book presents the fundamental principles of soil mechanics and shows how they can be applied in practical situations. Worked examples are included to reinforce the principles discussed. A solutions manual is available.

Das, B.M. (2001) *Principles of Geotechnical Engineering* (5th edn), Salt Lake City, UT: Thomson-Engineering. An overview of soil properties and mechanics along with a study of field practices and basic soil engineering procedures are presented. Numerous case studies are included.

Terzaghi, K., R.B. Peck, and G. Mesri (1996) *Soil Mechanics in Engineering Practice* (3rd edn), New York: John Wiley & Sons. This volume presents both theoretical and practical knowledge of soil mechanics in engineering. The third edition provides expanded coverage of vibration problems, mechanics of drainage, and consolidation.

Tomlinson, M.J. (2001) *Foundation Design and Construction* (7th edn), Upper Saddle River, NJ: Longman Group. This book covers such topics as site investigations, principles of foundation design, design of specific types of foundations, foundation construction, and shoring. Appendixes provide properties of materials and conversion tables.

Journals

Bulletin of the Seismological Society of America (0037–1106). Started in 1912, this journal of the SSA covers the general areas of seismology, seismicity, earthquake engineering, seismic hazards, and the effects of earthquakes.

Canadian Geotechnical Journal (0008–3674). Published by the National Research Council of Canada, this bimonthly journal was started in 1963. It presents papers on foundations, excavations, soil properties, dams, embankments, slopes, rock engineering, waste management, frozen soils, and offshore soils.

Earthquake Engineering and Structural Dynamics (0098–8847). Published since 1972, this Wiley publication covers all aspects of engineering related to earthquakes including soil amplification and failure, and ground motion characteristics.

Engineering Geology (0013–7952). Published since 1966, this Elsevier journal includes original studies, case histories, and comprehensive reviews in the field of engineering geology.

Geosynthetics International (1072–6349). This official journal of the International Geosynthetics Society, which was started in 1994, is now published by Thomas Telford. It is now available online and includes supplementary data and links.

Geotechnique (0016–8505). Started in 1948 and published by Thomas Telford, this journal includes papers in the fields of soil and rock mechanics, engineering geology, and environmental geotechnics. It publishes detailed case histories and theoretical research in these fields.

Geotechnical Engineering – Proceedings of the Institution of Civil Engineers (1353–2618). This journal, published by Thomas Telford, is part of the *Proceedings of the ICE* and the journal of the British Geotechnical Association. Published since 1848, it contains papers that are aimed at both the civil engineer and the geotechnical specialist.

Ground Improvement: Journal of the International Society for Soil Mechanics and Geotechnical Engineering (ISSMGE). Started in 1897 and published by Thomas Telford, this journal provides referred papers and original research on topics from foundation soils and ground reinforcement to ground treatment and admixtures.

International Journal of Rock Mechanics and Mining Sciences (1365–1609). Published by Elsevier since 1964, this journal includes papers concerned with original research, new developments, site measurement, and case studies in rock mechanics and rock engineering.

Journal of Cold Regions Engineering (0887–381X). Published by the American Society of Civil Engineering since 1987, this journal covers topics such as ice engineering, ice force, cold weather construction, environmental quality in cold regions, snow and ice control, and permafrost.

Journal of Earthquake Engineering (1363–2469). Started in 1998 by World Scientific Publishing, this journal publishes papers on research and development in analytical, experimental, and field studies of earthquakes from an engineering seismology as well as a structural engineering viewpoint.

*Journal of Geotechnical and Geoenvironmental Eng*ineering (1090–0241). Started in 1875 by the American Society of Civil Engineers, this journal covers a broad area of practice including areas such as foundations, retaining structures, soil dynamics, slope stability, dams, earthquake engineering, environmental geotechnics, and geosynthetics.

Soil Dynamics and Earthquake Engineering (0267–7261). Begun in 1982 by Elsevier, this journal publishes papers of applied mathematicians, engineers and other applied scientists involved in solving problems related to the field of earthquake and geotechnical engineering.

Soils and Foundations (0038–0806). Started in 1961 by the Japanese Geotechnical Society, this journal publishes research papers, reports of engineering experiences, state-of-the-art reports on certain themes, and discussions.

Websites

Compendium of On-Line Soil Survey Information <http://www.itc.nl/~rossiter/research/rsrch_ss.html> (accessed May 18, 2005). Compiled by D.G. Rossiter of the Department of Earth Systems Analysis, International Institute for Geo-information Science and Earth Observation, it brings together online information on soil survey activities, institutions, datasets, research, and teaching materials worldwide.

GeoIndex <http://www.geoindex.com/> (accessed May 18, 2005). Developed for the geo-environmental professional by Datasurge Company, this search engine covers geotechnical, environmental, hydrogeological, geological, mining, and petroleum subject areas. This information is divided into four categories: companies, associations, education, and government.

Geotechnical, Rock, and Water (GROW) <http://www.grow.arizona.edu/> (accessed May 18, 2005). The aim of this project is to develop a National Civil Engineering Resource Library (NCERL) that will encourage and promote interest, exploration, and learning in civil engineering. This library works to provide high-quality interactive digital learning objects that may be used for self-study or for instruction needs.

Mineral Resources Program <http://minerals.er.usgs.gov/> (accessed May 18, 2005). This program provides funding to communicate current information on available mineral resources.

Natural Resources Conservation Service Soils website <http://soils.usda.gov/> (accessed May 18, 2005). This site provides information on soil risks and hazards important to urban planners and construction contractors, soil data, and soil education.

Professional societies and other organizations

American Geological Institute (AGI) <http://www.agiweb.org/> (accessed May 18, 2005). Founded in 1948, the AGI provides information services to the geoscience community. The Institute also serves as a voice for shared interests in the profession, plays a major role in strengthening geosciences education, and works to increase public understanding of the role of the geosciences in society's use of natural resources.

American Rock Mechanics Association <http://armarocks.org/> (accessed May 18, 2005). This association provides research, education and training, and a digital library for those professionals, companies, teachers, and students in the field of rock mechanics and rock engineering.

American Soil and Foundation Engineers (ASFE) <http://www.asfe.org/> (accessed May 18, 2005). This trade association of earth engineering and related applied science services firms was founded in 1969. Its aim is to help its members prosper through professionalism. To do this, ASFE provides a variety of services, programs, and materials to help its members enhance their skills.

American Underground Construction Association <http://www.auca.org/> (accessed May 18, 2005). This organization is involved in all aspects of underground facilities: planning, design, development, construction, and use.

Association of Engineering Geologists <http://www.aeg.org/> (accessed May 18, 2005). This association provides leadership, advocacy, and applied research in environmental and engineering geology.

British Geotechnical Association (BGA) <http://www.britishgeotech.org.uk/> (accessed May 18, 2005). As the principal association for geotechnical engineers in the United Kingdom, the BGA plays a coordinating role in the UK with the International Society of Soil Mechanics and Geotechnical Engineering. It arranges programs to enable discussion of topics important to geotechnical engineers with the Institution of Civil Engineers.

Canadian Geotechnical Society (CGS) <http://www.cga.ca/> (accessed May 18, 2005). The CGS exists to serve and promote the geotechnical and geoscience community in Canada.

Deep Foundations Institute (DFI) <http://www.dfi.org/> (accessed May 18, 2005). Started in 1976, the DFI is a technical association of firms and individuals in the deep foundations and related industry. It offers its members international conferences, seminars, networking opportunities, and publications on the latest technologies.

Environmental and Engineering Geophysical Society (EEGS) <http://www.eegs.org/> (accessed May 18, 2005). The EEGS was formed in 1992 to promote the science of geophysics, to foster common scientific interests of geophysicists, and to maintain a high professional standing among its members. The Society holds an annual conference and produces various publications in the field.

Geo-Institute <http://www.geoinstitute.org/> (accessed May 18, 2005). Established by the American Society of Civil Engineers in 1996 as a specialty organization focused on the geo-industry, its major aims are to advance the state of the art, and to provide leadership on the professional, business, public policy, and educational aspects of the field. To accomplish these aims it sponsors conferences, seminars, and educational programs. The Institute is involved in the publication of journals, newsletters, manuals, and geotechnical special publications.

International Geosynthetics Society (IGS) <http://www.geosyntheticssociety.org/>

(accessed May 18, 2005). The IGS works for the scientific and engineering development of geosynthetics and related technologies. It publishes newsletters, journals, and sponsors conferences.

International Society for Rock Mechanics (ISRM) <http://www.isrm.net/gca/?id=51> (accessed May 18, 2005). The ISRM was founded in Salzburg in 1962. Its main objectives are to encourage international collaboration, to encourage teaching and research, and to promote high professional standards. The ISRM sponsors international congresses and publishes a news journal.

International Society for Soil Mechanics and Geotechnical Engineering <http://www.issmge.org/> (accessed May 18, 2005). This society promotes international cooperation among engineers and scientists for the advancement and dissemination of knowledge in the field of geotechnical engineering. In order to do this, its members may submit papers to many conferences, have access to the society's work in various fields, and have networking opportunities.

International Tunneling Association <http://www.ita-aites.org/> (accessed May 18, 2005). The Association holds meetings, organizes studies, and publishes reports and proceedings in order to promote advances in planning, design, construction, maintenance, and safety of tunnels and underground space.

National Geophysical Data Center (NGDC) <http://www.ngdc.noaa.gov/> (accessed May 18, 2005). The NGDC provides scientific leadership, products, and services for geophysical data describing the solid earth, marine, and solar-terrestrial environment. The Center contains more than 300 digital and analog databases.

National Information Service for Earthquake Engineering (NISEE) <http://nisee.berkeley.edu/> (accessed May 18, 2005). Sponsored by the National Science Foundation and the University of California, Berkeley, the NISEE provides access to technical research and development information in earthquake engineering and related fields.

Seismological Society of America <http://www.seismosoc.org/> (accessed May 18, 2005). Founded in 1906 in San Francisco, its aim is to advance earthquake science, and it represents a variety of technical interests.

Soil Science Society of America <http://www.soils.org/> (accessed May 18, 2005). The Society's purpose is to advance the discipline and practice of soil science.

United States Geological Survey (USGS) <http://www.usgs.gov/> (accessed May 18, 2005). Started in 1879, the USGS provides reliable scientific information to describe and understand the Earth and to manage water, biological, energy, and mineral resources.

WODA <http://www.woda.org/> (accessed May 18, 2005). The WODA incorporates the Western Dredging Association which serves the Americas, the Central Dredging Association which serves Europe, Africa, and the Middle East, and the Eastern Dredging Association which serves the Asian and Pacific regions.

Conferences

International Conference of the International Association of Computer Methods and Advances in Geomechanics <http://www.iacmag.org/index1.html> (accessed May 18, 2005). Started in 1972 and sponsored by IACMAG, these conferences, held every three years, cover such topics as computer modeling and their applications to a wide range of geomechanical problems.

International Conference on Soil Mechanics and Geotechnical Engineering <http://www.issmge.org> (accessed May 18, 2005). This four yearly conference, first held in 1936, is the most important in the geotechnical field.

Rapid Excavation and Tunnelling Conference <http://www.dfi.org/> (accessed May 18, 2005). Each year for thirty years, contractors, academics, suppliers, and manufacturers from the deep foundations construction industry have met to learn from each other and discuss the state of the art related to the design and construction of deep foundations.

Maritime engineering

Maritime engineering deals with all aspects of waterway environments. These engineers are concerned with such issues as ocean exploration, design of offshore structures, wave action on coastlines and ports, and the protection of wetlands (Britannica, 2004).

Searching the library catalog

When looking for materials related to ocean engineering, the following are some of the Library of Congress Subject Headings to search:

Coastal engineering
Harbors
Marine geotechnique
Ocean – atmosphere
Ocean engineering
Ocean waves
Offshore structures
Shore protection
Water masses

There are two main Library of Congress classification code areas for ocean engineering topics: GC and TC.

GC 1–1581 Oceanography
GC 96–97.8 Estuarine oceanography
GC 151–155 Density
GC 190–190.5 Ocean–atmosphere interaction
GC 229–296.8 Currents
TC 183–201 Dredging, submarine building
TC 203–380 Harbors and coast protective works
TC 530–537 River protective works

Abstracts and indexes

Aquatic Sciences and Fisheries Abstracts (ASFA) <http://www.csa.com/> (accessed May 18, 2005). ASFA provides coverage of the field since 1971. It covers publications on such topics as dynamical oceanography, underwater acoustics, marine meteorology, marine technology, offshore structures, and underwater vehicles. It is updated monthly by Cambridge Scientific Abstracts.

Oceanic Abstracts <http://www.csa.com/> (accessed May 18, 2005). This publication covers the worldwide technical literature related to the marine and brackish-water environment. It covers such topics as ecology, marine geology, marine pollution, non-living marine resources, and navigation.

Handbooks and manuals

Clark, J.R. (1996) *Coastal Zone Management Handbook*, Boca Raton, FL: CRC Press. This handbook covers such topics as natural resources, economics, development, productivity, and diversity of coastal zones. The book discusses three aspects of management: strategies, methods, and information. The final section is a collection of project histories.

El-Hawary, F. (ed.) (2000) *Ocean Engineering Handbook*, Boca Raton, FL: CRC Press. This is a comprehensive compilation of information on the theory and practice of oceanic/coastal engineering. It covers such topics as modeling considerations, marine hydro dynamics, and applications of computational intelligence. The handbook also includes over 200 tables and figures on ocean engineering.

Gerwick, B.C. Jr (1999) *Construction of Marine and Offshore Structures*, Boca Raton, FL: CRC Press. This handbook provides a comprehensive treatment of the construction aspects of offshore structures. It discusses floating structures, deep-water structures, ice-resistant structures, and bridge foundations. It also details all the particulars of building in the marine environment.

Herbich, J. (ed.) (2000) *Handbook of Dredging Engineering* (2nd edn), New York: McGraw-Hill Professional. Providing an up-to-date guide to dredging theory and practice, this handbook covers such topics as fluid mechanics, dredging equipment, sediment, dredging methods, environmental effects, and project planning.

Kennish, M.J. (ed.) (2001) *Practical Handbook of Marine Science* (3rd edn), Boca Raton, FL: CRC Press. This handbook contains information on physical oceanography, marine biology, marine chemistry, geology, and pollution. It contains over 800 tables and figures.

Mader, C.L. (2004) *Numerical Modelling of Water Waves* (2nd edn), Boca Raton, FL: CRC Press. This manual covers all aspects of this topic from basic fluid dynamics and the basic models to the most complex, including the compressible Navier-Stokes techniques to model waves generated in various ways.

Task Committee of the Waterways Committee (1998) *Inland Navigation: Locks, Dams, and Channels*, Reston, VA: American Society of Civil Engineers. Based on the experience of the U.S. Army Corps of Engineers, this manual provides information on planning, designing, construction, and operation of the U.S. waterways used by barge traffic.

Task Committee on Ship Channel Design (1993) *Report on Ship Channel Design*. Reston, VA: American Society of Civil Engineers. This manual provides an overview of the design process and identifies the studies usually needed for a ship channel project.

Task Committee on Underwater Investigations (2001) *Underwater Investigations: Standard Practice Manual*, Reston, VA: American Society of Civil Engineering. This manual presents guidelines in such areas as standards of practice, documentation and reporting, repair design, and the inspection of unique structures.

Monograph

Randall, R.E. (1997) *Elements of Ocean Engineering*, Jersey City, NJ: Society of Naval Architects and Marine Engineers. This text addresses the application of engineering principles for the analysis, design, development, and management of marine systems.

Journals

Coastal Engineering: an International Journal for Coastal, Harbour, and Offshore Engineers (0378–3839). Started in 1977, this Elsevier publication combines practical application with modern technological achievements. It presents studies and case histories on all aspects of coastal engineering such as waves and currents, coastal morphology, and estuary hydraulics.

International Journal of Offshore and Polar Engineering (1053–5381). The principal journal of The International Society of Offshore and Polar Engineers (ISOPE), it publishes research in the fields of offshore, ocean, polar, marine, environment, mechanics, and materials engineering.

Journal of Atmospheric and Oceanic Technology (0739–0572). This journal, published by the American Meteorological Society, publishes papers on the instrumentation and methodology used in atmospheric and oceanic research such as computational techniques and methods of data acquisition.

Journal of Waterway, Port, Coastal, and Ocean Engineering (0733–950X). Published by the American Society of Civil Engineers, this journal covers such topics as dredging, floods, ice, sediment transport, and wave action. It also publishes papers on the development and operation of ports, harbors, and offshore facilities.

Maritime Engineering: Part of the Proceedings of the Institution of Civil Engineers (1741–7597). Started in 2004 by the ICE, this journal publishes papers that focus on such topics as fixed and moving port and harbor developments, estuarine and coastal protection, habitat creation, and seabed pipelines.

Ocean Engineering: an International Journal of Research and Development (0029–8018). Published since 1968, this journal covers such areas as design and building of structures, submarine soil mechanics, coastal engineering, fabrication of materials, hydrodynamic properties of shells, ocean energy, propulsion systems, and underwater acoustics.

Websites

Coastal and Hydraulics Laboratory <http://chl.erdc.usace.army.mil/chl.aspx/> (accessed May 18, 2005). This laboratory conducts research in areas such as coastal structures, dredging, erosion control, fish studies, flood control, navigation, and watersheds. Besides conducting research, this site also lists their publications and available software.

Coastalmanagement.com <http://www.coastalmanagement.com> (accessed May 18, 2005). A portal site for resources related to coastal management. It provides links to news groups, research, courses, and to many different sectors such as marinas, navigation, dredging, ocean engineering, and coastal environments.

EUCC Coastal Guide <http://www.coastalguide.com/> (accessed May 18, 2005). A portal site aimed at professionals in coastal management, planning, conservation, and research in Europe. It provides access to such services as EUCC coastal news, publications, coastal guide country files, ecosystems, various relevant policies, and trends in the field.

Glossary of Coastal Terminology <http://www.usace.mil/inet/usace-docs/eng-manuals/em1110-2-1100/AppA/a-a.pdf> (accessed May 18, 2005). This glossary is for the Army Corps of Engineer's Coastal Engineering Manual.

NOAA Coastal Services Center <http://www.csc.noaa.gov/> (accessed May 18, 2005). This agency is devoted to serving the state and local coastal resource management programs. It provides training classes, fellowships, online mapping of coastal areas, and software tools.

USGS Science Center for Coastal and Marine Geology <http://woodshole.er.usgs.gov/> (accessed May 18, 2005). This center explores and studies many aspects of the underwater areas between shore-lines and the ocean. The Center has modeled various aspects of the coasts and ocean waters such as tidal flushing and circulation. It concentrates on the eastern coastal waters while other centers concentrate on the other coastlines.

Professional societies and other organizations

Association of Coastal Engineers <http://www.coastalengineers.org/> (accessed May 18, 2005). Started in 1999, the Association is dedicated to excellence in education, research, and the practice of coastal engineering. Among other activities, it holds an annual conference where such topics as the sciences of oceanography, design construct, and monitor coastal structures, beach nourishment, and harbor design are discussed.

Coasts, Oceans, Ports, and Rivers Institute (COPRI) <http://www.coprinstitute.org/> (accessed May 18, 2005). Established by the American Society of Civil Engineers in 2000 as a semi-autonomous institute, COPRI officially replaced the Waterway, Port, Coastal, and Ocean Division of ASCE to serve as the multi-disciplinary and international leader in improving the knowledge, teaching, development, and practice of civil engineering and other disciplines working in waterway environments. It supports conferences, publications, continuing education seminars, and workshops in all areas of maritime engineering.

Institute of Marine Engineering, Science and Technology <http://www.imarest.org/> (accessed May 18, 2005). Recognizing the need to bring together marine engineers, scientists, and technologists, the Institute's aim is to promote the scientific development of these three related areas by integrating their different interests into one organization. Its main activities are publishing specialized journals and books, and organizing conferences and seminars.

International Navigation Association (PIANC) <http://www.pianc-aipcn.org/> (accessed May 18, 2005). PIANC, founded in 1885, is an international organization made up of individuals, corporations, and national governments. Its object is to promote the maintenance and operation of both inland and maritime navigation by fostering progress in the planning, design, construction, maintenance, and operation of inland and maritime waterways, ports, and coastal areas.

International Society of Offshore and Polar Engineers (ISOPE) <http://www.isope.org/> (accessed May 18, 2005). International in scope, ISOPE is interested in promoting engineering and scientific progress in the fields of offshore and polar engineering. It does this by sponsoring technical conferences, publications, scholarship programs, continuing education, and international cooperation. The Annual International Offshore and Polar Engineering Conference is the largest conference held on these topics.

Maritime Technology Society <http://www.mtsociety.org> (accessed May 18, 2005). Since the 1960s, the Society has had a mission to disseminate marine science and technical knowledge, promote education, advance research in the field, and engender a broader understanding of the field in the general public.

Office of Coast Survey <http://chartmaker.ncd.noaa.gov/> (accessed May 18, 2005). A part of the National Oceanic and Atmospheric Administration (NOAA), the Coast Survey is the oldest scientific organization in the United States. The Office of Coast Survey produces navigational charts and related publications, provides information on navigational obstructions and hydrographic surveys, and supports research in the field.

Offshore Engineering Society <http://www.oes.org.uk/> (accessed May 18, 2005). The Society provides networking opportunities for engineers in all related disciplines and works to disseminate technical information through lectures and other events.

Scripps Institution of Oceanography <http://sio.ucsd.edu/> (accessed May 18, 2005). Founded in 1903 as a biological research laboratory, Scripps now encompasses physical, chemical, biological, geological, and geophysical studies of the oceans. Part of the University of California, it supports ongoing investigations into the topography and composition of the ocean bottom, waves and currents, and the interactions between seawater and the atmosphere.

Society of Naval Architects and Marine Engineers (SNAME) <http://www.sname.org/> (accessed May 18, 2005). Started in 1893, the Society works to advance the art, science, and practice of naval architecture, shipbuilding, and marine engineering. It sponsors applied research, offers career guidance, disseminates information, and supports educational programs. SNAME is interested in all types of marine ships from yachts to submersibles.

U.S. Army Engineer Waterways Experiment Station (WES) <http://www.wes.army.mil/> (accessed May 18, 2005). The largest civil engineering and environmental quality research and development complex in the United States, WES conducts projects in response to the requirements of Army Civil Works research and development covering such topics as flood control, materials and structures, environmental quality, and military research and development related to survivability and protective structures, environmental clean-up, and software engineering.

Conferences

Coastal Engineering in the Oceans <http://www.coprinstitute.org/> (accessed May 18, 2005). This conference is for engineers and scientists from governmental agencies, academia, and the private sector to share information in the fields of hydrodynamics, coastal and offshore structures, marine transportation, and marine geotechnical engineering.

International Conference on Coastal Engineering <http://www.coprinstitute.org/> (accessed May 18, 2005). This international conference covers fields such as coastal processes and climate change, flood and coastal defense engineering, flood risk management, ports and harbors, and legislation.

International Conference on Offshore Mechanics and Artic Engineering <http://www.asme.org/> (accessed May 18, 2005). Sponsored by the Ocean, Offshore and Artic Engineering Division of ASME International, this conference is held to advance the development and exchange of information regarding ocean, offshore, and artic engineering.

International Navigation Congress <http://www.pianc-aipcn.org/> (accessed May 18, 2005). An international congress for the presentation of papers on subjects of current significance to waterways and maritime interests, it is organized into two sections – one is for inland navigation issues, while the other concentrates on maritime navigation topics.

Offshore Technology Conference <http://www.otcnet.org/> (accessed May 18, 2005). This conference, which was started in 1969, is the foremost trade show related to the development of offshore resources in the fields of drilling, exploration, production, and environmental protection. In addition to company exhibits, volunteers develop a technical program in order to share information on advances in the field.

World Dredging Congress <http://www.woda.org> (accessed May 18, 2005). This congress brings researchers and practitioners together to explore the latest developments in all aspects of dredging, including the ability to ensure safe navigation while addressing environmental issues.

Construction management: Introduction

Management information has become a major priority for civil engineers, and the published literature has proliferated. With competition from other professional groups in the construction sector, managerial qualifications and expertise are a *sine qua non* for any civil engineer aspiring to reach the top. General management gurus such as John Adair, Tom Peters, and Alan Weiss are as relevant as those writing specifically on construction. Within the UK the impact of the Latham and Egan reports has been to encourage partnership rather than confrontation in construction contracts, whereas worldwide encouragement of private funding for public infrastructure projects has led to a reconsideration of traditional roles of client/consultant/contractor. There has been a proliferation of new forms of contract from relevant professional bodies and trade associations to reflect these changes, and the ICE's New Engineering Contract has gained widespread acceptance as a result of these developments in the UK. The World Bank is actively considering its use. Much construction is done under the auspices of the building contracts issued by the Joint Contracts Tribunal rather than ICE and its partners.

Searching the library catalog

Management involves industry-specific knowledge, and generic knowledge about law and regulations, as well as an understanding of human resources issues such as psychology and motivation. Material is likely to be scattered within a library in law sections, general management, industrial management, and construction management.

Examples are:

BF Psychology
HD28–70 Management
HD61 Risk management
K623–968 Civil law
T55–5.3 Industrial safety
T55.4–60.8 Management engineering
T57.6–57.97 Operations research
T58.4 Managerial control systems
T58.6–58.62 Management information systems
T59.7–T59.77 Human engineering in industry
T60–60.8 Work measurement
TA 166–167 Human engineering
TA177.4–185 Engineering economy
TA194–197 Management of engineering works
TS15–194 Operations management

Handbooks

Hicks, T.G. and J.F. Mueller (1996) *Standard Handbook of Consulting Engineering Practice* (2nd edn), New York: McGraw-Hill. Covers all aspects of getting started as a consulting engineer, marketing and expanding your business.

O'Brien, J.J. (ed.) (1996) *Standard Handbook of Heavy Construction*, New York: McGraw-Hill. Concerned with all aspects of the management of major civil engineering projects including estimating, scheduling, earth-moving, and value engineering, with case studies of techniques and projects.

Ratay, R.T. (1996) *Handbook of Temporary Structures in Construction: Engineering Standards, Designs, Practices and Procedures*, New York: McGraw-Hill. Written by a team of specialist authors with a U.S. background, this covers issues such as site safety, with more specific techniques such as underpinning, formwork, and bracing. It covers issues often ignored elsewhere such as loadings created by construction equipment.

Shenson, H.L. (1990) *The Contract and Fee-setting Guide for Consultants and Professionals*, New York: Wiley. Although published in 1990, most of the material is still relevant. It provides sample copies of proposals, interim reports, and final reports, and illustrates the marketing role of the proposal. The book is of practical help to the independent consultant.

Turner, J.R. and S.J. Simister (2000) *Gower Handbook of Project Management* (3rd edn), Burlington, VT: Aldershot, England: Gower. Produced with the support of the International Project Management Association and the Project Management Institute (USA), this covers the systems of project management, the context of projects including political, economic, social, technical, legal, and environmental issues, the management of performance including quality, time, cost, risk, and safety, the management of the project life cycle, commercial issues including appraisal and finance, and contracts, and human resources management.

Journals

Building Law Reports George Godwin (0141–5875). Although largely concerned with building contracts and contract case law, some cases relate directly to civil engineering contracts, and many have an indirect relevance.

Construction Industry Law Newsletter Informa (0269–0039). A newsletter updating legal developments, meetings, and so on.

Construction Law Journal Sweet & Maxwell (0267–2359). Largely concerned with UK practice, contains articles and law reports.

Construction Law Reports Architectural Press (0950–3889). Like the *Building Law Reports* (above), this journal is largely concerned with UK case law in construction.

Construction Management and Economics Spon (0144–6193). An academic journal which is international in coverage, and with frequent analysis of case studies.

Engineering, Construction and Architectural Management Blackwell (0969–9988). This journal has carried some key papers on developments in private finance, and so on.

International Construction Law Review Lloyds/Informa (0265–1416). Very useful on the International Federation of Consulting Engineers (FIDIC) and also on international development of the New Engineering Contract (NEC).

International Journal of Project Management Elsevier (0263–7863). With excellent international coverage of all aspects of project management, this journal discusses the latest techniques and issues for project managers.

Journal of Construction Engineering and Management ASCE (0733–9364). This journal has a more practical emphasis than many of the other ASCE titles which publish construction case studies.

Journal of Management in Engineering ASCE (0742–597X). Examines contemporary issues associated with leadership and management; the focus is the practicing consulting civil engineer.

Leadership and Management in Engineering ASCE (1532–6748). A publication of the Committee on Professional Practice, this journal examines contemporary issues and principles of leadership and management, including news, brief and concise leadership and management "nuggets," and short articles of interest to practicing professionals in a variety of roles and industry segments. The focus includes, but is not limited to, individuals and public and/or private entities, small and large projects, and organizations. Areas of interest are: leadership; teamwork; communications; decision-making; partnering; project management; mentoring; diversity; office management; professional practice and development; financial management; productivity management and tools; globalization; networking; change management; involvement in the political process; legislative and regulatory issues; and economic and environmental sustainability.

Monographs

Arup, O. and Partners (1996) *Green Construction Handbook*, Bristol: J T Design Build Limited. Advice for the construction industry on sustainable methods and materials.

Bartlett, A. (1992) *Emden's Construction Law*, London: Butterworths. Continuously updated compendium of contract documents.

Bennett, J. (1998) *Seven Pillars of Partnering*, London: Thomas Telford. Identifies the main issues in partnering for the UK construction industry.

Blockley, D. and Godfrey, P. (2000) *Doing it Differently: Systems for Rethinking Construction*, London: Thomas Telford. Discusses new ways of procurement philosophies based on the Egan report.

Broome, J. (2002) *Procurement Routes for Partnering: A Practical Guide*, London: Thomas Telford. A comprehensive treatment of partnering and procurement, covering issues such as incentives, target cost and cost reimbursable contracts, and strategic alliances, with discussion of the NEC.

Bunni, N.G. (1997) *The FIDIC Form of Contract* (2nd edn), Oxford: Blackwell Scientific. This comprises both a discussion of the principles underlying the FIDIC forms and also a commentary on the clauses.

CESMM3 *Price Database*, Available from Thomas Telford. A database based on ICE Civil Engineering Standard Method of Measurement.

Chung, H.W. (1999) *Understanding Quality Assurance in Construction*, London: Spon. Discusses the implementation of ISO quality systems in construction. Contains a chapter relating to small firms.

Construction Industry Board (CIB) (1997) *Partnering in the Team*, London: Thomas Telford. CIB encouragement for partnering in the UK construction industry.

Construction Industry Council (CIC) (2001) *Rethinking Construction Implementation Toolkit*, London: Local Government Task Force. A CIC response to Latham and Egan.

CIOB (2000) *Code of Practice for Project Management* (2nd edn), Harlow: Longman. Includes the UK Chartered Institute of Building's code of practice, which is widely used by UK project managers.

CIRIA (2001) *Sustainable Construction: Company Indicators*, London: CIRIA. Provides measures for civil engineering companies to measure up to sustainability targets.

Clough, R.H. (ed.) (2000) *Construction Contracting* (4th edn), New York: Wiley. A well-established guide to planning and scheduling construction projects, with CPM software, and consideration of problems caused by inclement weather.

Clough, R.H. (ed.) (2005) *Construction Contracting* (7th edn), New York: Wiley. Concerned with all aspects of the ownership or management of a contracting construction company, including the latest regulations, contract documents, and U.S. law.

Corbett, E.C. (1991) *FIDIC a Practical Legal Guide* (4th edn), London: Sweet & Maxwell. A clause-by-clause legal commentary on the FIDIC condition.

Day, D.A. and N.B.H. Benjamin (1991) *Construction Equipment Guide* (2nd edn), New York: Wiley. Aimed at the contractor, it summarizes the performance of commonly recognized equipment, and it deals with the physical concepts of the work, the surrounding conditions and equipment requirements, with an emphasis on controls governing the equipment's performance.

DETR (1998) *Rethinking Construction*, London: DETR (Egan report). Commissioned by the then UK Department for Environment, Transport and the Regions, this builds upon the Latham report to try to establish a way forward.

Eggleston, B. (2001) *The ICE Conditions of Contract* (7th edn), Oxford: Blackwell Scientific. A relatively simple guide to the ICE conditions.

Geddes, S. (1996) *Estimating for Building and Civil Engineering* Works (9th edn), Oxford: Butterworths. A long-established textbook on estimating.

Gladden, S.C. and A. Orlitt (1996) *Marketing and Selling A/E and other Engineering Services*, New York: ASCE. A practical guide to marketing and sales, describing seminars, and indirect marketing tools.

Godfrey, K.A. Jr (1996) *Partnering in Design and Construction*, New York: McGraw-

Hill. A multi-authored work describing the development of partnering, with case studies including non-project uses of partnering, discussions of alternative dispute resolution and with a look to the future.

Goodowens, J.B. (1996) *A User's Guide to Federal Architect–Engineer Contracts*, New York: American Society of Civil Engineers. This explains all facets of the architect–engineering contract from pre-selection to cost negotiation to architect–engineer liability. It is intended to assist professionals in making informed decisions about which jobs to pursue, and how to better present their qualifications in obtaining government contracts. The book features inside details of government negotiations strategy, and perspectives on various implementation documents of the Federal Departments of Defense, Logistics, Army, Navy, Air Force, and Mapping.

Halpin, D.W. and R.W. Woodhead (1997) *Construction Management* (2nd edn), New York: Wiley. Looks at the skills required by a construction manager, and all professional groups including engineers and contractors, the resources in terms of equipment and materials as well as financial and human. There are chapters on estimating and cost control.

Harris, F. and R. McCaffer (2001) *Modern Construction Management* (5th edn), Oxford: Blackwell Scientific. A textbook that provides an introduction to most aspects of construction management from a U.K. perspective, covering bid preparation, plant management, and so on.

Holroyd, T.M. (1999) *Site Management for Engineers*, London: Thomas Telford. Site management is an important aspect of any successful project, and requires skills and knowledge rarely obtained through academic study; this is a useful introduction for engineers going on-site for the first time to best practice.

Holroyd, T.M. (2000) *Principles of Estimating*, London: Thomas Telford. An introduction to estimating practice.

Holz, H. (1997) *Complete Guide to Consulting Contracts* (2nd edn), New York: Dearbourn. Indispensable for those new to consulting contracts and agreements.

Illingworth, J.R. (1987) *Temporary Works: Their Role in Construction*, London: Thomas Telford. Helpful in describing temporary works and invaluable for young engineers going on-site for the first time with little knowledge of this area.

ICE (1996) *Civil Engineering Procedure* (5th edn), London: Thomas Telford. Based on UK practice, it explains all the stages involved in a construction project.

ICE (1998) *RAMP. Risk Appraisal and Management of Projects*, London: Thomas Telford. This is a guide to risk appraisal, now a fundamental aspect of all project planning

ICE (2001) *Management Development in the Construction Industry* (2nd edn), London: Thomas Telford. Although aimed at civil engineers, the basic principles enunciated in this guide are applicable to anybody interested in developing as a manager.

Irvine, D.J. and R.J.H. Smith (1992) *Trenching Practice CIRIA Report 97* (rev. edn), London: CIRIA. Trenching is an area of construction that, if incorrectly carried out, can result in deaths and disasters; this provides guidance on good practice.

Joyce, R. (2001) *The CDM Regulations Explained* (2nd edn), London: Thomas Telford. This provides guidance on the most significant aspect of health and safety legislation in the UK.

Latham, M. (1994) *Constructing the Team*, London: HMSO. A seminal report aimed at changing the face of the UK construction process away from confrontation and toward cooperation between all those involved in the construction process.

McInnnis, A. (2001) *The New Engineering Contract: A Legal Commentary*, London: Thomas Telford. The first heavyweight commentary on the New Engineering Contract (NEC), discussing the principles behind the NEC rather than a line-by-line analysis of clauses. Heavily referenced throughout.

Mackay, E.B. (1986) *Proprietary Trench Support Systems Technical Note 95* (3rd edn), London: CIRIA. For any engineer dealing with trench support, the prevalence of proprietary systems is a challenge. This is helpful in identifying characteristics and guidance on use.

O'Reilly, M. (1999) *Civil Engineering Construction Contracts* (2nd edn), London: Thomas Telford. More narrowly focused than Uff's work, and a useful comparison of alternative contract forms.

Rogers, M. (2001) *Engineering Project Appraisal*, Oxford: Blackwell Scientific. Details methods of project appraisal in engineering.

Scott, W. and B. Billing (1998) *Communication for Professional Engineers* (2nd edn), London: Thomas Telford. This provides good common-sense advice to engineers working to develop their communication skills in all their aspects – spoken, written, and so on.

Snell, M. (1997) *Cost Benefit Analysis for Engineers*, London: Thomas Telford. CBA is a well-established technique and this serves as a basic introduction.

Spon's Civil Engineering and Highway Works Price Book (annual) New York: E & FN Spon. The main UK price book for civil engineers (NB: Spon produce a number of specialist estimating books covering railway engineering and foreign regions, but not all are updated annually).

Uff, J. (1999) *Construction Law* (7th edn), London: Sweet & Maxwell. Useful introduction to UK construction law.

Wallace, I.N.D. (1995) *Hudson's Building and Engineering Contracts* (11th edn) (2 vols), London: Sweet & Maxwell. Heavyweight commentary on British contracts. As with all such texts changes in case law, the introduction of new contract forms means that it can only provide guidance.

Woodward, J.F. (1997) *Construction Project Management*, London: Thomas Telford. An introduction to the subject.

Web resources

Blue Book of Building and Construction <http://www.thebluebook.com/> (accessed May 18, 2005). This is the US industry's leading directory of information on construction products and firms, categorized by area and specialization.

Construction Industry <http://www.construction.about.com/industry/construction.mbody.htm> (accessed May 18, 2005). Provides links to sites related to every area of the construction industry such as codes, industry news, software, equipment, and manufacturers.

Means Costworks. Kingston, MA: RS Means <http://www.rsmeans.com/cworks> (accessed May 18, 2005). Subscriber access to Means construction cost books, aimed at the US market.

Work Zone Safety Information Clearinghouse <http://wzsafety.tamu.edu/> (accessed May 18, 2005). This site provides information on the latest developments in U.S. and Canadian legislation, research, government agencies, and public and private organizations relevant to safety in traffic work zones. Spanish-language version available.

Institutions, organizations, and societies

Association of Consulting Engineers (ACE) <http://www.acenet.co.uk/> (accessed May 18, 2005). The British-based ACE produces conditions of engagement for consulting engineers' services, and works with ICE and the contractors to develop conditions of contract.

Association for Planning Supervisors <http://www.aps.org.uk/> (accessed May 18, 2005). This organization has grown up as a result of changes in health and safety legislation in the UK under the CDM regulations, and the need for all construction projects to address health and safety issues by having accredited personnel.

Association for Project Management <http://www.apm.org.uk/> (accessed May 18, 2005). The UK society for project managers with links to the international society. They publish a journal (*Project Management*), and a major guide to project planning. Their former trading arm, APM Group, is now independent <http://www.apmgroup.co.uk/> (accessed May 18, 2005).

Construction Best Practice Programme <http://www.cbpp.org.uk/> (accessed May 18, 2005). The UK's Construction Best Practice Programme provides support to individuals, companies, organizations, and supply chains in the construction industry seeking to improve the way they do business.

Construction Federation (formerly Building Employers Federation) <http://www.thecc.org.uk/> (accessed May 18, 2005). The UK Trade Federation for major building contractors.

Construction Industry Council <http://www.cic.org.uk/> (accessed May 18, 2005). A U.K. industry-wide body, given great impetus by the recommendations of the Latham and Egan reports. Full members are admitted into electoral colleges depending on their constitution: Chartered professional institutions holding a Royal Charter; Professional institutes/bodies with independent status qualifying individuals in various disciplines throughout the construction process; Business organizations which represent professional services, materials/product supply and contracting in private and public companies or partnerships; Research organizations that are dedicated to serving construction, the industry and its clients. Associate membership embraces organizations within the construction industry that speak for a defined group – or groups – of members, but are not eligible for admission as a Full Member of Council.

Construction Management Association of America (CMAA) <http://www.cmaanet.org/> (accessed May 18, 2005). The CMAA encourages excellence among construction managers. It can assist project owners by providing information about construction management practice and by helping identify qualified professional construction managers for projects. The website gives access to award-winning projects, and the construction management e-journal.

Construction Specifications Institute (CSI) <http://www.csinet.org/s_csi/index.asp> (accessed May 18, 2005). US-based organization with multi-disciplinary membership (17,000+). Produces Masterformat (2004) as a standard for specifications, and the Project Resources Manual – the CSI Code of Practice. The website has a members' area, and online access to the journal.

Construction Specifier: Design-Build Institute of America (DBIA) <http://www.dbia.org/> (accessed May 18, 2005). The DBIA is a member organization that promotes the philosophy of design and build in construction. The site offers a database of design-build projects, an e-commerce section containing publications detailing the model contract forms available, a mission statement, membership information, courses, a members' area, and a summary on the design-build process.

Forum on the Construction Industry <http://www.abanet.org/forums/construction/> (accessed May 18, 2005). A forum of the American Bar Association on the construction industry, the website contains past issues of the electronic newsletter, an index to *Construction Lawyer*, and an online bookshop referring to publications on design-build, which encompasses all U.S. states and Canada; fundamentals of construction law, and federal government construction contracts.

International Federation of Consulting Engineers (FIDIC) http://www.fidic.org/. An international body based in Switzerland responsible for a range of contract forms and procedures that are widely used in international civil engineering contracts.

Lean Construction Institute <http://www.leanconstruction.org/> (accessed May 18, 2005). An organization that aims to extend to the construction industry the "lean production" revolution started in manufacturing. This approach maximizes value delivered to the customer while minimizing waste. The website describes lean construction, and contains a link to the *Lean Construction Journal*, started in late 2004.

National Building Specification (NBS) <http://www.thenbs.com/> (accessed May 18, 2005). The main purpose is to write, revise and publish the National Building Specification. The NBS was first launched in 1973, and two years later an updating service was started with loose-leaf binders, followed shortly by a small jobs version. In 1988, the documents were completely revised to take account of the Common Arrangement of Work Sections, part of the CPI (Co-ordinated Project Information) initiative led by the Royal Institution of Chartered Surveyors. Its building industry focus has meant it is not entirely satisfactory for engineering work.

Society of Construction Law <http://www.scl.org.uk.> (accessed May 18, 2005). A UK society that has developed with the tremendous growth of construction and law and litigation.

World Bank <http://publications.worldbank.org/online> (accessed May 18, 2005). A major funder of projects in the developing world, the website provides a welter of economic information and other advice for those involved in international projects.

Municipal engineering: Introduction

The tradition of municipal engineering goes back to the mid-nineteenth century when engineers were appointed to deal with the health and sanitary problems of the fast-growing towns. Earlier, in the eighteenth century, some communities had appointed county surveyors to look after their bridges. Some aspects of the specialization are covered elsewhere, such as under transport engineering and water engineering, and increasingly such engineering services are provided directly by, or in partnership with, the private sector. The activity of community engineering is rapidly changing to one of community focus, sustainability, and best value. Government funding across the developed world has encouraged the development of brownfield sites (sites previously developed for other purposes and often contaminated by past use) and urban regeneration rather than expansion of urban areas and the planning of new communities so much in vogue in the 1950s and 1960s.

In the municipal engineering context transport engineering is concerned with the planning and management of transportation systems, and the construction, maintenance, and renewal of the infrastructure. While this includes knowledge of the moving loads operating in the system, vehicle design is the province of the mechanical engineer, while the electrical engineer interfaces with the civil engineer in the design of electrical power systems and traffic control/signaling. Transport affects people in their daily lives and, when transport systems fail to operate efficiently, as has been witnessed in Britain and elsewhere in recent years, the work of the municipal engineer is taken into the world of politics.

Investment in roads has characterized most Western economies since the 1930s, and at a local level has traditionally been the province of municipal engineers. There is an enormous literature on the subject which is generally readily accessible on subscription via the most successful cooperative database – International Transport Research Documentation (ITRD), and the freely available US-based Transportation Research Information Services (TRIS). Although the UK Department of Transport/Highways Agency has produced standards and codes through its various changes of practice for many years, ready access has only been made possible via the transfer of sales to The Stationery Office (TSO), and their publication on the Internet. It is an area when American practice has played a significant role, and despite the proliferation of Highways Agency documentation, a large number of other bodies produce important guidelines on best practice and specialist areas such as heavy-duty pavements.

Although generally associated with road traffic congestion in towns, traffic engineering has its origins in railway signaling in the nineteenth century. Today, with rapid developments in IT and GIS, the cutting edge of traffic engineering interfaces with vehicle design, and encompasses crude physical devices such as road humps and the most sophisticated smart card technology.

Many airports, although subject to (inter)national regulation, are municipally owned. Most countries have a research organization concerned with transport, generally government funded, and central governments issue regulations which may govern safe operation as well as minimum design standards.

Searching the library catalog

Municipal engineering forms a subset of sanitary engineering and environmental technology in the LC Classification scheme: TD 159–168. However, municipal engineers are likely to be interested in the legal aspects of their work as well the technical, and many are involved in other areas of civil engineering such as TE: Highway engineering; TF: Railroad engineering where it relates to metros; TG: Bridge engineering, as well as other aspects of sanitary engineering such as Municipal waste: TD 783–812.5, and Street cleaning: TD813–870.

Abstracts and indexing services

International Transport Research Database (ITRD) <http://www.stn-international.de/stndatabases/databases/itrd.html> (see Chapter 20 on Transportation engineering).

TRIS (TRB) <http://trisonline.bts.gov/search.cfm/http://www.nas.edu/trb> (see Chapter 20 on Transportation engineering).

Handbooks

American Railway Engineering and Maintenance of Way Association (AREMYA). *Manual of Railway Engineering*, Washington, DC: American Railway Engineering Association. A compilation of U.S. railway standards covering all aspects of the civil engineering of rail transport.

Atkinson, K. (1997) *Highway Maintenance Handbook* (2nd edn), London: Thomas Telford. Comprehensive treatment of the subject from a British perspective.

Brockenbrough, R.L. and K.J. Boedecke (2003) *Highway Engineering Handbook: Building and Rehabilitating the Infrastructure* (2nd edn), New York: McGraw-Hill. An American multi-authored handbook that covers highway planning, traffic engineering, pavement and highway structures design, street lighting and signage, as well as noise barriers and value engineering and life-cycle costs.

Cooper, A.R. (1998) *Properties of Hazardous Industrial Materials*, CRCnetBASE, Boca Raton, FL: Lewis Publishers. Available in digital format. Lists more than 25,000 hazardous materials. Each chemical is fully described.

Design Manual for Roads and Bridges, TSO <http://www.theso.co.uk/bookshop/bookstore.asp?FO=1164185> (accessed May 18, 2005). Multi-volume manual for highway design in the UK. Sections may be downloaded from <http://www.official-documents.co.uk/document/deps/ha/dmrb/> (accessed May 18, 2005).

Freeman, H. (1998) *Standard Handbook of Hazardous Waste Treatment and Disposal*, New York: McGraw-Hill. Summarizes U.S. laws and regulations, with an overview of the hazardous waste problem and state-of-the-art alternative treatment and disposal processes.

Grava, S. (2003) *Urban Transportation Systems: Choices for Communities*, New York: McGraw-Hill. Individual chapters discuss each mode in turn from walking to cable cars with a helpful guide to further reading. An excellent introduction for any covered subject.

Hess, K. (1998) *Environmental Site Assessment, Phase I: A Basic Guide*, Boca Raton, FL: Lewis Publishers. Useful introduction to U.S. practice.

Institute of Transportation Engineers (1997) *Traffic Signing Handbook*, Washington, DC.

Institute of Transportation Engineers (1997) *Trip Generation* (3 vols), Washington, DC.

Institution of Structural Engineers/Institution of Highways and Transportation (2002) *Design Recommendations for Multi-storey and Underground Car Parks* (3rd edn), London: Institution of Structural Engineers.

Jahm, R.K. (ed.) (2002) *Environmental Assessment* (2nd edn), New York: McGraw-Hill. Covers U.S. legislative background and the implications for the National Environmental Policy Act if applied to overseas projects involving U.S. firms. There are chapters on public participation and other issues affecting the environment globally.

Jane's Urban Transport Systems, Coulsdon: Jane's Information Systems. Comprehensive coverage of systems as existing and planned, with suppliers' details.

Karnofsky, B. (1997) *Hazardous Waste Management Compliance Handbook*, New York: Van Nostrand Reinhold. A handbook concerned with U.S. hazardous waste management regulations.

Keith, L.H. (1996) *Compilation of EPA's Sampling and Analysis Methods*, Boca Raton, FL: CRC. Digital format available. Information about related software programs is available at <http://www.instantref.com/inst-ref.htm> (accessed May 18, 2005).

Kindred Association (1994) *A Practical Recycling Handbook*, London: Thomas Telford. Despite the proliferation of initiatives to encourage local authority recycling programs there are relatively few textbooks on the subject.

Lamm, R. (ed.) (1999) *Highway Design and Traffic Engineering Handbook*, New York: McGraw-Hill. The specific objective of this work was its traffic safety focus, rather than engineering design per se.

Lewis, R.J. (1996) *Sax's Dangerous Properties of Industrial Materials*, New York: Van Nostrand Reinhold. Digital format available. The leading source on hazardous substances, including health and safety data, regulatory standards, toxicity, and physical properties.

Minsk, L.D. (1998) *Snow and Ice Control Manual for Transportation Facilities*, New York: McGraw-Hill.

Nelson, P.M. (1997) *Transportation Noise Reference Book*, London: Butterworths. *Practical Environmental Bioremediation: The Field Guide* (1998) Boca Raton, FL: Lewis Publishers. Bioremediation commonly uses micro-organisms as a biological activity to reduce the effects of a pollutant in the environment. This work is illustrated with numerous case studies.

SDU (1995–) *Leidraad Bodembescherming* (2 vols), The Hague: SDU. Contains widely quoted standards, which are a corollary for planning development.

Shell Bitumen Handbook (2003) London: TTL. Its association with a single company is misleading regarding the coverage of this handbook, which is a valuable up-to-date source on the use of bitumen.

Strong, D.L. (1997) *Recycling in America: A Reference Handbook*, Santa Barbara, CA: ABC-CLIO. Discusses legal issues surrounding recycling, and new recycling technology focusing on the role of recycling in solid waste management with reference to environmental issues such as global warming, conservation, and the depletion of natural resources. It reviews materials currently being recycled and those that have been overlooked, and the responsibility of industry to create new materials and products in the future.

Transportation Research Board (1994) *Highway Capacity Manual*, Washington, DC: National Research Council (TRB Special Report 209). Although other methods of traffic assessment are available and used in the UK, the methods employed in this manual are the most widely used internationally.

Journals

APWA Reporter (APWA) (0092–4873). Monthly magazine of the APWA, available via their website <http://www.apwa.net> (accessed May 18, 2005).

Association of Asphalt Paving Technologists Journal (US). The leading learned society journal in this area where most literature is trade driven.

International Journal of Pavement Engineering Taylor & Francis (1029–8436). An academic journal reflecting the body of university research now taking place in this area.

Journal of Infrastructure Systems ASCE (1076–0342). The only journal aimed at engineers that takes a holistic view of infrastructure.

Journal of Urban Planning and Development ASCE (0733–9488). The ASCE journal covering urban planning and municipal issues.

Municipal Engineer ICE (0965–0903). Part of the ICE Proceedings and published quarterly aimed at a municipal engineering audience, the majority of issues are based around a current theme.

Planner (RTPI). A weekly newsletter on UK planning issues.

Public Works <http://www.pwmag.com/> (0033–3840). A monthly journal with an annual directory/buyers' guide aimed at consulting engineers, and contractors involved with public works and infrastructure projects.

Surveyor Hemming Group (not supplied). The weekly for municipal engineers in the UK, available online to subscribers

Traffic Engineering and Control <http://www.tecmagazine.com/> (0041–0683). Editorially independent of all suppliers, with an international reputation, despite its UK base.

Transport ICE (0965–092X). The ICE's journal covers all aspects of transport, including public transport, so often the concern of municipal engineers.

United Nations *Annual Bulletin of Transport Statistics for Europe and North America* UNECE (1027–3093). Useful for comparative statistics on transport provision and development.

Monographs

Armstrong-Wright, A. (1993) *Public Transport in Third World Cities*, Crowthorne: TRL. A state-of-the-art review of public transport in the developing world.

Ashford, N. and P.H. Wright (1992) *Airport Engineering* (3rd edn), New York: Wiley. Probably the most comprehensive text on all aspects of airport engineering

Baier, R. (ed.) (2000) *Strassen und Plaetze Neu Gestaltet*, Bonn: Kirschbaum. Case studies of best practice in the design of streets for all users. The pictures tell the story.

Blow, G.J. (1996) *Airport Terminals* (2nd edn),Oxford: Butterworth-Heinemann. A worldwide review of developments in airport terminal design.

Bregman, J.I. (1999) *Environmental Impact Statements*, Boca Raton, FL: Lewis Publishers. Useful for understanding the development of environmental impact statements in the U.S.A.

Chrest, A.P. (ed.) (2001) *Parking Structures* (3rd edn), Kluwer. Written with reference to U.S. codes, this discusses all aspects of parking structures including structural design, signage, lighting, accessibility, maintenance, and repair.

Countryside Commission and ICE (2002) *Rural Routes and Networks: Creating and Preserving Routes that are Sustainable, Convenient, Tranquil, Attractive and Safe*, London: Thomas Telford. This report addresses concerns that increases in vehicular traffic are changing the character of rural roads, and how they can be made sustainable for all road users without destroying their essential character.

Croney, D. and P. Croney (1997) *Design and Performance of Road Pavements* (3rd edn), New York: McGraw-Hill. Packed full of informed advice on pavement design based originally on work at the UK TRL.

DETR (2001) *The Value of Urban Design*, London: Thomas Telford. Part of a series of reports aimed at raising the quality of urban design in the U.K.

DETR (2000) *Environmental Impact Assessment: A Guide to Procedures*, London: DETR. Promotes guidance on EIA in the UK; official publications may enlarge on this in specific areas.

Department of the Environment, Transport and the Regions (DETR) (2001) *By Design: Better Places to Live*, London: Thomas Telford.

Edwards, J.T. (ed.) (1990) *Civil Engineering for Underground Rail Transport*, London: Butterworths. A comprehensive review of the civil engineering aspects of metro construction.

Horonjeff, R. and F.A. McKelvey (1994) *Planning and Design of Airports* (4th edn), New York: McGraw-Hill. Very good on the civil engineering aspects of airport design.

ICE (1996) *Sustainability and Acceptability in Infrastructure Development*, London: Thomas Telford. Provides guidance on the main issues in the sustainability debate regarding infrastructure.

Kendrick, P. (ed.) (2004) *Roadwork, Theory and Practice* (5th edn), Oxford: Elsevier Butterworth-Heinemann. A very practical approach to road work, aimed at technicians rather than concerned with the intricacies of highway design and planning.

Laidler, D.W. (ed.) (2002) *Brownfields – Managing the Development of Previously Developed Land*, London: CIRIA (C578). Very much aimed at the management of the development of brownfield sites, rather than the detailed technical measures necessary to deal with issues such as contamination.

McClintock, H. (2002) *Planning for Cycling*, Boca Raton, FL: CRC Press. A multi-authored work, focused on UK experience, but with contributions on European, U.S. and Australian policies and experience. Very much state of the art.

McCluskey, J. (1987) *Parking: A Handbook of Environmental Design*, London: Spon. Comprehensive advice on all types of parking including multi-storey structures from an aesthetic/environmental viewpoint. Describes best practice by function and location.

Motorway Archive Trust (2002–) *The Motorway Achievement* (3 vols), London: Thomas Telford. A history of the building of the U.K. motorway system, written largely by the engineers responsible, with general interest for all involved in developing a road-based infrastructure.

O'Flaherty, C.A. (2002) *Highways* (4th edn), Butterworth-Heinemann. Focused on the engineering design of pavements, their substructure, and surface treatment. Based on UK practice but with many references to practice elsewhere.

Oppenheim, N. (1995) *Urban Travel Demand Modeling*, New York: Wiley. Discusses modeling demand under congested and uncongested conditions, transport system design, and an appendix on the mathematical background.

Ortuzar, J.D. and L.G. Willumsen (2001) *Modelling Transport*, Chichester: Wiley. Based on UK–Chilean cooperation, this is concerned with explaining the mathematical modeling techniques available, with relatively little discussion of the transport issues.

Salter, R.J. (1983) *Highway Traffic Analysis and Design* (2nd edn), London: Macmillan. An introduction to traffic engineering aimed at students.

Salter, R.J. (1988) *Highway Design and Construction* (2nd edn), London: Macmillan. An introduction to highway design aimed at students.

Sherwood, B. (ed.) (2002) *Wildlife and Roads*, London: ICP. Focuses on the ways in which roads can be integrated with nature rather than bulldozing it.

Sussman, J. (2000) *Introduction to Transport Systems*, Boston: Artech. This is intended as an introduction to the subject and discusses all aspects of transportation systems including labor, customers, networks, models, and modes. Recent developments such as Intelligent Transport Systems are covered as well as trucking, railroads, and urban transport systems.

Tyler, N. (ed.) (2002) *Accessibility and the Bus System: From Concepts to Practice*, London: Thomas Telford. Written by international specialists who form the Accessibility Research Group, this deals holistically with all aspects of bus systems and is required reading for anybody responsible for bus transport.

Wells, A.T. (2000) *Airport Planning and Management* (4th edn), New York: McGraw-Hill. Much more on the planning and management of airports than their engineering, but of value for municipal engineers with those responsibilities.

Whitelegg, J. and G. Haq (2003) *World Transport Policy and Practice*, London: Earthscan. Arranged by continent specialists, the authors describe the state of the art in urban, rural, and regional contexts around the world.

Whittles, M.J. (2003) *Urban Road Pricing: Public and Political Acceptability*, Aldershot: Ashgate. Based on a Ph.D. thesis, this discusses data on public attitudes to various road-pricing options and how they might be addressed.

Websites

Contaminated Land <http://www.contaminatedland.co.uk/> (accessed May 18, 2005). Excellent website for anybody in the UK interested professionally or otherwise in contaminated land issues.

Environment Agency (EA) <http://www.environment-agency.gov.uk/subjects/landquality/> (accessed May 18, 2005). The UK EA website provides important guidance on contaminated and derelict land, with many CLR reports available to download.

Hazardous Waste Clean-up Information (CLU-IN) <http://www.clu-in.org> (accessed May 18, 2005). This site, developed by the U.S. EPA, provides information about innovative hazardous waste treatment technology.

Highway Statistics <http://www.bts.gov/ntda/fhwa/prod.html> (accessed May 18, 2005). U.S. highway statistics.

Infoshare A free newsletter from the ICE on municipal issues; contact helen.hildon@ice.org.uk.

Light Rail Now <http://www.lightrailnow.org/> (accessed May 18, 2005). Grassroots group of volunteers (Austin, TX) supporting light rail transit. Offers transit links, light rail links, and contact details of volunteers presenting educational programs.

Motorway Archive Trust <http://www.ukmotorwayarchive.org/> (accessed May 18, 2005). An online encyclopedia of UK motorway development.

US Department of Transportation (DOT) National Transportation Library (NTL) <http://ntl.bts.gov> (see Chapter 20 on Transportation engineering).

US Environmental Protection Agency (EPA) *The Municipal Solid Waste Factboo* <http://www.epa.gov/epaoswer/non-hw/muncpl/factbook> (accessed May 18, 2005). An electronic reference manual containing information about U.S. household waste management practices including the complete text of EPA's regulations for municipal solid waste landfills.

Institutions

Advanced Transit Association (ATRA) <http://advancedtransit.org/> (accessed May 18, 2005). ATRA promotes investigation and development of advanced transit technologies and strategies. Site includes a large library, links, list of publications and conferences, and advanced technology descriptions.

American Association of State Highway and Transportation Officials (AASHTO) <http://transportation.org/aashto/home.nsf/frontpage> (accessed May 18, 2005) (see Chapter 20 on Transportation engineering).

American Public Works Association (APWA) <http://www.apwa.net/> (accessed May 18, 2005) (see Chapter 20 on Transportation engineering).

American Public Transit Association (APTA) <http://www.apta.com> (accessed May 18, 2005) (see Chapter 20 on Transportation engineering).

Asphalt Institute (US) <www.asphaltinstitute.org> (accessed May 18, 2005). The Asphalt Institute produces a number of manuals aimed at the practitioner, together with useful FAQS on their website. They played a major role in the establishment of the International Society for Asphalt Pavements which publishes a major series of conferences.

Association of Metropolitan Water Agencies (AMWA) <http://www.amwa.org> (accessed May 18, 2005). AMWA represents the interests of large, publicly owned drinking-water systems. The website provides current news, interactive bulletin boards, summaries of legislation and regulations, and links to member websites, Congress, EPA, and others.

Association of Pedestrian and Bicycle Professionals (APBP) <http://www.apbp.org> (accessed May 18, 2005). APBP promotes excellence in the field of pedestrian and bicycle transportation. The site contains membership info, links, newsletter, job listings, and an events calendar.

County Surveyors Society (CSS) <http://www.cssnet.org.uk/> (accessed May 18, 2005). The CSS acts as a forum and pressure group for county surveyors in the UK. It publishes useful guides on subjects of interest to members.

Chartered Institution of Wastes Management (CIWM) <http://www.ciwm.co.uk/> (accessed May 18, 2005). The UK society concerned with the fast-growing field of waste management and recycling.

CROW (Information and Technology Centre for Transport and Infrastructure) <http://www.crow.nl/engels/> (accessed May 18, 2005). A Dutch organization whose research is in the area of traffic, transport and infrastructure; standardization in this sector; transfer of knowledge and knowledge management. The Center's manuals and guidelines are widely respected across Europe.

Institute of Transportation Engineers (ITE) <http://www.ite.org> (accessed May 18, 2005). The Institute is one of five organizations designated by the U.S. Department of Transportation to develop standards. The site provides an online directory of transportation product and service providers. Other publications include the Trip Generation and the Traffic Signing Handbook.

Institution of Highways and Transportation (IHT) <http://www.iht.org/> (accessed May 18, 2005). IHT has been broadening its brief and appeal outside the road engineering community which first established the Institution. It publishes a journal, organizes conferences and produces important guidelines.

International Civil Aviation Organisation (ICAO) <http://www.icao.int/> (accessed May 18, 2005). Produces international standards and codes of practice for airport design.

International Society for Concrete Pavements <http://www.concretepavements.org> (accessed May 18, 2005. Organizes meetings and activities on rigid/concrete pavements.

National Association of County Engineers <http://www.countyengineers.org/> (accessed May 18, 2005). Across the U.S., county engineers have responsibility for enormous infrastructure assets, notably county bridges and roads. The former often have great historic value. Generally, states will also have their own statewide association. The website has an alerts service and training and career information.

The National Joint Utilities Group (NJUG) <http://www.njug.org.uk/> (accessed May 18, 2005). This is the UK trade grouping for the utilities in street works.

Portland Cement Association <http://www.cement.org/> (accessed May 18, 2005). Produces guidance and so on on road and airport pavements.

Transportation Research Board (TRB) <http://www.nas.edu/trb> (accessed May 18, 2005). The Transportation Research Board is part of the National Research Council. TRB publishes the TRIS database. This site provides full-text publications and free access to the database.

Transport Research Laboratory (TRL) <http://www.trl.co.uk/1024/mainpage.asp> (accessed May 18, 2005). Although largely concerned with highway/road-related research, this is the leading research body in the UK, a major publisher of research, and with comprehensive library holdings.

US Department of Transportation (DOT). Federal Highway Administration (FHWA) <http://www.fhwa.dot.gov/> (accessed May 18, 2005). The FHWA sponsors research and issue regulations. The website links to a series of subsidiary websites on issues such as bridge technology, with a whole range of downloadable pdf documents, and an electronic library with access to TRIS and regulations.

United States Federal Aviation Administration <http://www.faa.gov/> (accessed May 18, 2005). Responsible for an enormous volume of research and documentation in the sector.

World Road Association, AIPCR/PIARC <http://www.piarc.org/en/> (accessed May 18, 2005). PIARC deals with road infrastructure planning, design, construction, maintenance, and operation. Founded in 1909, currently has ninety-seven national or federal government members, 2,000 collective or individual members in 129 countries, and over 750 experts in twenty standing technical committees. In addition to organizing conferences, PIARC produces guidance on subjects such as road tunnels via its technical committees.

Structural engineering: Introduction

Structural engineering as an identifiable subdivision of civil engineering developed over the second half of the nineteenth century as increasingly sophisticated analytical tools were developed to analyze arched and more particularly framed structures of iron, and latterly steel and reinforced concrete. In the twenty years before World War I, the introduction of structural steel framing and reinforced concrete encouraged widespread use of these materials in tall buildings, and structures which had, in previous times, been largely the province of the architect, such as civic buildings, theaters, offices, and shops. Architects wishing to take full advantage of the economy and design potential of these new materials needed an understanding of the analytical tools that underpinned their safe use. A demand for specialist structural engineers grew which extended beyond the engineers who had specialized in the design of iron bridges and railway structures of the previous century. This was echoed in Britain by the creation of the Institution of Structural Engineers, and internationally by the First Congress on Bridge and Structural Engineering in 1926 (see IABSE below) and the establishment of the International Association of Bridge and Structural Engineering (see below).

The best introduction to the historical development of the subject is that by Timoshenko (1951), although Charlton is helpful for the nineteenth century. For a more academic approach one may refer to Benvenuto. Addis is very useful as an introduction to the philosophy of design, while the writings of David Billington convey an enthusiasm for the subject that is contagious.

Structural engineering is the discipline within civil engineering that seizes the imagination of the public and inspires youngsters to join the profession. It emerged as a discipline in the early twentieth century with the development of reinforced concrete and steel-framed structures, most spectacularly with the skyscrapers of Chicago and New York. While much of the work of the structural engineer is involved with more mundane questions relating to building regulations approval for domestic housing extensions, and so on, it is the interface of structural engineers with modern architecture that understandably attracts media attention. Since the nineteenth century, increasingly sophisticated methods of analysis have been developed to design structures, dominated today by computer programmers. Codes have changed to reflect these developments, and an increasing number of British structures are now based on Eurocodes <http://www.eurocodes.co.uk/.> (accessed May 18, 2005).

The design of modern building structures is closely allied with the provision of building services, and the concept of "intelligent buildings." These issues are dealt with in Chapter 5 on Architectural engineering.

Structural engineers' use of materials is overwhelmingly concerned with the use of concrete and mild steel. With the properties of these materials specified in standards, the need for civil engineers to concern themselves with materials science on a daily basis might be considered limited, but in fact higher performance materials are frequently sought for specialist applications, and to reduce the mass or cost of structures. Moreover, "new" materials such as plastics have gained an increasingly important role. More traditional materials such as timber and glass have enjoyed something of a revival, with jointing and joining methods facilitating timber construction, while a better understanding among engineers of the structural properties has facilitated the adventurous use of glass by architects. Materials such as cast iron are generally associated with older structures like mill buildings, but continue to have specialist uses, while a newer metal such as aluminum, which enjoyed a minor vogue in the 1950s, continues to be important where lightness and strength are required.

Engineering structures have a long design life – decades if not centuries –

and so the durability of materials, and their decay and corrosion mechanisms are ultimately as important as their strength. While standards and specifications for most materials are produced by the leading national standard organizations, when new developments take place engineers have to turn to industry standards and manufacturers' specifications for guidance until the "official" standards have caught up. The problem in seeking such information is that in many fields such specialist organization proliferate – more than sixty are involved in concrete in the UK alone. The Internet and the concept of the "one-stop-shop" can help, and what follows is intended to provide guidance on the most important of such organizations and sources.

Bibliography

Addis, W.A. (1990) *Structural Engineering: The Nature of Theory and Design*, Chichester: Ellis Horwood Ltd.

Benvenuto, E. (1991) *An Introduction to the History of Structural Mechanics* (2 vols), New York: Springer.

Billington, D.P. (1983) *The Tower and the Bridge: The New Art of Structural Engineering*, New York: Basic Books.

Billington, D.P. (1990) *Robert Maillart and the Art of Reinforced Concrete*, New York: Architectural History Foundation.

Billington, D.P. (1996) *The Innovators: The Engineering Pioneers Who Made America Modern*, New York: John Wiley.

Charlton, T.M. (1982) *A History of Theory of Structures in the Nineteenth Century*, Cambridge: Cambridge University Press.

Timoshenko, S.P. (1953) *History of Strength of Materials*, New York: McGraw-Hill.

Searching the library catalog

Structural engineering is generally one specialization which fits neatly into a Library of Congress subclass (i.e. TA 630–695); while there needs to be an awareness of other areas, for example, chemical technology namely TPO 751–762 clay industries, ceramics, glass, and TP875–888, and TP 1080–1185, this will be rare. There is a strong overlap in structural analysis with the classes TA329–348 engineering mathematics and analysis, and TA349–TA 359 applied mechanics, reflected in the works of world famous civil engineers such as S.P. Timoshenko. In terms of keyword searching one must bear in mind that an engineer may be seeking information on shell structures, which the architects and general public might describe as domes.

Abstracts and indexes

Building (Construction) References 1946–1990 (MPBW/PSA). A useful source for references to older buildings.

Building Science Abstracts 1926–1976. In part continued in ICONDA, this abstract series created by the UK Building Research Station is very useful for tracing articles on building science (e.g., properties of materials).

International Construction Database (ICONDA) (1974–) (Fraunhofer-Gesellschaft Informationzentrum Raum und Bau (IRB). Excellent on European coverage but poor on British coverage since 1990 due to lack of input. Available on CD-ROM from SilverPlatter.

Instruct (1923–). This database is accessible via the website of the Institution of Structural Engineers and covers a wide range of books, periodicals, and related publications (e.g., *BRE Digests*) held in the library. This includes all articles in *The Structural Engineer* (the Institution journal) from 1923 to date. These are digitized.

Databases

Barbour Index <http://www.barbour-index.co.uk/content/aboutservices/cfm.asp> (accessed May 18, 2005). *Barbour Index* has a range of information services for built environment professionals in the U.K. It is more relevant to the building and architectural professions than TI (below), which provides more information of relevance to engineers. The Index, like TI, has developed from a microfiche-based product library. Barbour produces a number of services: Building Product Expert, a building product database containing detailed technical and product information from manufacturers of building products; Specification Expert [Incorporating the National Engineering Specification], a specification writing tool for building professionals and building services engineers; BSRIA Detail Drawings, a library of over 850 drawings and symbols illustrating common building services design and installation details. Construction Expert provides access to technical and product information on CD or via the Web with full-text documents from over 200 technical publishers, a directory of UK manufacturers, 25,000 trade names, and 50,000 product ranges with performance characteristics.

Construction Information Service (TI) Technical Indexes Limited mktg@techindex.co.uk. Developing from a microfilm-based library and product information service TI is now arguably the most important UK provider of civil engineering documentation. Rapidoc is the document supply division. It acts as authorized distributor for the British Standards Institution, ETSU, and TSO (some of these do not have other names any more, e.g., TSO was the Stationery Office), and many other publishers. The RIBA.ti Construction Information Service is produced jointly with RIBA Enterprises Ltd. Its target audience includes civil and structural engineers. It is in use at networked locations as well as on stand-alone pcs. The service may be delivered online or on CD-ROM; subscription includes regular updates, technical support, and access to Technical Indexes helplines. Although it is aimed at UK civil engineers, there are certain important categories of documentation that are not available, such as ICE publications. On the other hand, TI agreements with Network Rail and the Highways Agency mean that a lot of key data are available from a single source.

Bibliographies

BRE Bibliography of Structural Failures 1850–1970, Garston, England: Building Research Establishment. While failures preceded 1850 and have continued to occur, this is a useful starting point for precedents.

ICE Bibliography on Prestressed Concrete 1920–1957. A useful introduction to the early literature on the subject although lacking particularly in early German work.

Jakkula, A.A. (1941) *A History of Suspension Bridges in Bibliographic Form 1941*, Washington, DC: Public Roads Administration. Despite its age this is a good starting point for anybody interested in the history of suspension bridges. It is weakest on European developments where the only literature is in non-English-language sources.

Handbooks and manuals

Bickford, J.H. and S. Nassas (1998) *The Handbook of Bolts and Bolted Joints*, New York: Marcel Dekker. The design of bolted connections requires specialist knowledge that is frequently lacking in more general handbooks.

Chen, W-F. (2002) *Structural Engineering Handbook* (2nd edn), Boca Raton, FL: CRC Press, 2002. The early chapters deal with basic concepts and aspects of structural analysis such as theory of plates and stability; structural analysis; seismic engineering; structural steel design and composite construction. More specialist topics include aluminum and timber structures; bridges; shell and space framed structures; multi-storey frames, cooling towers transmissions. Elements are dealt with such as stub-girder floor systems. Other aspects covered include shock loading underground pipes and structural reliability. There is a chapter on passive energy dissipation and active control. Available at<http://www.engnetbase.com/ejournals/books/book_summary/toc.asp?id=417>.

Chen, W-F. and L. Duan (1999) *Bridge Engineering Handbook*, Boca Raton, FL: CRC Press. In seven main sections. The first deals with fundamental issues such as bridge aesthetics, the second with types of bridge superstructure, including less familiar types such as floating and movable bridges, the third with foundations, the fourth seismic design, including retrofit, and the fifth with construction and repair, the sixth with specialist aspects such as ship collision. The final section deals with international design, including Russia and China. Given the overall coverage it is surprising how little there is on footbridges.

Doran, D.K. (1992) *Construction Materials Reference Book*, Butterworth-Heinemann. Written by leading authorities in the field and covering most engineering materials, including some more generally associated with historic structures such as cast and wrought iron.

Gaylord, E.H., C.N. Gaylord and J.E. Stallmeyer (1997) *Structural Engineering Handbook*, New York: McGraw-Hill. This is a comprehensive and well-established title that, with few exceptions, is as good a basic reference tool as one could desire. Some of the references to standards are out of date. Covers structural analysis, computer application, and earthquake-resistant design, geotechnical issues, structural elements by types and material, and specialist structures including suspended roofs, shells, industrial buildings, silos, steel tanks, and chimneys.

Levy, S.M. (2001) *Construction Building Envelope and Interior Finishes Databook*, New York: McGraw-Hill. This databook is intended to provide access to hundreds of tables, specifications, charts, diagrams, and illustrations covering materials and components most frequently used on a typical job.

Ryall, M.J. (ed.) (2000) *Manual of Bridge Engineering*, London: Thomas Telford. Covers all aspects of bridge design by material and form, including joints and foundations. Written largely by UK authors.

Tilly, G. (2002) *Conservation of Bridges*, London: Spon. Intended to encourage best practice in bridge conservation across the U.K.

Monographs

General works

Adler, D. (ed.) (1999) *Metric Handbook*, Oxford: Architectural Press. Very useful for dimensions and guidelines on all types of structures.

ASCE (1989) *Manual 52: Guide for Design of Steel Transmission Lines*; ASCE (1990) *Manual 72: Guide for Design of Steel Transmission Pile Structures*, Reston, VA: ASCE; ASCE (1991) *Manual 74: Guidelines for Electrical Transmission Line Structural Loading*. These three titles are the best source of published advice on the design of transmission line structures.

ASCE (1999) *Structural Design for Physical Security*, Reston, VA: ASCE. Security is now an international issue and this is a helpful reference point for all engineers.

Bangash, M.Y.H. (1993) *Impact and Explosion*, Oxford: Blackwell (new edn, Springer, October 2004). Contains many numerical examples, including modeling of aircraft impact on tall buildings, and nuclear and conventional weapons effects.

Berlow, L.H. (1998) *The Reference Guide to Famous Engineering Landmarks of the World*, Chicago, IL: Fitzroy Dearborn. Not nearly as comprehensive as the *Guinness Book* by Stephens (noted below), but useful for more modern structures.

Brown, G.J. and J. Neilsen (1990) *Silos*, London: Spon. Silo design and construction from a U.K. perspective.

CIRIA (2004) *Principles of Design for Deconstruction to Facilitate Reuse and Recycling*, London: CIRIA. Discusses all elements of a building in terms of life cycle and recycling, including fittings as well as structure, with a view to facilitating deconstruction and reuse through design (C607).

Cook, N.J. (1986) *Designer's Guide to Wind Loading of Building Structures* (2 parts), London; Boston, MA: Butterworths. Although not concerned with the latest codes, the basic principles of wind engineering are well covered here.

Cook, N.J. (1999) *Wind Loading: A Practical Guide to BS 6399, Part 2*, London: Thomas Telford. The title is somewhat misleading as it also covers BS 8100 (lattice structures). Not as comprehensive as Cook (1986).

Curwell, S. (ed.) (2002) *Hazardous Building Materials: A Guide to the Selection of Environmentally Responsible Alternatives* (2nd edn), London: Spon. The most helpful feature is the use of data sheets which form the bulk of this publication and provide a check-list both for a professional and home-owner chart.

Davis, J. and R. Lambert (2002) *Engineering in Emergencies* (2nd edn), London: ITDG. A useful guide to the appropriate technology for anybody dealing with emergencies.

Fernandez Troyano, L. (2003) *Bridge Engineering: A Global Perspective*, Madrid and London: TTL and CICCP. A magnificently presented compendium by one of the world's foremost bridge designers, which would inspire anybody to take up engineering.

Hambly, E.C. (1991) *Bridge Deck Behaviour* (2nd edn), London: E & FN Spon. A simple approach to a fundamental subject for bridge engineers.

Holmes, J.D. (2001) *Wind Loading of Structures*, London: Spon. Good general textbook with extensive references on all aspects of wind-resistant design.

Illston, J.M. and P.L.J. Podmore (eds) (2001) *Construction Materials: Their Nature and Behaviour* (3rd edn), London: Spon. A useful introduction covering fundamentals, metals, including aluminum and copper, concrete, bitumen, brickwork, polymers and fiber composites, and cements and timber. Aimed primarily at students.

IStructE (1996) *Appraisal of Existing Structures* (2nd edn), London: IStructE. Almost a check-list approach to appraisal and as such generally applicable for all involved in appraisal.

Jackson, N. and R.K. Dhir (eds) (1996) *Civil Engineering Materials* (5th edn), Basingstoke: Macmillan. Good basic introduction to materials used regularly by civil engineers. Covers metals, concrete, timber, bitumen, soils, polymers, and bricks and blockwork. Aimed at a student audience.

Jones, N. (1989) *Structural Impact*, Cambridge: Cambridge University Press. Investigation of the crashworthiness of structures has grown over recent years. This text provides a good starting point.

Mainstone, R. (1998) *Developments in Structural Form* (2nd edn), Oxford: Architectural Press. This is perhaps the best introduction to the art of structural engineering through all ages, and very useful for those unfamiliar with various structural types.

Mays, G.C. and Smith, P.D. (1995) *Blast Effects on Buildings*, London: Thomas Telford. A useful starting point for those new to blast-resistant design.

Petroski, H. (1994) *Design Paradigms: Case Histories of Error and Judgement in Engineering*, Cambridge: Cambridge University Press. This study of failures is as good a place as any to begin understanding the secret of good structural engineering design.

Plate, E.J. (1982) *Engineering Meteorology*, Amsterdam: Elsevier. No structure can be designed without an understanding of weather, and this book provides a useful background on this aspect of engineering.

Podolny, W. and J.B. Scalzi (1986) *Construction and Design of Cable Stayed Bridges* (2nd edn), New York: Wiley. There are a growing number of texts on cable-stayed bridges but this is the best for its historical insight, with lots of examples.

Ravenet, J. (1987) *Silos* (3 vols), Barcelona: EJA. The most comprehensive book on the subject, albeit in Spanish.

Reimbert, M.L. and A.M. Reimbert (1987) *Silos* (2nd edn), Paris: Lavoiser. The Reimberts have developed one of the most commonly used methods of silo design.

Ryall, M.J. (2001) *Bridge Management*, Oxford: Butterworth-Heinemann. A useful introduction to all aspects of bridge maintenance and assessment from a U.K. perspective.

Safarian, S.S. and E.C. Harris (1985) *Design and Construction of Silos and Bunkers*, New York: Van Nostrand Reinhold. U.S.-based silo design.

Stafford-Smith, B. and A. Coull (1991) *Tall Building Structures*, New York: Wiley. These authors wrote a classic on shear wall design in the 1960s and have carried their work through into a very readable text for graduate engineers involved in design.

Stephens, J.H. (1976) *Guinness Book of Structures*, Enfield: Guinness Superlatives. Despite its age, its comprehensive coverage, encompassing dams and tunnels as well as building structures, makes this a valuable source of reference data on important structures.

Taranath, B.S. (1998) *Steel, Concrete and Composite Design of Tall Buildings* (2nd edn), New York: McGraw-Hill. General introduction to tall building design covering the various loadings that must be embraced.

Young, W.C. and R.C. Budynas (2002) *Roark's Formulas for Stress and Strain* (7th edn), New York: McGraw-Hill. The main reference work for structural formulas.

Zaknic, I. (ed.) (1998) *100 of the World's Tallest Buildings*, Corte Madera, CA:

Gingko. Given that such a work is continually in need of updating, this is a useful compendium on some of the best-known skyscrapers.

Concrete

Bangash, M.Y.H. (2001) *Manual of Numerical Methods in Concrete*, London: Thomas Telford. Part of an introduction to numerical methods, an important feature is the case studies and worked examples including U.S., British, and European codes. Some understanding of numerical methods is required to make best use of this.

Dobrowolski, J A. (1998) *Concrete Construction Handbook*, New York: McGraw-Hill. A long established handbook on U.S. practice.

Nawy, E G. (1998) *Concrete Construction Engineering Handbook*, Boca Raton, FL: CRC. Deals with recent advances in materials including high-strength concrete, and fiber reinforcement as well as the main fields of reinforced and pre-stressed concrete. There are chapters on seismic and fire-resistant design, and specialist areas such as offshore structures and pre-stressed concrete bridges.

Neville, A.M. (1995) *Properties of Concrete* (4th edn), London: Prentice Hall. Neville's work is essentially concerned with concrete practice and provides the best introduction to the properties of the material.

Reynolds, C.E. and J.C. Steedman (1998) *Reinforced Concrete Designers Handbook* (10th edn), London: E & FN Spon. Despite criticism of its age, this handbook is a useful compendium of information on subjects not readily found elsewhere (e.g., weights of various materials).

Sutherland, R.J.M. (ed.) (2001) *Historic Concrete: Background to Appraisal*, London: Thomas Telford. Useful background for appraisal, with a U.K. bias, this work comprises a series of chapters by leading construction historians and engineers covering the use of concrete in all types of structures.

Iron and steel

AISC (1998) *Load and Resistance Factor Design (LRFD): Manual of Steel Construction, Metric* (2nd edn) (2 vols), Chicago, IL: American Institute of Steel Construction, 1998. The main U.S. design manual for structural steel, widely used internationally.

Angus, H.T. (1976) *Cast Iron: Physical and Engineering Properties* (2nd edn), London: Butterworth. The best guide to the properties of modern cast iron.

Bangash, M.Y.H. (2000) *Structural Detailing in Steel*, London: Thomas Telford. Deals with joints, welding, components, with examples of buildings and bridges from U.K., U.S., and European practice.

Brockenburgh, R.L. and Merritt, F.S. (1994) *Structural Steel Designer's Handbook* (3rd edn), New York: McGraw-Hill. Multi-authored guide, useful for American practice, although all traditional units, this handbook deals with properties, structural theory and analysis and design of all types of buildings and bridges, including truss bridges that are often omitted from bridge designer books.

Galambos, T.V. (1998) *Guide to Stability Design Criteria for Metal Structures*, New York: Wiley. The best monograph on this specialist aspect of structural steel design.

Hayward, A. and Oakhill, A. (2002) *Steel Detailer's Manual* (2nd edn), Oxford: Blackwell. Useful for detailing UK standards with examples of buildings and bridges.

Hoffman, E.S. (ed.) (1996) *Structural Design Guide to the AISC (LRFD) Specification for Buildings*, New York: Chapman and Hall. Useful guide to the AISC specification.

SCI (1997) *Appraisal of Existing Iron and Steel Structures*, Berkshire: Steel Construction Institute. This work covers all aspects of structural appraisal, although the reference points are generally to British practice.

SCI (2003) *Steel Designers Manual* (6th edn), Oxford: Blackwell Scientific. This work encompasses most aspects of structural steel design in the U.K. context, with a chapter on Eurocodes. Bridges and pile design are covered.

Timber

Baird, J.A. and Ozelton, E.C. (2002) *Timber Designers Manual* (3rd edn), Oxford: Blackwell. This provides the manual to British Standards practice in the U.K. while incorporating references to Eurocode 5. Some chapters reflect developments in North America which have now influenced U.K. practice.

Breyer, D., Fridley, K.J., and Cobeen, K.E. (2003) *Design of Wood Structures* (ASD) (5th edn), New York: McGraw-Hill. Widely used in teaching timber engineering at undergraduate level in the U.S., the 5th edition has references to the latest International Building Code and 2001 National Design Specification for Wood Construction. There is extensive treatment of joints and fixings, laminated, and sandwich construction.

Centrum Holst (1995) *Timber Engineering: STEP 1 and STEP 2*, Netherlands: Centrum Holst. This is an excellent introduction to timber design with Eurocodes in design.

Faherty, K.F. and Williamson, T.G. (2003) *Wood Engineering and Construction Handbook* (3rd edn), New York: McGraw-Hill. Multi-authored work which deals with materials and connections, and chapters on all kinds of structures, including trusses, columns, domes and diaphragms, and foundations.

McKenzie, W.M.C. (2000) *Design of Structural Timber*, Basingstoke: Macmillan. Aimed at undergraduates, this serves as an introduction to designs by British Standards and Eurocodes.

Sunley, J. and Bedding, B. (eds) (1985) *Timber in Construction*, London, Batsford: TRADA. American Institute of Timber Construction (1994) *Timber Construction Manual* (4th edn), Comprehensive manual for American practice.

Yeomans, D. (1992) *The Trussed Roof*, Aldershot, Hants: Ashgate. For anybody interested in the subject this is a readable introduction and more.

Masonry

The term "masonry," particularly in the U.S., is used to describe mass concrete (without reinforcement), as well as natural stone and brick and other ceramics; for concrete generally see above.

Amrhein, J.E. (1998) *Reinforced Masonry Engineering Handbook Clay and Concrete Masonry* (5th edn), Los Angeles, CA: Masonry Institute of America. The most comprehensive source available on the subject.

Brick Development Association (1994) *Brick Development Association Guide to Successful Brickwork*, London: Edward Arnold. This guide provides practical advice on brick construction.

Curtin, W. (ed.) (1995) *Structural Masonry Designers Manual* (2nd edn revised), Oxford: Blackwell Scientific. Comprehensive approach, covering relevant British standards.

Sowden, A.M. (ed.) (1990) *The Maintenance of Brick and Stone Masonry Structures*, New York: E & FN Spon. A series of papers by specialists on all types of masonry structures including brick lined tunnels.

Other materials

The Aluminum Association (2000) *Aluminum Design Manual*, Washington, DC: The Association. The basic source for engineers in the English-speaking world.

ASCE (1985) *Structural Plastics Selection Manual*, New York: ASCE. ASCE (1984) *Structural Plastics Design Manual*, New York: ASCE. These two ASCE manuals remain the best source for the structural use of plastics, although there have been many advances in applications since they were written (e.g., with the use of carbon fiber bonded plates for strengthening bridges) but such applications can be traced through conference and journal papers.

Amstock, J.S. (1997). *Handbook of Glass in Construction*, New York: McGraw-Hill. The first comprehensive guide for American audiences.

Amstock, J.S. (2000) *Handbook of Adhesives and Sealants in Construction*, New York: McGraw-Hill. Chapters on the various sealant materials are followed by discussions of specifications, joint design, and performance and repair.

CIRIA (1987) *TN128: Civil Engineering Sealants in Wet Conditions – Review of Performance and Interim Guidance on Use*, London: CIRIA. A little dated, but the guidance remains helpful.

Dutton, H. and Rice, P. (1999) *Structural Glass* (2nd edn), London: Spon. Peter Rice was probably the most important figure in popularizing the use of structural glass in Britain and France. This book originally appeared in French.

Dwight, J. (1999) *Aluminium Design and Construction*, London: E & FN Spon. Useful for those designing with British codes; as Eurocode 9 is introduced more texts are likely to appear.

Eekhout, M. (1989) *Architecture in Space Structures*, Rotterdam: Uitgeverij OIO Publishers. Eekhout, alongside Rice, is the great pioneer of structural glass.

Institution of Structural Engineers (1999) *Guide to the Structural Use of Adhesives*, London: IStructE. Useful for those with little knowledge intending to use adhesives structurally for the first time.

Institution of Structural Engineers (1999) *Structural Use of Glass in Buildings*, London: IStructE. For any engineer short of knowledge or experience of the structural use of glass, this provides excellent guidelines.

Mays, G.C. and Hutchinson, A.R. (1992) *Adhesives in Civil Engineering*, Cambridge: University Press. Reasonably comprehensive source of applications in civil engineering.

Schittich, C. (1999) *Glass Construction Manual*, Wiesbaden: Birkhauser. A useful introduction to European practice.

Journals

In Britain and the U.S. the institutions produce the leading journals. The ICE now produces *Structures and Buildings*, while The Institution of

Structural Engineers has published *The Structural Engineer* since 1923. The ASCE produces the *Journal of Structural Engineering*, and a number of more specialist titles listed below.

General titles

Annales de l'Institut Technique du Bâtiment et Travaux Publics (Institute Technique du Bâtiment et Travaux Publics) (France). Mostly research papers on structures and the performance of materials.

Bauingenieur (Association of German Engineers – VDI) (Springer Verlag) (0005–6650). Contains papers on analysis, design, and construction.

Bautechnik (Ernst und Sohn) (Germany) (0932–8351). Similar to Bauingenieur but complements rather than duplicates its coverage.

Bridge Design and Construction. The first international journal devoted to bridge engineering in all its glory.

Bridge Engineering: Proceedings of the Institution of Civil Engineers (1478–4629). One of the latest ICE titles, with an emphasis on projects rather than research.

Bulletin of the International Association for Shell and Spatial Structures (IASS) (0304–3622). Well-presented specialist journal with case studies.

Computers and Structures (Elsevier) (0045–7949). The leading academic journal in this area, really intended for the specialist.

Construction and Building Materials (Butterworth-Heinemann) (0950–0618). Mostly research papers into building materials.

Engineering Structures (Elsevier) (0141–0296). Concerned with designing structures for live loads.

Industria Italiana del Cemento (Italian Cement Association) (0019–7637). Its specialist title gives no clue to the inspirational quality of its drawings and photographs. A journal that would sit easily in an art bookshop.

International Journal of Solids and Structures (Elsevier) (0020–7683). An academic journal concerned with the mechanics of engineering materials.

International Journal of Space Structures (Multi-Science Publishing) (0956–0599). This is much more academic in its approach than the IASS bulletin (above)

Journal of Bridge Engineering (ASCE) (1084–0702). More research papers than construction examples.

Journal of Engineering Mechanics (ASCE) (0733–9399). One of the oldest of the ASCE specialty journals, and very academic.

Journal of Performance of Constructed Facilities (ASCE) (0887–3828). This is one ASCE title that is of obvious interest to the practitioner rather than the academic.

Journal of Structural Engineering (ASCE) (0733–9445). The leading title on structural analysis and design.

Materials and Structures (RILEM) (France) (1359–5997). The leading international journal on materials testing for civil engineers.

National Hazards Review (ASCE). One of the new ASCE journal titles, with a more practical or state-of-the-art approach than some other titles in what is becoming an important area for civil engineers.

Practice Periodical on Structural Design and Construction (ASCE) (1084–0680). This journal is aimed at the practitioner rather than the academic and research audience of the journal of structural engineering.

Structural Design of Tall Buildings (Wiley) (1541–7794). This was the first academic journal to be concerned with the design of tall buildings.

Structural Engineering International (International Association for Bridge and Structural Engineering) (1016–8664). IABSE's magazine, with international contributions.

Structural Safety (Elsevier) (0167–4730). Concerned with analyzing probability and risk associated with structures.

Concrete

ACI Materials Journal (American Concrete Institute) (1987–) (0889–325X). Bimonthly, the world's leading journal on the material properties of concrete. Abstracts are searchable <http://www.concrete.org/PUBS/JOURNALS/ABSTRACTSEARCH.ASP> (accessed May 18, 2005).

ACI Structural Journal (American Concrete Institute) (1987–) (0889–3241). Bimonthly, the world's leading journal on the analysis of concrete structures. Abstracts are searchable <http://www.aci-int.net/journals/oljsearch.asp> (accessed May 18, 2005).

Advances in Cement Research (TTL) (0951–7197). International journal on cement research; claims to be more selective than *Cement and Concrete Research.*

Beton Armé (Societe des Editions Andre Guerrin (France)). A French title on reinforced concrete.

Beton-Kalendar Ernst. Written in German this provides annual reviews of developments in concrete practice and abstracts (in some cases full text) of German and European Standards.

Beton und Stahlbetonbau (Ernst) (0005–9900). One of several German journals on concrete, this produces more on structural design and practice than do others.

Cement and Concrete Aggregates (ASTM) (0149–6123). ASTM's research journal on cement and concrete.

Cement and Concrete Research (Elsevier) (0008–8846). In terms of papers published this is the leading research journal on the subject in the world. Probably too academic to be readily beneficial for most practicing engineers.

Concrete (Concrete Society) (0010–5317). Published for the UK Concrete Society this journal contains project descriptions and brief papers on design topics; similar to the ACI's *Concrete International.*

Concrete Construction (formerly Aberdeen's concrete construction) (1533–7316). The leading U.S. trade journal, which is often the first to announce new developments in concrete construction techniques.

Concrete International (ACI) (not supplied). The ACI monthly which often has "special issues" featuring recent developments in concrete technology.

Concrete Science and Engineering (RILEM) (1295–2826). Until recently, RILEM published all its concrete papers in *Materials and Structures*; the launch of this journal reflects the volume of academic research in this area.

Indian Concrete Journal <http://www.icjonline.com>. Although generally concerned with India, a major area of civil engineering activity, this journal reports on developments elsewhere.

Magazine of Concrete Research (TTL) (0024–9831). The longest established U.K. research journal with an international circulation; all back issues are digitized.

PCI Journal (PCI) (0887–9672). The leading international source on the use of prestressed and post-tensioned concrete.

Structural Concrete (TTL for FIB) (1464–4177). A relatively new entrant, but given kudos by the FIB (International Concrete Federation for Prestressed Concrete) brand.

Steel

Construction Metallique (CTICM) (France) (not supplied). The leading French-language journal on steel structures, blending research and practice.

Costruzione Metalliche (ACS-ACAI) (Italy). Similar approach to Rivista, but more restricted in its coverage.

Engineering Journal (AISC) (0013–8029). Design and research papers; back issues are now all available on one CD-ROM.

Modern Steel Construction (AISC) (0026–8445). Monthly magazine on structural steel developments in the U.S.

New Steel Construction (Steel Construction Institute/BCSA) (0968–0098). The U.K.'s monthly magazine on steel construction.

Stahlbau (Ernst and Sohn) (Germany) (0038–9145). The leading German journal on structural steel analysis and practice.

Websites

Advanced Buildings Technologies and Practices <http://www.advancedbuildings.org./> (accessed May 18, 2005). Site guides building professionals to more than ninety examples of environmentally appropriate design for commercial, industrial, and multi-unit residential buildings.

Bridge Building – Art and Science <http://www.brantacan.co.uk/bridges.htm> (accessed May 18, 2005). Covers all aspects of bridge-building from arches to stress with lots of links and images.

Bridge Construction and Engineering <http://bridgepros.com/> (accessed May 18, 2005). Site presents construction and history of bridges. Covers past, current, and planned bridge projects worldwide. Learning Center discusses bridge types, and has models and links to lesson plans. Extensive links.

Bridge Site <http://www.bridgesite.com/> (accessed May 18, 2005). Something for everybody. Site intends to provide a way to bring people of different communities together. Lots of links and discussion forums including one used chiefly by schoolchildren seeking help with their projects.

Council on Tall Buildings and Urban Habitat <http://www.ctbuh.org/> (accessed May 18, 2005). Includes the online journal *CTBUH Review* <http://www.ctbuh.org/journal/index.htm> and subscription area.

Greatbuildingsonline <http://www.greatbuildings.com/search.html> (accessed May 18, 2005). Very much an architectural resource (one cannot search by name of engineer) with close links to *Architecture Week*, but it still provides access to brief details of some well-known structures. Structurae (below) is better from an engineering viewpoint.

High-performance Buildings Research <http://www.nrel.gov/> (accessed May 18, 2005). The National Renewable Energy's research site encompasses high-performance whole-building design to develop very low-energy, environmentally sensitive buildings.

Office of Bridge Technology <http://www.fhwa.dot.gov/bridge/index.htm> (accessed May 18, 2005). Department of Transportation, Federal Highway Administration Office of Bridge Technology contains an electronic library, training courses, NBI information.

Spreadsheet Solutions for Structural Engineers <http://www.yakpol.net/> (accessed May 18, 2005). A series of links to free software sites for structural engineers.

Spreadsheets for Structural Engineering <http://www.structural-engineering.fsnet.co.uk/> (accessed May 18, 2005). Free software for all types of civil engineering problems.

Steel Bridges in the World <http://www.sbi.se/default_en.asp> (accessed May 18, 2005). Bridge statistics compiled by the Swedish Institute of Steel Construction; the site is generally of use to anyone interested in steel construction.

Structurae <http://www.structurae.de> (accessed May 18, 2005). Multilingual website which is growing rapidly in its coverage of engineers and engineering feats, with pictures and bibliography; all the work of informed amateurs.

West Point Bridge Designer <http://bridgecontest.usma.edu/> (accessed May 18, 2005). Bridge Design Contest developed by the US Military Academy (West Point) for U.S. students in grades K-12 aimed at promoting math, science, and technology education. Open to anyone, but U.S. students (K-12) may compete for prizes. Downloadable contest information.

Associations, organizations, and societies

Aluminum Association (AAI) <http://www.aluminum.org> (accessed May 18, 2005). The Aluminum Association is the U.S. trade association. It publishes *Aluminum Standards*, which are widely used internationally, and the *Aluminum Statistical Review*.

Aluminium Federation <http://www.alfed.org.uk/> (accessed May 18, 2005). The UK equivalent body of the AAI, with more limited publications programmed.

American Ceramic Society <http://www.ceramics.org/acers/acers.asp> (accessed May 18, 2005). A 100-year-old non-profit organization that serves the informational, educational, and professional needs of the international ceramics community. The leading such organization in the world, although many of its publications will have no direct relevance to civil engineering.

American Concrete Institute (ACI) <http://www.aci-int.org> (accessed May 18, 2005). The world's leading learned society for concrete construction, publishing in English and Spanish. The *ACI Manual of Concrete Practice* (MCP) includes all the ACI standards, specifications, recommendations, and guides used for the construction, construction management, inspection, and design of concrete structures (available on CD-ROM and online).

American Institute of Steel Construction (AISC) <http://www.aisc.org> (accessed May 18, 2005). Standards activities and publications include the AISC Specifications and Codes. The Institute publishes the *Engineering Journal* and *Modern Steel Construction* (1964 to present), the former dealing with analysis and design while the latter is more concerned with practice. Useful FAQs dealing with issues such as progressive collapse. There is also information on all kinds of long span bridges. Available at http://www.aisc.org/Content/ContentGroups/Documents/NSBA5/20_NSBA_LongestSpans.PDF (accessed May 18, 2005).

American Institute of Timber Construction http://www.aitc-glulam.org/ (accessed May 18, 2005). Organization produces manuals and other key publications on U.S. practice, particularly on laminated timber.

American Iron and Steel Institute (AISI) <http://www.steel.org> (accessed May 18, 2005). AISI represents Canadian, U.S., and Mexican steel-makers. Produce standards, including a compilation of out-of-date sections.

American Welding Society (AWS) <http://www.aws.org> (accessed May 18, 2005).

The AWS site includes a buyer's guide and research supplements to the *Welding Journal*.

American Wood Council <http://www.awc.org/index.html> (accessed May 18, 2005). The American Forest and Paper Council is the engineered wood products division of AFPA. The website contains a large number of downloadable technical advice notes; the Council publishes standards and design manuals.

Association for Specialist Fire Protection (ASFP) http://www.asfp.org.uk/. The ASFP produces guidelines and organizes activities for UK specialists in structural fire protection.

Brick Development Association (BDA) <http://www.brick.org.uk/Index.html> (accessed May 18, 2005). The BDA publishes design guides, notes, and technical information papers.

British Cement Association (BCA) <http://www.bca.org.uk/> (accessed May 18, 2005). The British Cement Association, formerly the Cement and Concrete Association, publishes a large number and range of guides, construction guides, interim technical notes, bibliographies, reprints, project profiles, and videos. The database *Concquest*, an online service for publications relating to concrete design and construction, is being redeveloped with BRE and the Concrete Society as Concrete Information Ltd, a one-stop shop for UK concrete information <http://www.concreteinfo.org/default.asp> (accessed May 18, 2005).

British Constructional Steelwork Association Ltd (BCSA) <http://www.steelconstruction.org/steelconstruction/guestLogin> (accessed May 18, 2005). Publications include: Handbook of structural steelwork (with SCI), Historical structural steelwork handbook (1984), "The red book" – Handbook of structural steelwork, "The green book" – Joints in simple construction, "The black book" – National structural steelwork specification for building construction, "The orange book" – The contractual handbook, "The blue book" – Erectors' manual. Their "Steel buildings" (2003) is a useful introduction to all aspects of the design and procurement of steel buildings, useful for engineers and clients.

British Masonry Society, c/o CERAM Research (see below). This society publishes a journal and organizes regular conferences.

Building Research Establishment (BRE) <http://www.bre.co.uk/> (accessed May 18, 2005). Publishes a range of material including *BRE Digests*, *BRE Reports*, *BRE Newsletter*, *BRE Information Directory*, *Good Building Guides* (which have superseded the *Defect Action Sheets*), *Overseas Building Notes*, and *Information Papers*. Although library services are now minimal, this is the UK national source for building and geotechnical information.

Centre for Window and Cladding Technology <http://www.cwct.co.uk> (accessed May 18, 2005). A leading information provider in the field of building envelopes and glazing.

CERAM Research <http://www.ceram.co.uk/> (accessed May 18, 2005). CERAM has published research papers and special publications, and specifications such as the model specification for clay and calcium silicate structural brickwork (1988).

Civieltechnisch Centrum Uitvoering Research en Regelgeving (CUR) <http://www.cur.nl/index.asp?l=eng> (accessed May 18, 2005). A Dutch body responsible for a number of important design manuals, now available in English.

Concrete Reinforcing Steel Institute (CRSI) <http://www.crsi.org> (accessed May 18, 2005). CRSI publishes a Design Handbook and Manual of Standard Practice that relate to ACI Standards. Their site includes several full-text publications.

Concrete Society <http://www.concrete.org.uk/> (accessed May 18, 2005). Publishes many technical reports, guides, and digests. Now taking the lead with the British Cement Association in developing Concrete information, a one-stop shop for U.K. concrete information.

Corus <http://www.corusconstruction.com/index.asp> (accessed May 18, 2005). Provides a wide range of information (largely free) such as the guide to Sections: Structural Sections and corrosion protection guides.

Council on Tall Buildings and Urban Habitat <http://www.ctbuh.org/> (accessed May 18, 2005). Based at Lehigh University, the Council has organized many conferences and produced a significant body of research.

Fédération Internationale de Beton (FIB) (International Concrete Federation) <http://fib.epfl.ch> (accessed May 18, 2005). An international organization formed by the merger of Fédération Internationale de la Precontrainte and Comité Eurointernationale du Beton for the development of structural concrete; it publishes (via Thomas Telford) the journal *Structural Concrete*, many recommendations, and state-of-the-art reports, on both reinforced and pre-stressed concrete.

Institution of Structural Engineers (IStructE) <http://www.istructe.org.uk/> (accessed May 18, 2005). Founded in 1908 (as the Concrete Institute) IStructE publishes *The Structural Engineer*, technical reports, guidance, conference and symposium papers. In addition, it organizes CPD events. The Library is an important specialist resource.

International Association for Bridge and Structural Engineering (IABSE) <http://www.iabse.ethz.ch/> (accessed May 18, 2005). IABSE organizes regular international conferences on current issues in structural engineering as well as a journal and state-of-the-art reports.

International Association for Earthquake Engineering (IAEE) <http://www.iaee.org.jp/> (accessed May 18, 2005). Although based in Japan, this is an international society with national and regional activities which organizes major international conferences, supports a journal (*Earthquake Engineering and Structural Dynamics*), and produces the definitive Earthquake resistant regulations world list.

International Association for Shell and Spatial Structures <http://www.iass-structures.org/> (accessed May 18, 2005). Organizes international and regional conferences and publishes a journal.

International Council for Building Research Studies and Documentation (CIB) <http://www.cibworld.nl/> (accessed May 18, 2005). A number of conferences and commercially published journals are produced under CIB auspices, normally related to CIB working groups. CIB covers all aspects of building and construction.

Masonry Institute of America <http://www.masonryinstitute.org/> (accessed May 18, 2005). Produces manuals for engineers, standards, and guidance for builders. Online bookshop and industry links.

National Fire Protection Association (NFPA) <http://www.nfpa.org.> (see Chapter 5 on Architectural engineering).

Portland Cement Association (PCA) <http://www.cement.org/> (accessed May 18, 2005). The PCA is a major publisher of reports on cements and their structural applications. The website has a number of specialist sub-sites aimed at practitioners.

Precast/Prestressed Concrete Institute <http://www.pci.org/> (accessed May 18, 2005). Publishes a journal and a range of design manuals, making this the foremost source of information on pre-stressed concrete. The designers' knowledge bank is an online service intended to help answer designers' questions by referral to PCI publications and elsewhere.

RILEM <http://www.rilem.org/> (accessed May 18, 2005). Réunion Internationale des Laboratoires d'Essais et de Recherches sur les Matériaux et Constructions (RILEM), the premier international body for research and testing materials, organizes conferences and publishes a growing number of journals, either directly or in partnership with other publishers such as Materials and Structures.

Standing Committee on Structural Safety (SCOSS) <http://www.scoss.org.uk/> (accessed May 18, 2005). This is an independent body established by the Institution of Civil Engineers and the Institution of Structural Engineers and others in 1976 to maintain a continuing review of building and civil engineering matters affecting the safety of structures. The website contains full text of all their reports.

Steel Construction Institute (SCI) <http://www.steel-sci.org/> (accessed May 18, 2005). A membership-based research institute that produces design guides, commentaries on codes of practice, and manuals relating to the use of steel in bridges, buildings, and foundations. Although aimed largely at a British audience, as European Codes and Standards are introduced, its publications will reflect this focus.

The Welding Institute (TWI) <http://www.twi.co.uk> (accessed May 18, 2005). An excellent specialist source on welding whose database, WELDASEARCH, is marketed by CSA.

Timber Research and Development Association (TRADA) <http://www.trada.co.uk/> (accessed May 18, 2005). TRADA, a U.K. trade and research body, publishes books, reports, and wood information sheets. The website has an ask TRADA feature.

United States Corps of Engineers <http://www.usace.army.mil/> (accessed May 18, 2005). The ASCE is publishing a growing number of the Corps' engineering manuals, but its codes for blast-resistant construction and military specifications are used worldwide, and are available to download freely.

Water engineering: Introduction

From extant remains of early civilizations in China, Mesopotamia and the Nile Basin, it is evident that water resources have been an important consideration for humans for millennia. The design of works such as the aqueduct feeding Nines in southern France shows evidence that the Romans had great practical knowledge of hydraulics, and indeed speculations on the theory of the flow of water may be found among the earlier printed books on engineering. Water engineers had access to serviceable hydraulic formulae from the eighteenth century, but it was not until the nineteenth century, when engineers began to address issues of urban water supply and drainage and irrigation on a large scale, that water resources engineering may be said to emerge. In this regard the publication of Nathaniel Beardmore's *Manual of Hydrology* and papers by the Irish engineers William Thomas Mulvany and Robert Manning may be regarded as milestones, the one for coining the term "hydrology" for a scientific specialization, the others for developing a servicable formula for run-off in catchments. Since then our understanding of the hydrological style and its implications for water supply have improved through observations.

Dealing with hydraulics and hydrology is only one aspect of the water engineers' work. Those involved in supplying potable drinking-water need to have an understanding of chemistry and biology, and this has a strong interface with the handling and treatment of waste water whether human sewage or industrial effluent. Structures have to be designed to improve and convey water, and also to control it to ensure that it can act as a channel for navigation and not pose a threat through flooding.

Bibliography

Biswas, A.K. (1970) *History of Hydrology*, Amsterdam; London: North-Holland Publishing.

Fahlbusch, H. (ed.) (2001) *Historical Dams: Foundations of the Future Rest on the Achievements of the Past*, New Delhi: International Commission on Irrigation and Drainage.

Rouse, H. and S. Ince (1957) *History of Hydraulics*, Iowa City: Iowa Institute of Hydraulic Research.

Wikander, O. (ed.) (2000) *Handbook of Ancient Water Technology*, Leiden: Brill.

Searching the library catalog

Library searching can be complicated since, although the main texts will be found in water resources, related material will be found in other sections such as chemistry of water, meteorology, and other related earth sciences.

Within the Library of Congress Subject Headings, relevant classes are as follows:

TC 160–181 Technical hydraulics
TC 401–506 River, lake, and water-supply engineering
TC 530–537 River protective works, regulation, flood control
TC 540–558 Dams barrages
TC 601–791 Canals and inland navigation, waterways
TC 801–978 Irrigation engineering, reclamation of wasteland, drainage
TD 201–500 Water supply for domestic and industrial purposes including:
 TD 419–428 Water pollution
 TD 429.5–480.7 Water purification and treatment
 TD 481–493 Water distribution systems
 TD 511–780 Sewerage

and there is already a strong relationship with:

TD 169–171.8 Environmental protection
TD 172–193.5 Environmental pollution
TD 194–195 Environmental effects of industries and plants

as well as subjects such as water chemistry, rainfall, and other branches of the earth sciences. Probably the civil engineers' most significant contribution to modern society is the provision of safe drinking-water. Closely allied with this is the treatment and disposal of waste water – both from industry and in the form of sewage and agricultural wastes. Water supply involves the design of dams, reservoirs, and water towers, as well as pipelines and aqueducts. These involve knowledge of hydraulics and fluid mechanics, as well as geotechnical and structural engineering. Underlying it all is a careful husbanding of water resources. Water purification requires knowledge of chemistry and biology, a clear overlap with science.

Water engineering may also be seen in other aspects of the control of the natural environment (e.g., river engineering to control flooding and make navigation possible). Artificial navigation is seen in the construction of canals, and man-made channels are also required for irrigation schemes, also the work of the civil engineer. These are all closely allied with drainage engineering.

Flood control is often associated with land reclamation, and most dramatically in coastal protection schemes. The cost of such works is causing reconsideration of hard engineering in the form of sea-walls, and an acceptance of some loss of land to flooding and the sea, and an effort to use natural processes to protect the coasts. This area is dealt with in maritime engineering (above).

Abstracting and indexing services

Aqualine The Aqualine database provides comprehensive coverage from the 1950s, concerning water resources, supply and treatment, waste water and sewage treatment, and ecological and environmental effects of water pollution. Articles are drawn from approximately 300 journals as well as from conference proceedings, scientific reports, books, and theses. The Aqualine Thesaurus is available for aiding in online searching. Previously published by WRc in England, Aqualine is now produced in joint cooperation with WRc and Cambridge Scientific Abstracts, e-mail: sales@csa.com*ICID Bibliography on Irrigation, Drainage and Flood Control* (1954–). Only available in paper format but comprehensive.

National Ground Water Association (NGWA) <http://www.ngwa.org> (accessed May 18, 2005). NGWA provides subscriber access to the database *Ground Water On-Line*

Water Pollution Research Abstracts 1927–1975 (continued in *Aqualine*). A printed abstracting service focused on research

Water resources abstracts

A total of 1,094 journals indexed. Almost perfectly complements *Aqualine*, dealing with water resources in its broadest sense, rather than engineering aspects of water supply and treatment.

Dictionaries

APHA/ASCE/AWWA/WPCF (1981) *Water and Wastewater Control Engineering: Glossary*, Washington, DC: APHA. A comprehensive dictionary, going well beyond a focus on water supply and treatment.

ICID Multilingual Technical Dictionary on Irrigation and Drainage (produced by ICID national committees) (English/French/Italian/Spanish/Turkish/Japanese/German). A series of multilingual dictionaries; these are frequently the only specialist civil engineering dictionaries in the languages concerned available for translation into English.

ICOLD Technical Dictionary on Dams, Paris: ICOLD, various edns (French/English/German/Spanish/Italian/Portuguese). Multilingual with diagrams; one of the best dictionaries of this type.

Meinck, F. and H. Mohle (1963) *Dictionary of Water and Sewage Engineering*, Amsterdam: Elsevier (English/German/French/Italian). Useful multilingual dictionary.

Smith, P.G. and J.S. Scott (2002) *Dictionary of Water and Waste Management*, Oxford: Butterworth-Heinemann, for IWA. This builds on the work of J.S. Scott, well known for his Penguin dictionary of civil engineering and now sadly deceased. Sparsely illustrated but good definitions, up-to-date and comprehensive.

Van der Tuin, J.D. (1997) *Elsevier's Dictionary of Water and Hydraulic Engineering*, New York: Elsevier (available digitally). Reasonable coverage of the field.

Handbooks

The American Society of Civil Engineers, Water Environment Federation, and American Water Works Association, sometimes in combination with other partners, produce a series of handbooks on design of municipal waste-water treatment plants, odor control in waste-water plants, waste-water and storm-water pumping stations, gravity sewer design, ground-water management, sewer evaluation and rehabilitation, aeration, urban run-off quality, urban storm-water management systems, sewer system overflows, manhole inspection, urban subsurface drainage, steel pipe design, and so on, and are the obvious first source of reference for U.S. practice.

Alley, E.R. (2000) *Water Quality Control Handbook*, New York: McGraw-Hill. From a U.S. regulatory perspective this deals with water pollution, its treatment and control.

Corbitt, R.A. (1998) *Standard Handbook of Environmental Engineering*, New York: McGraw-Hill. Although this handbook is concerned with all aspects of the environment it has chapters on water quality, water supply, storm-water management, and waste-water treatment, with general environmental information.

Davis' Handbook of Applied Hydraulics (1993) (4th edn), New York: McGraw-Hill. Written for the practicing engineer, this book is organized in the order usually followed in making regional plans for water use and control. Coverage includes basic hydraulics, reservoir hydraulics, and natural channels, river diversion and the construction of reservoirs and dams, pumped storage, navigation locks, irrigation and water, and waste-water systems. Twelve sections are completely new to this edition; the remaining sixteen sections have been condensed and updated to reflect current practice. More tables and illustrations have been included than in previous editions. They provide examples of many actual projects.

Delleur, J.W. (1999) *The Handbook of Groundwater Engineering*, Boca Raton. FL: CRC Press. Comprehensive treatment dealing with aquifers, well-sinking, contamination and landfill design, and water quality. Included in EngNetBase.

De Zuane, J. (1997) *Handbook of Drinking Water Quality* (2nd edn), New York: Van

Nostrand Reinhold. Describes U.S. and WHO (World Health Organization) water quality guidelines, problems of water quality treatment and control, and related water supply engineering issues.

Gallagher, L.M. and L.M. Miller (2003) *Clean Water Handbook* (3rd edn), Rockville, MD: Government Institutes. Each chapter focuses on different aspects of the (U.S.) Clean Water Act with up-to-date coverage of the latest enactments and EPA regulations.

Golze, A.R. (1977) *Handbook of Dam Engineering*, New York: Van Nostrand Reinhold. Comprehensive, if dated. Very helpful for anybody unfamiliar with dam engineering.

Hunt, T, and N. Vaughan (1996) *Hydraulic Handbook*, New York: Elsevier.

Institute of Hydrology (1999) *Flood Estimation Handbook* (5 vols), Wallingford, and CD-ROM of data. Based on UK hydrology, a detailed handbook for flood calculations.

National Engineering Handbook. Washington, DC: Natural Resources Conservation Services <http://www.mi.nrcs.usda.gov/technical/engineering/neh.html> (accessed May 18, 2005). Produced for the U.S. Department of Agriculture; sections deal with design, construction, hydrology, irrigation, and drainage. Very much aimed at the practitioner in the field. A general engineering handbook aimed at those involved in agricultural engineering.

WRc (1990) *Design Guide for Marine Treatment Schemes* (4 vols), Swindon: WRc. Design manual for waste-water treatment for marine disposal.

WRc (2001) *Sewer Rehabilitation Manual* (4th edn), Swindon: WRc. Comprehensive U.K. guidance on sewer rehabilitation, including specifications.

Journals

There are a large number of trade journals of interest to the industry, but with little intellectual content.

American Water Works Association Journal (0003–150X). Useful for AWWA and water industry news with lots of advertisements, issues are often themed around latest developments.

Aqua (IWA) (0003–7214). Although mostly concerned with research, this does contain papers on practical applications and operational aspects of water supply worldwide.

Circulation (BHS). News of British Hydrological Society meetings and articles.

Hydrological Sciences Journal (IAHS) (0262–6667). Contains research papers and news of IAHS meetings.

ICID Bulletin Wiley (0971–7412). The *ICID Bulletin* is the longest established international journal in the field.

Irrigation and Drainage Systems (Kluwer) (0168–6291). A relative newcomer, with an academic flavor.

Journal of Contaminant Hydrology (Elsevier) (0169–7722). The leading academic journal in this specialist field.

Journal of Environmental Engineering (ASCE) (0733–9372). This ASCE title covers waste-water treatment.

Journal of Hydraulic Engineering (ASCE) (0733–9429). A long-established ASCE specialty journal that has published many key papers on hydraulic design.

Journal of Hydroinformatics (IWA/IAHR) (1464–7141) An international journal intended to be cross-disciplinary dealing with the application of information technology to aquatic sciences; a preponderance of articles deal with numerical flow modeling.

Journal of Hydrologic Engineering (ASCE) (1084–0699). The ASCE journal for engineering hydrologists.

Journal of Hydrology (Elsevier) (0022–1694). The cost of this journal has been the cause of comment among hydrologists, but it remains the leader in the field.

Journal of Irrigation and Drainage Engineering (ASCE) (0733–9437). The leading U.S. journal in the field.

Journal of Water and Health (IWA/WHO) (1477–8920). A multi-disciplinary approach to water, recognizing its basic relationship to public health.

Journal of Water Resources, Planning and Management (ASCE) (0733–9496). The ASCE's journal on water resources.

Water 21 (IWA) (1561–9508). The IWA's monthly magazine with news and feature articles on water issues worldwide.

Water and Environment Magazine (CIWEM) (1362–9360). U.K. news, feature articles. and CIWEM activities.

Water Environment and Technology (Water Environment Federation) (1044–9493). News and feature articles on recent developments in water treatment.

Water Environment Research (Water Environment Federation) (1061–4303). Perhaps the best source, aside from the IWA, on the latest research into waste-water treatment.

Water Management (ICE) (1741–7589). The latest ICE proceedings title, bringing together all the papers on water supply, treatment, and resources management in the broadest sense.

Water Research (Elsevier) (0043–1354). Mostly academic research papers on all aspects of water quality and treatment.

Water Resources (AWRA). Perhaps the most widely referenced journal in the field in the U.S.

Water Resources Research (American Geophysical Union) (0043–1397). Broad issues of climate are published alongside papers on water chemistry.

Water Science and Technology, 2 parts (IWA) (0273–1223). Effectively two conference series on water supply and waste-water treatment; frequently these are state-of-the-art volumes from international contributors.

Monographs

American Public Health Association. *Standard Methods for the Examination of Water and Wastewater*, New York: American Public Health Association. These are probably the most widely used methods for water analysis, both in the U.S. and overseas. Irregular serial.

Applegate, G. (2002) *The Complete Guide to Dowsing*, London: Vega. For those interested in the use of water divining this is the most up-to-date source on the technique.

ASCE (1980) *Manual 57: Operation and Maintenance of Irrigation and Drainage Systems*, New York: ASCE. For the developed world this provides comprehensive advice of managing irrigation systems.

AWWA (1999) *Water Quality and Treatment* (5th edn), New York: McGraw-Hill.

An introductory chapter deals with U.S. standards and regulations, and successive chapters deal succinctly with treatment methods and issues of corrosion and microbial contamination in distribution systems.

Barr, D.I.H. (1998) *Tables for the Hydraulic Design of Pipes, Sewers and Channels* (2 vols), London: Thomas Telford. A useful compendium of tables for pipe and channel design.

Bernstein, L.B. (1996) *Tidal Power Plants*, Seoul: KORO. In many ways the seminal text on this aspect of power from water.

BRE (1999) *Engineering Guide to the Safety of Embankment Dams in the UK*, London: CRC Ltd. An essential guide to safety assessment of earth dams.

CIRIA (2004) *Sustainable Drainage Systems: Hydraulic, Structural and Water Quality Advice*, London: CIRIA. Engineering advice on the design of SUDS – intended to mimic as far as possible natural drainage systems and thus limit the environmental impact of a drainage scheme.

Escarameia, M. (1998) *River and Channel Revetments: A Design Manual*, London: Thomas Telford. Revetment design is an essential aspect of river control and this work covers the latest European practice.

Hansen, V.E. (ed.) (1980) *Irrigation Principles*, New York: Wiley. Despite its age, this book remains on many course reading lists, and covers most aspects of irrigation engineering.

ICE (1975) *Guide to the Reservoirs Act, 1975*, London: Thomas Telford. An essential guide to the legislation governing the safety assessment of dams in the U.K.

ICE (2001) *Learning to Live with Rivers*, London: Thomas Telford. Intended to encourage engineers to use rivers as an environmental asset, rather than controlling and hiding them.

Jansen, R.B. (ed.) (1988) *Advanced Dam Engineering: For Design, Construction and Rehabilitation*, New York: Van Nostrand Reinhold. A comprehensive overview of advances in dam design in the second half of the twentieth century.

Metcalf and Eddy Inc. (2003) *Wastewater Engineering: Treatment Disposal and Reuse* (4th edn), New York: McGraw-Hill. This work has been in print in some form or other for nearly a century. This edition incorporates SI units. It covers all aspects of water treatment and reuse, and is fully referenced. Data tables and worked design examples are included.

Nienhuis, P.H. (ed.) (1998) *New Concepts for Sustainable Management of River Basins*, Leiden: Backhuys Publishers. A series of papers exploring sustainable management of river basins, illustrating international practice.

Novak, P. (2001) *Hydraulic Structures*, London: Spon. Covers most aspects of hydraulic structures (e.g., dams and weirs at a basic level), suitable for undergraduates or those seeking a basic understanding of the subject.

Parker, D.H. (ed.) (1987) *Urban Flood Protection Benefits*, Aldershot: Gower. Cost benefit analysis of flood alleviation in an urban context.

Penning Rowsell, E.C. and J B. Chatterton (1977) *Benefits of Flood Alleviation*, Westmead, Hants: Saxon House. An early and classic investigation of the cost benefit analysis of flood alleviation schemes.

Read, G.C. (ed.) (1997) *Sewers*, London: Arnold. A multi-authored work, largely based on British practice, dealing with sewer repair and reconstruction.

Read, G.C. (ed.) (2004) *Sewers: Replacement and New Construction*, Amsterdam: Elsevier. Largely complementing the above title, with state-of-the-art contributions on the latest sewer construction techniques.

Sanks, R.L. (1998) *Pumping Station Design*, London: Butterworth. A comprehensive work on a subject where mechanical and civil engineers are regularly brought together.

Schiechtl, H.M. and Stern, R. (1997) *Water Bioengineering Techniques for Watercourse Bank and Shoreline Protection*, Oxford: Blackwell Science. Explores natural alternatives to hard engineering solutions such as concrete for slope protection.

Skogerboe, G.V. and G.P. Merkley (1996) *Irrigation Maintenance and Operations: Learning Process*, Colorado: Water Resources. Provides guidelines and procedures for improved maintenance and operations of conveyance and distribution systems for irrigation water.

Stein, D. (2001) *Rehabilitation of Drains and Sewers*, Berlin: Ernst. A comprehensive treatment of sewer maintenance and repair based on German practice.

Twort, A.C. (2000) *Water Supply* (5th edn), London: Arnold. The best U.K. textbook on the subject with coverage of the organization of the U.K. water industry, as well as storage, distribution and treatment of water and related aspects of hydraulics.

U.S. Bureau of Reclamation (1987) *Design of Small Dams*, Washington, DC: USGPO. Probably the most comprehensive treatment of dam design in a single book, now on the USBR website.

U.S. Bureau of Reclamation (1997) *Water Measurement Manual*, Washington, DC: USGPO. A useful handbook on methods of water flow measurement, and so on, available for downloading <http://www.usbr.gov/pmts/hydraulics_lab/pubs/wmm/> (accessed May 18, 2005).

U.S. Bureau of Reclamation (1983) *Safety Evaluation of Existing Dams*, Washington, DC: USGPO. U.S. procedures for the safety inspection of dams.

U.S. Environmental Protection Agency (EPA) (1997) *EPA Methods and Guidance for the Analysis of Water*, Rockville, MD: Government Institutes. U.S. Government Agency's guidelines for water analysis; available in digital format.

Web resources

Dam decommissioning in France <http://rivernet.org/welcome.htm> (accessed May 18, 2005). Describes the political processes and engineering challenges of decommissioning three dams in France, with references, links, and images.

GRDC <http://grdc.bafg.de/servlet/is/857> (accessed May 18, 2005). GRDC is an international cooperative venture under the auspices of the World Meteorological Organization and hosted within the German Federal Institute of Hydrology to provide an international databank on surface water hydrology. Although access to data from the bank is fee-based, the website provides free access to many other publications.

Ground-water Remediation Technologies Analysis Centre (GWRTAC) <http://www.gwrtac.org> (accessed May 18, 2005). The GWRTAC site includes links to other environmental sites, full-text technical reports, and a list of U.S. regulations.

Hydrology Web <http://hydrologyweb.pnl.gov/index.asp> (accessed May 18, 2005). Site hosts a comprehensive list of hydrology and hydrology-related resources.

National inventory of dams <http://crunch.tec.army.mil/nid/webpages/nid.cfm> (accessed May 18, 2005). Compiled by the U.S. Army Corps of Engineers and other agencies, this is a database giving data on 75,000 dams in the U.S. and Puerto Rico.

NWS Hydrologic Information Center <http://www.nws.noaa.gov/oh/hic> (accessed May 18, 2005). Provides historical and current data on streamflow, drought, soil moisture, flood damage, and so on.

Safe Drinking Water <http://www.safedrinkingwater.com/> (accessed May 18, 2005). This is a resource for the drinking-water community providing news and information, with an emphasis on California issues.

United Kingdom Environment Agency <http://www.environment-agency.gov.uk/subjects/flood/> (accessed May 18, 2005). Responsibility for dealing with flooding has traditionally been a divided responsibility in the U.K. Around five million people, in two million properties, live in flood risk areas in England and Wales. The Environment Agency's website has a lot of helpful information: Floodline information homepage – before, during and after; Information on how to prepare for floods and cope with the inevitable clear-up operation in the aftermath; Current flood warnings in force – allows you to view flood warnings in place and flood warnings issued. Updated every fifteen minutes; Flood management and R&D – details of the agency's flood management roles, including research and development (joint with Defra) and the National Flood and Coastal Defence Database (NFCDD).

U.K. DEFRA (Annual) Digest of environmental statistics <http://www.defra.gov.uk/environment/statistics/des/index.html> (accessed May 18, 2005). Statistics on all aspects of the U.K. environment including water supply and the state of resources.

USGS SPARROW <http://water.usgs.gov/nawqa/sparrow/> (accessed May 18, 2005). The SPARROW (Spatially Referenced Regressions on Watershed Attributes) model was developed to estimate the origin and fate of contaminants in streams, based on regional water quality monitoring data. Site gives examples of model's application.

Water Librarian's Homepage <http://www.interleaves.org/~rteeter/waterlib.html> (accessed May 18, 2005). This site contains a variety of useful links in the field of water resources with recommended websites.

Water Strategist Community (WSC) <http://www.waterchat.com> (accessed May 18, 2005). Water Strategist Community is a water news website that provides daily updates on press releases from U.S. environmental and water agencies as well as news on water investment. WSC provides e-mail alerts and a trade directory.

Wateright <http://www.wateright.org/> (accessed May 18, 2005). An educational resource for irrigation water management for three audiences: home-owners, commercial turf growers, and agriculture. Presents a series of advisories and tutorials on multiple topics.

World Commission on Dams <http://www.dams.org> (accessed May 18, 2005). A UN Commission to review the development of large dams, and to develop guidelines and standards on their use. The report is downloadable.

Associations, organizations, and societies

American Institute of Hydrology (AIH) <http://www.aihydro.org> (accessed May 18, 2005). The U.S. society for hydrologists.

American Society of Agronomy (ASA) http://www.agronomy.org/ (accessed May 18, 2005). The ASA has published a number of monographs on irrigation and drainage.

American Water Resources Association (AWRA) <http://www.awra.org/> (accessed May 18, 2005). The AWRA's mission is to be the pre-eminent multi-disciplinary association for information exchange, professional development, and education about water resources and related issues. It publishes the leading U.S. journal (*Water Resources*) in the field. The site contains conference information, publications, employment news, e-learning courses, links, and membership information.

American Water Works Association (AWWA) <http://www.awwa.org> (accessed May 18, 2005). The AWWA produces standards and specifications, a journal, and conference proceedings, including an important series on water reuse.

Association of State Dam Safety Officials <http://www.damsafety.org/> (accessed May 18, 2005). The Association site provides dam safety information for regulators, with links, training programs, and an electronic bibliography.

CEH Wallingford (CEH) <http://www.nwl.ac.uk/ih> (accessed May 18, 2005). The Centre for Ecology and Hydrology (formerly the Institute of Hydrology) carries out research into the effects of land use, climate, topography, and geology on the volume and character of surface water resources. A new Joint Centre for Hydrometeorological Research has been established in partnership with The Met Office.

Chartered Institution of Water and Environmental Management (CIWEM) <http://www.ciwem.org.uk/> (accessed May 18, 2005). Formed as the result of a series of mergers of societies for engineers and scientists in the water and environmental field, CIWEM produces two journals, organizes conferences, and publishes state-of-the-art manuals. The multi-volume *Manual of British Water Engineering Practice*, although no longer up to date, remains the best general source of information on topics such as water supply and river engineering

Colorado Water Resources Research Institute <http://cwrri.colostate.edu/links.html> (accessed May 18, 2005). Provides links to sources for specialty water data and information on the worldwide web.

Computational Fluid Dynamics (CFD)-Online <http://www.cfd-online.com/Resources/> (accessed May 18, 2005). Nicely arranged directory of websites on the topic of fluid dynamics.

ESCAP (UN) <http://www.unescap.org/> (accessed May 18, 2005). ESCAP supports the efforts of members and associate members in Asia and the Pacific regions to achieve their desired development goals in a sustainable manner. Contains sections on energy resources; environment; water and mineral resources; and space technology. It has a sub-site on sustainability: <http://www.unescap.org/esd/index.asp>.

Food and Agriculture Organization of the United Nations <http://www.fao.org> (accessed May 18, 2005). Responsible for several series of publications relevant to the work of civil engineers.

HR Wallingford <http://www.hrwallingford.co.uk> (accessed May 18, 2005). Developed from the former Hydraulics Research Station, providing research, modeling, and consultancy services on hydraulic engineering in its broadest sense.

Hydraulic Engineering Research Unit <http://www.ars.usda.gov/main/site_main.htm?modecode=62171000> (accessed May 18, 2005). Unit is part of the USDA Plant Science and Water Conservation Research Laboratory. Research is performed to develop criteria for design and analysis of structures and channels for control, conveyance, storage, and disposal of run-off water.

Hydrological Sciences Branch <http://hsb.gsfc.nasa.gov/> (accessed May 18, 2005). NASA-based organization dedicated exclusively to the understanding, quantification, and analysis of the different components of the hydrological cycle, with emphasis on land surface hydrological processes and their interaction with the atmosphere.

International Association of Hydrogeologists <http://www.iah.org/> (accessed May 18, 2005). An international body for scientists and engineers; the site provides information on conferences, publications and so on, with a member's area.

International Association of Hydrological Sciences <http://www.cig.ensmp.fr/~iahs/index.html> (accessed May 18, 2005). The society publishes a journal, and produces two series of publications, a series of symposia, and specialist reports. The Association has national associated societies.

International Commission on Irrigation and Drainage (ICID) <http://www.icid.org/> (accessed May 18, 2005). The ICID was established in 1950 as a non-governmental international organization (NGO) with headquarters in New Delhi. The Commission is dedicated to enhancing the worldwide supply of food for all people by improving water and land management and the productivity of irrigated and drained lands through appropriate management of water, environment and application of irrigation, drainage and flood management techniques. "Irrigation and drainage in the world" provides nation-by-nation summaries of major schemes.

International Commission on Large Dams (ICOLD) <http://www.icold-cigb.org> (accessed May 18, 2005). Created in 1928, ICOLD has membership in eighty-two countries. It holds regular conferences and produces bulletins of good practice, and a world register of large dams. Its publications are indexed on the website. Many national societies have produced national directories and histories of dams.

International Water Association (IWA) <http://www.iwap.co.uk/> (accessed May 18, 2005). IWA sponsors a number of journals, and organizes major conferences related to water supply and waste-water treatment.

International Water Management Institute (IWMI) <http://www.iwmi.cgiar.org/about/intro.htm> (accessed May 18, 2005). IWMI is a non-profit scientific research organization focusing on the sustainable use of water and land resources in agriculture, and on the water needs of developing countries. IWMI works with partners in the South to develop tools and methods to help these countries eradicate poverty through more effective management of their water and land resources. The website provides links, guidance notes, and news of international developments.

Irrigation Association (IA) <http://www.irrigation.org/> (accessed May 18, 2005). The Irrigation Association's mission is to improve the products and practices used to manage water resources and to help shape the worldwide business environment of the irrigation industry. Site includes certification, water conservation policy, and brochures.

Irrigation Association of Australia <http://www.irrigation.org.au> (accessed May 18, 2005). This site provides comprehensive information about irrigation industry and practices in Australia.

Low Impact Development Center <http://www.lowimpactdevelopment.org/> (accessed May 18, 2005). The Center provides information to individuals and organizations dedicated to protecting the environment and water resources through proper site design techniques replicating pre-existing hydrologic site conditions.

National Ground Water Association <http://www.ngwa.org> (accessed May 18, 2005). Searchable bibliographic database is restricted to subscribers. Free access areas include information on maintaining domestic water supplies plus ground water in water supplies and product specifications. The site contains membership information and activities.

National Water Resources Association (NWRA) <http://www.nwra.org> (accessed May 18, 2005). NWRA is concerned with appropriate use of water and land resources. Works closely with Congress and Executive Branch in the U.S.

New Mexico Water Resources Research Institute (NMWRRI) <http://wrri.nmsu.edu/> (accessed May 18, 2005). NMWRRI (at New Mexico State University) is part of New Mexico's Rio Grande Research Corridor and participates in water resources planning. Site contains a resource data and information system, with access to publications.

Office of Water Services (OFWAT) <http://www.ofwat.gov.uk/> (accessed May 18, 2005). The regulator for the privatized water industry in the UK, producing a number of reports, industry reviews and codes of practice, mostly available free, or downloadable from the website.

UNESCO <http://upo.unesco.org> (accessed May 18, 2005). UNESCO publishes a number of series and reports relevant to water engineers including studies in hydrology, and discharge records for selected (major) rivers.

USGS-Water Resources of the US <http://water.usgs.gov/> (accessed May 18, 2005). Provides maps concerning water conditions for U.S. and local areas.

U.S. River Systems and Meteorology Group (RSMG) <http://www.usbr.gov/pmts/rivers> (accessed May 18, 2005). Part of the US Bureau of Reclamation, RSMG provides expertise in meteorology, NEXRAD radar, estimates of precipitation, hydrology, hydraulic engineering, water management, water rights, statistical analysis, and user interface design and implementation.

United States NWIS-W Data Retrieval <http://waterdata.usgs.gov/usa/nwis/> (accessed May 18, 2005). Home page links to U.S. and international water gauging stations.

Water, Engineering and Development Centre, Loughborough University <http://www.lboro.ac.uk/departments/cv/wedc/> (accessed May 18, 2005). The U.K.'s leading institute involved in infrastructure development for the Third World.

Water Environment Federation (WEF) <http://www.wef.org> (accessed May 18, 2005). WEF compliments AWWA in its activities. The website provides weekly water environmental news in WEF Reporter, a product locator search tool, regulations, and legislation details, member and affiliated organizations, research information and events calendar. There are outside links, and forum discussions. Publications include journals, conferences, manuals, and a CD-ROM on operations training.

Water Research Commission (South Africa) <http://www.wrc.org.za/> (accessed May 18, 2005). Promotes water research and application of research findings in South Africa. Site contains research information, publications, databases, calendar, software, reference information, and WRC information.

WRc Group <http://www.wrcplc.co.uk/> (accessed May 18, 2005). The former Water Research Centre provides consultancy services related to water, waste-water, and the environment, and publishes numerous reports and standards/codes of practice for the U.K. water industry.

Water UK <http://www.water.org.uk> (accessed May 18, 2005). Membership comprises the water and waste-water suppliers of the U.K. Funded by the members who all have representation on the council, it publishes reports on the water industry, and statistics.

WateReuse Association <http://www.watereuse.org> (accessed May 18, 2005).

National organization dedicated to increasing the beneficial use of recycled water. Site contains meetings calendar, technical, and membership information. Membership includes public agencies, water suppliers, with local, state and federal groups. The site has links to other sites and documentation.

Waterways Experiment Station (WES) <http://www.wes.army.mil/> (accessed May 18, 2005.) WES is responsible for a large amount of hydraulics research with the majority of its recent publications available to download.

World Health Organization (WHO) <http://www.who.int/water_sanitation_health/> (accessed May 18, 2005). WHO produces drinking-water guidelines and sponsors a number of initiatives linking water, sanitation, and health.

Standards: Introduction

Like all engineers, civil engineers depend on local, national, and international official standards and specifications. They can be produced by a national standards body such as the British Standards Institution, or by a professional body, or sometimes by a trade association. The status of the originating body is the key to the authority of the standard. Particularly with developing areas, such as occurred with reinforced earth in the 1970s, the patentee/product developer is often the chief source of design information. Engineers evidently need to exercise engineering judgment in applying all standards and codes.

Standards and specifications provide minimum performance requirements for all aspects of a construction project. A standard can involve an entire project such as the American Society of Civil Engineer's Minimum Design Loads for buildings and other structures or one as specific as the American Society of Mechanical Engineers' *International Boiler and Pressure Vessel Code.*

In addition to technical standards, civil engineering projects are generally governed by local, state, or national regulations. The U.K. building regulations are available from TSO <http://www.tso.co.uk/bookshop/bookstore.asp> and on subscription via Knights/Tolley. The following are some of the major standards-producing organizations about which civil engineers should be familiar.

Societies, associations, and organizations

ASM International <http://www.asminternational.org> (accessed May 18, 2005). Founded in 1913, ASM International serves the technical interests of metals and materials professionals.

ASTM International <http://www.astm.org> (accessed May 18, 2005).

American Association of State Highway and Transportation Officials (AASHTO) <http://www.transportation.org/aashto/home.nsf/FrontPage> (accessed May 18, 2005). AASHTO represents all five transportation modes. It works to develop specifications related to highways and public transportation (see Chapter 20 on Transportation engineering).

American Concrete Institute (ACI) <http://www.concrete.org/general/home.asp> (accessed May 18, 2005). Founded in 1904, the ACI develops codes and specifications related to the use of concrete in structures.

American Institute of Steel Construction (AISC) <http://www.aisc.org> (accessed May 18, 2005). Established in 1921, the AISC aims to serve the structural steel design community by developing technical information including codes and specifications related to steel construction.

American Iron and Steel Institute (AISI) <http://www.steel.org> (accessed May 18, 2005). AISI comprises North American steel producers and plays a lead role in the development and application of new steels and steel-making technology.

American Railway Engineering and Maintenance of Way Association <http://www.arema.org/> (accessed May 18, 2005). Develops standards on all aspects of railway engineering including permanent way and bridges.

American Society of Civil Engineers (ASCE) <http://www.asce.org> (accessed May 18, 2005). ASCE, founded in 1852, is America's oldest national engineering society. Through its technical institutes and committees it produces codes and standards related to all areas of civil engineering from building design to waste-water treatment.

American Society of Heating, Refrigeration, and Air-Conditioning Engineers (ASHRAE) <http://www.ashrae.org> (accessed May 18, 2005). ASHRAE develops standards and guidelines in its field. The Society addresses such areas as indoor air quality, thermal comfort, and energy conservation in buildings.

American Society of Mechanical Engineers (ASME) <http://www.asme.org> (accessed May 18, 2005). Founded in 1880, the ASME focuses on technical, educational, and research issues related to the field of mechanical engineering. It is responsible for the development of the International Boiler and Pressure Vessel Code that establishes rules of safety governing the design, fabrication, and inspection of boilers and pressure vessels (see Chapter 16 on Mechanical engineering).

American Water Works Association (AWWA) <http://www.awwa.org> (accessed May 18, 2005). The largest organization of water professionals, the AWWA develops codes and standards related to water quality and supply.

British Standards Institution (BSI) <http://www.bsi.org.uk/index.xalter> (accessed May 18, 2005). BSI, initially established by the ICE, was the first national standards body in the world. The U.K. contact point for other international and national standards organizations, including: International Standards Organisation (ISO), Comité Européen de Normalisation (CEN, European Committee for Standardisation), those of European Union nations including France (AFNOR) and Germany (DIN), Standards Association of Australia (SAA), and Standards Council of Canada (SCC). The monthly periodical *Update Standards* details development in U.K. standards. BSI's work is increasingly involved with quality assurance, while European and ISO standards are replacing the national standards.

Canadian Codes Centre <http://codes.nrc.ca> (accessed May 18, 2005). The Centre provides technical and administrative support to the Canadian Commission on Building and Fire Codes which is responsible for the development of the national model construction codes of Canada. Some of these codes are the National Building Code, National Fire Code, and National Plumbing Code.

Deutsches Institut fuer Normung e.V. <http://www2.din.de/index.php?lang=en> (accessed May 18, 2005). The German Standards Institute's standards are probably the most widely used internationally aside from U.S. standards. The most important are available in English. Those relating to structural loads and reinforced and pre-stressed concrete and structural steelwork are widely used.

Eurocodes website <http://www.eurocodes.co.uk/> (accessed May 18, 2005). The ICE has developed a website for all those interested in the development of Eurocodes that was launched in March 2003.

European Committee for Standardisation (CEN), rue de Stassart 36, B-1050 Bruxelles, Belgium. Tel 32 2 519 6811; Fax: +32 2 519 6819. CEN, in combination with European national standards bodies, is gradually developing product standards in civil engineering to replace existing national standards.

German Geotechnical Society http://www.dggt.de/ (accessed May 18, 2005). Produces a number of internationally recognized design guidelines, in areas such as waterfront structures and contaminated land.

International Code Council (ICC) <http://www.iccsafe.org> (accessed May 18, 2005). The ICC was established in 1994 by the Building Officials and Code Administrators International, Inc., International Conference of Building Officials, and Southern Building Code Congress International, Inc. These three organizations used to develop separate regionalized building codes. With the founding of the ICC, they work together to develop a single set of codes without regional limitations.

International Union of Railways (UIC) <http://www.uic.asso.fr/> (accessed May 18, 2005). Internationally recognized railway standards. Although each national railway system has its own standards and regulations, UIC standards are those which generally govern international routes.

National Fire Prevention Association (NFPA) <http://www.nfpa.org/catalog/home/index.asp> (accessed May 18, 2005). The NFPA develops safety codes and standards, such as the Fire Prevention Code, that influence buildings, processes, services, and design.

U.S. Department of Commerce, National Institute of Standards and Technology (NIST) <http://www.nist.gov> (accessed May 18, 2005). NIST, established by Congress, is responsible for an enormous variety of standards-related research programs, including publishing a series on building, and a cooperative program with Japan on wind and seismic engineering.

References

American Society of Civil Engineers (2004) *2004 Official Register*, Reston, VA: ASCE.

Chrimes, M. and A. Bhogal (2001) "Civil Engineering – A Brief History of the Profession: The Perspective of The Institution of Civil Engineers," in J.R.F. Rogers and J. Augustine (eds) *International Engineering History and Heritage: Improving Bridges to ASCE's 150th Anniversary*, Reston, VA: ASCE.

Goodman, R.E. (2002) "Karl Terzaghi's Legacy in Geotechnical Engineering," *Geostrata* www.geoengineer.org/terzaghi2.html (accessed August 10, 2004).

Merdinger, C.J. (1953) *Civil Engineering Through the Ages*, Washington, DC: Society of American Military Engineers.

http://www.britannica.com/eb/print/eu=70311 (accessed August 9, 2004).

http://www.britannica.com/eb/print?eu=108130 (accessed August 9, 2004).

http://www.geoengineer.org/terzaghi2.html (accessed August 8, 2004).

9 Computer engineering

Hema Ramachandran and Renée McHenry

Introduction

According to *Webster's New World Dictionary of Computer Terms* (3rd edn 1988), computer engineering is defined as "the field of knowledge that includes the design of computer hardware systems." As simple as it is, this definition is a good starting point for the general reader, but it ignores its relationship to computer science which the *Dictionary* describes as "the field of knowledge embracing all aspects of the design and use of computers."

The Joint Task Force on Computer Engineering Curricula, consisting of members from the Association of Computing Machinery (ACM) and the Institute of Electrical and Electronics Engineers (IEEE) Computer Society, provides us with a more detailed working definition as outlined in its *Curriculum Guidelines for Undergraduate Degree Programs in Computer Engineering* published in December 2004:

> Computer engineering as an academic field encompasses the broad areas of computer science and electrical engineering. Computer engineering is defined . . . as follows:
>
> Computer engineering is a discipline that embodies the science and technology of design, construction, implementation, and maintenance of software and hardware components of modern computing systems and computer-controlled equipment. Computer engineering has traditionally been viewed as a combination of both computer science (CS) and electrical engineering (EE).

The Internet encyclopedia, *Wikipedia*, describes a computer engineer as "an electrical engineer with a focus on digital logic systems, and less emphasis on radio frequency or power electronics. From a computer science perspective, a computer engineer is a software architect with a focus on the interaction between software programs and the underlying hardware components."

Another way to consider this is that the computer engineer's primary goal is to implement theories developed by computer scientists to solve practical problems and create viable systems. This comparison is analogous to science and engineering in general: science often provides results that are significant to engineering. Both computer science and computer engineering owe their foundation to mathematics.

Computer engineering, also sometimes known as "Computer Systems Engineering" or even "Computer Science Engineering," has been viewed in the past as just "designing computers." Now, more often than not, computer engineers are an integral part of professional teams who design computer-based systems throughout industries such as aerospace, telecommunications, power production, manufacturing, defense as well as the computer and electronics industry.

Professional specialty areas include system architecture, system integration, layout, RTL/chip design, package/board design, and the fabrication process while the more theoretical areas are computer architecture, digital control theory, digital signal processing, semiconductors, and software engineering.

Overview of the literature

Prior to 1947, the sparse computing literature was concerned with analog computers or analyzers and punched card machines. Simple desk calculators existed but usually calculations were made by hand. Journals in computing barely existed and pertinent literature, mostly technical, was scattered throughout a variety of disciplines – mathematics, statistics, physics, electrical engineering, and even sciences such as astronomy.

How different the situation is today! In under sixty years, computers are ubiquitous, and literature about them in all shapes and forms is everywhere. From literature for the professional computer engineer to the average consumer, the volume of literature is almost breathtaking. The average consumer encounters chain bookstores with large sections devoted to computer books for the general public as well as the professional practitioner. In the bookstore's "How to" section are books (and book series) devoted to the computer neophyte – the familiar "Dummies" and "Idiots" series is known to everyone. The popular computer magazine is part of our everyday culture and found in supermarkets side-by-side with grocery items. The general news media frequently reports on the computing industry and those who work in it. This was particularly evident during the rise and fall of the dotcoms. The entrepreneurs in California's Silicon Valley are our modern-day technological heroes and their rise has almost become part of folklore. The computing industry, however it is defined, is indeed big business and any economic shifts have global ramifications – as seen with the controversial subject of "outsourcing" to India and other countries.

The ACM's *Guide to Computing Literature*, published annually since 1960, may be considered a guide or bibliography of the computing literature. Indexed by author, subject, and keyword using the ACM Computing Classification System, the *Guide* includes references to scholarly and trade magazines, and books from major publishers in the computing field.

Of course, from time to time librarians will also produce bibliographies on topics of interest. This is a particular challenge in computer engineering where a subject compilation can be out of date as soon as it is published but it is certainly useful for retrospective literature searching. This is certainly evident with the older compilations listed below:

Hildebrandt, D.M. (1996) *Computing Information Directory* (13th edn), Federal Way, WA: Pedaro. It is a shame that this annual series, which ceased in 1996, has not been updated. It has been a very useful directory for librarians and information-seekers alike due to its comprehensive coverage of the computing literature. The author, Darlene Myers Hildebrandt, Head of the Science Libraries of Washington State University, provides a treasure trove of information. Even though it is outdated, hold on to your editions, since it still provides a good overview of the literature. The directory has listings and descriptions of journals, newsletters, books, dictionaries, indexes, abstracts, software resources, review sources, information on special issues in journals, directories, computer languages, standards, and publishers.

Rousseau, R. (compiler) (1985) *Selective Guide to Literature on Computer Science*, Washington, DC: American Society for Engineering Education, Engineering Libraries Division. A new book devoted to computing resources by the computer science librarian at Carnegie Mellon University, Missy Harvey, is due to be published by Libraries Unlimited <http://lu.com> in 2006.

It is interesting to compare the computer engineering literature with that of other engineering disciplines. Due to the pervasive nature of computing in society today, this may be the only engineering field where publishers cater to the needs of a whole spectrum of users hungry for information (from neophyte to expert): academicians, researchers, teachers, practitioners, students, hobbyists, and consumers. Given the fast-paced nature of the computing industry at large, material is churned out at a phenomenal rate; the literature has become a lucrative business for many publishers. The range of publishers who meet these needs covers the gamut.

The standard commercial publishers – Springer, Wiley, Cambridge University Press, Oxford University Press, Elsevier, Newnes, and Morgan Kaufmann as well as the society publishers IEEE Press and ACM – carry many monographic titles catering to the research and educational communities. Many of these publishers also publish journals – for more information, see the Journals and Current awareness sections. Then we have the "practitioner" market served by such publishers as O'Reilly (the major player), Peachtree Press, Prentice Hall PTR, Que and Sam's Publishing (see Tools for the Practitioner section).

With the development of the Internet – a medium that computer engineers not surprisingly embraced from the beginning – there is a vast amount of information freely available. The challenge now, and for the authors of this chapter, is how to make sense of the overabundance of literature and offer guidance to the computer engineer and information specialist. This chapter aims to be as comprehensive as possible within the confines of this publication. The primary audiences are the computer engineering professional and information specialist with a secondary audience of the educated layperson. The task of identifying basic resources – for instance, for the high school student – is left to others. At the very least, this chapter will give pointers for further investigation and serve as a reminder of sources that may have been overlooked. The most important resources for the computer engineering field have been marked with an asterisk.

Fast and rapid dissemination of information is the hallmark of computer engineering, and for this reason access to conference proceedings is relatively more important than journals (the reverse is true in other science and engineering disciplines) – although this is changing with the development of online journals. The concept of the "technical report" emerged in computer engineering as a way of communicating early research results. With the advent of the Internet, it became even easier to provide access to technical reports via FTP servers and then later, the World Wide Web. The Internet has been a boon for the computer engineer with many tools freely available, and it is always a challenge to decide which sites to include in a publication such as this requiring one to critically evaluate each site for content, authority, usefulness, and stability.

Classic texts

A brief list of some of the classic texts in the discipline includes:

Abelson, H., Sussman, G.J., and Sussman, J. (1996) *The Structure and Interpretation of Computer Programs* (2nd edn), Cambridge, MA: MIT Press; New York: McGraw-Hill.

Brooks, F.P. Jr (1995) *The Mythical Man-month: Essays on Software Engineering* (Anniversary edn), Reading, MA: Addison-Wesley.

Dahl, O-J., Dijkstra, E.W., and Hoare, C.A.R. (1972) *Structured Programming*, London; New York: New Academic Press.

Dijkstra, E.W. (1976) *A Discipline of Programming*, Englewood Cliffs, NJ: Prentice-Hall.

Kernigham, B.W. and Plauger, P.J. (1978) *The Elements of Programming Style* (2nd edn), New York: McGraw-Hill.

Knuth, D.E. (1998) *The Art of Computer Programming* (3rd edn) (3 vols), Reading, MA: Addison-Wesley (Vol. 1. Fundamental algorithms; Vol. 2. Seminumerical algorithms; Vol. 3. Sorting and searching).

Note: As a continuation of this volume series (and to update previous volumes), Knuth has begun writing a series of books called "fascicles" to be published at regular intervals. The ultimate goal is for these new contributions to become part of the fourth edition.

Knuth, D.E. (2005) *The Art of Computer Programming, Volume 1, Fascicle 1. MMIX – A RISC Computer for the New Millennium*, Reading, MA: Addison-Wesley Professional.

Knuth, D.E. (2005) *The Art of Computer Programming, Volume 4, Fascicle 2. Generating All Tuples and Permutations*, Reading, MA: Addison-Wesley Professional.

Knuth, D.E. (2005) *The Art of Computer Programming, Volume 4, Fascicle 3. Generating All Combinations and Partitions*, Reading, MA: Addison-Wesley Professional.

Sammet, J.E. (1969) *Programming Languages: History and Fundamentals*, Englewood Cliffs, NJ: Prentice-Hall.

Weinberg, G. (1998) *The Psychology of Computer Programming* (Silver Anniversary edn), New York: Dorset House.

Wirth, N. (1986) *Algorithms and Data Structures*, Englewood Cliffs, NJ: Prentice Hall.

Searching library catalogs

For those wishing to identify sources within the field of computer engineering, the access points provided by Library of Congress Subject Headings (LCSH) are well defined. The subject heading Computer engineering is used for the field and Computer engineers is used for individuals within the field. The LCSH scope note states that:

> Here are entered works on the design of computer hardware and circuitry. Works on the logical structure that determines the way a computer executes programs are entered under Computer architecture. Works on the way a computer is constructed to implement its architecture, including what components are used and how they are connected, are entered under Computer organization.

Library of Congress Subject Headings

Computer Algorithms	Computer Programming
Computer Architecture	Computer Programs
Computer Graphics	Computer Science
Computer Hardware	Computer Software
Computer Interfaces	Computers
Computer Logic	Programming Languages (also specific names of languages such as C++, Java, XML)
Computer Networks	
Computer Organization	Software Engineering

Where the computer engineering literature may be shelved in a library's collection depends on whether its focus is on computer science, electrical engineering, systems security, or Internet-related technologies.

Library of Congress call numbers

Q 300	Artificial intelligence
QA 75.5–76.95	Electronic computers. Computer science
QA 76.75–76.765	Computer software
T58.5–T58.64	Information technology

TA 157	Computer engineers
TK 5100	Computer networks
TK 7800–8360	Electronics
TK 7885–7895	Computer engineering. Computer hardware
Z 102.5–104	Cryptography
ZA 3201–3250	Information superhighway
ZA 4201–4251	Internet

History of computing

The birth and development of computer engineering in its relatively short life has been both phenomenal and dramatic. The 1930s marked the beginning of the discipline – the era of Turing and Church – which gave us the fundamental mathematical concepts in computing. Dominant issues in the current arena include the Internet, and the growth of network technologies such as wireless, security and research on human–computer interfaces. The specialized field of computational science lends its techniques to all the other scientific disciplines as they try to manage huge amounts of data.

The discipline has resolved the identity issues it had in the 1950s. Does it belong in electrical engineering or is it a subdivision of mathematics? Will it develop into a science in its own right or will it fade away? The former question still sometimes poses a challenge for the field but the latter question is no longer an issue, as it has become a well-established subject. In its article on computer engineering, *Wikipedia* states:

> computer engineering degrees have been added to a number of schools' degree programs since the early 1990s. Some schools have integrated computer engineering, along with software engineering, into their electrical engineering departments, while others such as MIT have chosen to merge electrical engineering and computer science departments instead. Since computer engineers are mainly focused on electronics and computers, their course loads tend to involve fewer courses on natural sciences such as statics or dynamics than traditional engineering programs. Instead, courses on fundamentals of computer sciences are taught.

The first Ph.D. in computer science was awarded in 1965 to Richard L. Wexelblat at the University of Pennsylvania.

It is interesting to note, as this book goes to press, that the computing field is at an interesting crossroads – young enough that some of the early pioneers of the field are still practicing, either teaching and/or working in industry, and yet the "second generation" has been around long enough to have heard stories from the early pioneers directly.

Luckily, students of computing history have several key resources at their disposal. The Institute of Electrical and Electronics Engineers (IEEE) and Association for Computing Machinery (ACM), the two major professional associations in computer science, have published a vast amount of information on the topic through sponsorship of conferences, journals, and books. A search of the ACM's *The Guide to the Computing Literature*, Inspec and Compendex databases (see Databases section) will enable the user to do a more exhaustive search of the literature.

The journal IEEE *Annals of the History of Computing* is worth a special mention, since it provides a constant output of scholarly articles. It is one of the few periodicals to concentrate exclusively on the history of a scientific subject.

The study of computer engineering is a popular topic not only for the history of science scholar but also for the general public. A list of some of the significant, seminal, and recently published studies on the topic is provided below. It should not be considered comprehensive but a good starting point for further exploration.

Bibliographies

Cortado, J.W. (compiler) (1983) *Annotated Bibliography on the History of Data Processing*, Westport, CT: Greenwood Press.

Cortado, J.W. (compiler) (1990) *Bibliographic Guide to the History of Computing, Computers and the Information Processing Industry*, Westport, CT: Greenwood Press.

Cortado, J.W. (compiler) (1996) *Bibliographic Guide to the History of Computer Applications, 1950–1990*, Westport, CT: Greenwood Press.

These three titles provide a complete bibliographic history of computer engineering from 1800 to 1990 and cover contributions made by individuals and institutions in hardware, computing concepts, and software. The second title in the above list is a supplement to the first title. Each work is well organized with annotated entries and indexed by author and subject. There are about 10,000 entries in total over the three volumes.

Monographic series

The MIT Press "History of Computing Series" provides significant book-length studies of the subject. The series began in 1984 and has about thirty titles as of 2005. They may be viewed on the MIT Press home page <http://mitpress.mit.edu>; click on the topic "Computer Science and Intelligent Systems," then "Series" and "History of Computing." All of the titles in the series represent important in-depth studies and, for the sake of brevity, only a few of the well-known titles are listed below:

Aspray, W. (1990) *John von Neumann and the Origins of Modern Computing*, Cambridge, MA: MIT Press.

Ceruzzi, P.E. (2003) *A History of Modern Computing* (2nd edn), Cambridge, MA: MIT Press.

Cohen, I.B. (1999) *Howard Aiken: Portrait of a Computer Pioneer*, Cambridge, MA: MIT Press.

Cohen, I.B. and Welch, G.W. (1999) *Makin' Numbers: Howard Aiken and the Computer*, Cambridge, MA: MIT Press.

Pugh, E.W. (1984) *Memories that Shaped an Industry: Decisions Leading to IBM System 360*, Cambridge, MA: MIT Press.

Rojas, R. and Hashagen, U. (2000) *The First Computers – History and Architectures*, Cambridge, MA: MIT Press.

Stein, D. (1985) *Ada: A Life and Legacy*, Cambridge, MA: MIT Press.

Wilkins, M.V. (1985) *Memoirs of a Computer Pioneer*, Cambridge, MA: MIT Press.

Other significant texts

Agar, J. (2001) *Turing and the Universal Machine: The Making of the Modern Computer*, Cambridge: Icon Books; Lanham, MD: Totem Books; distributed to the trade in the U.S. by National Book Network.

Aspray, W. (ed.) (1990) *Computing Before Computers*, Ames, IA: Iowa State University Press.

Burks, A.R. (1988) *The First Electronic Computer: The Atanasoff Story*, Ann Arbor, MI: University of Michigan Press.

Dubbey, J.M. (1978) *The Mathematical Work of Charles Babbage*, Cambridge: Cambridge University Press.

Goldstine, H.H. (1993) *The Computer from Pascal to von Neumann* (reprint edn) Princeton, NJ: Princeton University Press. Available as part of ACLS History E-Book Project if library subscribes <http://www.hti.umich.edu/cgi/b/bib/bibperm?q1=HEB01140>.

Ifrah, G. (2001) *The Universal History of Computing: From the Abacus to the Quantum Computer* (translated from the French: *Histoire Universelle des Chiffres.* (1994). Paris, France: Editions Robert Laffont). New York: John Wiley.

Lukoff, H. (1979) *From Dits to Bits: A Personal History of the Electronic Computer*, Portland, OR: Robotics Press.

Mollenhoff, C.R. (1988) *Atanasoff: Forgotten Father of the Computer*, Ames, IA: Iowa State University Press.

Reilly, E.D. (2003) *Milestones in Computer Science and Information Technology*, Westport, CT: Greenwood Press.

Slater, R. (1987) *Portraits in Silicon*, Cambridge, MA: MIT Press.

Stern, N.B. (1981) *From ENIAC to UNIVAC: An Appraisal of the Eckert-Mauchly Computers*, Bedford, MA: Digital Press.

Wexelblat, R.L. (ed.) (1981) *History of Programming Languages: Proceedings of the History of Programming Languages Conference, Los Angeles, California, June 1–3, 1978*, New York: Academic Press.

Williams, M.R. (1997) *A History of Computing Technology* (2nd edn), Los Alamitos, CA: IEEE Computer Society Press.

Serials

IEEE Computer Society (1992–) *IEEE Annals of the History of Computing*, Los Alamitos, CA: IEEE Computer Society <http://www.computer.org/annals/> (accessed May 18, 2005). Begun in 1979, the *Annals* are the primary source for documenting the history of the discipline. They have articles by leading scholars as well as computer pioneers. It is available in print, via *IEEE Xplore* and the IEEE Computer Society. One of its useful features is the publication of special issues spotlighting topics. For instance, forthcoming issues will highlight the following: computing in Japan, creating the Internet, foundations of computer science.

Websites

Alan Turing.net: The Turing Archive for the History of Computing <http://www.cs.usfca.edu/www.AlanTuring.net/turing_archive/index.html> (accessed May 18, 2005). "Largest web collection of digital facsimiles of original documents by Turing and other pioneers of computing in addition to articles about Turing and his work (including Artificial Intelligence)" (source: Home page).

Charles Babbage Institute: Center for the History of Information Technology <http://www.cbi.umn.edu/collections/search.html> (accessed May 18, 2005). "The Charles Babbage Institute is an historical archives and research center of the University of Minnesota. CBI is dedicated to promoting study of the history of information technology and information processing and their impact on society. CBI preserves relevant historical documentation in all media, conducts and fosters research in history and archival methods, offers graduate fellowships, and sponsors symposia, conferences, and publications" (source: About CBI page). Of particular note is the *CBI Reprint Series for the History of Computing*.

Computer Oral History Project. The Lemelson Center for the Study of Invention and Innovation (National Museum of American History, Smithsonian Institution) <http://invention.smithsonian.org/resources/fa_comporalhist_index.aspx> (accessed May 18, 2005). "The Computer Oral History Collection (1969–1973, 1977) was a cooperative project of the American Federation of Information Processing Societies (AFIPS) and the Smithsonian Institution. This project began in 1967 with the main objective to collect, document, house, and make available for research source material surrounding the development of the computer. The project collected taped oral interviews with individuals who figured prominently in developing or advancing the computer field and supplemental written documentation – working papers, reports, drawings, and photographs" (source: Home page). Some interviews are available full-text online.

Virtual Museum of Computing <http://vmoc.museophile.com/> (accessed May 18, 2005). This virtual museum, which "opened" in 1995, is a comprehensive site specializing in the history of computing compiled by Professor Bowen of South Bank University, London, U.K.

Book reviews

In addition to the standard book review journals such as *Choice* and *Library Journal*, here are some specialized resources for computer engineering:

**Computing Reviews* <http://www.reviews.com> (accessed May 18, 2005). *Computing Reviews* has been published for over forty years by ACM and is the major review journal for the discipline. Each review is authored by experts and covers the book and journal literature. In 2000, ACM entered into a partnership with Reviews.com and the journal metamorphosed into an online database. The database, which now has over 20,000 entries, continues the tradition of excellent expert reviews with presentation of up-to-date material and some value-added features such as personalized alerts, customized searching and browsing. The aim is to create a virtual community. The print version of *Computing Reviews* is still being published, and remains a good tool for those who cannot afford the database but want to keep up with the literature. The print is included in the ACM Journal print package (but not the ACM Digital Library). The sources are selected from the ACM *Guide to Computing Literature*, which includes both ACM and non-ACM material, and is organized according to the ACM Computing Classification System (CCS). This database, either in the printed or online version, is an invaluable tool for book selection. Considered a "consumer reports" journal for information specialists, *The Charleston Advisor* in its October 2004 issue gave it an overall ranking of 4.25 out of 5 possible stars. The quality of its search engine was given 5 stars <http://www.charlestonco.com/review.cfm?id=200> (accessed May 18, 2005).

Electronic Review of Computer Books <http://www.ercb.com/> (accessed May 18, 2005). ERCB is a collection of reviews on books about computer hardware, software, and networking that have appeared in *Dr. Dobb's Journal*.

TechBookReport <http://www.techbookreport.com/> (accessed May 18, 2005). "*TechBookReport* aims to provide independent, interesting and informative book reviews for developers, technologists and the plain geeky" (source: Home page). Reviews are written "by and for" the practitioner.

Dictionaries

Anyone brave enough to compile a dictionary of computing terms is faced with many challenges. In short, by the time the dictionary is published it may already be out-of-date. However, every library collection or computing professional should have a few choice titles at their fingertips to clarify terms and usage. Online dictionaries offer obvious advantages for this reason. The bulk of the literature in computer engineering is published in English but it may be prudent to have a couple of good dictionaries to translate terms from other languages. When the discipline was young (not very long ago), many dictionaries were published to explain basic and complex terms to the neophyte. However, publishing in this area has slowed down somewhat. A current trend is for dictionaries to be published in niche or new areas. One example is the dictionary on embedded systems listed below.

We offer a hand-picked list of well-respected and established titles with the expectation that, if desired, the serious researcher can search *Books in Print* and Amazon.com to find more titles.

Print dictionaries

*Daintith, J. (ed.) (2005) *A Dictionary of Computing* (5th edn), Oxford; New York: Oxford University Press. Fully revised by a team of computer scientists, this edition provides comprehensive, up-to-date coverage of the language of computing, including hardware and software applications, programming languages, networks and communications, the Internet, and e-commerce. In addition to the clearly explained 10,000 terms, many of which are new for this edition, the dictionary offers words used in context, and includes quotations from computing magazines and useful tables of programming codes and languages. It is an ideal resource for any general computer user.

Downing, D.A., Covington, M.A., and Covington, M.M. (2003) *Dictionary of Computer and Internet Terms* (8th edn), Hauppauge, NY: Barron's. Completely revised, this pocket reference provides more than 2,500 definitions for terms and concepts relating to computers (including business software and computer architecture) and Internet functions with illustrations, photographs, diagrams or tables for many of the applications described. Useful for the new to intermediate computer user.

Freedman, A. (2001) *The Computer Glossary: The Complete Illustrated Dictionary* (with CD-ROM) (9th edn), New York: AMACOM. *The Computer Glossary* was considered one of the best one-volume computer dictionaries on the market and an essential resource for many computer users over the past decade. Sadly, there will be no more print editions [see entry on author's *Computer Desktop* Encyclopedia for more details]. This edition contains 6,000 definitions along with 175 illustrations. Alan Freedman is President of the Computer Language Company – an organization dedicated to computer education and training.

Ganssle, J. and Barr, M. (2003) *Embedded Systems Dictionary*, San Francisco, CA: CMP Books. This is a good example of a recent dictionary that deals with a segment of computer science and engineering. It is intended for engineers (on both the hardware and software side) who design embedded systems. It covers the history, technologies, tools, abbreviations, and definitions for the terms most used in embedded systems. Serving both technical and non-technical audiences, the dictionary contains 2,800 terms.

*LaPlante, P.A. (ed.) (2001) *Dictionary of Computer Science, Engineering and Technology*, Boca Raton, FL: CRC Press. Written by an international team of over eighty contributors, this dictionary provides detailed definitions (including illustrations where appropriate) and practical information spanning various disciplines and industry sectors. Its 8,000 terms cover all aspects of computing and computer technology from multiple vantage points – including academic, applied, and professional. Unfortunately, this edition has not been updated.

McGraw-Hill (2003) *McGraw-Hill Dictionary of Computing and Communications*, New York: McGraw-Hill. Compiled from the *McGraw-Hill Dictionary of Scientific and Technical Terms* (6th edn 2003), this dictionary contains 11,000 entries essential in the field of computer science, information technology, and communications (relating to both digital and analog data). Its definitions indicate which field the term belongs to together with synonyms, pronunciations, cross-references, acronyms, and abbreviations. Includes an appendix containing charts and tables for measurement conversion and other useful data. Suitable for the general reader up to the professional. *Note:* Depending on their clientele and budget, libraries may not wish to duplicate content in their collection if they already own the larger sixth edition. Content from the larger sixth edition (plus the *McGraw-Hill Encyclopedia of Science and Technology* (9th edn) is available via annual subscription to *Access Science* <http://www.accessscience.com/> (accessed May 18, 2005).

McGraw-Hill (2004) *McGraw-Hill Dictionary of Electrical and Computer Engineering*, New York: McGraw-Hill. This dictionary, published in late 2004, covers the areas of electrical, computer, and electronics engineering and related areas of mathematics and communications. It features 15,000 entries based on terms from the *McGraw-Hill Dictionary of Scientific and Technical Terms* (6th edn) 2003). In addition to definitions, it includes synonyms, acronyms, and abbreviations and pronunciations for all terms as well as an extensive appendix. *Note:* Depending on their clientele and budget, libraries may not wish to duplicate content in their collection if they already own the larger sixth edition. The January 2005 *Choice* review states that "practicing engineers could benefit from the present dictionary because of its tighter focus and lengthy appendixes." Content from the larger sixth edition (plus the *McGraw-Hill Encyclopedia of Science & Technology* (9th edn) is available via annual subscription to *AccessScience* <http://www.accessscience.com/>.

Microsoft (2002) *Microsoft Computer Dictionary* (5th edn), Redmond, WA: Microsoft Press. Now minus the accompanying CD-ROM, this fifth edition defines computing terms (including acronyms, jargon, and slang) and concepts pertaining to hardware, networks, programming, applications, and databases. Unlike other dictionaries, it does not tackle information about companies, commercial products, or proprietary technologies (except from Microsoft itself). The more than 10,000 entries explain concepts and technical terms in easy-to-understand language, making this a good resource to have for either the beginning or advanced computer user.

Narins, B. (ed.) (2002) *World of Computer Science* (2 vols), Detroit: Gale Group/Thomson Learning. A good introductory reference on the subject with an impressive list of contributors and advisors. Alphabetically arranged, the 800+ entries discuss pioneers, discoveries, theories, concepts, issues, and ethics. Of particular interest are the biographies of living computer personalities. Not aimed at an academic audience, *Choice* remarks that it will help new computer science students and professionals alike.

Newton, H. (2005) *Newton's Telecom Dictionary: Covering Telecommunications, Networking, Information Technology, the Internet, Fiber Optics, RFID, Wireless, and VoIP* (21st edn), San Francisco, CA: CMP Books. Long considered the telecom "Bible," Newton's newest dictionary editions continue to cover technical terms and acronyms in the ever-changing telecom, network, and IT industry for technology and business professionals. Each revised edition includes updated and expanded definitions including any new standards, technologies, and vendor-specific terms. With its use of non-technical language to explain concepts, *Newton's Telecom Dictionary* is considered an "essential reference" (*PC Magazine*) for anyone involved with telecom and IT systems and services.

Pfaffenberger, B. (ed.) (2003) *Webster's New World Computer Dictionary* (10th edn), Indianapolis, IN: Wiley. Useful for the novice or the professional, this updated dictionary contains over 4,750 definitions for computer terms including current coverage of standards and protocols for storage, memory, and peripherals (with excellent cross-referencing of terms). Bryan Pfaffenberger is the author of more than seventy-five books on personal computing (including the *HTML 4 Bible*) and teaches at the University of Virginia's Division of Technology, Culture, and Communication. This is a good reference work for any collection.

Free online dictionaries

BABEL: A Glossary of Computer Oriented Abbreviations and Acronyms <http://www.geocities.com/ikind_babel/babel/babel.html> (accessed May 18, 2005). Available on the Internet since 1989, Babel covers computer-related abbreviations and acronyms. It is alphabetically arranged and includes three appendices. It offers no built-in search capability.

Dictionary of Algorithms and Data Structures <http://www.nist.gov/dads/> (accessed May 18, 2005). "This is a dictionary of algorithms, algorithmic techniques, data structures, archetypical problems, and related definitions. . . . We do not include algorithms particular to business data processing, communications, operating systems or distributed algorithms, programming languages, AI, graphics, or numerical analysis" (source: Home page). Use the alphabetical arrangement or the search feature to find a definition.

*FOLDOC: Free On-line Dictionary of Computing <http://foldoc.doc.ic.ac.uk/foldoc/index.html> (accessed May 18, 2005). "FOLDOC is a searchable dictionary of acronyms, jargon, programming languages, tools, architecture, operating systems, networking, theory, conventions, standards, mathematics, telecoms, electronics, institutions, companies, projects, products, history, in fact anything to do with computing" (source: Home page).

Mathematical Programming Glossary© <http://carbon.cudenver.edu/~hgreenbe/glossary/index.php> (accessed May 18, 2005). Professor Harvey J. Greenberg from the Mathematics Department at University of Colorado at Denver began this glossary in 1996. It "contains terms specific to mathematical programming, and some terms from other disciplines, notably economics, computer science, and mathematics, that are directly related" (source: Home page). Search the alphabetical index or jump to a letter to begin. No sources are cited.

Netlingo <http://www.netlingo.com/inframes.cfm> (accessed May 18, 2005). "NetLingo is an online dictionary about the Internet. It contains thousands of words and definitions that describe the technology and community of the World Wide Web" (source: Home page).

Wĕbopēdia™ <http://www.pcwebopaedia.com/> (accessed May 18, 2005). "Webopedia is a free online dictionary for words, phrases and abbreviations that are related to computer and Internet technology. . . . Full-time experienced editors gather information from standards bodies, leading technology companies, universities, professional online technical publications, white papers and professionals working in the field. The sources used are often listed in the links section below. . . . Every definition is verified among multiple sources; definitions are never based on just one source. . . . New terms are added on a daily basis. . . . Webopedia is part of the internet.com network of Web sites" (source: Home page).

Encyclopedias

Dictionaries provide a quick answer to the meaning or definition of terms but encyclopedias will give you an overview of the topic. Encyclopedias are particularly useful when one is embarking on research in a new discipline or topic. They will provide you with background information, brief historical context within the discipline, written by experts in the field, and a short bibliography for further exploration. However, these short articles are a "snapshot" in time and it is up to information-seekers to use more up-to-date resources (such as books, periodical indexes, and web resources) to augment their knowledge. Librarians in particular will find a good encyclopedia a godsend when faced with reference questions on a new topic.

As with the section on Dictionaries, we offer a few select titles for your computer engineering science collection. Encyclopedias are expensive, so replacing copies with new editions or new titles every year might not be economical or necessary, but replacing titles every five years is more realistic. Our highly selective list below includes the well-established and respected comprehensive encyclopedias in computer science plus some newcomers to the arena who concentrate on a particular aspect of the discipline.

Bainbridge, W.S. (ed.) (2004) *Berkshire Encyclopedia of Human–Computer Interaction* (2 vols), Great Barrington, MA: Berkshire Group. This encyclopedia, edited by the Deputy Director of the National Science Foundation's Division of Information and Intelligent Systems, consists of 186 signed articles on human–computer interaction (HCI). It is the first major reference resource for a new and fast-changing field that draws upon many branches of social, behavioral, and information sciences, as well as computer science, medicine, engineering, and design. The encyclopedia covers all aspects of HCI including applications, breakthroughs, challenges, interfaces, and methods. The articles are accompanied by a comprehensive bibliography, glossaries, and a "popular culture database." This would complement any of the standard encyclopedias listed in this section for its unique content.

Belzer, J., Holzman, A.G., and Kent, A. (eds) (1975–2001) *Encyclopedia of Computer Science and Technology* (43 vols), New York: Marcel Dekker. Updated by supplements. This comprehensive reference provides access to articles on state-of-the-art computer technology. Approximately 700 articles written by 900 international authorities feature current developments and trends in computers, software, vendors, and applications as well as in-depth analysis of future directions. There are extensive bibliographies of leading figures in the field. Subscribers to the encyclopedia can stay up-to-date with book-length supplements that cover a variety of timely topics, keeping readers aware of the newest developments, the latest buzz-words, hardware and software changes, and individuals making noteworthy contributions. Volume 15 (Supplement 1) begins the Supplement series, then continues with Volume 17 (Supplement 2) – Volume 16 is the index for the previous volumes. Libraries may be interested in selecting specific supplement titles of interest for their collection instead of purchasing the entire set of available forty-three volumes which carries a hefty price tag.

Bidgoli, H. (ed.) (2004) *The Internet Encyclopedia* (3 vols), Hoboken, NJ: John Wiley & Sons. This substantive three-volume set covers every aspect of fast-moving Internet technology for computer professionals, offering a broad perspective on the Internet as a business tool, an IT platform, and a medium for communications and commerce. Experts from institutions such as Stanford University and Harvard University and leading corporations such as Microsoft and Sun Microsystems describe leading-edge technology and recent developments in lengthy signed articles. Articles are supplemented with flow charts, programming samples, and technical diagrams as appropriate as well as bibliographies. The editor is a tenured professor at California State University Bakersfield, School of Business, Management Information System Department. Volume 3 is also available as a Wiley Interscience OnlineBook <http://www.interscience.wiley.com/online-books>.

*Freedman, A. (ed.) *Computer Desktop Encyclopedia: the Indispensable Reference on Computers*, Point Pleasant, PA: Computer Language Co. CD-ROM product (quarterly updates) <http://www.computerlanguage.com/index.htm> (accessed May 18,

2005). No longer published as a print edition (the last one (9th edn 2001) was published by McGraw-Hill/Osborne), it is now available as a CD-ROM product containing more than 18,000 definitions about the computer field and IT industry accompanied by 2,500 images. Content is available through various technical websites (one example is TechEncyclopedia <http://www.techweb.com/encyclopedia/> (accessed May 18, 2005)), but this lacks the full functionality of the CD-ROM.

Marciniak, J.J. (ed.) (2002) *Encyclopedia of Software Engineering* (2nd edn) (2 vols), New York: Wiley. Considered the most comprehensive reference in software engineering by *Computing Reviews*, this is an important encyclopedia for practitioners who design, write, or test computer programs covering all the issues and principles of software design and engineering. This completely revised and updated edition contains 500+ entries in thirty-five taxonomic areas as well as biographies for over 100 important contributors to the field. Among the issues discussed are the Software Engineering Body of Knowledge Project, software engineering ethics, licensing and certification of software engineering personnel, and education and training in software engineering. The terminology used complies with the standard set by IEEE. Also available as a Wiley Interscience OnlineBook <http://www.interscience.wiley.com/onlinebooks>.

*Ralston, A., Reilly, E.D., and Hemmendinger, D. (eds) (2003) *Encyclopedia of Computer Science* (4th edn), Chichester, West Sussex; Hoboken, NJ: Wiley. Since 1976, this has been the definitive and comprehensive reference work on computers, computing, and computer science. With over 2,000 pages, the newly revised edition contains over 100 new articles and more than 600 completely updated articles by internationally known experts. The articles are signed and the author affiliations included at the front of the book. Alphabetically arranged and classified into broad subject areas, the entries cover hardware, computer systems, information and data, software, the mathematics of computing, theory of computation, methodologies, and applications and the computing milieu. The encyclopedia skillfully combines historical perspective with practical reference information. This work is a must-have for all academic and public libraries and is an essential resource for computer professionals, engineers, mathematicians, students, scientists, and librarians. Short bibliographies, cross-references to other articles in the volume plus a name and subject index further increase the usefulness of this publication. The nine appendices at the back of the book are worth a special mention. They cover Abbreviation and acronyms; Notations and units; Computer journals and magazines; Ph.D. granting departments of computer science and engineering; Presidents of major computing societies; Key high-level languages; Glossary of major terms in five languages; Articles deleted from previous editions; and a Timeline of significance computing milestones.

Reilly, E.D. (ed.) (2004) *Concise Encyclopedia of Computer Science*, Chichester, West Sussex; Hoboken, NJ: Wiley. This is the perfect encyclopedia for the non-specialist and is based on the fourth edition of the *Encyclopedia of Computer Science* (Ralston *et al.* above), with shorter versions of 60 percent of the entries in the fourth edition.

Rojas, R. (ed.) (2001) *Encyclopedia of Computers and Computer History* (2 vols), Chicago, IL: Fitzroy Dearborn. This encyclopedia aims to cover the complete subject of computers and their history from personal computing to main-frames to robotics and artificial intelligence as well the theoretical foundations of computer science. The 600+ entries cover facts, definitions, biographies, histories, and explanations of diverse topics. Contributors are scholars in computer science and computer history from around the world. This source will only serve as a starting point for more sophisticated users.

Sheldon, T. (2001) *McGraw-Hill Encyclopedia of Networking & Telecommunications*, Berkeley, CA: Osborne. Accompanied by CD-ROM. An encyclopedic reference of information with 1,400 entries and over 3,000 links on computer networking and telecommunications, covering subjects ranging from Bluetooth to mobile computing. The fully searchable CD-ROM consists of an e-version of the text with hyperlinks between related topics (the complete set of Internet RFCs are accessible via hyperlinks throughout the e-version of the book), external hyperlinks to websites, and illustrations of complex networking topics. A free website <http://www.linktionary.com/about.html> (accessed May 18, 2005) allows you to search for and view reduced versions of the definitions found in the book but the searching capabilities are not as robust as the CD-ROM's. The website is also the "home" for book addendums and recent topic updates.

Wiley Encyclopedia of Computer Science and Engineering (6 vols), forthcoming title from a publisher with a reputation for excellent technological encyclopedias. Please consult the publisher's website for more information <http://www.wiley.com/>.

Handbooks and manuals

Engineering handbooks are ready-reference tools – compact, comprehensive sources of data and information frequently needed by engineers. They can help you find quick, factual information to support your ideas or reacquaint you with theories, formulas, and other data that may be scattered throughout the literature.

Chen, W-K. (ed.) (2003) *The Circuits and Filters Handbook* (2nd edn), Boca Raton, FL: CRC Press. Written for practicing electrical engineers, the second edition has been thoroughly updated to provide the most current, comprehensive information available in both the classical and emerging fields of circuits and filters, both analog and digital. This edition contains twenty-nine new chapters, with significant additions in the areas of computer-aided design, circuit simulation, VLSI circuits, design automation, and active and digital filters. The volume begins with an overview of mathematics as it relates to the subject. The articles are written by experts in the field with a bibliography at the end of each chapter. This handbook is part of the "Electrical Engineering Handbook Series." Contents of CRC Press engineering handbooks may be searched online at <http://www.engnetbase. com/> (accessed May 18, 2005); access to CRC full-text resources is available with annual subscription.

Freeman, R.L. (2002) *Reference Manual for Telecommunications Engineering* (3rd edn) (2 vols), New York: Wiley. For over fifteen years, the *Reference Manual for Telecommunications* has been regarded as an essential design tool for engineers and technicians who deal with communications technology. Now expanded to a two-volume set, this completely revised and updated third edition features over 3,500 pages of the latest information (organized into forty-one subject areas) on designing, building, purchasing, using, and maintaining telecommunications systems. Gathered from industry, government, and academic sources, the manual contains all the technical material a telecom professional might need on a daily basis. It also includes a wealth of tables, figures, nomograms, formulas, statistics, standards, regulations, and explanatory text. Also available as a Wiley Interscience OnlineBook <http://www.interscience.wiley.com/onlinebooks>.

Lyons, R.G. (2004) *Understanding Digital Signal Processing* (2nd edn), Upper Saddle River, NJ: Prentice Hall PIR. In this updated and expanded edition, the author demonstrates how engineers and other technical professionals can master and apply DSP techniques. This edition adds extensive new coverage of quadrature signals for digital communications, recent improvements in digital filtering, and contains more than twice as many "DSP Tips and Tricks."

Oklobdzija, V.G. (ed.) (2002) *The Computer Engineering Handbook*, Boca Raton, FL: CRC Press. The author (who hails from IEEE) along with other industry experts have created a comprehensive, state-of-the-art review of the most recent achievements and new directions and developments for the field of computer design and engineering. The handbook organizes information into three top-level sections covering fabrication and technology; computer systems and architecture; and reliability and testability. Unfortunately, this edition has not been updated. Contents of CRC Press engineering handbooks may be searched online at http://www.engnetbase.com/; access to CRC full-text resources is available with annual subscription.

Smith, S.W. (2003) *Digital Signal Processing: A Practical Guide for Engineers and Scientists*, Amsterdam; Boston, MA: Newnes. Accompanied by CD-ROM. This guide explains DSP design, algorithms, and techniques as well as the operation and usage of DSP chips. Even though this reference work is aimed at engineers and scientists, it avoids abstract, theoretical, and mathematical explanations. The accompanying CD-ROM contains information from various DSP processor manufacturers, DSP software tools, and the code used in the applications examples.

*Tucker, A.B. (ed.) (2004) *Computer Science Handbook* (2nd edn), Boca Raton, FL: Chapman & Hall/CRC Press. Aimed at computer scientists, software engineers, and IT professionals, this edition has broadened its scope, emphasizing a more practical and applied approach to computing. The seventy-plus new or revised chapters are written by over 150 recognized experts. It provides coverage across all eleven subject areas of the discipline as defined in ACM/IEEE 2001 Computing Curricula 2001. Also available as a CD-ROM.

Van Leeuwen, J. (ed.) (1990) *Handbook of Theoretical Computer Science* (2 vols), Amsterdam; New York: Elsevier; Cambridge, MA: MIT Press. This handbook provides professionals and students with a comprehensive overview of the main results and developments in this rapidly evolving field. Volume A covers models of computation, complexity theory, data structures, and efficient computation in many recognized subdisciplines of theoretical computer science. Volume B takes up the theory of automata and rewriting systems, the foundations of modern programming languages, and logics for program specification and verification, and presents several studies on the theoretic modeling of advanced information processing. The two volumes comprise thirty-seven chapters, with extensive chapter references and individual tables of contents for each chapter. The editor points out in the preface that although "the volumes can be used independently, there are many interesting connections that show that the two areas really are highly intertwined." There are 5,387 entry subject indexes that include notational symbols and a list of contributors and affiliations in each volume. Even though this book was published more than a decade ago, it is still useful given the nature of the subject area – the theoretical and mathematical underpinnings of computer science.

Technical reports

Technical reports, usually numbered report series, are often published by university departments, government agencies, and companies (to a lesser extent) to report on current research in a timely fashion. Technical reports are highly technical in nature, not peer-reviewed, and describe unsuccessful as well as successful research. Given the nature of computer science, technical reports remain an important vehicle for reporting early results prior to presentation at conferences and publication in the journal literature.

The future of computer science technical reports is uncertain at present as most organizations include information on current research projects on their websites, and it remains to be seen if the genre will disappear altogether. However, from time to time, there is a need to track down computer science technical reports – retrospective as well as current reports.

Before the Internet, organizations, mostly academic computer science departments, exchanged hard copies and microfiche of computer science technical reports; later these were placed on FTP servers for easy access. During that time, directories of FTP servers were compiled, but a quick survey of these listings shows that they are, not surprisingly, out-of-date. Now just a quick review of an organization's website will reveal if the technical report is available for download or a contact address to help you track down a copy. Some institutions, such as California Institute of Technology, have digitized their entire computer science technical reports collection and made them available globally. These reports may be searched via the CaltechCSTR website <http://caltechcstr.library.caltech.edu/> (accessed May 18, 2005) or Google's search engine <http://www.google.com/>.

Below are some starting points that may be used to find computer science technical reports:

Carnegie Mellon University Engineering and Science Library <http://www.library.cmu.edu/Research/EngineeringAndSciences/Eng/techrpts.html> (accessed May 18, 2005). This library houses one of the largest physical collections of computer science technical reports in the nation dating back to 1966 (approximately 68,000 reports in 2000). The Library's Technical Reports web page provides links to CMU departmental technical reports as well as other technical report sites. Or you can find a technical report in the library collection by searching the library catalog CAMEO at <https://unicorn.library.cmu.edu/> (accessed May 18, 2005).

*Networked Computer Science Technical Report Library (NCSTRL) <http://uther.dlib.vt.edu/oaincstrl/index.pl> (accessed May 18, 2005). NCSTRL (pronounced "Ancestral") is a distributed library of computer science technical reports developed in the mid-1990s using the Dienst protocol. The architecture was developed with DARPA funding and intended for developing NCSTRL and other similar digital libraries. By 1999, the collection consisted of papers from over 100 research institutions residing in servers distributed over the U.S., Europe, and Asia. With the advent of the Open Archives Initiative (OAI) <http://www.openarchives.org/> (accessed May 18, 2005), NCSTRL has been moving toward an OAI-compliant environment with data being harvested from source archives and then stored in a Union Catalog. A search engine has been developed for demonstration purposes. *D-Lib Magazine* <http://www.dlib.org/>, a free online journal, has several articles about the development of NCSTRL. In addition to its own articles, *D-Lib Magazine* also tracks other articles, books, news, conferences, and links to other sites on digital libraries.

University of Maryland's Virtual Technical Reports Collection <http://www.lib.umd.edu/ENGIN/TechReports/Virtual-TechReports.html> (accessed May 18, 2005). This metasite provides links to full-text or searchable extended abstracts of technical reports, preprints, reprints, dissertations, theses, and research reports in all disciplines.

Online bibliographies, preprints, and repositories

In computer science, we find a proliferation of free online resources that augment the traditional means of publishing papers in journals and conference proceedings for fast dissemination and exchange of information. In many cases, these resources are experimental and provide a testbed for emulation not only in computer science but also in other disciplines.

This section provides a list of the major resources in this category mostly limited to those covering the discipline as a whole. Technical reports have been covered in the previous section. For an excellent listing of current resources in subspecialties (such as cryptology, computer vision) please see the computer science metasite maintained by the State University of New York at Albany Library <http://library.albany.edu/subject/csci.htm> (accessed May 18, 2005).

Please note that it is not advisable to use any of these resources as the sole source for bibliographic research but in conjunction with the other traditional databases (listed in the section on Databases). All these resources are freely available.

BibFinder <http://kilimanjaro.eas.asu.edu/> (accessed May 18, 2005). BibFinder is an innovative bibliographic search engine for computer science literature which simultaneously searches a variety of databases including the *ACM Digital Library*, *IEEE Xplore*, *ScienceDirect*, *CiteSeer*, and Google. These sources are partially overlapping and the developers of BiBFinder are using this as a testbed.

*CiteSeer <http://citeseer.ist.psu.edu/cs/> (accessed May 18, 2005). CiteSeer (formerly known as Research Index) was developed by the NEC Research Institute and is now hosted by Pennsylvania State University. "CiteSeer is a scientific literature digital library and search engine that focuses primarily on the literature in computer and information science. CiteSeer aims to improve the dissemination and feedback of the scientific literature and to provide improvements in functionality, usability, availability, cost, comprehensiveness, efficiency, and timeliness in the access of scientific and scholarly knowledge" (source: Home page). The archive provides access to journal articles, conference papers, and technical reports in computer science. In addition to providing a list of documents by author or subject, the search engine also generates a form of citation indexing similar to the *Science Citation Index* (produced by the Institute for Scientific Information) called in this case "autonomous citation indexing."

CoGPrints <http://cogprints.org/> (accessed May 18, 2005). Using Eprints.org software and in compliance with the Open Archives Initiative (OAI) <http://www.openarchives.org/> (accessed May 18, 2005), CoGPrints allows self-archiving of papers in psychology, neuroscience, linguistics, philosophy, biology, and computer science (e.g., artificial intelligence, robotics, vision, learning, speech, neural networks). Anyone can deposit papers for consideration but registration is required.

The Collection of Computer Science Bibliographies <http://liinwww.ira.uka.de/bibliography/index.html> (accessed May 18, 2005). This resource, updated monthly, is a collection of references from almost 1,500 bibliographies covering most of computer science and related areas in mathematics. The database provides access to journal articles, conference papers, and technical reports. Searching can be limited by publication type and date.

*Computing Research Repository (CoRR) <http://arxiv.org/corr/home> (accessed May 18, 2005). The goal of CoRR is to be the single repository for computer science preprints. It is sponsored by ACM, the arXiv archive, NCSTRL, and the American Association for Artificial Intelligence (AAAI). Established in 1998, CoRR allows researchers to search the repository, browse by year or subject class as well as download papers.

Digital Bibliography and Library Project (DBLP) <http://dblp.uni-trier.de/> (accessed May 18, 2005). This was originally a bibliography for database systems and logic programming but later widened its coverage to include more of the discipline. Hosted by several servers around the world, it lists over 500,000 references from major computer science journals and conference papers.

E-Print Network: Research Communications for Scientists and Engineers Computer Technologies and Information Sciences <http://www.osti.gov/eprints/pathways/computertech.shtml> (accessed May 18, 2005). "The *E-print Network* is a set of powerful tools that facilitate access to and use of scientific and technical e-prints communicating the results of a wide range of research activities of interest to the Department of Energy" (source: Home page). However, there is a section on "Computer Technologies and Information Sciences." The site also includes links to scientific societies, and an alerting feature notifies users when new items are added in their areas of interest.

Patents and standards

For information on how to locate patents, please refer to the Patents section in Chapter 2. In the computing sector, software has always been a patent "problem child." The rather unique website listed below helps to sort out how to research software prior art.

Software Patent Institute Database of Software Technologies <http://www.spi.org/> (accessed May 18, 2005). "The Software Patent Institute is dedicated to providing information to the public and assisting the United States Patent and Trademark Office and others by providing technical support in the form of educational and training programs and providing access to information and retrieval resources concerning software prior art" (source: Home page).

The following two sites provide good links to sources of computer standards (e.g., for hardware) or standards organizations. Please also refer to the Standards section in Chapter 10 on Electrical engineering and electronics for additional information.

CompInfo – The Computer Information Center: Computer Standards <http://www.compinfo.co.uk/itman/computer_standards.htm> (accessed May 18, 2005).

Yahoo! Directory > Computers and Internet > Standards <http://dir.yahoo.com/Computers_and_Internet/standards/> (accessed May 18, 2005).

Some of the important standards organizations for computing-related standards are IEEE (e.g., for wireless – see entry below), International Standards Organization (ISO) <http://www.iso.org/> (accessed May 18, 2005), and Internet Engineering Task Force (IETF) (e.g., for networking and Internet – see entry below).

IEEE Standards Online: IT Standards <http://standards.ieee.org/catalog/olis/allit.html> (accessed May 18, 2005). IEEE offers an information technology subscription package which includes all of the current IEEE standards in the following areas: Bus architecture/Microprocessor/Microcomputer, Communications, Design automation, Local area networks/Metropolitan area networks (LAN/MAN 802) plus Drafts, portable applications, and Software engineering (SE). Also included in this package are the technical collections which provide standards on computer simulation, Internet, learning technology, storage systems, test and diagnosis for electric systems and test technology. Access to full-text standards is available through IEEE Xplore <http://ieeexplore.ieee.org/xpl/standards.jsp> (accessed May 18, 2005) as an IEEE or subscribing member, or one may purchase individual computer engineering standards from Shop IEEE <http://shop.ieee.org/ieeestore/default.aspx> (accessed May 18, 2005).

Internet Engineering Task Force (IETF) <http://www.ietf.org/> (accessed May 18, 2005). "The Internet Engineering Task Force (IETF) is a large open international community of network designers, operators, vendors, and researchers concerned with the evolution of the Internet architecture and the smooth operation of the Internet. . . . It is the principal body engaged in the development of new Internet standard specifications" (source: Overview page and Tao page).

Conferences

This section provides guidance on locating information on forthcoming conferences. More information on locating papers published in conference proceedings is listed in the Databases section.

Face-to-face exchange of information at conferences is still as important as ever in computer science even though communicating via electronic media has become commonplace. This is evidenced by the large number of conferences sponsored annually by the major associations in computer engineering: ACM, IEEE Computer Society, IEE and the British Computer Society, to name a few of the key players. A list of the major organizations in computer engineering is in the Associations, societies and other organizations section of this chapter. The websites usually compile listings not only of their own forthcoming sponsored conferences but also others that would be of interest to their research community. ACM and IEEE also include information on how to organize a conference for their respective organizations. These are two sites to bookmark and monitor on a regular basis:

ACM Events and Conferences <http://www.acm.org/events/> (accessed May 18, 2005).
IEEE Computer Society Conferences <http://www.computer.org/conferences/> (accessed May 18, 2005).

Below are some additional sites specializing in conference announcements:

All Conferences.Com <http://www.allconferences.com/> (accessed May 18, 2005). Online directory focusing on conferences, conventions, trade shows, exhibits, workshops, events, and business meetings with a category for "Computers and Internet." In addition to listing sites, the company also provides services to conference organizers.

Atlas Conferences Inc. <http://atlas-conferences.com/index.html> (accessed May 18, 2005). Founded in 2000, this website maintains a database of forthcoming academic conferences, meetings, and events. Browse conferences by subject, such as "Computer science," date, or country. Also provides conference support services.

EEVL: Internet Guide to Engineering, Mathematics, and Computing Current Awareness: Conferences, Symposia and Conference Papers Page <http://www.eevl.ac.uk/current_awareness_services.html#conferences> (accessed May 18, 2005). This metasite complements the two society sites listed above and provides a good starting point for tracking conference announcements.

Tech Expo: A One-Stop Technical Conference Information Center <http://www. techexpo.com/events/evnts-p1.html> (accessed May 18, 2005). Events are listed in chronological order. To find upcoming technical events, click on an appropriate date range or search by conference name, topic, sponsor, country, city, or state. A link to Amazon.com is also provided to search for books on the same topic as the conference.

Journals

The phenomenal growth of the field has given rise to a plethora of journals. We can categorize the journals into the following main types:

Academic/scholarly Journals published mainly by associations, societies, and commercial publishers

Any list of computer-related academic journals is dominated by the IEEE Computer Society and ACM titles. Researchers who monitor appropriate journals from these two societies are keeping up fairly well with the professional literature in their field. Some of the academic journals are published commercially. Please also see Publisher resources under the Current awareness section. Two of the major commercial publishers are:

Elsevier: Computer Science Journal Titles <http://www.elsevier.com/wps/find/journal_browse.cws_home/P05?pseudotype=&sortBy=Title&SH1Code=P05&letter=A> (accessed May 18, 2005).

Springer: Computer Science Journal Titles <http://www.springeronline.com/sgw/cda/frontpage/0,11855,5-40100-2-70920-0,00.html> (accessed May 18, 2005).

Springer also publishes the important series *Lecture Notes in Computer Science* (LNCS) <http://www.springeronline.com/sgw/cda/frontpage/0,11855,4-164-12-72397-0,00.html> (accessed May 18, 2005) and its two sub-series (to date) *Lecture Notes in Artificial Intelligence* (LNAI) and *Lecture Notes in Bioinformatics* (LNBI). These series, which have a well-deserved reputation in the computer science research community, provide an important venue for publication of new developments in the field. Online access has greatly enhanced their usability.

A special subset in this group is journals from industrial organizations. Three of the well-known titles in this category are:

Bell Labs Technical Journal <http://www.lucent.com/minds/techjournal/> (accessed May 18, 2005).
IBM Journal of Research and Development <http://www.research.ibm.com/journal/> (accessed May 18, 2005).
IBM Systems Journal <http://www.research.ibm.com/journal/sj/> (accessed May 18, 2005).

Trade magazines are published for and read by members of a particular trade. They are usually aimed at design engineers, developers, technical, or IT managers. These range from the "how-to" magazines based on particular system, tool, or language to those for the broader electronics industry such as *EE Times* <http://www.eetimes.com/> (accessed May 18, 2005) to those for the enterprise market.

Dr. Dobbs Journal <http://www.ddj.com/> (accessed May 18, 2005) is a journal of choice for many development managers together with the more specialized magazines, *Linux Journal* or *Software Development*. *Dr. Dobbs Journal* is written by and for professional software developers, and reports on all languages, platforms, tools, gives practical tricks of the trade and examples of code. By becoming a paid member of the publisher CMP Media's Developer Network you will have access not only to *Dr. Dobb's Journal* but also *BYTE.com*, *C/C++ Users Journal*, *The Perl Journal*, and *Software Development* magazine.

Trade journals aimed at IT professionals (see also Tools for the practitioner section) usually carry a combination of technical, trade, and business information and are often sent free to "qualified" subscribers (usually individuals within a company who make major IT-related budget or purchase decisions). Some examples of these are:

CIO <http://www.cio.com/> (accessed May 18, 2005).
Computerworld <http://www.computerworld.com/> (accessed May 18, 2005).
InformationWeek <http://www.informationweek.com/> (accessed May 18, 2005).

Consumer magazines for the PC end-user market

These magazines often begin life serving the "hobbyist" on a small scale but become magazines with healthy circulation figures. In this category, we have titles such as *PC Magazine* <http://www.pcmag.com/>, *PC World* <http://www.pcworld.com/>, and *Smart Computing* <http://www.smartcomputing.com/> to name a few – each of which is accompanied by informative websites. *BYTE* died in 1998 but lives on in an online-only version <http://www.byte.com> (all accessed May 18, 2005).

Magazines with broad appeal to the general public

A good example of this type is the magazine *Wired* <http://www.wired.com/wired/> (accessed May 18, 2005).

Finding comprehensive listings of computer journals is quite a challenge. The ACM *Guide to Computing Literature* provides a good starting point. Under "Browse" select "Journals" as publication type. You can retrieve a list of journals in alphabetical order or by publisher – this is free for all users. For more information, you need a personal or institutional membership to the ACM Digital Library. Members can retrieve the full citation and link to the table of contents (TOCs). Keep in mind that ACM-published TOCs are kept up-to-date but non-ACM titles may not be as up-to-date.

Core journals

The *Encyclopedia of Computer Science* (2003) is an excellent source for journal titles, providing us with as comprehensive a list as is possible given that the number of titles is growing at a rapid pace. In the Encyclopedia's Appendix III, the editors list the titles under these broad headings:

- Society journals (namely from ACM, IEEE, British Computer Society, the Institution of Electrical Engineers-UK, Society for Industrial and Applied Mathematics)
- Industrial journals
- Journals of specific countries and regions
- International journals
- Journals and magazines in specialized areas (including related fields such as mathematics)
- Magazines
- Major publishers in computing

A different approach to determining the leading journals in the field is to look at those which are most frequently cited. The alphabetical list below is based on the Institute of Scientific Information's *Journal Citation Reports* (a Web of Knowledge database) under the category "Computer science, hardware and architecture." *Journal Citation Reports*, available through a subscription, is a unique tool that gives users the ability to compare journals using citation data from over 7,000 scholarly and technical journals. IEEE and ACM titles are removed from the list. In effect, this listing, plus the IEEE and ACM titles, forms the major scholarly journals in this category. See the Publisher resources section below for detailed information about ACM and IEEE Computer Society and other key publishers.

Advances in Computers (0065–2458)
Analog Integrated Circuits and Signal Processing (0925–1030)
Canadian Journal of Electrical and Computer Engineering Revue (0840–8688)
Computer Standards and Interfaces (0920–5489)
Computer Communications (0140–3664)
Computers and Electrical Engineering (0045–7906)
Computer Journal (0010–4620)
Computer Networks – The International Journal of Computer and Telecommunications Networking (1389–1286)
Computer Systems Science and Engineering (0267–6192)
Design Automation for Embedded Systems (0929–5585)
Displays (0141–9382)
IBM Journal of Research and Development (0018–8646)
International Journal of High Performance Computing Applications (1094–3420)
Integration – The VLSI Journal (0167–9260)
Journal of Circuits Systems and Computers (0218–1266)
Journal of Computer Science and Technology (1000–9000)
Journal of Computer and System Sciences (0022–0000)
Journal of High Speed Networks (0926–6801)
Journal of Information Storage and Processing Systems (1099–8047)
Journal of Network and Computer Applications (1084–8045)
Journal of Supercomputing (0920–8542)
Journal of Systems Architecture (1383–7621)
Microprocessors and Microsystems (0141–9331)
Mobile Networks and Applications (1383–469X)
Networks (0028–3045)
New Generation Computing (0288–3636)
Performance Evaluation (0166–5316)
VLDB Journal (1066–8888)
VLSI Design (1065–514X)

Directories

Computer Science Journals <http://www.informatik.uni-trier.de/~ley/db/journals/> (accessed May 18, 2005). This listing from the DBLP (Digital Bibliography and Library Project) concentrates on scientific journals in computer science – trade journals are excluded.

Directory of Computing Science Journals <http://elib.cs.sfu.ca/Collections/CMPT/cs-journals/> (accessed May 18, 2005). This comprehensive list emanates from the

Internet Electronic Library Project at SFU. The project leader Professor Rob Cameron's aim was "to develop a comprehensive Electronic Library in Computing Science" and "to include information about both on-line and print materials" (source: About page). According to the website, a total of 522 journals are listed. The page offers a keyword index of title words as well as an alphabetical journal list. However, it is not clear how often the list is updated.

Directory of Open Access Journals: Computer Science <http://www.doaj.org/ljbs?cpid=114> (accessed May 18, 2005). "Open Access Journals" are defined as journals that do not charge individuals or their institution for access. This directory aims to be the "one-stop shop" listing for these peer-reviewed scientific and scholarly journals.

Top 100 Magazines: Computer and Software WWW Magazines and Journals <http://www.netvalley.com/top100mag.html> (accessed May 18, 2005). This is a subjective view of the top IT-related journals from San Francisco-based Internet consulting and publishing company NetValley. Although not explicitly dated, the page links to late 2002 to 2003 representative journal stories. Journal "leaders" (some are more portals than journals) are arranged by resource title.

Publisher resources

*ACM Journals and Magazines <http://www.acm.org/pubs/journals.html> (accessed May 18, 2005). This page points to the editorial home page of each of the ACM journals.

Elsevier: Computer Science <http://www.elsevier.com/wps/find/P05.cws_home/main> (accessed May 18, 2005). Choose "Journals" as product type or browse a more specific subject area within computer science. Only basic information is available about each title as well as a free sample issue. Access to journal full-text articles is only available via ScienceDirect (must be a subscriber).

*IEEE Computer Society Periodicals <http://www.computer.org/publications/index.htm#Magazines> (accessed May 18, 2005). Choose a magazine or transaction by title or browse periodicals by subject.

SpringerLink (Online Libraries: Computer Science). <http://springerlink.metapress.com/> (accessed May 18, 2005). Click on "Browse by Online Libraries," then choose "Computer Science." Free access is provided to search functions, tables of contents, as well as keyword and tables of contents alerts. Subscribers to a journal title receive online access to that title whether by a print-plus online or an electronic subscription. KluwerOnline (formerly Kluwer Academic Journals) is now part of SpringerLink.

Database digital collections

*ACM Digital Library <http://portal.acm.org/dl.cfm> (accessed May 18, 2005). Established in 1947, the Association of Computing Machinery (ACM) is the first and foremost computing association in the world. The ACM publishes twenty-two journals, thirty-two SIG newsletters, and twenty-four transactions series. ACM sponsors over eighty-five SIG-related conferences worldwide every year. Access to its journals, conferences, and newsletters is available in print and via its "Portal," making it the single most important full-text resource in computer engineering and science. The "Portal" consists of two main databases: the ACM Digital Library (DL) and the Guide to Computing Literature. The former contains bibliographic information, abstracts, and full text of all the ACM journals, conference proceedings and newsletters, and the latter is a collection of bibliographic citations and abstracts published by ACM and other major publishers in computer science. The Guide provides access to more than 750,000 citations from 3,000+ publishers (including ACM) covering the entire literature of computer science in books, journal articles, conference proceedings, doctoral dissertations, Masters' theses, and technical reports.

The ACM has two types of subscriptions: institutional or individual via ACM membership. Institutional and individual members receive access to the resources mentioned above. However, individual members have additional "Personalized" services: table of content (TOC) e-mail alerts when a new issue of an ACM journal, magazine, newsletter, or proceedings has been posted in the DL and the ability to create "Binders" where searches and queries may be saved and shared with colleagues. For those who do not have access to the ACM Portal via a library, becoming an ACM member provides a cost-effective way to access this vast storehouse of information. *Library Journal*'s August 2002 review calls the ACM Portal "a bonanza of computing literature at a bargain price . . . a valuable resource for academic use but a powerful tool kit for IT professionals."

A unique Portal feature available since March 2004 is CrossRef Search, a pilot program to implement full text interpublisher searchability <http://www.crossref.org/crossrefsearch.html> (accessed May 18, 2005). Google has indexed the full-text of ACM journal articles along with scholarly research content from nine other participating publishers. A Google search filters set results to content from participating publishers in an attempt to reduce the "noise" produced by more general web searches.

*IEEE Computer Society Digital Library <http://www.computer.org/publications/dlib/> (accessed May 18, 2005). A subscription to the IEEE Computer Society Digital Library includes online access to twenty-two Society periodicals and over 1,400 conference publications. This resource is available only to IEEE Computer Society members and library/institution customers.

Plans for online, print, or combination subscriptions are available to libraries. OPAC links may be used for publication titles within the Digital Library. In addition, the IEEE Computer Society and the ACM have agreed to exchange bibliographic data and abstracts. One may now search across both digital library collections from the IEEE Computer Society and link directly to either publisher's content.

Non-members will always have free access to abstracts and tables of contents in the Digital Library and may purchase individual documents.

Commercial database packages

EBSCO Technical Package <http://www.ebsco.com/home/whatsnew/ipca.asp> (accessed May 18, 2005). EBSCO offers a collection of databases which provide a "comprehensive compilation of abstracts and indexing for the top journals in various fields of electronic information management and computer science.... This collection is comprised of five databases including the following: Inspec, *IPCA* [*Internet and Personal Computing Abstracts* – new 2004 addition], *Information Science and Technology Abstracts*™ (ISTA), *Computer Science Index*™ (CSI), and Computer Source [according to EBSCO sales representative], as well as a complementary full-text component" (source: Home page). EBSCO acquired the IPCA and ISTA files in July 2003 from Information Today, Inc.

Engineering Village 2 (*EV 2*) <http://www.ei.org/eicorp/ev2.html> (accessed May 18, 2005). Launched in 2003, EV 2 is a "one-stop shop" for engineers providing access to resources important for computer engineering research – namely, the *Compendex* database, an Engineering Village cornerstone from its inception, plus additional subscription content (*Engineering Index Backfile, Inspec, Inspec Archive* and *NTIS*). An important advantage of this arrangement to the end user is that with one common interface you can search the *Compendex, Inspec* and *NTIS* databases simultaneously (without duplicate search results). In addition, Engineering Village 2 offers the ability to access other free and add-on subscription content such as Referex Engineering (Electronics and electrical collection is of particular interest), CRC ENGnetBASE (engineering handbooks), patents from the U.S. Patent and Trademark Office and European Patent Office, specifications from GlobalSpec, daily engineering news briefs from LexisNexis, Scirus, EEVL and much more. For more detailed information on these sources, see other entries within this chapter or in Chapter 2.

ISI Web of Knowledge <http://www.isiwebofknowledge.com> (accessed May 18, 2005). See Chapter 2.

Scopus <http://www.info.scopus.com/> (accessed May 18, 2005). Competing directly with ISI's Web of Science, the Elsevier product Scopus, in development since 2002, is an abstracts database comprising twenty-seven million records with articles from 14,000 peer-reviewed titles (including a number of open access journals) from 4,000+ publishers around the globe. Through the Scopus interface, users may also search Scirus. Includes computer science-related content from publishers such as ACM, IEEE Computer Society, CRC Press, MIT Press, Kluwer, and Springer.

Individual commercial databases

CMP Computer Full-text (Dialog File 647) <http://library.dialog.com/bluesheets/html/bl0647.html> (accessed May 18, 2005). "CMP Computer Full-text provides timely, relevant information about the computer, communications, and electronic industries. The file includes full-text, cover-to-cover coverage of twelve top rated newspapers and magazines published by CMP Media LLC" (source: Home page).

*Compendex <http://www.ei.org/eicorp/compendex.html> (accessed May 18, 2005). See Chapter 2.

*Compumath Citation Index <http://www.isinet.com/products/citation/specialty/cmci/> (accessed May 18, 2005). This CD-ROM product from the Institute for Scientific Information covers "current and retrospective bibliographic information and author abstracts from over 600 of the world's leading scholarly journals, in computers and mathematics. It also reaches outside the core literature to provide coverage of related articles from the multidisciplinary ISI collection of over 8,000 of the world's premier scholarly journals" (source: Home page). This database includes cited reference search – a unique feature of ISI databases.

*Computer Abstracts International Database <http://www.csa.com/csa/factsheets/computab.shtml> (accessed May 18, 2005). Published by Emerald Abstracts (a part of MCB University Press), this database "provides access to the latest developments in computing and computer science for practitioners and scholars." The database is particularly targeted to those working in the "production, processing, use and fabrication of hardware and software" (source: Home page). For more than forty years, Computer Abstracts International Database has been an important tool in the field and online access back to 1987 is provided via Cambridge Scientific Abstracts. Content within the database is classified using the ACM Classification Scheme. Major topics covered include: artificial intelligence, communications and networks, computer theory, data, database and information systems applications, hardware, human–computer interaction, mathematics of computing, programming, and systems organization.

*Computer and Information Systems Abstracts <http://www.csa.com/csa/factsheets/computer.shtml> (accessed May 18, 2005). This database, and its print equivalent of the same name, published by Cambridge Scientific Abstracts, "provides a comprehensive monthly update on the latest theoretical research and practical applications around the world" as covered in the journal and conference literature. One of the oldest databases in the field, this database "provides international coverage with the monitoring of over 3,000 serial titles as well as numerous non-serial publications" (source: Home page). Topics include: artificial intelligence, computer applications, computer programming, computer systems organization, computing milieu, hardware, information systems, mathematics of computing, and software engineering.

Computer Science Index <http://www.epnet.com/academic/computersci.asp> (accessed May 18, 2005). "Computer Science Index (formerly Computer Literature Index) offers abstracting and indexing of academic journals, professional publications, and other reference sources at the highest scholarly and technical levels of computer science. The collection covers more than 6,500 periodicals and books, with coverage going back to the mid-1960s. Computer Science Index focuses on subjects such as artificial intelligence, expert systems, system design, data structures, computer theory, computer systems and architecture, software engineering, human–computer interaction, new technologies, social and professional context, and much more. Enhancements to the original database include hundreds of new titles and searchable cited references for key academic journals. This database also includes editor-selected articles from magazine and journal titles in related areas of study" (source: Home page).

Computer Source <http://www.epnet.com/academic/computersource.asp> (accessed May 18, 2005). Computer Source (available via EBSCO) provides researchers with the latest information and current trends in high technology with abstracts and indexing for over 330 periodicals. This database features over 250 full-text periodicals covering valuable market information in computers, telecommunications, electronics, and the Internet.

Computer Technology Database <http://www.csa.com/csa/factsheets/ird-CS.shtml> (accessed May 18, 2005). Offered by Cambridge Scientific Abstracts, "the Computer Technology database is a collection of almost 15,000 high-quality websites that are hand-picked and indexed by CSA editors. We choose only sites containing specific, technical information of interest to a college-level audience, from respected, nonbiased sources such as educational institutions, government agencies, and scientific organizations. We review all links on a monthly basis, and average a phenomenal rate of less than 2% dead links" (source: Home page). Subject coverage includes artificial intelligence, communications and networks, computer graphics, computer theory, database and information systems, hardware, human–computer interactions, programming languages, security, and software applications.

Gale Group Computer Database (Dialog File 275) <http://library.dialog.com/bluesheets/html/bl0275.html> (accessed May 18, 2005). This database "provides comprehensive information about the computer, electronics, and telecommunications industries. Coverage includes detailed information about the evaluation, purchase, use, and support of computer and other electronic products. Gale Group Computer Database is designed to answer the questions of business and computer professionals about hardware, software, networks, peripherals, and services. Lengthy abstracts are available for most records from 1983 to present. Complete text is fully searchable for many records from 1988 to present" (source: Home page).

**Inspec* <http://www.iee.org/Publish/INSPEC/> (accessed May 18, 2005). *Inspec*, one of the most respected scientific databases in the world, was formed in 1967, based on the Science Abstracts service which has been provided by the UK Institution of Electrical Engineers since 1898. *Inspec* is the online version of the three print indexes: *Physics Abstracts*, *Electrical and Electronics Abstracts*, and *Computer and Control Abstracts* which together form *Science Abstracts*. It provides comprehensive access to the world's leading scientific and technical literature in computers and computing, information technology, electrical engineering, electronics, communications, control engineering, and physics. The database is international in scope, and indexes (full citations with lengthy abstracts) journal articles, conference proceedings, reports, dissertations, reports, and books. In addition to its depth of coverage, its strength lies in consistent indexing by subject specialists. The scope of indexing has grown since its inception to include not only the usual Thesaurus Term and Classification Code indexes, but also the Uncontrolled Index Terms (Identifiers), Treatment Codes, Chemical Substance Indexing, Numerical Data Indexing and Astronomical Object Indexing. This indexing helps guide the user to related key terms and subject areas, and to gain a better understanding of topics.

Internet and Personal Computing Abstracts <http://www.epnet.com/academic/internet&personal.asp> (accessed May 18, 2005). Formerly *Microcomputer Abstracts*, this database "provides abstracts and indexing for literature related to personal computing products and developments in business, the Internet, the home, and all other applied areas. This resource contains content coverage that extends back to the 1980s. Over 400 of the most important trade publications, mainstream computer magazines, and professional journals are covered, including those that focus on specific topics such as Macintosh and Windows platforms, programming, web development and more. Special emphasis is also given to hardware and software reviews. The product includes content from such titles as *Byte.com, PC World, Macworld*, and *Linux Journal*. The product also includes editor-selected articles from hundreds of popular magazine titles" (source: Home page). *Note:* EBSCO purchased the back file from Information Today, Inc. in July 2003.

MathSciNet <http://www.ams.org/mathscinet/> (accessed May 18, 2005). MathSciNet is the most comprehensive database for mathematics, produced by the American Mathematical Society (AMS), and covering the world's mathematical literature since 1940. It is analogous to the print publication *Mathematical Reviews* and also includes recent issues of *Current Mathematical Publications*. The database indexes journal articles, conference proceedings, and books. The database is organized by the Mathematics Subject Classification (MSC). This database is important to computer engineers who want to research fundamental theories in mathematics related to their discipline.

TecInfoSource (Dialog File 256) <http://library.dialog.com/bluesheets/html/bl0256.html> (accessed May 18, 2005). Use this file to find information about products and companies in the Information Technology industry. Formerly known as *Softbase: Reviews, Companies, and Products*.

Current awareness

Managing and keeping up with new information and literature in computer engineering is a challenge. However, the good news is that now more than ever there are automatic ways to keep abreast of the literature. For instance, e-mail alerts from the major databases (listed below) allow you to set up single or multiple search strategies on your research topic, and every week (or whenever the database is updated) you will receive relevant hits in your e-mail. To keep up with new books in the discipline, sign up with the various major publishers mentioned below and you will receive timely alerts (see also Chapter 2).

Metasites

EEVL: Internet Guide to Engineering, Mathematics, and Computing Computing Current Awareness Services <http://www.eevl.ac.uk/computing/currentawarenesscomp.htm> (accessed May 18, 2005). Lists new Internet resources, books, articles, conferences, and conference papers.

Databases

ACM Digital Library <http://portal.acm.org/> (accessed May 18, 2005). At present, the TOC (Table of Contents Service) is only available to personal members (not institutional subscribers). The service enables a member to receive tables of contents of issues and proceedings.

Cambridge Scientific Abstracts databases: *Computer Abstracts International, Computer Information Systems Abstracts*, and *Computer Technology.* E-mail notification service is available for SDIs and when a saved search is re-run periodically.

*Current Contents/Engineering, Computing & Technology Edition <http://www.isinet.com/products/cap/ccc/editions/ccect/> (accessed May 18, 2005). Updated daily, this database provides access to tables of contents, abstracts and bibliographic information from scholarly journals and books as well as evaluated websites and full-text web documents.

Subject Category: Computer Science and Engineering <http://www.isinet.com/journals/scope/scope_ccect.html> (accessed May 18, 2005). "Includes resources on computer hardware and architecture, computer software, software engineering and design, computer graphics, programming languages, theoretical computing,

computing methodologies, broad computing topics, and interdisciplinary computer applications" (source: Home page). Indexes 116 journal titles. Other related subject categories are *Information Technology and Communications Systems* and *AI, Robotics and Automatic Control.*

Computer Science and Engineering Journal Titles <http://sunweb.isinet.com/cgi-bin/jrnlst/jlresults.cgi?PC=T&SC=CSE> (accessed May 18, 2005).

Engineering Village 2 <http://www.engineeringvillage2.org/> (accessed May 18, 2005). With an institutional subscription and registration, you can set up weekly alerts in Engineering Village 2.

IEEE Xplore <http://ieeexplore.ieee.org/Xplore/DynWel.jsp> (accessed May 18, 2005). IEEE's e-mail alert service is free. Select journal titles of choice and you will receive regular e-mail notification of recently posted IEEE and IEE journals and magazines. Each e-mail alert will give you a direct link to the issue's latest table of contents in *IEEE Xplore*. Access to abstracts and full text of the articles depends on whether you have personal or institutional membership.

Inspec <http://www.iee.org/Publish/> (accessed May 18, 2005). As has been noted in the Databases section, *Inspec* is available from many vendors. Depending on the vendor, current awareness alerts are provided. For instance, Ovid has "Autoalerts" and the ISI (Web of Knowledge) platform provides e-mail alerts with registration. To use the alerting feature, one needs to have access to an institutional subscription, usually through one's library. With either of these platforms one may set up a search strategy and every week, when the database is uploaded, alerts will be sent to your e-mail.

Web of Science <http://isi17.isiknowledge.com/portal.cgi/> (accessed May 18, 2005). Web of Science, described in detail elsewhere in this book, is a comprehensive database covering science and engineering and includes the major journals in each discipline. Although this is not a database specializing in engineering, let alone computer engineering, it is useful for tracking interdisciplinary topics. Web of Science, updated weekly, has three types of alerting services: a table of contents alert, citation alert (helps you track references to particular citations), and an alert by topic. One needs to have institutional access and registration is a perquisite.

Publishers

Listed below are the major book publishers in computer engineering and science. Each has an e-mail alert service, making it easy for the librarian and practitioner alike to keep up with the latest publications in this subject area.

Cambridge University Press Computer Science <http://us.cambridge.org/computerscience/> (accessed May 18, 2005). Receive monthly updates of special offers and new titles at <http://us.cambridge.org/cais>.

CRC Press: E-mail Alerts <https://www.crcpress.com/mailing_lists/default.asp> (accessed May 18, 2005). Subscribe to e-mail updates on specials and promotions.

CRC Press: Computer Engineering <http://www.crcpress.com/shopping_cart/categories/categories_products.asp?parent_id=388&> (accessed May 18, 2005). Review publications by date to see forthcoming and new titles.

CS Press Alert <http://www.computer.org/cspress/csp-alert/index.html> (accessed May 18, 2005). Monthly e-mail bulletin of IEEE Computer Society's latest product offerings.

Digital Press <http://books.elsevier.com/us//digitalpress/us/subindex.asp?maintarget=&isbn=&country=United+States&srccode=&ref=&subcode=&head=&pdf=&basiccode=&txtSearch=&SearchField=&operator=&order=&community=digitalpress> (accessed May 18, 2005). Part of Elsevier. Browse new and forthcoming titles or click on "Sign up here" link for special offer notification.

Elsevier <http://books.elsevier.com/us//computerscience/us/subindex.asp?maintarget=&isbn=&country=United+States&srccode=&ref=&subcode=&head=&pdf=&basiccode=&txtSearch=&SearchField=&operator=&order=&community=computerscience> (accessed May 18, 2005). Browse new computer science titles or click on "Opt-in here" link to subscribe to a targeted e-mail list. See also imprints Digital Press, Morgan Kaufmann and Newnes.

MIT Press: E-mail Alerts <https://mitpress.mit.edu/shared/mlist/default.asp?sid=3AFD4B02-DE02-4BCD-8E6E-AFEDA36E40EA&> (accessed May 18, 2005). Must sign up and create a profile.

MIT Press: Computer Science and Intelligent Systems <http://mitpress.mit.edu/catalog/browse/default.asp?sid=3AFD4B02-DE02-4BCD-8E6E-AFEDA36E40EA&cid=5> (accessed May 18, 2005). See in particular links for "New Releases" and "Coming Soon."

Morgan Kaufmann: Computing Books <http://books.elsevier.com/us//mk/us/subindex.asp?maintarget=&isbn=&country=United+States&srccode=&ref=&subcode=&head=&pdf=&basiccode=&txtSearch=&SearchField=&operator=&order=&community=mk> (accessed May 18, 2005). Part of Elsevier. From this page, sign up for a targeted e-mail list.

Newnes: Electronics and Computer Engineering <http://books.elsevier.com/us//computereng/us/subindex.asp?maintarget=&isbn=&country=United+States&srccode=&ref=&subcode=&head=&pdf=&basiccode=&txtSearch=&SearchField=&operator=&order=&community=computereng> (accessed May 18, 2005). Part of Elsevier. Browse new and forthcoming titles or click on "Sign up here" link for special offer notification.

Prentice-Hall Professional Technical Reference <http://www.phptr.com/> (accessed May 18, 2005). "Stay up to date on our new publications and special promotions by subscribing to our newsletters. You can choose to hear about all of our computer titles, or choose specific topics based on your areas of interest including: Engineering, Java, Security, Operating Systems, Business, and more" (source: Home page).

Oxford Science Publications: Mathematics, Statistics, and Computer Science <http://www.oup.co.uk/academic/science/e-mailnews/> (accessed May 18, 2005). Subscribe to MATHCOMPNEWS-L for notification of new titles in mathematics, statistics and computer science.

Springer: SpringerAlerts for Computer Science <http://www.springeronline.com/sgw/cda/frontpage/0,11855,4-103-610-0-0,00.html> (accessed May 18, 2005). Select an area of interest for delivery of TOC or keyword alerts. Should now include Kluwer publications.

Springer: SpringerAlerts for Librarians <http://www.springeronline.com/sgw/cda/frontpage/0,11855,4-103-2-74426-0,00.html> (accessed May 18, 2005). "Receive monthly e-mail updates about Springer News Previews, a PDF version of our monthly catalog of new books and journals. Also get news on our library bestsellers, new releases, and special promotions" (source: Home page).

What's News @ IEEE: Computing Monthly Newsletter <http://whatsnew.ieee.org/> (accessed May 18, 2005). Select "Computing." Newsletters are monthly e-mail updates "with the latest news regarding IEEE activities, industry trends, career development tips, and new IEEE product releases" (source: Home page).

Wiley Computing <http://www.wiley.com/WileyCDA/Section/id-2925.html> (accessed May 18, 2005). Browse all forthcoming titles for computing or select a subject subcategory. You can also sign up for e-mail alerts to be notified of promotions and new Wiley titles at <http://www.wiley.com/enewsletters> (accessed May 18, 2005).

Other

National Technical Information Service (NTIS) <http://www.ntis.gov/new/alerts_printed.asp?loc=5-0-0> (accessed May 18, 2005). Alerts cover Computer hardware; Computer software; Control systems and control theory; Information processing standards; Pattern recognition and image processing; Application software; Information theory; and Data files. An annual subscription to the printed version is available and averages about eighty summaries per issue. Please note that the weekly e-mail alerts service has been suspended (NTIS order number: SUB-9062).

Science.gov Alerts (launched February 18, 2005) <http://www.science.gov/helpalerts.html> (accessed May 18, 2005). Science.gov, the Web portal for federal science information, has just announced an ALERTS service that will provide weekly notifications of new Science.gov information in a specific area of interest. Relevant topics are computer hardware, computer security, computer networking and computer software.

Associations, organizations, and societies

Professional associations and scholarly societies provide the foundation for structured peer-to-peer exchanges. Listed below are the most important of these organizations. A more exhaustive list of computer science scholarly societies around the world is compiled by the Scholarly Societies Project <http://www.scholarly-societies.org/compsci_soc.html> (accessed May 18, 2005). Non-profit and government organizations are listed in the Virtual Computer Library <http://web.austin.utexas.edu/vclib/nonprofit.cfm> (accessed May 18, 2005). A comprehensive list of computer science departments worldwide is available from Mathtools.net <http://www.mathtools.net/Learning_and_Education/Universities/Computer_Science_Departments/> (accessed May 18, 2005). Research institutes related to computer science have been compiled by the DMOZ Open Directory Project <http://dmoz.org/Computers/Computer_Science/Research_Institutes/> (accessed May 18, 2005).

American Association for Artificial Intelligence (AAAI) <http://www.aaai.org/> (accessed May 18, 2005). "Founded in 1979, the American Association for Artificial Intelligence (AAAI) is a non-profit scientific society devoted to advancing the scientific understanding of the mechanisms underlying thought and intelligent behavior and their embodiment in machines. AAAI also aims to increase public understanding of artificial intelligence, improve the teaching and training of AI practitioners, and provide guidance for research planners and funders concerning the importance and potential of current AI developments and future directions.

Major AAAI activities include organizing and sponsoring conferences, symposia, and workshops, publishing a quarterly magazine for all members, publishing books, proceedings, and reports, and awarding grants, scholarships, and other honors" (source: Home page). AAAI members have full access to most association publications in digital form.

*Association for Computing Machinery (ACM) <http://www.acm.org/> (accessed May 18, 2005). Founded in 1947, ACM is an international scientific and educational organization dedicated to advancing the arts, sciences, and applications of information technology. With 78,000 members around the world, ACM is a leading resource for computing professionals and students working in the various fields of information technology, and for interpreting the impact of information technology on society. ACM's Special Interest Groups (SIGs) in thirty-four distinct areas of information technology address varied interests: programming languages, graphics, computer–human interaction, and mobile communications, to name a few. Each SIG organizes itself around specific activities that best serve both its practitioner- and research-based constituencies. Many SIGs sponsor conferences and workshops, and offer members reduced rates for registration and proceedings. SIGs also produce newsletters and other publications, or support lively e-mail forums for information exchange. ACM and its SIGs sponsor more than 100 conferences around the world every year, attracting over 100,000 attendees in total. Each conference publishes a proceeding, and many have exhibitions. Many of ACM's conferences are considered "main events" in the IT industry. The ACM Press Books program covers a broad spectrum of interests in computer science and engineering.

Computing Research Association (CRA) <http://cra.org/. (accessed May 18, 2005). "The Computing Research Association (CRA) is an association of more than 200 North American academic departments of computer science, computer engineering, and related fields; laboratories and centers in industry, government, and academia engaging in basic computing research; and affiliated professional societies . . . CRA's mission is to strengthen research and advanced education in the computing fields, expand opportunities for women and minorities, and improve public and policymaker understanding of the importance of computing and computing research in our society" (source: Home page).

*IEEE Computer Society <http://www.computer.org/> (accessed May 18, 2005). "With nearly 100,000 members, the IEEE Computer Society is the world's leading organization of computer professionals. Founded in 1946, it is the largest of the 37 societies of the Institute of Electrical and Electronics Engineers (IEEE). The Computer Society's vision is to be the leading provider of technical information and services to the world's computing professionals. The Society is dedicated to advancing the theory, practice, and application of computer and information processing technology. Through its conferences, applications-related and research-oriented journals, local and student chapters, distance learning campus, technical committees, and standards working groups, the Society promotes an active exchange of information, ideas, and technological innovation among its members. . . . With over 40 percent of its members living and working outside the U.S., the Computer Society fosters international communication, cooperation, and information exchange" (source: About the Computer Society page). Members receive *Computer Magazine* free and access to the Digital Library with twenty-two periodicals available online by subscription. Conference Proceedings online is included with an IEEE Computer Society Digital Library subscription. Members have access to special online collections of unabridged books on a variety of technology topics.

International Federation for Information Processing (IFIP) <http://www.ifip.org/> (accessed May 18, 2005). "IFIP is a non-governmental, non-profit umbrella organization for national societies working in the field of information processing. It was established in 1960 under the auspices of UNESCO as an aftermath of the first World Computer Congress held in Paris in 1959. Today, IFIP has several types of Members and maintains friendly connections to specialized agencies of the UN system and non-governmental organizations. Technical work, which is the heart of IFIP's activity, is managed by a series of Technical Committees" (source: About IFIP: General information page).

Society for Industrial and Applied Mathematics (SIAM) <http://www.siam.org/> (accessed May 18, 2005). SIAM is dedicated to advancing "the application of mathematics and computational science to engineering, industry, science, and society" (source: About SIAM page). It is very active in sponsoring conferences and publishing high-quality books, book series and a suite of significant journals. All the journals are available electronically as well as in print. The product "Locus" introduced in 2005 contains the full text for every SIAM journal article published from the journal's inception through 1996, making this a valuable resource for retrospective research in applied mathematics.

USENIX <http://www.usenix.org/> (accessed May 18, 2005). "USENIX, the Advanced Computing Systems Association, fosters technical excellence and innovation, supports and disseminates research with a practical bias, provides a neutral forum for discussion of technical issues, and encourages computing outreach into the community at large. Since 1975, the USENIX Association has brought together the community of engineers, system administrators, scientists, and technicians working on the cutting edge of the computing world. The USENIX conferences have become the essential meeting grounds for the presentation and discussion of the most advanced information on the developments of all aspects of computing systems" (source: About USENIX page). The best papers from these conferences are available in their site's Compendium.

Careers and education

Undergraduate and graduate programs in computer engineering as well as two-year IT education programs can all be identified by searching the well-known *Peterson's Guides* on the Web. Another popular resource, GradSchools.com, allows one to search for specific graduate programs in computer science. For more information see Chapter 11 on Engineering education.

Peterson's IT Channel <http://www.petersons.com/itchannel/> (accessed May 18, 2005). Search two-year information technology education programs. Contains information on computer-based training resources for the IT professional and information on, and online registration for, various IT certification tests.

GradSchools.com: Computers and Information Technology Graduate Program Directories <http://www.gradschools.com/computers_info.html> (accessed May 18, 2005). The program directories are categorized by curriculum and subdivided by geography. Choose from topics such as computer science, information technology, or software engineering.

Technical societies remain the first and best point of entry for career-related resources, namely ACM and IEEE Computer Society. Keep in mind that other IT professional organizations, journals, or web portals will likely have their own career-related resources of potential interest to computer engineers. See our sections on Journals and Tools for the practitioner for more information.

*ACM Career Resource Center <http://campus.acm.org/crc/> (accessed May 18, 2005). Search and apply for jobs, find out information about CS/IS/MIS careers and industry trends. Read member recommendations for technical resources such as books, conferences, courses, and websites. Members can use self-assessment tools and participate in community discussion forums. In its Professional Development Centre <http://pd.acm.org/> (accessed May 18, 2005), ACM offers a wide variety of online courses free to its professional and student members and member-discounted courses offered through Stevens Institute of Technology.

*IEEE Computer Society Career Services Center <http://www.computer.org/ careers/> (accessed May 18, 2005). In addition to career resources, this site also provides links to the Society's education and certification resources. The Society offers its members 100 online training courses through its Distance Learning Campus <http://www.computer.org/distancelearning/> (accessed May 18, 2005). Its Certified Software Development Professional (CSDP) Program <http://computer.org/certification> (accessed May 18, 2005) is a certification program for software engineers. IEEE also maintains its own job site <http://careers.ieee.org/> (accessed May 18, 2005).

World Lecture Hall <http://web.austin.utexas.edu/wlh/browse.cfm> (accessed May 18, 2005). Browse by "Computer science" or "Electrical and computer engineering." "World Lecture Hall publishes links to pages created by faculty worldwide who are using the Web to deliver course materials in any language. Some courses are delivered entirely over the Internet. Others are designed for students in residence" (source: About WLH page).

Tools for the practitioner

Interpreting practitioners as "real world" engineers in the area of information systems and technologies, they can range from educators to programmers and analysts to management. In some cases, practitioners may have completed professional certification programs.

For relevant journals, see the Journals or Web portals section (where many of the trade print journal content is available online).

Book collections

*Books24x7: ITPro Collection <http://marketing.books24x7.com/browseabout.asp?item=itpro> (accessed May 18, 2005). Available as an individual, library, or corporate subscription, ITPro "provides both broad and deep coverage of over 100 different technology topics.... Premier industry publishers, such as Wrox, McGraw-Hill, Microsoft Press, and many more, contribute front-list, best selling, classic and niche titles. Popular book series, such as The Complete Reference, Inside Out, Bibles and many others provide multifaceted, multi-skilled approaches to topics" (source: Home page). Access to this collection is also made available through a partnership with ACM for its members with a subscription <http://pd.acm.org/books/faq.cfm> (accessed May 18, 2005).

FreeTechBooks <http://www.freetechbooks.com/> (accessed May 18, 2005). This site provides links to free online computer books and documentation. There are over 100 books covering programming languages, scripting languages, operating systems, and other computer science topics such as data structures, algorithms, object-oriented programming, logic programming, compiler design, and software development.

Numerical Recipes Books On-line <http://lib-www.lanl.gov/numerical/> (accessed May 18, 2005). This site provides access to the complete text of the following Numerical Recipes books from Cambridge University Press: *Numerical Recipes in C: The Art of Scientific Computing, Numerical Recipes in Fortran 77: The Art of Scientific Computing*, and *Numerical Recipes in Fortran 90: The Art of Parallel Scientific Computing*.

*O'Reilly Books <http://oreilly.com> (accessed May 18, 2005). "O'Reilly Media is the premier information source for leading-edge computer technologies. The company's books, conferences, and websites bring to light the knowledge of technology innovators. O'Reilly books, known for the animals on their covers, occupy a treasured place on the shelves of the developers building the next generation of software. O'Reilly conferences and summits bring alpha geeks and forward-thinking business leaders together to shape the revolutionary ideas that spark new industries" (source: About O'Reilly page). To keep up with O'Reilly news and application specific newsletters, sign up at the home page.

*Safari Books Online <http://www.safaribooksonline.com/> (accessed May 18, 2005). O'Reilly and the Pearson Technology Group have joined forces to create Safari Books Online. Safari is a continuously updated online library that features the best IT titles not only from O'Reilly but also Adobe Press, Addison Wesley Professional, Cisco Press, New Riders, Peachpit Press, Prentice Hall PTR, Que, and Sam's Publishing. Individual, enterprise, and institutional subscriptions are available. Users can access the full text of hundreds of top-selling IT books. Safari lets you search specific topics, pinpoint the specific chapter or section of the book relevant to your issue, and download the chapter. You can update the titles in your "collection" every month or expand your subscription to access the whole collection. One of the major advantages of this is copying and pasting code to eliminate typographical errors. Safari Books Online is a good investment for companies and organizations that depend on teams of developers for their livelihood.

Market research

In addition to association, society, trade, and commercial publications, IT practitioners may also seek out intelligence from the leading providers of technology market research in order to make better decisions regarding the buying and selling of technologies for a company. Companies may subscribe to some of these offerings through enterprise subscriptions. Otherwise, complete reports (or specific sections) can usually be downloaded from commercial online services such as Dialog Profound or Thomson Research (see your librarian). Some of the larger IT analyst firms are listed below:

Forrester Research <http://www.forrester.com/> (accessed May 18, 2005) "is an independent technology research company that provides pragmatic and forward-thinking advice about technology's impact on business" (source: Home page). *Note:* Acquired Giga Information Group in February 2003.

Gartner, Inc. <http://www4.gartner.com/> (accessed May 18, 2005) "is the leading provider of research and analysis on the global IT industry" (source: About Gartner page). Gartner provides "advice for IT professionals, technology companies and technology investors in the form of research reports, briefings or events" (source: About Gartner > Our Business page). *Note:* Acquired Meta Group in December 2004.

IDC <http://www.idc.com/> (accessed May 18, 2005) "is the premier global market intelligence and advisory firm in the IT and telecommunications industries. We analyze and predict technology trends so that their clients can make strategic, fact-based decisions on IT purchases and business strategy" (source: About IDC page).

InfoTech Trends <http://infotechtrends.com/> (accessed May 18, 2005), "formerly Computer Industry Forecasts, provides market data on computers, peripherals, software, storage, the Internet, and communications equipment" (source: Home page).

Web portals

Representative examples of some of the most useful web portals for IT practitioners include the following.

*CNET Networks <http://www.cnetnetworks.com> (accessed May 18, 2005). With a presence in the U.S., Asia and Europe, CNET Networks supplies "content in the personal technology, games and entertainment, and business technology categories" (source: About CNET Networks page). Selected CNET portals are listed below which may be of more interest to practitioners.

Builder.com <http://www.builder.com> (accessed May 18, 2005). "Created by developers, for developers, Builder.com brings software developers fresh, real-world perspective on topics from programming to architecture to management" (source: About CNET Networks page).

DevX.com <http://www.devx.com> (accessed May 18, 2005). "DevX.com is the leading provider of technical information and services that enable corporate application development teams to efficiently address development challenges and projects" (source: Jupiterweb.com Home page).

EarthWeb <http://www.earthweb.com> (accessed May 18, 2005). "EarthWeb.com's sites are organized into five "channels" targeting the needs of IT management, hardware & systems professionals, networking & communications administrators, web and software developers" (source: Jupiterweb.com Home page). One of its best-known sites is Developer.com which has a wealth of resources for software development.

itmWeb <http://www.itmweb.com/> (accessed May 18, 2005). Established in 1996, itmWeb is a "source for information technology reference, methodology, and technical content focused on IT departmental management, technology support, and project leadership" (source: Home page). Designed for CIOs, project managers, IT educators, and students, the site also publishes a monthly IT eZine.

*JupiterWeb Network internet.com <http://www.internet.com/> (accessed May 18, 2005). "internet.com provides enterprise IT and Internet industry professionals with the news, information resources and community they need to succeed in today's rapidly evolving IT and business environment" (source: Jupiterweb.com Home page). It is comprised of technology-specific Websites, e-mail newsletters, announcement lists, and discussion lists.

*O'Reilly Network <http://www.oreillynet.com/> (accessed May 18, 2005). "The O'Reilly Network is the essential portal for developers interested in open and emerging technologies, including new platforms, programming languages, and operating systems. Just like O'Reilly & Associates' books, this hub site provides in-depth technical information, clearly and consistently, for expert developers. Beyond that, it creates a forum for the O'Reilly developer community. . . . The O'Reilly Network hub is a high-quality aggregator of information from affiliate sites, as well as provider of original content and FAQs on new technologies" (source: About the O'Reilly Network page).

OSTG: Open Source Technology Group Network <http://www.ostg.com/> (accessed May 18, 2005). "OSTG is the leading network of technology sites for today's IT managers and development professionals and provides a unique combination of news, original articles, downloadable resources, and community forums to help IT buyers, influencers and users make critical decisions about information technology products and services" (source: Home page). Its network of websites focuses on IT (Slashdot, IT Manager's Journal, NewsForge, Linux.com) and developers (SourceForge.net, freshmeat.net, DevChannel.org). Around since 1997, Slashdot <http://slashdot.org> (accessed May 18, 2005) remains a "go-to" website for all things technical – their mission "news for nerds – stuff that matters."

SE Online: Software Engineering Online <http://www.computer.org/portal/ site/seportal/> (accessed May 18, 2005). "The IEEE Computer Society has launched this new online resource for researchers and practitioners to learn and find information about software engineering." Their objective is to make the site the "most trusted, non-commercial, and objective source of SE knowledge available online." Regardless of source, the information found in SE Online "has been selected by knowledge area editors based on its quality and value to the community" (source: Home page).

TechOnLine <http://www.techonline.com> (accessed May 18, 2005). TechOnLine serves the information, demonstration, training, and support needs of design engineers. Some access requires site registration. Resources include featured articles, technical papers, courses, and on-demand webcasts. Technology groups include Analognet, Communicationsnet, DSPnet, EDAnet, Embeddnet, SOCnet, and T&Mnet. Tech Topics include Bluetooth, CSoC, Design Services, Internet/Network, Java, Linux, MEMS, Power Supplies, RTOS, Timing Closure, UML, USB/Firewire, VoIP, and Web-enabled Design.

TechRepublic <http://techrepublic.com.com> (accessed May 18, 2005). "TechRepublic serves the needs of the professionals representing all segments of the IT industry, providing information and tools for IT decision support and professional advice by job function" (source: About CNET Networks page). Requires sign-up for free membership which allows access to thousands of how-to articles, largest collection of technical white papers (separately accessible via <http://itpapers.com/>), downloads, technical Q&A, discussions, and newsletters.

TechTarget Network <http://www.techtarget.com> (accessed May 18, 2005). TechTarget's primary audience are enterprise IT professionals. The Network "family" comprises twenty-six IT-related websites for applications, core technologies, Windows, enterprise IT management, and platforms. Search across all network sites at <http://searchtechtarget.techtarget.com/> (accessed May 18, 2005). *Note:* TechTarget acquired Bitpipe, Inc. in December 2004.

*TechWeb: The Business Technology Network <http://www.techweb.com> (accessed May 18, 2005). "TechWeb.com adds its own IT reporting to the best-of-breed news and information from *InformationWeek, InternetWeek, Network Computing, Network Magazine, Optimize Magazine, The Open Enterprise*, and the Financial Technology Network, thus creating the complete online resource for technology managers" (source: About Us page).

ZDNet <http://www.zdnet.com> (accessed May 18, 2005). ZDNet operates a "worldwide network of websites that offer content, services, and commerce opportunities that enable IT professionals and business influencers to gain an edge in business" (source: About CNET Networks page).

Other web resources

Algorithms

Collected Algorithms of the ACM (CALGO) <http://www.acm.org/calgo/contents/> (accessed May 18, 2005). CALGO contains the "software associated with papers published in the *Transactions on Mathematical Software*" (TOMS) which are available in print and through the ACM Digital Library. "This software is refereed for originality, accuracy, robustness, completeness, portability, and lasting value" (source: Home page). Until December 2004, CALGO was also published in loose-leaf form, but this has now ceased. The associated free website will continue to be maintained and updated.

The Stony Brook Algorithm Repository <http://www.cs.sunysb.edu/~algorith/> (accessed May 18, 2005). Professor Steven S. Skiena has mounted this site, based on his book with the same title, "to serve as a comprehensive collection of algorithm implementations for over seventy of the most fundamental problems in combinatorial algorithms" (source: Home page).

Compilers and interpreters

Catalog of Free Compilers and Interpreters <http://www.idiom.com/free-compilers/> (accessed May 18, 2005). This directory, aimed at developers (rather than researchers), provides links to compilers, compiler generators, interpreters, translators, important libraries, assemblers, and so on, and is maintained by Idiom Computers. It is not clear when the list was last updated; nevertheless, it is a useful source.

Compilers.net <http://www.compilers.net/> (accessed May 18, 2005). Based in Hungary, Avalon IT hosts this site of links to free compilers. To stay abreast of the latest news, sign up for their mailing list which will provide you with information on free compilers, tutorials, and new links added to their page.

Free Compilers and Interpreters for Programming Languages <http://www.thefreecountry.com/compilers/index.shtml> (accessed May 18, 2005). This site almost disappeared when the administrator, Christopher Heng, could no longer maintain it (due to costs and time involved in maintenance), but his "fans" inundated him with e-mails begging him to continue with offers of free hosting on their servers. So he took up one of the offers and the site has been resurrected with a new look. This example indicates how sites develop and then disappear, but also shows the power of the medium to bring parties together. It is indeed an excellent site linking not only to free compilers and interpreters but also to the whole gamut of programming tools – a site to bookmark certainly.

Programming

There are many sites on the Internet which provide programming tools for individual languages and groups of languages. It would be impossible to list them all in this chapter. The Computer Science Resource Guide at the State University of New York at Albany, maintained by Michael Knee of the Science Library, does a magnificent job in tracking programming sites <http://library.albany.edu/subject/csci.htm>. The site below could be considered a metasite for programming tools:

Programmers Heaven <http://www.programmersheaven.com/> (accessed May 18, 2005). Maintained by Synchron (located in Sweden), Programmers Heaven contains a wealth of information for programmers. It is not just a website but more of a "virtual community" for programmers. In addition to links to programming sites, it also has links to other resources such as tutorials, magazines, sample chapters in books, and a bulletin board.

Internet resources subject guides

*Computer Science (State University of New York at Albany). Prepared by Michael Knee, Science Bibliographer and Reference Librarian. Both print and web resource guides listed below are organized by subtopic with good annotations:

- Computer Science Subject Guide <http://library.albany.edu/subject/csci.htm>
- Computer Science: A Brief Guide to Reference Resources <http://library.albany.edu/subject/guides/csciguid.htm>

*Information Sources: Computer Science (University of Manitoba Sciences and Technology Library) <http://www.umanitoba.ca/libraries/units/science/compsci.html>. Compiled by Ryan Schultz, Subject Bibliographer for Computer Science. Extensive listing of links for guides to the literature, dictionaries and glossaries, encyclopedias, handbooks and manuals, abstracting and indexing services, and full-text e-journals.

*Research Guide for Computer Science and Electrical and Computer Engineering (Carnegie Mellon University Engineering & Science Library) <http://www.library.cmu.edu/Research/EngineeringAndSciences/CS%2BECE/index.html>. Prepared by Missy Harvey, Computer Science Librarian. Resource guide organized by sections for fundamentals, other resources, local special collections, and related CMU library research guides. All sections have links to relevant sources with scattered, brief annotations.

Metasites

Computer science and engineering

*EEVL: Internet Guide to Engineering, Mathematics, and Computing.

Computing Section <http://www.eevl.ac.uk/computing/index.htm>.

Electrical, Electronic and Computer Engineering Section <http://www.eevl.ac.uk/engineering/eng-browse-page.htm> (accessed May 18, 2005). Choose "Engineering" then "Electrical, electronic and computer engineering." Termed an "exceptional guide" by Gary Price in *SearchDay*, EEVL's award-winning not-for-profit site is created and run by librarians and information professionals in the U.K. It provides access with annotations to selected high-quality networked engineering, mathematics, and computing resources. It is the most comprehensive resource guide for engineering.

HCI Bibliography: Human-Computer Interaction Resources <http://www.hcibib.org/> (accessed May 18, 2005). Comprehensive metasite on HCI with over 29,000 entries (as of July 2004). In addition to the book and journal literature, the bibliography covers history of the field, conferences, columns and news, and developer resources. Individual webliographies allow access at the subcategory level (e.g., accessibility links).

WWW Computer Architecture Page <http://www.cs.wisc.edu/~arch/www/> (accessed May 18, 2005). A cooperative from computer science faculty at University of Wisconsin-Madison, University of Pennsylvania, and University of Texas-Austin, this page gathers together resources relating to computer architecture which they define as "the science and art of selecting and interconnecting hardware components to create computers that meet functional, performance and cost goals" (source: Home page). New links are frequently added for calls for papers and tools.

Computers (general)

CompInfo: The Computer Information Center <http://www.compinfo-center.com/> (accessed May 18, 2005). CompInfo uses a very broad interpretation of IT. Its extensive listings are organized by topic and subtopic with brief annotations.

Open Directory: Computers <http://dmoz.org/Computers/> (accessed May 18, 2005). DMOZ uses a very encompassing definition of computer-related resources. Its extensive listings currently total 149,349 and are subdivided by topic with brief descriptions provided. There is a separate category for "Business: Information technology."

Virtual Computer Library (University of Texas at Austin) <http://www.utexas.edu/computer/vcl/> (accessed May 18, 2005). The VCL provides general computing resource links organized by topic.

Specialty search engines

*Scirus <http://www.scirus.com/srsapp/advanced/index.jsp> (accessed May 18, 2005). For more information see Chapter 2. Specialty sources relevant to computer science include full-text articles from ScienceDirect, Project Euclid, Scitation, and SIAM in addition to USPTO patents, e-prints from ArXiv.org and CogPrints and NASA technical reports. Access to some resources requires registration or subscription.

Sourcebank: The Search Engine for Developers <http://archive.devx.com/sourcebank/> (accessed May 18, 2005). "DevX's Sourcebank is a directory of links to source code and script posted around the Web. Use the Search option to find terms within the source code" (source: Home page). In addition to code, resources include articles, white papers, and websites.

Mailing lists, newsgroups, and online forums

In the early days of the Internet, mailing lists and newsgroups (such as Usenet) were important channels of communication for scientists and engineers. With the advent of interactive online forums associated with organizations, journals, or portals, these "older" tools are not used so often. Members of professional associations or societies may still participate in a mailing list or discussion group at the technical committee or SIG level. For example, ACM-related lists may be viewed at <http://www.lsoft.com/scripts/wl.exe?XH=LISTSERV.ACM.ORG> (accessed May 18, 2005).

Google has integrated the past twenty years of Usenet archives into its Google Groups site. Computer-related Usenet groups (or more properly, the comp.* hierarchy) may now be browsed via <http://groups-beta.google.com/groups/dir?sel=33584255&expand=1>. One noteworthy resource for software engineers continues to be COMP.SOFTWARE-ENG <http://groups-beta.google.com/group/comp.software-eng> with FAQs and archive available from <http://www.cs.queensu.ca/Software-Engineering/> (all accessed May 18, 2005).

IT professionals also seek out one another in many online communities based on interest in platform, operating system, language, product, or technology (too numerous to provide a comprehensive listing here). An example of an online community for developers is Java.net Forums <http://forums.java.net/jive/index.jspa> (accessed May 18, 2005).

Weblogs and webfeeds

A representative sampling of weblogs and webfeeds of broad interest to the computer engineer has been included here. The most useful and substantive resources are those from reputable research organizations, established technical journals, or professional web portals. Blogs and feeds are also aimed at narrower audiences such as association SIG group members (e.g., ACM SIGGRAPH blog <http://www.siggraph.org/blog/> (accessed May 18, 2005)) and faculty and students of computer engineering academic departments.

CNET Networks: CNET maintains a news.blog <http://news.com.com> as well as a directory of RSS feeds for all CNET Networks <http://www.cnet.com/4520-6022-5115113.html?tag=rss>. ZDNet also hosts blogs <http://blogs.zdnet.com/?tag=hdr> (all accessed May 18, 2005).

Computing Research Policy Blog: Advocacy and Policy Analysis for the Computing Community <http://www.cra.org/govaffairs/blog/index.php> (accessed May 18, 2005). Available since January 2004, this blog from the Computing Research

Association provides "the latest take on issues affecting the computing research community" (source: Home page).

*IDG Publications: *Computerworld* <http://www.computerworld.com/news/xml/index/> (accessed May 18, 2005). Offers numerous feeds for breaking news, *Computerworld* departments, seventy-plus topics, forty-plus special coverage topics, ten industries, and sixteen knowledge centers.

IDG Publications: *InfoWorld* <http://weblog.infoworld.com/> (accessed May 18, 2005). Offers columnist blogs as well as RSS feeds on top news, columnists, test center reviews and topics such as "applications, application development, e-business solutions & strategies, end-user hardware, networking, operating systems, platforms, security, standards, storage, telecom, wireless, and web services" (source: Home page).

IDG Publications: Network World Fusion <http://www.nwfusion.com/rss/> (accessed May 18, 2005). *Network World* offers feeds for "breaking news, reviews, opinion columns, Weblogs and original-content newsletters" (source: Home page).

*IEEE Computer Society Digital Library <http://bell.computer.org/rss/index.jsp> (accessed May 18, 2005). "IEEE Computer Society now offers the availability of the latest magazines and transactions content through RSS (Really Simple Syndication) feeds using XML (or eXtensible Markup Language) to automatically deliver new abstracts to your desktop" (source: Home page).

ITtoolbox Blogs <http://blogs.ittoolbox.com/> (accessed May 18, 2005). "ITtoolbox now features professional blogs, providing viewers with a revealing look into the daily challenges faced by real world IT professionals" (source: Home page).

Java.net: Offers developer weblogs at <http://weblogs.java.net/> and Java-related RSS feeds at <http://today.java.net/pub/q/rsschannels> (accessed May 18, 2005).

*JupiterWeb Network: Offers no blogs but the websites below offer more traditional news commentary, e-mail discussion lists, e-mail newsletters, opt-in announcements, and sometimes online forums.

- *EarthWeb:* RSS feeds – "IT Management News, Tips and Tutorials Delivered Straight to Your Desktop" – <http://www.jupiterweb.com/rss/earthweb.html> (accessed May 18, 2005).
- internet.com: RSS feeds – "IT News and Developer Tools Delivered Straight to Your Desktop" – <http://www.jupiterweb.com/rss/internet.html> (accessed May 18, 2005).
- *DevX:* RSS feeds – "Application Development Tools and Tips Delivered Straight to Your Desktop" – <http://www.jupiterweb.com/rss/devx.html> (accessed May 18, 2005).

*O'Reilly Network: Offers developer weblogs at <http://weblogs.oreilly.com/>. An O'Reilly original, Meerkat <http://www.oreillynet.com/meerkat/> (accessed May 18, 2005) is an RSS-based "syndicated content reader" which focuses on "stories of interest to developers, programmers, Web designers, intranet/extranet administrators" with a large variety of categories and channels on special topics (source: Meerkat Home page).

*TechWeb Network <http://www.techweb.com/rss/index.html> (accessed May 18, 2005). RSS feeds are available for TechWeb.com, *InformationWeek, InternetWeek, Network Computing*, and topical Pipeline sites.

Acknowledgments

Our appreciative and heartfelt thanks go to those who kindly took the time to review the content of our chapter: Jay Bhatt, Engineering Information Services Librarian, Drexel University Library; Joe Ellison, Document Delivery Assistant, Northwestern University Transportation Library; Joseph R. Kiniry, Ph.D., Lecturer in Computer Science, University College Dublin; Sandy Lewis, Electrical Engineering Collection Specialist, University of California Santa Barbara Sciences-Engineering Library; Eve Schooler, Intel Corporation.

I would like to dedicate this chapter to Anne Buck, University Librarian, California Institute of Technology from 1995 to April 2003 when she lost her battle to cancer. I know that Anne would have been pleased and proud to know that I had taken on this challenging project. This one is for you Anne! I will always regret that I did not stop and spend more quality time with Anne but was always running to the next task. (HR)

References

Hurt, C.D. (1998) *Information Sources in Science and Technology* (3rd edn), Englewood, CO: Libraries Unlimited.

Immroth, J.P. (1971) *A Guide to the Library of Congress Classification* (2nd edn), Littleton, CO: Libraries Unlimited.

Joint Task Force on Computer Engineering Curricula (2004) *Computer Engineering 2004: Curriculum Guidelines for Undergraduate Degree Programs: A Report in the Computing Curricula Series*, Washington, DC: IEEE Computer Society <http://www.eng.auburn.edu/ece/CCCE/CCCE-FinalReport-2004Dec12.pdf> (accessed February 22, 2005).

Knee, M. (2001) "Computer science: a guide to selected resources on the Internet," *C&RL News* 62 (6): 609–615 <http://www.ala.org/ala/acrl/acrlpubs/crlnews/backissues2001/june1/computerscience.htm> (accessed February 27, 2005).

Library of Congress (2001) *Library of Congress Subject Headings* (24th edn), Washington, DC: Library of Congress.

Lord, C.R. (2000) *Guide to Information Sources in Engineering and Technology*, Englewood, CO: Libraries Unlimited.

Mildren, K.W. and Hicks, P.J. (eds) (1996) *Information Sources in Engineering* (3rd edn), London; New Jersey: Bowker-Saur.

Webster's New World Dictionary of Computer Terms (1988) (3rd edn), New York: Webster's New World.

Wikipedia "Computer engineering [definition entry]" http://en.wikipedia.org/wiki/Computer_engineering (accessed February 22, 2005).

10 Electrical and electronics engineering

Larry Thompson

Introduction

There are various ways to describe electrical engineering. The following scheme (Irwin, 1995: 5–10) divides the discipline into seven areas:

1 *Power engineering* – Generating and transferring electrical energy from one location to another, and transforming it into forms that can do useful work. Power is most often generated by conversion of mechanical energy from a rotating shaft to electric energy in a generator. Power can also be produced by solar cells that convert solar energy into electrical energy or from chemical reactions, such as a battery. Power is distributed by high-voltage power lines, and is used for purposes such as heating, illumination, and driving electric motors.
2 *Electromagnetics* – Concerning the interaction between magnetic fields, electric fields, and the flow of current.
3 *Communications and signal processing* – Transmitting information from one place to another via unconfined electromagnetic waves, or telephone wires, cables or optical fibers. It includes modulating (encoding) information, and demodulation (decoding) information.
4 *Computers* – Designing and developing computer hardware and software.
5 *Electronics* – Using materials in special configurations to make devices that control current flow. These devices, such as transistors or diodes, can be interconnected to make circuits. Electronics are used in devices such as circuit boards for computers, engine monitors in cars, radio receivers, and radar systems.
6 *Systems* – Using mathematical principles to model and describe complex systems.
7 *Controls* – Providing fast and accurate adjustments or placements upon command of mechanical systems (robotic arms, airplane autopilots).

A simpler scheme (Sarma, 2001: xxi) divides the discipline into only two areas: (1) information systems (electrical means are used to transmit, store, and process information); and (2) power and energy systems (bulk energy is transmitted from one place to another and power is converted from one form to another).

It is impossible to specify an exact date when electrical engineering began. Discoveries involving electricity were made throughout the centuries, and electrical engineering gradually evolved into a distinct discipline. One of the first mentions of phenomena related to electrical engineering (Martin, 1919: 9) is credited to Thales, a Greek philosopher, who in about 600 BC recorded that a piece of amber, when rubbed against clothing, would attract and repel light objects brought close to it. In later years (Martin, 1919: 11–12), various individuals experimented with magnetism and the use of compasses to determine direction. In 1729, Stephen Grey pointed out the differences between conductors and non-conductors, and in 1745 the Leyden jar, an early capacitor, was discovered by Pieter van Musschenbroek and Ewald Jurgens von Kleist. Benjamin Franklin, famous for his 1742 kite and key experiment, invented the lightning rod as a result of that experiment.

The 1800s ushered in the more traditional beginnings of the field, with many well-known experimenters setting the foundations: Andre Marie Ampere (1775–1836) – electromagnetism and Ampere's Law; Greg Simon Ohm (1787–1854) – Ohm's Law; Michael Faraday (1791–1867) – electromagnetic induction; Samuel F.B. Morse (1791–1872) – telegraph; Alexander Graham Bell (1847–1922) – telephone; and Thomas Alva Edison (1847–1931) – incandescent light bulb and phonograph.

The electrical engineering accomplishments of the 1900s began with Guglielmo Marconi's (1874–1937) wireless transatlantic telegraph transmission in 1901. During the rest of the century, developments in areas such as electronics, computers, communications, and energy shaped the world in unprecedented ways.

- *Electronics:* The rapid development of electronics has been the driver behind much of the progress in the past century. Devices that were formerly difficult to transport and were large consumers of electric power have been reduced in size and increased in efficiency so that they can be taken almost anywhere.
- *Computers:* Although mechanical calculating machines have been available for hundreds of years, and ENIAC was built using over 19,000 vacuum tubes, it was not until the advent of low-cost electronics that computers became feasible for widespread use. Since that time they have become ubiquitous, not only as stand-alone products, but also incorporated into products ranging from automobiles to military weapons.
- *Communications:* One of the biggest benefactors of electronics developments has been communications. Electronics have made possible the development of cell phones, satellites, and other devices that enable instantaneous audio and video communication to and from any point on the globe. They have also made it possible to store huge quantities of data in readily transportable devices such as audio and video players.
- *Energy:* Coupled with the development of electronics has been the development of energy sources to power the devices. Advances in recharge-

able battery technology along with more efficient photovoltaic cells have enabled the users of electronic devices to take them almost anywhere.

Combining all these devices together, it is now possible for an individual sailing on a solo circumnavigation of the globe to power up a laptop, establish a satellite link, and send out daily updates to millions of people throughout the world. This is quite a change from Marconi's 1901 transmission of the letter "s" from Poldhu, Cornwall to St. John's, Newfoundland.

More recently, electrical engineers have been involved in such areas as nanotechnology, e-textiles, and biomedicine. It is impossible to predict exactly how research in these areas will be applied to products in the future. However, some possibilities are to use nanotechnology in the development of smaller and more portable devices, e-textiles to detect bio-hazardous chemicals used during warfare, and developments in biomedicine to produce more sophisticated diagnostic imaging machines.

The formation of professional societies paralleled the development of the discipline. In 1871 the Institution of Electrical Engineers (U.K.) was established, followed in 1884 by the American Institute of Electrical Engineers (AIEE), and in 1912 by the Institute of Radio Engineers (IRE). In 1963 the AIEE and IRE merged to form the Institute of Electrical and Electronics Engineers.

Searching the library catalog, keywords, LC Subject Headings, LC call numbers

Because of the rapid advances being made in the field, it is difficult to search a library catalog using LC Subject Headings. These headings often lag behind and do not reflect the most current research. Therefore, in most cases, using the keyword function of the catalog is the best way to start a search. This will allow items to be retrieved without relying on the out-of-date subject headings.

After relevant entries have been retrieved using keywords, the Library of Congress Subject Headings can be reviewed, and if they appear to target the desired area, a search may be done using them. In many cases, though, the keyword will retrieve satisfactory results.

The primary Library of Congress call numbers for electrical engineering are found from TK1–TK9971. The important subdivisions are as follows:

TK301–TK399 Electric meters
TK452–TK454.4 Electric circuits, electric networks
TK1001–TK1841 Production of electric energy or power
TK2000–TK2891 Dynamoelectric machinery including generators, motors, transformers
TK2896–TK2985 Production of electricity by direct energy conversion
TK3001–TK3521 Distribution or transmission of electric power
TK4001–TK4102 Applications of electric power

TK4125–TK4399 Electric lighting
TK4601–TK4661 Electric heating
TK5101–TK6720 Telecommunication including satellites, computer networks, telephones, television, radio
TK7800–TK8360 Electronics including electronics apparatus, computer engineering, computer hardware, optoelectronic devices
TK9001–TK9401 Atomic power
TK9900–TK9971 Electricity for amateurs

Also related to electrical engineering are selected areas in mathematics, computing, and physics:

Q300–Q390 Cybernetics
Q350–Q390 Information theory
QA75–QA76 Calculating machines, electronic computers, computer science, computer software
QC501–QC721 Electricity including electromagnetic theory, radio waves (theory), electric discharge, plasma physics, and ionized gases
QC750–QC766 Magnetism

In the social sciences, call numbers that cover industries related to electrical engineering may be of interest:

HD9684–HD9685 Lighting industries and electric utilities
HD9696 Electronic industries
HD9697 Electric industries
HE7601–HE8700.9 Telecommunications industry, including telegraph, wireless telegraph, radiotelegraphy, radio and television broadcasting
HE8701–HE9680.7 Telephone industry
HE9713–HE9714 Cellular telephone services industry and wireless telephone industry
HE9719–HE9721 Artificial satellite telecommunications

Article indexes and full-text resources

During most of publishing's history, there has been a division between article indexes and the articles indexed. Someone searching for an article by a particular author or on a particular subject would first consult the appropriate index, and then use the citation information in the index to retrieve the article.

The indexes that were first commercially available, such as *Engineering Index* (1884) and *Science Abstracts* (1898), were one-stop shops that allowed users to search for appropriate material across several publishers. This pattern of cross-publisher indexing continued for a hundred years, first in print format and then through online database searching.

However, in the late 1990s, a significant change occurred. Publishers began web-based journal distribution, and, to enable users to find the desired online article, they developed indexes for their own journals. In addition to being publisher-centric, these indexes had the added disadvantage of covering only a few years of publication, rather than the decades of coverage offered by the commercial indexes.

Still, users were attracted to the new indexes because they provided a quick, free, and easy way to search, and link to, the full-text online articles. Commercial databases did not yet have links to the online full text.

During the past five years, the publisher-based indexes have increased in influence. ACM, Elsevier, IEEE, and other publishers have not only expanded their online full-text holdings; they have also increased the search capabilities of their in-house indexes. Users find it convenient to go directly to a product such as the *ACM Digital Library*, *Elsevier Science Direct*, or *IEEE Xplore* to access online full-text resources. Undergraduate students, and others who do not need a comprehensive view of a topic, are understandably reluctant to search a multi-publisher database when the publisher-specific database is much more efficient. After all, why go through *Compendex* or *Inspec* with their questionable links to online full text, when a search of the *IEEE Xplore* or *ScienceDirect* sites will yield "enough" full-text articles for their project or research paper?

However, there is a danger in this. For the graduate or faculty researcher, the publisher-based search imposes an unacceptable limitation on a single publisher. It is not sufficient to get "enough" articles to write a paper; rather, the objective is often to retrieve all resources on a topic. For this type of research, it is necessary to search the comprehensive indexes such as *Inspec* and *Compendex*, as well as subject-specific indexes, which not only search multiple publishers, but also search beyond journal articles to include conference proceedings, technical reports, and government documents.

It has always been necessary for librarians and users of the resources to be aware of the limitations inherent in a resource. However, it is even more necessary now that publisher-specific indexes are freely available and compete for the user's attention. For many users there is no separation between the index and the full text. With the intertwining between the two, the online index is simply a gateway to the online full text, and each collection of full text contains its own specific index. For this reason, it is necessary to look at indexes and full text together.

ACM Digital Library <http://www.acm.org> (accessed May 18, 2005) (1954–) New York: Association for Computing Machinery. The home page of this resource states: "Full text of every article ever published by ACM." From Volume 1, Issue 1 (January, 1954) of the *Journal of the ACM* through the latest SIG newsletter, this site offers excellent coverage, but unfortunately not *every* article ever published by ACM. A comparison between the hard copy and online offerings shows incomplete full text in titles such as *Electronic Art and Animation Catalog* and *Conference Abstracts and Applications*. Still, it is an excellent effort, and all publishers should strive to produce the extent of full-text coverage that ACM has. The site offers basic, advanced, and browse modes to its subscribers, while the basic and browse modes are freely available to the general public.

Applied Science and Technology. Within the EE subject area, AS&T provides good coverage of the journals published by the major societies such as ACM, IEE, and IEEE. In addition, a variety of trade journals and selected commercial titles are included. This product is suitable for undergraduate searching (see Chapter 2 on General engineering).

Cambridge Scientific Abstracts Products. Within its array of indexes, CSA offers three that are of interest to the electrical engineer: *Computer and Information Systems Abstracts, Electronics and Communications Abstracts*, and *Solid State and Superconductivity Abstracts*. Until recently, these three databases were produced in conjunction with Engineering Information. Since 2003 they have been produced solely by CSA (see Chapter 2).

Compendex. This product has thorough coverage in the EE area and when used alone it provides adequate indexing for undergraduate studies. However, for graduate studies or when a more comprehensive search is needed it should be used in conjunction with *Inspec*. Neither *Compendex* nor *Inspec* used alone provides a comprehensive search in EE (see Chapter 2).

Infotrac <http://www.galegroup.com)> (accessed May 18, 2005) (1980–) Farmington Hills, MI: Thomson Gale. *Infotrac* provides two indexes helpful to those searching the literature in EE: *Expanded Academic Index ASAP* and *Infotrac OneFile*. Both index a good selection of IEEE and ACM journals, while OneFile also has additional coverage in IEE and trade journals. Either index would meet the needs of many undergraduate EE students, but *Expanded Academic Index* would probably be better due to the ready availability of most of its sources.

IEEE Xplore (1988–) Piscataway, NJ: IEEE. *IEEE Xplore* is the online delivery system that provides access to IEEE and IEE publications. Within *IEEE Xplore*, the public may use the free search-and-browse interface to access tables of contents and abstract records of IEEE journals, magazines, conference proceedings, and standards, as well as IEE journals and conference proceedings. IEEE members have the same browse-and-search access to all records in the database as well as access to selected full text activated in their IEEE web account. Corporate, government, and university subscribers can search and browse all records in the database and access the full-text documents permitted in their subscription agreements. Various options for accessing the full text are available. These include flat rate subscriptions to the total library or to selected portions, as well as individual article purchase. All materials included in the collection are available from 1988 to the present, and IEEE has an ongoing program to increase the backfile length, with some titles already available to the 1950s.

Inspec <http://www.iee.org> (accessed May 18, 2005) (1898–) London: IEE. *Inspec* is the combined, electronic version of *Physics Abstracts*, *Computer and Control Abstracts*, and *Electrical and Electronics Abstracts*. This has been the standard database in the field of EE, and with the recent extension of the backfile from 1969 back to 1898, it encompasses the very earliest EE literature. No literature search in the field would be complete without consulting this resource, and in most cases this database would be the first choice for a literature search. However, *Inspec* does not provide blanket coverage of all EE resources, and in order to make a comprehensive search it is necessary to use it in conjunction with *Compendex*. Several vendors offer subscription and/or pay-as-you-go access to the database.

ISI – Science Citation Index. Within EE, the primary reason for using this resource is to access the citation indexing. For a subject search, *Inspec* or *Compendex* will usually provide superior results (see Chapter 2).

NTIS. This resource indexes technical reports not covered in any of the other indexes. It is valuable for those who need to do a comprehensive subject search, but is probably of little use to most undergraduates in the field (see Chapter 2).

ScienceDirect. Elsevier and its various imprints have many titles in EE, and many of them have extended backfile online access to Volume 1, Issue 1 (see Chapter 2).

Scopus. This product has the advantage of offering citation indexing for conferences, which are very important in EE, giving it an advantage over ISI's Science Citation Index, which only covers journals. The disadvantage of the product is the loss of a controlled vocabulary which can be very helpful when doing EE subject searches (see Chapter 2).

Databooks and integrated circuits

In the past, libraries supporting an electrical engineering curriculum were expected to maintain collections of databooks from various manufacturers. This could be accomplished either through obtaining the hard copy from the manufacturers, or subscribing through a commercial vendor to obtain collections either on microfiche, or later through CD or the Web. Now, although some vendors still produce hard copy of their databooks, most researchers access the information online. In addition to the online databooks produced by specific manufacturers, there are several websites that have compiled the data from many resources.

CAPS Expert – Information Handling Services I(HS<) http://www.ihs.com> (accessed May 18, 2005). IHS is a long-time commercial provider of component information, first with microfiche and now through an online service. This is a subscription-based service.

Electronic Engineers Master Catalog (EEM) <http://www.eem.com> (accessed May 18, 2005)

IC Master <http://www.icmaster.com> (accessed May 18, 2005). Both of these resources are provided by Hearst Business Communications. Registration is required, but is free, and the same registration will work for both resources. The EEM provides information on over 6,000 manufacturers and suppliers. Various search options lead to data sheets, inventory listings, and the option to purchase components. In addition to the online version, Hearst Business Communications also publishes a multi-volume print version of the EEM. IC Master offers a searchable collection of over fifty million parts. Searches may be done by part number and there is the capability to compare parts.

UIUC Data Sheet Links <http://www.crhc.uiuc.edu/databookshelf/> (accessed May 18, 2005). This list, from the University of Illinois at Urbana-Champaign, skips the manufacturers' corporate home page and links directly to their online specifications. Very comprehensive.

Handbooks, encyclopedias, and dictionaries

Just as the availability of full-text journals online has blurred the distinction between indexes and the full-text, the availability of online full text handbooks, encyclopedias, and dictionaries has blurred the distinction between the three. In the hard copy world, it was not unusual to have subject-specific encyclopedia sets relegated to a distinct section of the reference area, while handbooks and dictionaries might be found more close at hand around the reference desk. Times have changed.

Handbooks are now less of a ready reference tool, and more of a subject overview, arranged by subdisciplines within the subject. Encyclopedias cover similar material, but in an alphabetical arrangement. For users who search an online site for a particular topic, it matters little if the needed information was gleaned from a handbook or an encyclopedia, and sites may contain both types of resources, as well as dictionaries. Because the resources have been combined online, and because users are more interested in finding information than specifying a type of resource from which to find it, all three resources will be considered in this section.

There are hundreds of handbooks available within EE. Some volumes give an overview of the entire field, while others concentrate on a specific area such as lasers or semiconductors. Whatever the scope of the title, the common characteristic is an author-devised arrangement of sub-topics which comprise the subject at hand. Although handbooks are useful additions to the collection, it is difficult to recommend specific titles. First, the content of handbooks differs, even among handbooks with similar subject emphases. Most librarians have had the experience of going to a handbook expecting to find a specific piece of information, only to discover that the handbook does not provide it. However, another handbook with a similar title but from a different publisher may provide just the piece of data needed. Is one handbook better than the other? Usually not; they are simply different. Second, faculty and students (as well as librarians) are becoming accustomed to searching collections rather than individual titles. In the not too distant past, librarians and users developed a general knowledge of what books to consult for certain information. When confronted with a heretofore unknown situation, the librarian consulted the *Composite Index for CRC Handbooks* in order to determine what CRC handbook held the desired bit of information. Now, however, with the advent of products such as CRC ENGnetBASE and Knovel, the knowledge of individual titles is of less importance. One search made over dozens of titles will retrieve hits across the collection.

With that in mind, and if feasible, subscribe to an online collection of reference books such as CRC ENGnetBASE, Knovel, netLibrary, Elsevier's Referex, Wiley Interscience, or McGraw-Hill's Digital Engineering Library. Students and faculty will appreciate the convenience of desktop access and the power of searching across multiple titles to retrieve the information they need.

In addition to an online subscription, or if an online subscription is not within the budget, selected hard copy editions should be in the collection. Using the reviews in *Choice* or conducting a search in WorldCat can be helpful to determine recent and popular titles in particular areas. For example, a WorldCat title search on "handbook or reference or manual" and "optics," limited to the latest three years, gives a selection of titles that could be added to the collection.

The number of encyclopedias and dictionaries available in electrical engineering is much less than the handbooks. However, they are published by many of the same publishers that produce the handbooks, and are often included in the online handbook collections. For instance, Elsevier's Referex collection is composed primarily of handbooks, but also contains the *Dictionary of Video Television Technology*, while the CRC ENGnetBASE contains the *Electrical Engineering Dictionary* in its collection.

There are distinctions between handbooks and encyclopedias/dictionaries in the arrangement of the material. Rather than presenting the sub-topics as part of a larger whole, encyclopedias and dictionaries present the topics alphabetically. However, when a user searches online across a publisher's collection of reference books, and is linked to the particular page of a book containing the keyword, the context of the article and the page is lost. Whether the volume is arranged topically or alphabetically is of little consequence as long as the needed information is found.

However, every library will not be subscribing to electronic sets of handbooks and other references. Because of this, a selected list of print format titles that could be considered is given below.

Bovik, A. (ed.) (2005) *Handbook of Image and Video Processing*, San Diego: Academic Press. This volume provides a technical context for the images and videos which have become ubiquitous in our daily lives.

Cadick, J., Capelli-Schellpfeffer, M., and Neitzel, D.K. (2006) *Electrical Safety Handbook*, New York: McGraw-Hill. This handbook takes a very practical approach, and is geared for the safety training and reference needs of a company. Electrical safety codes from agencies such as NFPA, NEC, and OSHA are referenced.

Chang, K. (ed.) (2003) *Handbook of RF/Microwave Components and Engineering* (2nd edn), Hoboken: Wiley. Provides principles, methods, and design data for practicing engineers in the field of radio frequency and microwave engineering.

Chen, W. (ed.) (2005) *The Electrical Engineering Handbook*, San Diego: Academic Press. A comprehensive volume giving an overview of all aspects of electrical engineering.

Chen W-K. (2003) *The Circuits and Filters Handbook* (2nd edn), Boca Raton, FL: CRC Press. Focuses on practical applications for the practicing engineer. Covers circuits and filters, both analog and digital.

Christiansen, D. and Alexander, C. (2005) *Standard Handbook of Electronic Engineering* (5th edn), New York: McGraw-Hill. In its latest edition, with over thirty years in existence, this handbook no longer places its emphasis on computers. Although the handbook still contains all the basic material, the applications now reflect a change of direction in EE, and focus on communications, media, and medicine.

Diggers, R. (ed.) (2003) *Encyclopedia of Optical Engineering*, New York: Dekker. This three-volume set provides comprehensive coverage of the topic including digital image enhancement, holography, radiometry, and lasers in medicine.

Dorf, R.C. (ed.) (1997) *The Electrical Engineering Handbook* (2nd edn), Boca Raton, FL: CRC Press. In over 2,700 pages, this handbook provides broad and comprehensive coverage of the major topics of electrical engineering. Although it cannot provide the specificity found in more narrowly defined volumes, it is a good choice.

Harper, C. (2005) *Electronic Packaging and Interconnection Handbook* (4th edn), New York: McGraw-Hill. When the phrase "electronic packaging" was first used, some thought that it referred to the box within which a device was shipped. Not so. Electronic packaging refers to the process by which an electronic component is placed within a device so that it is protected as well as enabled to connect with other electronic components.

IEEE (2000) *The Authoritative Dictionary of IEEE Standards Terms* (7th edn), Piscataway, NJ: IEEE. Within the standards authored by IEEE, it is essential that the technical terms be defined. This volume gives the definitions for terms that have been defined within the IEEE standards. In addition to giving the definition of a word or phrase, an index gives the IEEE standard in which the word or phrase is defined.

Kaiser, K.L. (2004) *Electromagnetic Compatibility Handbook*, Boca Raton, FL: CRC Press. In order to function properly, it is necessary for electronic devices to operate without interfering with other devices in the area. This handbook provides guidelines for ensuring this electromagnetic compatibility in an increasingly electronic environment.

Kaplan, S.M. (2004) *Wiley Electrical and Electronics Engineering Dictionary*, Hoboken: Wiley-IEEE Press. This is one of the many monographs that are co-published by Wiley and IEEE. Containing over 35,000 terms, this is an up-to-date resource for definitions of electrical engineering terms and acronyms.

Kent, A. and Williams, J.G. (2001) *Encyclopedia of Computer Science*, New York: Dekker. This set contains fourteen core volumes and an index. Since its initial publication, thirty supplement volumes have been added for a total of forty-five volumes.

Linden, D. and Reddy, T.B. (2002) *Handbook of Batteries* (3rd edn), New York: McGraw-Hill. This comprehensive volume covers dozens of battery types. Separate sections cover batteries used in electric vehicles, as well as portable fuel cells which may be a competitive alternative to battery systems.

McGraw-Hill (2004) *McGraw-Hill Dictionary of Electrical and Computer Engineering*, New York: McGraw-Hill. This volume is derived from the *McGraw-Hill Dictionary of Scientific and Technical Terms* (6th edn). Therefore, if you have the larger, more comprehensive work in your collection, this volume would be superfluous.

Miller, M.A. (2004) *Internet Technologies Handbook: Optimizing the IP Network*, Hoboken: Wiley. This book is application oriented, not just focusing on protocol theory, but primarily on the practical management issues facing network professionals.

Petersen, J.K. (2002) *The Telecommunications Illustrated Dictionary* (2nd edn), New York: CRC Press. This volume contains over 10,000 terms covering many aspects of telecommunications. Biographies of telecommunications pioneers, timelines, and charts enhance the volume.

Short, T.A. (2003) *Electric Power Distribution Handbook*, Boca Raton, FL: CRC Press. This book focuses on the distribution of electricity, including reliability, equipment, safety, and distributed generation.

Skvarenina, T.L. (2001) *The Power Electronics Handbook*, Boca Raton, FL: CRC Press. Power electronics is a key component in building more energy-efficient devices such as appliances and heat pumps. This book emphasizes the practical aspects of this growing technology.

Toliyat, H.A. and Kilman, G.B. (2004) *Handbook of Electric Motors*, New York: Dekker. Gives details on motor types as well as guidelines for motor selection.

Warne, D.F. (ed.) (2005) *Newnes Electrical Power Engineer's Handbook* (2nd edn), Oxford: Newnes. As the name implies, this book covers all aspects of electrical power, including generators, transformers, motors, batteries, and fuel cells.

Webster, J.G. (2001) *Wiley Encyclopedia of Electrical and Electronics Engineering*, Hoboken: Wiley. This is the most comprehensive encyclopedia available in the area of electrical engineering. Only one supplemental update has been printed at this point. An online version is available, which Wiley says is updated "regularly." Because the print is rapidly losing its currency, the online version may be a better choice.

Whitaker, J. (2005) *Standard Handbook of Broadcast Engineering*, New York: McGraw-Hill. With the advent of digital TV and radio, engineers have need of an up-to-date source of information. This volume provides the data and equations necessary for understanding these new technologies.

Whitaker, J.C. (ed.) (2005) *The Electronics Handbook* (2nd edn), New York: CRC Press. Provides both the basic theory and the practical applications of electronics. Its 2,640 pages are divided into twenty-three sections covering not only electronics, but also safety and reliability.

Wilson, J. (2005) *Sensor Technology Handbook*, Amsterdam; Boston: Elsevier. The purpose of this handbook is to assist engineers and designers in the selection of sensors for their applications. It covers various sensor types, manufacturers, guidelines for selecting and specifying sensors, as well as information on MEMS and nanotechnology applications.

Monographs: professional and textbooks

It is often difficult within EE to make a clear distinction between professional and textbook monographs. Imagine a continuum that has introductory circuit theory textbooks at one end and scholarly professional volumes at the other. The introductory circuit theory books are clearly oriented toward the student, and the focused scholarly works are written for the researcher or practicing professional. As one moves toward the middle of the continuum, the distinctions become blurred. Titles used as textbooks for upper undergraduate courses may be quite specific and useful as resource materials for the professional. Titles that were written primarily for the researcher may form the basis for a graduate-level course.

Publishers sometimes add to the confusion by mingling professional monographs and textbooks within the same series. For instance, the Power System series from Springer contains *Control of Electrical Devices*, classified as a textbook for advanced students of engineering, but also the book *Insulation of High-Voltage Equipment* for experts in power and high-voltage engineering.

Book vendors (e.g., BNA and Yankee Book Peddler) try to establish some order by classifying monographs with descriptors such as lower undergraduate, upper undergraduate, graduate, and professional in order to distinguish between the academic levels. Even though the vendor categories are not exact, they can be useful for libraries that are setting up monograph collection development policies.

Libraries vary in how they do this. Some routinely exclude lower-level undergraduate textbooks from their acquisitions plans and only purchase upper-level undergraduate and graduate-level textbooks, along with professional volumes. Other libraries take the opposite tact and specifically purchase copies of textbooks being used in classes, sometimes placing them in the library reserve collection.

Two factors combine to make textbook collection development in EE particularly challenging. First, the electrical/computer engineering field has the largest student enrollment of all engineering disciplines. This creates a demand for a large number of textbooks. Second, the field is also one of the most dynamic in engineering, with changes occurring at a rapid pace. This creates a demand for frequent revisions containing the latest material. Publishers recognize this situation and have accommodated academia with a wealth of textbooks geared to every academic level.

Because of the large number of EE textbooks available in the marketplace, it would be outside the scope of this chapter to try to select those that would be most appropriate for a collection. Indeed, as noted above, some libraries have decided that undergraduate texts are not appropriate for an academic collection.

Two articles in the *IEEE Spectrum* (Nebeker, 2003a, 2003b) discuss classic textbooks in electrical engineering. Further articles may be found through searches in *IEEE Xplore* or *Inspec*.

With regard to the professional-level monographs, it is best to start with professional societies such as IEEE, ACM, and SPIE (International Society for Optical Engineering). From there, expand into the commercial publishers and select according to the research needs of the graduate students and faculty. These groups will be the primary users of professional-level books. Most undergraduate EE students will have neither the time nor the expertise to read monographs of this type.

Be wary of standing order plans for a publisher series. They can expand and consume ever larger portions of the acquisitions budget. In addition, as noted above, the academic level may vary within a series, with a mixture of textbooks and professional monographs. If a standing order seems appropriate, be sure to check the circulation statistics for the series after a year or two to see if the volumes are being used. In many cases usage will vary greatly across the series, and it may be financially advantageous to buy selected volumes rather than the entire series.

In addition to the traditional paper format, electronic textbooks and monographs are becoming increasingly available. This format enables distance or e-learners to enjoy access to the same titles as their on-campus counterparts. Most major publishers have sites for their own titles, and several third-party vendors are providing cross-publisher packages. Some of the packages offer the advantage of automatically updating to the latest edition of a title when it becomes available, which is of significant importance in this rapidly changing field.

A slightly different type of monograph falls into the category of tutorial, user's manual, programming manual, cookbook, and so on. Although these books are not "scholarly" in nature, they cover computing topics in a practical manner and are heavily used by undergraduates and graduates in EE. Unfortunately, their popularity makes them prime candidates for theft, and their frequent updates make it difficult to keep the newest editions on the shelf. They are, however, perfectly suited for an online collection. Providers such as NetLibrary, Safari Tech Books Online, and books24x7 have collections worth considering.

Journals and conference proceedings

As is true with all other engineering disciplines, the journal literature is the foundation of scholarly communication. In electrical engineering, the conference literature also plays an important part.

In electrical engineering, the place to begin is IEEE. In terms of impact, quality, and cost-effectiveness, it is the leader. In the 2003 Impact Factor rankings in the Institute for Scientific Information's (ISI) *Journal Citation Reports* (JCR), the IEEE published eighteen out of the top twenty journals in the category of Electrical and Electronic Engineering.

Although IEEE offers various journal package plans that cover a subset of its total journal collection, if the budget will allow, a subscription to the IEEE All Society Periodicals Package (ASPP) will provide a better foundation for a quality collection. This package includes all 100+ journal titles from IEEE with a 1998 backfile. Many established engineering programs have cancelled the print format of the IEEE journals in favor of online-only. For new programs, the online-only strategy also makes the most sense because it immediately gives researchers access to the 1998 backfile.

For programs with larger budgets, the next serials addition should be one of the IEEE Proceedings Order Plans (POP). This will give students and faculty access to the important IEEE conference proceedings. Again, the best choice would be the online format because it gives immediate access to the backfile.

For the largest programs, the *IEEE/IEE Electronic Library (IEL)*, also known as *IEEE Xplore*, is the best choice. This collection offers all IEEE and IEE journals and proceedings, as well as IEEE standards. A major advantage of the IEL is its extended backfile, which increases to 1988 for all titles and as far back as the 1950s for some titles.

Once the foundation has been laid with the appropriate IEEE titles, the collection may be expanded with titles from other publishers. The ISI's *JCR* may be used to identify other important titles in the field. In addition to the category of Electrical and Electronic Engineering mentioned above, the *JCR* also ranks journals in the categories of Automation and control systems, Remote sensing, Robotics, and Telecommunications. Any or all of these categories may be appropriate, depending on the research interests of the institution.

Another strategy to determine journal additions for the collection is to identify the journals in which faculty are publishing. Talking directly with faculty may give valuable information about what they read and where they publish. A less direct method is to search *Inspec* or *Compendex* by faculty name to determine past publication patterns, or search by author affiliation.

It is difficult to recommend specific journals, publishers, or package plans other than IEEE, because the needs of institutions vary greatly. For smaller undergraduate programs, the IEEE ASPP along with a very few additional titles will probably meet the needs of the user group. For the largest institutions, the complete IEL, along with individual titles or packages from Elsevier, Wiley, Springer, and others may be necessary.

Regardless of the size of the electrical engineering program or the collection, online access statistics have made it possible to fine-tune the collection to an extent that was heretofore impossible. By analyzing the number of article downloads from each online journal, it is possible to calculate the cost per article for each journal and determine if it is more cost-effective to subscribe to the journal or to purchase articles on demand. It is no longer necessary to rely on inaccurate reshelving counts or other means to determine usage, and funds that are being spent on low-use journals may be used to subscribe to other journals that have the potential for higher usage.

As noted above, a journal collection should start with the IEEE and IEE publications. Some of the major titles outside of the IEEE and IEE publishers are listed below.

Automatica (0005–1098) Pergamon – Elsevier. Covers all aspects of theoretical and experimental control theory.

Autonomous Robots (0929–5593) Springer Science. Theory and applications of autonomous robotic systems with preference given to papers that include data from actual robots in the real world.

ETRI Journal (1225–6463) Electronics Telecommunications Research Institute. This is an English-language journal published by the Electronics and Telecommunications Research Institute of South Korea. The journal publishes articles that report on research in information, telecommunications, and network technology.

International Journal of Robotics Research (0278–3649) Sage Publications. Covers many aspects of robotics including applied mathematics, computer science, and electrical and mechanical engineering.

Optical Fiber Technology (1068–5200) Academic – Elsevier. Theoretical and experimental papers with an emphasis on practical applications.

Robotics and Autonomous Systems (0921–8890) North Holland – Elsevier. Experimental and theoretical aspects of robotics with an emphasis upon autonomous systems.

SIAM Journal on Control and Optimization (0363–0129) SIAM Publications. Research articles on the mathematics and applications of control theory.

Systems and Control Letters (0167–6911) North Holland – Elsevier. Specializes in concise papers and the rapid dissemination of information in the area of system and control.

Wireless Networks (1022–0038) Springer Science. Focuses on the networking and user aspects of mobile communications and computing.

With regard to conference proceedings, the IEEE and IEE conferences have already been mentioned above as the first choice. For most programs the next collection to consider would probably be SPIE – The International Society for Optical Engineering <http://www.spie.org> (accessed May 18, 2005). These volumes are available in both print and online formats. The online format has the advantage of offering an immediate backfile of several years. The print format offers the advantage of customization, and therefore cost savings, because it is only necessary to purchase those volumes which are actually needed by students and faculty.

Other society proceedings which may be of interest to electrical engineers are those sponsored by ACM and SIAM. The *Lecture Notes in Computer Science* series by Springer may also be of interest.

Standards and codes

Standards are an important part of the electrical engineer's world. In the past, it was common for libraries to maintain just-in-case hard copy collections of standards. Later, the standards became available on CD-ROM through commercial vendors, and are now available from vendors through immediate download from the Web.

The result is that libraries no longer need to invest large sums of money for collections and collection maintenance in order to provide just-in-case access to standards. These flat-rate, subscription-based packages will provide a cost advantage only for standards from the most heavily used organizations. For most standards collections, a just-in-time strategy may be used to deliver downloaded standards to clients on demand.

In most cases, standards requests are for a specific standard, rather than a request for a standard on a specific topic. Researchers will have found a reference to a specific standard in the literature, or be required to meet a specific standard in a proposal or project. The need is simply to retrieve the standard. There will be a need to identify a standard related to a particular topic in only a few instances.

The major commercial standards suppliers have been identified in Chapter 2, along with a discussion of the online indexes. As noted therein, many societies do not sell standards directly, but have contracted with third-party vendors to do this.

Several organizations produce standards relevant to electrical engineering:

IEC – <http://www.iec.ch> (accessed May 18, 2005). International Electrotechnical Commission. The IEC prepares and publishes international standards in the area of electrical engineering. In doing this it works closely with organizations such as the IEEE and ISO.

IEEE – <http://www.ieee.org> (accessed May 18, 2005). Standards may be purchased directly through the IEEE website, as well as through third-party vendors. All current IEEE standards are included in the IEL/IEEE *Xplore* package described above.

ISO – see Chapter 2.

ITU – <http://www.itu.int> (accessed May 18, 2005). International Telecommunication Union. The ITU publishes two major sets of recommendations, the ITU-T and the ITU-R. These may be purchased at the ITU site as well as through third-party vendors.

NFPA – <http://www.nfpa.org> (accessed May 18, 2005). National Fire Protection Association. The NFPA publishes the National Electrical Code, updated every three years, which gives guidelines for electrical installations within buildings.

UL – <http://www.ul.com> (accessed May 18, 2005). Underwriters Laboratories. Available through the UL website or from third-party vendors.

In addition to the above organizations, a more extensive list may be found at the WWW Virtual Library – Electrical and Electronics Engineering <http://www.cem.itesm.mx/vlee/Standards> (accessed May 18, 2005).

Websites

Many of the most important sites related specifically to EE have already been mentioned in two previous sections of this chapter: (1) Article indexes and Full-text resources, and (2) Standards and codes.

A compilation of sites specific to EE is found at the WWW Virtual Library – Electrical and Electronics Engineering <http://www.cem.itesm. mx/vlee> (accessed May 18, 2005). This contains an assortment of EE-related links grouped under the categories of Academic and research institutions, Products and services, Information resources, Journals and magazines, and Standards. The site also allows keyword searches in order to find relevant entries.

In addition to EE-specific sites, there are sites that cover many engineering disciplines and have EE components. Two of the most helpful are the Scout Report and Engineering Village, both of which have been covered in Chapter 2. The Scout Report, which is freely available, has an archival search feature that retrieves a listing of reviewed websites matching the search criteria. Examples of some of the valuable sites found in the Scout Report follow.

BOWest Pty Ltd <http://www.bowest.com.au/library.html> (accessed May 18, 2005). This site, sometimes listed as the Electrical Theorems and Formulas site, is sponsored by an Australian consulting company. It contains information helpful to both analog circuit designers and students.

Electrical Engineering Circuits Archive <http://www.ee.washington.edu/circuit_archive/> (accessed May 18, 2005). The site is sponsored by the University of Washington, and provides an e-mail address so that users can submit new materials for inclusion in the database.

Electric Power Research Institute (EPRI) <http://www.epri.com> (accessed May 18, 2005). Founded in 1973, EPRI is a non-profit collaborative organization established for the purpose of research and development in electric generation, delivery, and use.

Energy Information Administration <http://www.eia.doe.gov/fuelelectric.html> (accessed May 18, 2005). This contains a wealth of statistical data including power generation capacities, pricing, and state-by-state information, as well as full text of many reports.

In addition to the above compilations which are oriented toward EE or engineering, many general web directories have EE sections within them. These EE sections contain information on topics such as discussion groups, professional societies, conferences, and university programs.

Galaxy Electrical Engineering <http://www.galaxy.com/galaxy/Engineering-and-Technology/Electrical-Engineering/> (accessed May 18, 2005).
Google Electrical Engineering <http://www.google.com/Top/Science/Technology/Electrical_Engineering/> (accessed May 18, 2005).
Yahoo Electrical Engineering <http://dir.yahoo.com/science/engineering/Electrical_Engineering/> (accessed May 18, 2005).

Associations and societies

There are numerous associations and societies related to the field of electrical engineering. Some have already been mentioned in previous sections of this chapter, but a listing of the more important includes:

Association for Computing Machinery <http://www.acm.org> (accessed May 18, 2005). Founded in 1947, this association is oriented heavily toward the computer literature, but is used by electrical engineers.
Audio Engineering Society <http://www.aes.org> (accessed May 18, 2005). Founded in 1948, this society consists of those who work with recording and reproducing equipment.
Institute of Electrical and Electronics Engineers (IEEE) <http://www.ieee.org> (accessed May 18, 2005). Founded in 1963 by the merger of the American Institute of Electrical Engineers (1884) and the Institute of Radio Engineers (1912), it is one of the premier organizations for publishing and conferences in electrical engineering. There are numerous societies within the IEEE that meet the needs of the specific subdisciplines of electrical engineering.
Institute of Electrical Engineers (IEE) <http://www.iee.org> (accessed May 18, 2005). Founded in 1871, the IEE is the publisher of *Inspec*, one of the most important indexes in the field.
International Microelectronic and Packaging Society (IMAPS) <http://www.imaps.org> (accessed May 18, 2005). Founded in 1967, IMAPS holds an annual national conference as well as international, national, and regional seminars.

Both the Associations Unlimited <http://www.galegroup.com> (accessed May 18, 2005) and the Scholarly Societies Project <http://www.scholarly-societies.org> (accessed May 18, 2005) are described in Chapter 2, and are valuable for locating additional groups in the field of electrical engineering.

Conclusion

Electrical engineering continues to be one of the most dynamic branches of engineering. Advances in telecommunications, biotechnology, nanotechnology, and other areas of research all guarantee that the field will continue to expand. As the research becomes more interdisciplinary, it will become increasingly important for electrical engineers to access efficiently the information they need.

References

Irwin, J.D. and Kerns, D.V. Jr (1995) *Introduction to Electrical Engineering*, Englewood Cliffs, NJ: Prentice-Hall.

Martin, T.C. and Coles, S.L. (eds) (1919) *The Story of Electricity: Volume One: A Popular and Practical Historical Account of the Establishment and Wonderful Development of the Electrical Industry*, New York: The Story of Electricity Company, M.M. Marcy.

Nebeker, F. (2003a) "Treasured Texts," *IEEE Spectrum* 40 (4): 44–49.

Nebeker, F. (2003b) "More Treasured Texts," *IEEE Spectrum* 40 (7): 31–36.

Sarma, M.S. (2001) *Introduction to Electrical Engineering*, New York: Oxford University Press.

11 Engineering education

Jill H. Powell

Engineering education is the education which students receive at university level that prepares them for a career as an engineer. A certain number of courses in mathematics, basic sciences, engineering sciences, engineering design, and other studies are required. Engineers differ from scientists in that engineers are concerned with design; scientists with fundamentals (Ruth, 2004).

Engineering education is also the study of problems and issues surrounding the education of the engineering student. While it is more social than technical in nature, such study is crucial to remain competitive and anticipate future needs. *Educating the Engineer for the 21st Century* (Weichert *et al.*, 2001) contains thirty-five papers presented at a conference about many global issues from students, faculty, staff, and practitioners from around the world. *Teaching the Majority* (Rosser, 1995) discusses how to teach in a way that is more inclusive to women and minorities. Both would be good books to study in a course on engineering education.

One of the best books describing the history of engineering education is Lawrence Grayson's *The Making of an Engineer* (Grayson, 1993). This book was written to celebrate the centennial of the American Society for Engineering Education (ASEE), and contains some 350 photographs as well as an exhaustive textual history of engineering education. Research in engineering education is not complete without consulting the publications and extensive web pages of this society. ASEE's mission is "furthering education and engineering technology" and "promoting excellence in instruction, research, public service, and practice" <http://www.asee.org/about/missionAndVision.cfm> accessed March 9, 2005).

Engineering education encompasses a number of categories, including students, faculty, practitioners and alumni, courses and programs, and assessment and evaluation. Topics relating to students include learning resources and practices, admissions and graduation requirements, advising, research opportunities, retention, and minority groups. Topics relating to faculty include specific teaching methods, publications and research methods, hiring, promotion, tenure, and strategies and tools used in the classroom. Practitioners and alumni address subjects relating to industry and collaborations. Assessment and evaluation refer to improving specific courses as well as overall engineering programs, plus distance learning, design, and ethics courses. Other topics include history of engineering education, and education and learning theory. Taken together they comprise the foundational subjects of engineering education.

History of engineering education

In 1802 West Point was established as the first engineering school in the U.S. (Hollister, 1966: 144). Engineering programs then began at Rensselaer Polytechnic Institute in 1824, Norwich University in 1825, and offered in the 1850s at University of Michigan, Harvard, Yale, Union College, and Dartmouth (Reynolds, 1991: 20). The number of engineering schools grew from three in 1900 to thirty schools by 1909 (Grayson, 1993: 77). In 2004 there are over 358 accredited schools (<http://www.asee.org/publications/> accessed May 18, 2005). Various engineering societies developed in the late 1800s, including the American Society of Civil Engineers (ASCE) in 1852, American Society of Mechanical Engineers (ASME) in 1880, American Institute of Electrical Engineers in 1884, and later the American Institute of Chemical Engineers (AIChE) in 1908 (Reynolds, 1991: 24, 355).

Engineering became established in the academic community with the passage of the Morrill Land Grant Act in 1862. This act provided public lands to the states to establish colleges of agriculture and mechanics arts, an early term to describe engineering. During the 1870s and 1880s, inventions, such as the internal combustion engine, steam-engine, electric generator, incandescent lamp, phonograph, and telephone helped advance the country toward the Industrial Revolution. The World Columbian Exposition was held in 1893 in Chicago, showcasing many of these inventions. A meeting of engineers at this exposition created the ASEE, which provided a forum for engineering educators to identify and address issues, such as entrance standards, curricula, course content, and requirements for graduation. Previously, engineering education had developed in an unplanned and uncoordinated manner, but after the creation of ASEE, schools had a resource for advice on curricula, teacher development, and ways to affect national policy (Grayson, 1993).

Wilhelm von Humboldt, Prussian Minister of Education, oversaw the system of Technische Hochschule and Gymnasium that made Prussia a strong scientific and intellectual power in Europe in the nineteenth century. He introduced two features that influenced many universities in Europe and North America. One of the requirements was that research and teaching be done by the same people. Another was freedom for the professor to decide what to teach and research, and for students the freedom to design their own program. The Humboldt University education is neither practical nor focused on technical skills for a specific profession. It aims to produce educated, well-rounded citizens and scholars for the long term. The large increase in the student/faculty ratio has weakened this ideology somewhat, as has the undergraduate programs in the United States, with its required set of courses. However, Humboldt's ideas are still apparent in America's graduate schools. Comprehensive exams replace credit point systems, and the exams are not always linked to the content of specific courses. Students work with faculty to design their own research. American graduate schools still embody the principles of Humboldt, and educate those who take a deep interest in science and research (Doepke, 2003).

Challenges in engineering education that continue to this day include the debate about the best way to prepare graduates. Should they be trained as specialists or generalists? Should they be taught in industrial shops for immediate usefulness in the workplace, or educated in a laboratory, studying fundamental principles? During various times in history each of these sides has become dominant.

Until around 1950 engineering education was vocationally rather than scientifically oriented. After World War II the emphasis shifted to fundamental principles, as physicists were valued more for their role in weapons development. Later, companies in a competitive marketplace began to demand practically trained engineers. The circle continues, as schools responded by emphasizing engineering practice, manufacturing techniques, and concepts such as reliability and quality (Grayson, 1993: x). In the late 1990s and continuing into the twentieth-first century information technology has driven the U.S. economy, and courses on entrepreneurial skills are included in the curriculum. Women, minorities, and large numbers of foreign students also began to receive degrees and join the workforce. This education landscape produces engineers with the skills to improve all aspects of engineering practice and to strengthen the economy.

World War II played several remarkable roles in the engineering education: it infused large sums of money into universities for the purposes of research, and the government paid for many returning draftees to train as engineers. The competition from the Cold War with the Soviet Union also increased the U.S.'s desire to educate more engineers and scientists.

Searching the library catalog

While keyword searching will find some useful books, one can find more complete information with the use of subject headings. Listed below are some of the most useful Library of Congress Subject Headings.

Engineering – study and teaching
Engineers – education
Electric engineering – study and teaching
Mechanical engineering – study and teaching
Civil engineering – study and teaching
Chemical engineering – study and teaching
Technical education
Science – study and teaching

The headings may be divided further by country, type of study, and type of format. Since many conference papers are published in conference proceedings, the word *Congresses* may be present:

Engineering – study and teaching (graduate) – United States
Engineering – study and teaching – periodicals
Engineers – education – United States
Engineering – study and teaching (Higher) – computer-assisted instruction – congresses
Chemical engineering – study and teaching – congresses

These last two headings may also be useful, when combined with the keyword "Engineers" or "Engineering":

Education, cooperative
Industry and education

Abstracts and indexes

Research databases, also called abstracts and indexes, offer access to citations to journal articles, conference papers, and theses. They are vital instruments in the early steps of research. The best indexes for recent engineering education are *Compendex*, *ERIC*, and *Applied Science and Technology Abstracts*. For early historical information going back as far as the eighteenth century, *Making of America* and the *History of Science, Technology, and Medicine* are useful tools. Depending on the type of engineering studied, additional indexes should be added.

ERIC (1966–) Boston, MA: SilverPlatter. ERIC abstracts journal articles and reports on education. Many materials, such as curriculum guides, instructional materials, conference papers, and project reports are available in the ERIC microfiche collection. ERIC indexes materials on education, child development, classroom techniques, computer education, counseling, testing, communication skills, career education, science education, teacher education, evaluation, disabled children, gifted children, and library and information science. An online thesaurus shows relevant subject headings: engineering education, professional education, engineering technology, engineers, land grant universities, science education, and technical education.

History of Science, Technology, and Medicine (HST) (1975–) Palo Alto, CA: Research Libraries Information Network. HST is the resource to go to for the earliest information on engineering education. It comprises four bibliographies: *Bibliografia Italiana di Storia Della Scienza, Current Bibliography in the History of Technology (Technology and Culture), Current Work in the History of Medicine* (Wellcome Library), and *Isis Current Bibliography of the History of Science*. It indexes journal articles, conference proceedings, books, book reviews, and dissertations on the history of science and technology. Relevant subject headings are education – engineering, technical societies, technical education; education, technical.

Making of America: The Cornell University Library MOA Collection <http://cdl.library.cornell.edu/moa/index.html> (1996–) Ithaca, NY: Cornell University Library. A digital library of primary sources in American history from the antebellum period through reconstruction. The collection is strong in education, psychology, American history, sociology, religion, and science and technology. Full-page images of journals are included.

When researching engineering education, one should go beyond the general science and education indexes given above. Specific disciplines are best covered by their own databases. Society publications are also indexed in these sources. For example, engineering education for computer science would be well covered by *Inspec, IEEE Xplore*, and *ACM Portal* (which indexes publications by the Association for Computing Machinery).

Bibliographies and guides to the literature

There are several annotated bibliographies that describe research in the field of engineering education. Currency is always a concern, and one needs to follow up with database searches from the entries in the Research Databases section; however, these bibliographies can save time in reviewing historical papers.

Channell, D.F. (1989) *The History of Engineering Science: An Annotated Bibliography*, New York: Garland [Bibliographies of the History of Science and Technology; vol. 16]. A chapter titled "Institutions" lists over 100 annotated references on the history of various engineering colleges around the world. Research institutes and professional organizations are also covered.

Cooper, J. and Robinson, P. (1997) *Small-group Instruction: An Annotated Bibliography of Science, Mathematics, Engineering, and Technology Resources in Higher Education*, Occasional Paper. University of Wisconsin, ED472334. This source addresses research, theory, and practice in small groups and cooperative instruction in higher education.

Dyrud, M.A. (2004) "2003 engineering technology education bibliography," *Journal of Engineering Technology*, 21 (2): 26–40. Published yearly in this journal, this resource lists books and articles under the following subject headings: administration; aviation; architectural; assessment; biomedical; civil, computers, curriculum; distance education; electrical/electronics; industrial; industry/government/employers; information technology; instructional technology; international; laboratories; liberal studies; manufacturing; mechanical; minorities; nanotechnology; photonics; teaching methodology; tech prep; telecommunications; technical graphics.

Seymour, E. (2002) "Tracking the processes of change in US undergraduate engineering education in science, mathematics, engineering, and technology," *Science Education* 86 (1): 101–105. A review article that discusses the changes in engineering education over the past decade and cites over seventy articles on the subject. Excellent coverage of many hard-to-find articles.

Directories

Directories help prospective students to evaluate the nature, scope, requirements, and statistics of various engineering programs. Many are published yearly, and most libraries will keep the current year only. Many directories will have online equivalents, and the URLs are noted where found. This is not an exhaustive list. For specific disciplines, consult library catalogs using appropriate subject headings; for example, Chemical engineering – study and teaching (Higher) – directories.

America's Best Graduate Schools (1994–)Washington, DC: U.S. News and World Report <http://www.usnews.com/usnews/edu/grad/rankings/rankindex_brief.php> (accessed May 17, 2005). Controversial, subjective, and popular ranking of graduate schools in engineering. Ranks by program. Subscription portion lists average GRE scores, enrollments, research expenditures, acceptance rates, recruiter scores, and more. This title is usually published during the month of March. The subscription section provides more in-depth information.

American Council on Education. American Universities and Colleges (1928–) New York: Walter de Gruyter. Contains detailed descriptions of over 1,900 institutions in the U.S. Narratives on structure of higher education, foreign students, undergraduate, graduate, and professional education. Descriptions include history, structure, admission requirements, statistical characteristics of freshmen, degrees offered, fees, enrollment, student life, library collections, and more. Indices, tables, and appendices are included. One of the most useful indexes is the listing of colleges by degree program.

ASEE Directory of Engineering and Engineering Technology Colleges (1997–) Washington, DC: American Society for Engineering Education <http://www.asee.org/about/publications/profiles/index.cfm> (accessed May 17, 2005). This directory is for prospective engineering students wishing to compare over 358 engineering and engineering technology colleges in the United States and Canada. Profiles include institutional information, undergraduate information, and graduate information. Statistical tables and profiles include information on enrollment, degrees granted, student expenses, faculty numbers, and research expenditures. Indexes by degree program, geographic location, and alphabetical by institution. Online version contains much more information, including research centers, student support programs, dual degrees, and department areas of expertise.

ASEE Directory of Graduate Engineering and Research Statistics (1999–) Washington, DC: American Society for Engineering Education. Partially online at <http://www.asee.org/colleges/> (accessed May 17, 2005). This directory provides extensive statistical information on graduate engineering departments. It compares programs in areas such as enrollments, degrees awarded by gender and ethnic groups, student appointments by department with average monthly stipend, and faculty statistics. Includes index by subject areas of graduate engineering-related research.

ASEE Directory of Undergraduate Engineering Statistics (1997–) Washington, DC: American Society for Engineering Education. Partially online at <http://www.asee.org/colleges/> (accessed May 17, 2005). This directory provides extensive statistical information on undergraduate engineering departments. It compares programs in areas such as enrollments, degrees awarded by gender and ethnic groups, faculty numbers, student expenses, and number of coop participants.

Bear, J., Bear, M., McQueary, L., and Head, T.C. (2001) *Bears' Guide to the Best Computer Degrees by Distance Learning*, Berkeley, CA: Ten Speed Press. This book helps prospective students compare 100 computer science degrees programs by including profiles and statistics.

Chemical Engineering Faculty Directory (2000–) <http://www.che.utexas.edu/che-faculty/> (accessed May 17, 2005). Previously available as Chemical Engineering Faculties, 1900s–1999. Basic faculty directory maintained by members at the University of Texas, Austin.

DGRweb (1997–) Washington, DC: American Chemical Society. Available by subscription only <http://pubs.acs.org/dgrweb/> (accessed June 7, 2004). Includes ACS Directory of Graduate Research, publications, doctoral and Master's theses in departments of chemistry, chemical engineering, biochemistry, polymer science, materials science, forensic science, marine science, toxicology, and environmental science at U.S. and Canadian universities.

Directory of Human Factors/Ergonomics Graduate Programs in the United States and Canada (2002) Santa Monica, CA: Human Factors and Ergonomics Society <http://www.hfes.org/Publications/2002Gradschools/TofC.html> (accessed May 17, 2005). The information in this directory should be helpful for prospective graduate students in assessing the nature, scope, and requirements of various programs.

Engineering and Technology Degrees (1967/68–) New York: Engineering Workforce Commission of American Association of Engineering Societies. This annual report identifies degrees awarded by race, gender, and type of degree covering 339 engineering institutions and 284 engineering technology institutions.

Engineering and Technology Enrollments (1974–) New York: Engineering Workforce Commission of American Association of Engineering Societies. This annual directory covers 351 engineering and 282 engineering technology institutions. Enrollment data are by institution, field, gender, race, and part-time or full-time status.

Fiske, E.B. (1988–) *The Fiske Guide to Colleges*, Naperville, IL: Sourcebooks, Inc. Contains lengthy narratives on more than 300 colleges, discussing strengths and weakness. Includes student quotes on academics and social life. A survey helps students select the best college.

GradSchools.com: The most comprehensive online source of graduate school information <http://www.gradschools.com/listings/menus/mech_eng_menu.html> (accessed May 17, 2005). This URL provides access specifically to engineering graduate school programs that offer Master's and Ph.D. degrees. Provides access to universities by subject, by state, and to engineering programs available worldwide. Provides links directly to the engineering department of universities.

Graduate Programs in Engineering and Applied Sciences (1986–) Princeton, NJ: Peterson's Guides. Book 5 in the six-volume set published by Peterson's, this directory describes more than 4,000 graduate programs in sixty-seven disciplines. Includes admission requirements, expenses, financial support, programs of study, and faculty research specialties. Typeface is very small.

Keane, C.M. (1992–) *Directory of Geoscience Departments*. Alexandria, VA: American Geological Institute. Brief information on 1,047 geoscience departments in the United States and the rest of the world. Includes faculty names, contact information, and enrollments. Federal agencies of interest to geoscientists are also included.

Occupational Outlook Handbook (1949–) U.S. Department of Labor, Bureau of Labor Statistics. Indianapolis, IN: JIST Publications <http://www.bls.gov/oco/home.htm> (accessed May 17, 2005). From the website: "The *Occupational Outlook Handbook* is a nationally recognized source of career information, designed to provide valuable assistance to individuals making decisions about their future work lives. Revised every two years, the *Handbook* describes what workers do on the job, working conditions, the training and education needed, earnings, and expected job prospects in a wide range of occupations."

Research-doctorate Programs in the United States: Continuity and Change (1995) Committee for the Study of Research-Doctorate Programs in the United States, National Research Council. Washington, DC: National Academy of Sciences. Contains information on descriptive statistics of selected characteristics of Ph.D. granting institutions and faculty views on program quality. Extensive narratives, tables, and figures compare institutions.

World of Learning (1947–) London: Europa Publications. Available online to subscribers at <http://www.worldoflearning.com> (accessed May 17, 2005). Lists contact information, publications, faculty, and statistics on colleges and universities worldwide.

Books

Grayson, L.P. (1993) *The Making of an Engineer: An Illustrated History of Engineering Education in United States and Canada*, New York: John Wiley & Sons. This book was written to celebrate the centennial of the American Society for Engineering Education (ASEE) and has some 350 photographs as well as an exhaustive textual history of engineering education in the United States and Canada. Highly recommended.

Hoag, K. (2001) *Skills Development for Engineers: An Innovative Model for Advanced Learning in the Workplace*, London: IEE. Contains many practical examples of successful approaches, with diagrams. Contents include "Moving beyond the classroom," "Management roles," "Mechanisms for advanced learning," "Communicating the information," "The supervisor's role," and "There is no such thing as a free lunch."

Kline, R.R. (1992) *Steinmetz: Engineer and Socialist*, Baltimore, MD: Johns Hopkins University Press. Chapter 7, entitled "Reforming a profession" (pp. 165–199) covers the history of engineering education, the first colleges to offer engineering degrees, and Charles Steinmetz's role as college educator, professional society leader, and reformer.

Layton, E.T. Jr (1986) *The Revolt of the Engineers: Social Responsibility and the American Engineering Profession*, Baltimore, MD: Johns Hopkins University Press. Examines the politics of engineering societies.

Leslie, S.W. (1993) *The Cold War and American Science: The Military-industrial-academic Complex at MIT and Stanford*, New York: Columbia University Press. Discusses the changes World War II and the Cold War funding of engineering had on the workings of these two universities.

Oldenziel, R. (1999) *Making Technology Masculine: Men, Women, and Modern Machines in America, 1870–1950*, Amsterdam: Amsterdam University Press. Excellent coverage of women engineers up to the founding of the Society of Women Engineers in the 1950s.

Project 2061, American Association for the Advancement of Science (1993) *Bench-*

marks for Science Literacy, New York: Oxford University Press. While focused on what students should know about science, mathematics, and technology by the end of grades 2, 5, 3, and 12, this book provides well-thought-out competencies, discussions, and perspectives. Engineering educators may use this book for K-12 collaborations and also as a tool to what students should know before they come to college.

Project 2061, American Association for the Advancement of Science (2000) *Designs for Science Literacy*, New York: Oxford University Press. Contains curriculum designs, specifications, and suggestions for improvement.

Reynolds, T.S. (1991) *The Engineer in America: A Historical Anthology from Technology and Culture*, Chicago, IL: University of Chicago Press. Contains essays appearing in the journal *Technology and Culture*; this volume presents a thorough history of engineering and engineers from its earliest beginnings to the twentieth century.

Rosser, S.V. (1995) *Teaching the Majority: Breaking the Gender Barrier in Science, Mathematics, and Engineering*, New York: Teachers College Press. Discussions by various authors on how to alter teaching to include feminist and other diverse approaches into the science and engineering classroom. Chapters focus on various disciplines, such as chemistry, physics and engineering, mathematics, computer science, environmental sciences, and geosciences.

Scarl, D. (1998) *How to Solve Problems: For Success in Freshman Physics, Engineering, and Beyond*, Glen Cove, NY: Dosoris Press. Teaches problem-solving methods that experienced scientist and engineers use to define a problem, solve it, and present their solution to others.

Wankat, P.C. and Oreovicz, F.S. (1993) *Teaching Engineering*, USA: McGraw-Hill, Inc. An excellent practical guide for engineering instructors, covering all aspects of teaching college-level engineering, from the component of good teaching to efficiency, designing lectures, teaching with technology, laboratories, testing, ethics, and evaluation.

Weichert, D., Rauhut, B., and Schmidt, R. (eds) (2001) *Educating the Engineer for the 21st Century*, Dordrecht, the Netherlands: Kluwer Academic. Provides published papers from the Proceedings of the 3rd Workshop on Global Engineering Education. Topics include European, Asian, and American views of engineering education, developing personal skills, programs, curricula, educational concepts, successful university–industry partnerships, and design projects for the global engineer.

Reports

The following selected reports (many produced by committees) are recommended as good places to start familiarizing oneself with engineering education. Most were found from searching the ERIC database, Worldcat (OCLC), and National Academies of Press Publications <http://www.nap.edu/>.

Building a Workforce for the Information Economy (2001) National Research Council. Committee on Workforce Needs in Information Technology, Board on Testing and Assessment, Board on Science, Technology, and Economic Policy. Office of Scientific and Engineering Personnel. Washington, DC: National Academy Press <http://books.nap.edu/catalog/9830.html> (accessed May 17, 2005). This report presents the results of a study on the supply and demand for information technology workers over the next ten years. It discusses the nature of IT work, employee demographics, perceived worker shortages, older IT workers and age discrimination, foreign workers, and long-term recommendations for the future.

Educating Scientists and Engineers: Grade School to Grad School (1988) United States Congress. Office of Technology Assessment. Washington, DC: Congress of the U.S. O.T.A. (available May 17, 2005). This report presents statistics and demographics that detail the supply and demand for engineers. Many tables on enrollment, degrees, majors, employment, and spending are included. Extensive discussion about foreign students and immigration policies. Presents policy options to improve science and engineering education, and retention and recruitment strategies at all levels of schooling.

Enhancing the Postdoctoral Experience for Scientists and Engineers: A Guide For Postdoctoral Scholars, Advisers, Institutions, Funding Organizations, and Disciplinary Societies (2000) Committee on Science, Engineering, and Public Policy (U.S.). National Academy of Sciences, National Academy of Engineering, Institute of Medicine. Washington, DC: National Academy Press. This comprehensive guide addresses issues in the postdoctoral experience, including rights, opportunities, responsibilities, funding, advisers, and recommendations for improvements. Information is supplied from meetings with thirty-nine focus groups at eleven universities, seven national laboratories, and five private research institutes or industrial firms; plus a day-long workshop with 100 postdocs and administrators.

Fox, M.A. and Hackerman, N. (eds) (2003) *Evaluating and Improving Undergraduate Teaching in Science, Technology, Engineering, and Mathematics*, Committee on Recognizing, Evaluating, Rewarding, and Development Excellence in Teaching of Undergraduate Science, Mathematics, Engineering, and Technology. National Research Council. Washington, DC: National Academy Press. This report examines effective teaching practices, and offers both methodologies and practical ways of evaluating teachers and academic programs. Included are recommendations on how to achieve change, and sample evaluation instruments.

From Analysis to Action: Undergraduate Education in Science, Mathematics, Engineering, and Technology (1996) Report from the National Research Council/National Science Foundation Convocation on Undergraduate Science, Mathematics, and Engineering Education. Washington, DC: National Academy Press <http://www.nap.edu/readingroom/records/NI000012.html> (accessed May 17, 2005). Synopsis of a convocation held in April 1995 with students, faculty, and administrators to discuss such issues as access, literacy, competency, curriculum, teaching, reward systems, and evaluation systems.

Engineering Education: Designing an Adaptive System (1995) National Research Council. Board of Engineering Education, Commission on Engineering and Technical Systems, Office of Scientific and Engineering Personnel. Washington, DC: National Academy Press <http://www.nap.edu/catalog/4907.html> (accessed May 10, 2004). Board of Engineering Education's report on the strengths and weaknesses of engineering education, and ways to achieve progress amid global political, societal, and economic changes. Includes outlines for a new engineering curriculum and specific recommendations for institutions, faculty, industry, and NSF. A key theme is "think globally, act locally."

Grant, H. (1993) "The role of quality concepts in engineering education," *Frontiers in Education Conference. Twenty-Third Annual Conference Proceedings. Engineering Education: Renewing America's Technology*, New York: IEEE, pp. 535–539. Following a survey of Deans of Engineering and other research, a task force presents action items that NSF can implement to improve the quality of education, a bibliography of basic research in quality in engineering education, a workshop on quality of engineering education, and alliance proposals with other groups including ABET, ASEE, and others.

Jackson, S.A. (2003) *Envisioning a 21st Century Science and Engineering Workforce for the United States: Tasks for University, Industry, and Government*, National Academy of Sciences, National Academy of Engineering, Institute of Medicine <http://books.nap.edu/catalog/10647.html> (accessed May 10, 2004). Written by the President of Rensselaer Polytechnic Institute, this paper finds the competitiveness of the nation's science and engineering talent to be eroding and reviews options to manage risks to U.S. technological innovation.

Reshaping the Graduate Education of Scientists and Engineers (1995) National Academy of Sciences, National Academy of Engineering, Institute of Medicine. Committee on Science, Engineering, and Public Policy. Washington, DC: National Academy Press <http://books.nap.edu/catalog/4935.html> (accessed May 10, 2004). The Committee convened a panel of experts from academia, government, industry, and foundations to answer questions such as: Is there an oversupply of Ph.D.s? What are the current employment conditions and trends? What impact do foreign students have? What changes should we make that will benefit students as well as society? Extensive statistics on enrollments, degrees, fields, and employment are included.

Science and Engineering Indicators – 2002 (2002) National Science Foundation. Division of Science Resource Statistics. Arlington, VA: National Science Foundation (NSB 02-01) <http://www.nsf.gov/sbe/srs/seind02/start.htm> (accessed May 10, 2004). This report contains comprehensive quantitative analyses of the scope and quality of the nation's science and engineering programs and enterprises. The report presents material on science education from the elementary level through graduate school; the scientific and engineering workforce; R&D performers, U.S. competitiveness in high technology; public understanding of science and engineering; and the significance of information technologies.

Shaping the Future. Volume I: New Expectations for Undergraduate Education in Science, Mathematics, Engineering, and Technology (SMET) (1996) Advisory Committee to the National Science Foundation (NSF) Directorate for Education and Human Resources. Arlington, VA: National Science Foundation. ERIC document ED404158, NSF 96–139 <http://www.nsf.gov/cgi-bin/getpub?nsf96139> (accessed May 10, 2004). This report is the product of more than a year of intensive work studying problems hindering undergraduate education at both two- and four-year institutions. It addresses the state of education at the time (1995), barriers to improvement, and ways to meet new expectations, with specific recommendations for colleges, universities, businesses, state and federal governments, and the National Science Foundation.

Shaping the Future. Volume II: Perspectives on Undergraduate Education in Science, Mathematics, Engineering, and Technology (SMET) (1998) Advisory Committee to the National Science Foundation (NSF) Directorate for Education and Human Resources. Arlington, VA: National Science Foundation <http://www.nsf.gov/cgi-bin/getpub?nsf98128> (accessed May 10, 2004). ERIC document ED433882, NSF 98–128. Volume II discusses activities that have taken place since the publication of Volume I, and contains testimony by university deans, professors, and industry leaders presented at committee hearings. Remarks focus on developing and implementing strategies to improve undergraduate SMET education. The report also includes focus group findings, extensive background data on enrollment, expenditures, and faculty practices; and an eighteen-page bibliography.

Transforming Undergraduate Education in Science, Mathematics, Engineering, and Technology (1999) Committee on Undergraduate Science Education. Center for Science, Mathematics, and Engineering Education. National Research Council. Washington, DC: National Academy Press. Presents six visions for raising scientific awareness of all students, not just science and engineering students. These include requiring courses, establishing admission standards, evaluating these courses, having higher education professionals share responsibility for K-12 teacher education, and having institutions provide the necessary infrastructure to achieve these visions. Extensive references.

Journals

Below is a listing of journals that publish articles on engineering education, and are indexed in the databases mentioned in the Abstracts and Indexes section. Published articles may be categorized under such topics as students, faculty, practitioners, courses and programs, assessment and evaluation, and history of engineering education.

ASEE Prism (1991–) Washington, DC: American Society for Engineering Education (1056–8077).

Computer Applications in Engineering Education (1992–) New York: John Wiley & Sons, Inc. (1061–3773).

European Journal of Engineering Education (1975–) Abingdon, Oxon: Carfax, English, French, and German (0304–3797).

Global Journal of Engineering Education (1997–) Melbourne, Australia: UNESCO International Centre for Engineering Education, English and German (1328–3154).

IEEE Potentials (1982–) New York: Institute of Electrical and Electronics Engineers (0278–6648).

Innovations in Science and Technology Education (1986–) Paris: UNESCO.

International Journal of Engineering Education (1992–) Hamburg, Germany: Tempus Publications (formerly *International Journal of Applied Engineering Education* 1985–1991. New York: Pergamon) (0949–149X).

International Journal of Science Education (1987–) London; New York: Taylor & Francis (formerly *European Journal of Science Education*, 1979–1986) (0950–0693).

Journal of Engineering Education (1910–1969; 1993–) Other title *Engineering Education*, 1970–1992. Washington, DC: American Society for Engineering Education (1069–4730).

Journal of Engineering Technology (1984–) Washington, DC: Engineering Technology Division, American Society for Engineering Education (0747–9964).

Journal of Materials Education (1983–) University Park, PA: Materials Research Laboratory, Pennsylvania State University (0738–7989).

Journal of Professional Issues in Engineering Education and Practice (1991–) New York: American Society of Civil Engineers (formerly *Journal of Professional Issues in Engineering*, 1983–1990; *Issues in Engineering*, 1979–1982; *Engineering Issues*, 1958–1978) (1052–3928).

Journal of Science Education and Technology (1992–) New York: Plenum Press (1059–0145).

Journal of the Learning Sciences (1991–) Hillsdale, NJ: Lawrence Erlbaum Associates (1050–8406).

Technos (1972–2002). Fort Collins, CO: American Society for Engineering Education. International Division (0363–308X).

Conferences

Conference papers are indexed by certain databases, particularly Engineering Village 2 (*Compendex*) and *Inspec*.

ASEE Annual Conference Proceedings (1978–) (formerly *Proceedings. Papers, Reports, Discussions*, printed in the *Journal of Engineering Education* 1893–1966). Also called *American Society for Engineering Education ASEE Annual Conference.* Washington, DC: American Society for Engineering Education <http://www.asee.org/conferences/> (accessed May 10, 2004).

ASEE International Colloquium on Engineering Education (2002–) American Society for Engineering Education. European Society for Engineering Education (SERI). Technical University Berlin (TUB). Chinese Academy of Engineering. Locations vary <http://www.asee.org/conferences/international/> (accessed May 10, 2004).

ASEE Engineering Deans Institute. ASEE Engineering Deans Council. Includes programs such as "Engineering Education 2020: What a College of Engineering may Look Like in 20 Years" <http://www.asee.org/about/events/conferences/edi.cfm> (accessed May 17, 2005).

ASME International Mechanical Engineering Congress (1989–) (also called *International Mechanical Engineering Congress and Exposition*). New York: American Society of Mechanical Engineers <http://www.asme.org> (accessed May 10, 2004).

Frontiers in Education Conference (1975–) New York: Institute of Electrical and Electronics Engineers (1975–). Earlier names: *IEEE Conference on Frontiers in Education, International Conference on Frontiers in Education* <http://www.fie-conference.org/> (accessed May 17, 2005).

Intertech 2000: Proceedings of the Interamerican Conference on Engineering and Technology Education (1999–) Cincinnati, OH: University of Cincinnati <http://listserv.uc.edu/archives/intertech-2000.html> (accessed May 17, 2005).

Proceedings/Conference for Industry and Education Collaboration (1997–) Washington, DC: American Society for Engineering Education (formerly *College Industry Education Conference*, 1976–1996).

World Congress on Engineering Education (1975–) World Federation of Engineering Organizations. Washington, DC: American Society for Engineering Education, International Division.

Search engines and important websites

American Society for Engineering Education <http://www.asee.org> (accessed May 17, 2005). The most important society for engineering education in the United States provides:

- Engineering Resources – <http://www.asee.org/resources/index.cfm> – points to websites on distance education, multimedia course, research centers, textbooks.

- ASEE EngineeringK12 Center – <http://www.engineeringk12.org/> – contains lesson plans and experiments, data on outreach programs, career guidance materials, and links to readings related to engineering education.

Eisenhower National Clearinghouse for Mathematics and Science Education <http://www.enc.org> (accessed May 17, 2005). Located at the Ohio State University, the Eisenhower National Clearinghouse for Mathematics and Science Education (ENC) is funded through a contract with the U.S. Department of Education. It provides a searchable collection of thousands of curriculum materials for K-12 math and science teachers.

NASA <http://www.nasa.gov/home/> (accessed May 17, 2005). Click on "For educators." Includes many teaching resources, including downloadable multimedia, curriculum materials, and so on, separated by age group.

National Science Digital Library <http://nsdl.org> (accessed May 17, 2005). From the website: "NSDL is a digital library of exemplary resource collections and services, organized in support of science education at all levels. Starting with a partnership of NSDL-funded projects, NSDL is emerging as a center of innovation in digital libraries as applied to education, and a community center for groups focused on digital-library-enabled science education."

NEEDS <http://www.needs.org/needs/> (accessed May 17, 2005). From the website: "The National Engineering Education Delivery System (NEEDS) is a digital library of learning resources for engineering education. NEEDS provides web-based access to a database of learning resources where the user (whether they be learners or instructors) can search for, locate, download, and comment on resources to aid their learning or teaching process." NEEDS and its sponsor, John Wiley & Sons, Inc support the Premier Award, which recognizes high-quality, non-commercial courseware designed to enhance engineering education.

PBS TeacherSource <http://www.pbs.org/teachersource/> (accessed May 17, 2005). Rich in educational resources, this site provides access to 4,500 free lesson plans and activities.

Teachers' Domain: Multimedia Resources for the Classroom and Professional Development <http://www.teachersdomain.org/> (accessed May 17, 2005). A multimedia digital library for K-12 teachers and students. You will find an extensive collection of classroom-ready resources, as well as media-rich lesson plans and professional development resources. Each resource is catalogued by grade level and correlated to national and state standards. Teachers' Domain is produced by WGBH, with major funding from the National Science Foundation.

Societies

Accreditation Board for Engineering and Technology (ABET) <http://www.abet.org/> (accessed May 17, 2005). Sets standards for degree programs providing qualifications for engineering practice.

American Society for Engineering Education <http://www.asee.org> (accessed May 17, 2005). Publishes several journals and directories: *Engineering College Research and Graduate Study*, *Directory of Graduate Programs in Engineering*, *Journal of Engineering Education*. Sponsors conferences. The most important society in engineering education in the United States.

American Technical Education Association <http://www.ateaonline.org> (accessed May 17, 2005).

Association for Media-based Continuing Education for Engineers <http://www.amcee.org/> (accessed May 17, 2005).

Engineering Institute of Canada <http://www.eic-ici.ca/> (accessed May 17, 2005).

European Society for Engineering Education <http://www.ntb.ch/SEFI/> (accessed May 17, 2005). Société Européenne pour la Formation des Ingenieurs. Europaische Gesellschaft für Ingenieur-Ausbildung.

International Association for Continuing Education and Training <http://www.iacet.org/> (accessed May 17, 2005).

National Academy Press Publications (many are online and full text) <http://www.nap.edu/> (accessed May 17, 2005).

National Council of Examiners for Engineering and Surveying <http://www.ncees.org/> (accessed May 17, 2005).

National Science Foundation <http://www.nsf.gov/> (accessed May 17, 2005).

National Society of Professional Engineers <http://www.nspe.org/> (accessed May 17, 2005).

Society for the History of Technology <http://shot.press.jhu.edu/> (accessed May 17, 2005).

Society of Women Engineers <http://www.swe.org/> (accessed May 17, 2005).

Tau Beta Pi Association <http://www.tbp.org/pages/main.cfm> (accessed May 17, 2005).

Triangle Fraternity <http://www.triangle.org/home/> (accessed May 17, 2005).

In addition, consult the websites of many discipline-specific engineering societies, such as the Institute of Electrical and Electronics Engineers (IEEE), American Society of Civil Engineers (ASCE), American Society of Mechanical Engineers (ASME), Association for Computing Machinery (ACM), American Institute of Chemical Engineers (AIChE), American Society of Heating, Refrigerating and Air-conditioning Engineering (ASHRAE), and Society of Automotive Engineers International (SAE).

A complete list of engineering societies may be found in the *International Directory of Engineering Societies and Related Organizations* (1993–2000) Washington, DC: American Association of Engineering Societies. Formerly *Directory of Engineering Societies and Related Organizations* (1900s-1989). Besides contact information, it lists the organizations' officers, publications, objectives, budget, geographic, and student chapter data.

Academic discussion lists and newsgroups

Academic discussion lists and newsgroups are e-mail forums where people discuss issues of common interest. Newsgroups messages are posted to a website. One can search newsgroups at <http://www.onelist.com> and <http://www.google.com/groups>. Groups that discuss the issues of engineering education include sci.engr.education, misc.education.science, sci.engr.civil, sci.engr.mech, sci.engr.chem., and sci.edu.

Users subscribe to mailing (or discussion) lists when they want to read and receive messages via their personal e-mail. They may be searched at the *Directory of Scholarly Electronic Journals and Academic Discussion Lists* <http://dsej.arl.org/>; Tile.Net <http://tile.net/lists/>; *CataList*

<http://www.lsoft.com/lists/listref.html> and National Academic Mailing List Service <http://www.jiscmail.ac.uk/mailinglists/index.htm>. Examples include civil-l@unb.ca and cheme-l@ulkyvm.louisville.edu.

Conclusion

In fall 2002, the Franklin W. Olin College of Engineering opened a new college in Needham, Massachusetts. In an effort to attract students, every admitted student receives a four-year, full tuition scholarship valued at approximately $125,000. It aims to provide a new state-of-the-art engineering college and provides a unique opportunity to innovate.

According to a study done in 1998, the desired attributes of an engineering graduate include the following:

- a good grasp of engineering science fundamentals in math, physical and life sciences, and information technology;
- a basic understanding of the design and manufacturing process;
- a basic understanding of the context (economics, history, environment, customer) in which engineering is practiced;
- good communication skills;
- high ethical standards;
- ability to think creatively and critically;
- flexibility;
- curiosity and the desire to learn;
- the understanding of the importance of team work (*Shaping the Future*, Volume II, 1998).

Keeping in mind these attributes, this chapter provides a starting point for researching the history and current state of engineering education in the United States. Whether the future lies in new programs such as the Olin College in Massachusetts or in the older established colleges, or both, there is much to be learned by studying their examples and the resources enumerated here.

The following quote from a former engineering dean who lived from 1891 to 1982 is still relevant today:

> It [Engineering] is incessantly probing the unknown and untried for new methods and improved results. Its procession of innovations have their impact on man's whole way of life; on his ease of movement, on the range of instant communication, on the health, wealth, and happiness of people.
>
> (Hollister, 1966: 17)

Acknowledgments

I would like to thank Jeffrey Harris from the National Science Foundation, and Cornell Professors Ronald Kline, Simpson Linke, Anthony Ingraffea, and Michel Louge for reading this chapter and contributing useful entries and comments.

References

American Society for Engineering Education <http://www.asee.org/about/missionAndVision.cfm> (accessed March 9, 2005).

American Society for Engineering Education <http://www.asee.org/publications/> (accessed May 29, 2004).

Doepke, M. (2003. *Humboldt's University – Now and Then* <http://www.rundertisch-usa.de/chicago/site/statements/matthiasdoepke.html> (accessed September 1, 2004).

Grayson, L.P. (1993) *The Making of an Engineer: An Illustrated History of Engineering Education in United States and Canada*, New York: John Wiley & Sons, Inc.

Hollister, S.C. (1966) *Engineer*, New York: Macmillan.

Reynolds, T.S. (1991) *The Engineer in America: A Historical Anthology From Technology and Culture*, Chicago, IL: University of Chicago Press.

Rosser, S.V. (1995) *Teaching the Majority: Breaking the Gender Barrier in Science, Mathematics, and Engineering*, New York: Teachers College Press.

Ruth, D. (2004) *What is Engineering Education? A Personal Perspective*, University of Manitoba, <http://www.umanitoba.ca/engineering/deans_office/about_us/what_is_engineering_education.shtml> (accessed June 7, 2004).

Shaping the Future. Volume II: Perspectives on Undergraduate Education in Science, Mathematics, Engineering, and Technology (SMET) (1998) Advisory Committee to the National Science Foundation (NSF) Directorate for Education and Human Resources. Arlington, VA: National Science Foundation <http://www.nsf.gov/cgi-bin/getpub?nsf98128> (accessed, May 5, 2004). ERIC document ED433882, NSF 98–128.

Weichert, D., Rauhut, B., and Schmidt, R. (eds) (2001) *Educating the Engineer for the 21st Century*, Dordrecht, the Netherlands: Kluwer Academic.

12 Environmental engineering

Linda Vida and Lois Widmer

Introduction, history, and scope of discipline

The history of environmental engineering is extensive, although the discipline was not called that until recently. Early concerns about water quality appear in human history dating back to 2000 BC. The ancient Romans built an extensive aqueduct system to provide fresh water to their cities. Some evidence indicates that they had waste-water treatment collection systems as well.

The field of environmental engineering developed from a branch of civil engineering when the public became concerned about the spread of diseases. In the United States in the 1850s, engineers were primarily dedicated to providing infrastructure to transport water or to build drinking-water distribution systems. Drinking-water treatment became more widespread in the 1900s. The field of waste-water treatment engineering advanced more slowly until the 1950s when treatment systems were standardized.

Health concerns about air pollution can easily be documented as early as the seventeenth century, attributable to industrial development as well as to the fuel sources for domestic heating and cooking. It took such twentieth-century incidents as the 1952 London smog case and the 1948 Donora, Pennsylvania air pollution inversion, both of which produced increased particulate matter pollution leading to increased mortality, to stimulate regulatory action and steps to control air quality.

In the United States, a broad awareness of environmental issues began in the 1960s. In the 1970s, President Nixon authorized the creation of the U.S. Environmental Protection Agency and the National Oceanic and Atmospheric Administration to establish and enforce environmental protection standards to improve the environment. These agencies were empowered to conduct research on the adverse effects of pollution and to develop new methods and equipment for controlling pollution. This was the beginning of an effort to gather information on pollution, and to use this information to strengthen environmental programs and recommend policy changes.

A discipline that had begun to address drinking-water and waste-water treatment issues expanded and now encompasses ground water, storm water, estuarine environments, aquatic ecology, desalination, hazardous wastes and solid wastes, acid deposition, smog, chlorofluorocarbons, indoor air quality and hazardous air pollutants, and global climate change.

Today, environmental engineering focuses on complex and interdisciplinary issues and is concerned with a broad range of environmental contaminants and their impacts on the environment. Environmental engineering requires the understanding of natural processes as well as an understanding of waste streams; the development of applied technologies; and knowledge of past and current engineering practices. Environmental engineers need to understand the fundamentals of contaminants in water, air, soils, and hazardous waste and strive to use their understanding to develop and apply technologies that will improve the environment.

Searching the library catalog, keywords, LC Subject Headings, LC call numbers

Broad terms for searching this topic and broad call numbers:

Environmental engineering, (Library of Congress call numbers TA170 through TA171)
Pollution (Library of Congress call numbers TD172 through TD193.5)
Air quality (Library of Congress call number TD883)
Air – Pollution (Library of Congress call numbers TD881 through TD890 – for the technology)

LC call numbers – Water

The Library of Congress classification system places environmental engineering materials primarily in Class T – Technology. Some of the most important areas are as follows:

GB621–628	Wetlands
GB1199	Aquifers
RA591–598.5	Water supply in relation to public health
TA170–171	Environmental engineering
TC401–526	Water supply engineering
TD1–1066	Environmental technology. Sanitary engineering
TD159–167	Municipal engineering
TD169–171.8	Environmental protection
TD172–193.5	Pollution
TD201–500	Water supply for domestic and industrial purposes
TD419–428	Water pollution
TD429.5–480.7	Water purification. Water treatment and conditioning. Saline water conversion
TD480.92–493	Water distribution systems
TD657	Urban run-off
TD511–780	Sewage collection and disposal systems

TD783–812.5	Municipal refuse. Solid wastes
TD878–894	Special types of environment including soil pollution, air pollution, and noise pollution
TD895–899	Industrial and factory sanitation
TD896–899	Industrial and factory wastes
TD920–934	Rural and farm sanitary engineering
TD940–949	Low temperature sanitary engineering
TD1020–1066	Hazardous substances and their disposal
TH6014–6081	Environmental engineering buildings. Sanitary engineering of buildings
TJ212–225	Control engineering systems. Automatic machinery (general)
TP155–156	Environmental chemistry

As with most environmental issues, finding information on environmental engineering for air quality control requires searches of the specialized environmental literature as well as the engineering and technical literature.

Some useful Library of Congress Subject Headings (LCSH) for searching library catalogs are listed below. They are arranged hierarchically with the narrower subject headings indented under the broader headings to show the relationship among terms; however, each term may be searched independently.

Tips:
All these subject headings have a range of subdivisions such as a geographic location or publication type that can be appended to the subject heading, e.g.,
Air pollution – Canada
Air pollution – Handbooks

For information on a specific substance or pollutant, search the substance name and qualify it by including the subdivision "Environmental aspects," e.g.,
Dioxins – Environmental aspects

LC Subject headings – Air

- air – analysis
  - air – pollution – measurement
  - air-sampling apparatus
- air quality
- air quality indexes
- air quality management
  - air – purification
- atmospheric chemistry
  - atmospheric carbon dioxide

  - atmospheric nitrogen compounds
  - atmospheric ozone
  - precipitation scavenging
  - sinks (atmospheric chemistry)
- atmospheric deposition
- automobiles – motors – exhaust gas
- automobiles – motors – exhaust systems
- automobiles – pollution control devices
- environmental chemistry – industrial applications
- environmental engineering
  - buildings – environmental engineering
  - dwellings – environmental engineering
  - environmental auditing
  - environmental impact statements
  - environmental monitoring
  - environmental testing
  - noise control
  - work environment
- environmental health
  - environmental toxicology
  - industrial hygiene
- environmental indicators
- environmental protection
  - environmental impact analysis
  - pollution prevention
    - best management practices (pollution prevention)
    - dust control
    - gas vapor control
    - greenhouse gas mitigation
    - waste minimization
- environmental law
  - air – pollution – law and legislation
- environmental monitoring
  - air quality monitoring stations
  - environmental sampling
- global warming
- greenhouse effect, atmospheric
- pollutants
  - acid-forming emissions
  - acid deposition
    - acid precipitation
      - acid rain
- pollution
  - air – pollution
    - acid precipitation
      - acid rain

air pollution control industry
radon control industry
asbestos abatement industry
automobiles – motors – exhaust gas
automobiles – motors (diesel) – exhaust gas
automobiles – pollution control devises
dust
fume control
indoor air pollution
tobacco smoke pollution
motor vehicles – pollution control devices
odor control
odors
smog
photochemical smog
VOG
smoke
smaze
zoning, emission density
air – pollution potential
pollution control industry
air pollution control industry
space pollution
transboundary pollution
urban pollution
sick building syndrome

LC Subject headings – Water

aquifers (see also ground water basin)
contaminated sediments
desalination of water (use saline water conversion)
drinking-water
contamination
law and legislation
standards
environmental engineering (related term *environmental health*)
environmental monitoring
factory and trade waste (used for industrial wastes)
ground water (a variant form is ground water) (the former term was water, underground)
basin (*see also* aquifers)
flow
quality
hazardous substances
hazardous waste management industry

- hazardous waste sites
- hazardous wastes (used for hazardous waste disposal)
- hydraulic engineering
- hydrogeology
- hydrology
- industrial effluent (use sewage)
- industrial wastes (use factory and trade waste)
- injection wells
- marine pollution
- municipal engineering
- municipal water supply
- pollutants
- pollution
    - environmental aspects
    - law and legislation
    - measurement
    - prevention (use pollution control industry)
- pollution control industry
- pollution control equipment
- reverse osmosis
- run-off
- saline water conversion
- saline water barriers
- saline waters
- salinity
- sanitary engineering
- sanitary landfills
- sedimentation and deposition
- sewage
    - purification
        - activated sludge process
        - aeration
        - biological treatment
        - flocculation
        - flotation
        - reverse osmosis process
- sewage disposal
- sewage lagoons
- sewage sludge
    - irraditation
- sewage sludge digestion
- sewerage (used for house drainage, sewers)
- sludge bulking
- solid wastes
- storage tanks
- storm sewers

storm water retention basins
underground storage
underground waste disposal (Use waste disposal in the ground)
urban hydrology
urban run-off
waste disposal in the ground
water
  equipment and supplies
  pollution
  purification
water quality
water reuse
water supply engineering
water treatment plants
wells
wetland ecology
wetlands

Keyword searching

For database and web searching, these same Library of Congress Subject Headings can be useful. However, for thorough coverage synonyms should be used as well. Several resources in the Article indexes sections have online thesauri that can suggest additional terms. Another resource for search terms is the *Thesaurus of Sanitary and Environmental Engineering* (12th edn) (February 2003) from the World Health Organization and Pan American Health Organization <http://www.cepis.ops-oms.org/bvsair/i/manuales/tesa/teses.pdf> (accessed May 18, 2005).

Other useful online resources for additional keywords and search terms are:

ChemFinder <http://chemfinder.cambridgesoft.com/> (accessed May 18, 2005), for chemical names and CAS registry numbers.

EPA Controlled Vocabulary, available in two formats:

PDF <http://www.epa.gov/webguide/metadata/EPA_Controlled_Vocab.pdf>version 2.7.5;
HTML <http://www.epa.gov/epahome/topics.html> (accessed May 18, 2005).

Article indexes

In the following list of indexes and abstracts for environmental engineering, the producers and their websites are listed as resources for additional information. In many cases, the producers both offer their own subscriptions for access and license the indexes and abstracts to a number of service providers, each of which then offers its own subscription options and search interfaces. The major service providers, likely to offer access to these indexes and abstracts, include:

Dialog, a Thomson business <http://www.dialog.com/>
STN, <http://www.cas.org/stn.html>
OVID (including SilverPlatter) <http://www.ovid.com/site/index.jsp>
Cambridge Scientific Abstracts <http://www.csa.com>
OCLC FirstSearch <http://www.oclc.org/firstsearch/default.htm>

Most of these indexes are international in their coverage of the literature.

ASFA: Aquatic Sciences and Fisheries Abstracts (Other names: ASFA 1: Biological Sciences and Living Resources; ASFA 2: Ocean Technology, Policy and Non-living Resources; ASFA 3: Aquatic Pollution and Environmental Quality; ASFA Marine Biotechnology Abstracts; ASFA Aquaculture Abstracts – all print counterparts), Cambridge Scientific Abstracts <http://www.csa.com> (accessed May 18, 2005) (1971–) Updated monthly. Fee based. Types of publications indexed: Serial publications, books, reports, conference proceedings, translations, and limited distribution literature. A single database incorporating five subfiles, the ASFA series indexes research and policy on the contamination of oceans, seas, lakes, rivers, and estuaries. Major subjects covered include: aquaculture, aquatic organisms, aquatic pollution, brackish water environments, conservation, environmental quality, fisheries, freshwater environments, limnology, marine biotechnology, marine environments, meteorology, oceanography, policy and legislation, wildlife management. Cambridge Scientific Abstracts serves as the publishing partner for Aquatic Sciences and Fisheries Information System (ASFIS), formed by four United Nations agency sponsors of ASFA and a network of international and national partners.

Cambridge Scientific Abstracts <http://www.csa.com> (accessed May 18, 2005) (1982–) Updated bimonthly. Types of documents indexed: papers and poster sessions of major scientific meetings. Since 1995 conference papers emphasizing the life sciences, environmental sciences, and aquatic sciences. Older material also covers physics, engineering, and materials science. Information is derived from final programs, abstracts, booklets, and published proceedings, as well as from questionnaire responses.

Chemical Abstracts (CAPlus, SciFinder) (see Chapter 2 on general engineering).

Conference Papers Index (Conference Papers Index – print counterpart).

CSA Engineering Research Database (see Chapter 2).

CSA Engineering Research Database contains the following subfiles, which may be searched separately:

ANTE: Abstracts in new technologies and engineering
Civil engineering abstracts
Environmental engineering abstracts (see also separate entry under this name)
Earthquake engineering abstracts
Mechanical and transportation engineering abstracts

Content includes basic and applied research, design, construction, technological and engineering aspects of air and water quality, environmental safety, energy production, and developments in new technologies. Topics of interest to environmental engineers include: automotive design and engineering, bridges and tunnels, buildings, towers, and tanks, coastal and offshore structures, construction materials, design and properties of substructures, electric and hybrid vehicles, engineering for electric power generation, flood analysis, fuels and propellants, geotechnical engineering, hazardous materials, industrial waste and sewage, internal combustion engines, land development, irrigation and drainage, pollution, waste and water engineering, seismic engineering, seismic phenomena, site remediation and reclamation, storm water.

Ei Compendex (see Chapter 2).

Energy Science and Technology (other name: Energy Research Abstracts – print counterpart). U.S. Department of Energy – Office of Scientific and Technical Information <http://grc.ntis.gov/energy.htm> (accessed May 18, 2005) (1976–) Updated biweekly. Types of publications indexed: journal literature, books, conference proceedings, papers, patents, dissertations, engineering drawings. Scientific and technical reports of the U.S. Atomic Energy Commission, U.S. Energy Research and Development Administration and its contractors, other agencies, universities, and industrial and research organizations. A comprehensive source of worldwide energy-related information, this file contains references to basic and applied scientific and technical research literature. Subject coverage of interest to this audience includes: energy sources, use, and conservation; environmental effects; waste processing and disposal; hazardous waste management; conservation technology; energy conversion; renewable energy sources; energy policy; synthetic fuels; engineering; and environmental science.

EnergyFiles: Environmental Sciences, Safety, and Health. U.S. Dept. of Energy – Office of Scientific and Technical Information <http://www.osti.gov/energyfiles/Environmental/main.html> (accessed May 18, 2005). The environmental sciences, safety and health subject area is defined as information on the effects of any energy-related activity on the environment, on methods for mitigating or eliminating adverse effects, and on technical aspects of ensuring that energy-related activities are environmentally safe and socially acceptable. This area covers all aspects of global climate change. Monitoring and transport of chemicals, radioactive materials, and thermal effluents within the atmospheric, terrestrial, and aquatic environs are covered.

EnergyFiles is an umbrella interface that permits searching of multiple government databases. Of potential interest to environmental engineers are:

Atmospheric Radiation Measurement Program (ARM) – Global change research supported by the Department of Energy.

EPA Full-text Reports – National Environmental Publications Internet Site (NEPIS) EPA full-text documents online, archival and current.

EPA Technical Reports – Searchable access to the Environmental Protection Agency website which includes technical reports, publications, and other information.

Office of Biological and Environmental Research (OBER) Abstracts Database – Abstracts and other information on OBER-funded R&D including abstracts for Environmental Sciences from FY 94–96.

USGS Publications Warehouse – The publications center of the U.S. Geological Survey, a federal source for science about the Earth, its natural and living resources, natural hazards, and the environment.

DOE Information Bridge – Searchable and downloadable bibliographic records and full text of DOE research report literature from 1995 onward.

Energy Citations Database – Bibliographic records for energy and energy-related STI from the DOE and its predecessor agencies, ERDA and AEC, from 1948 to the present.

Environment Abstracts (Other names: Enviroline; Environment Abstracts – print counterpart). LexisNexis – Congressional Information Service, Inc. <http://web.lexis-nexis.com> (accessed May 18, 2005) (1975–) Updated monthly. Types of publications indexed: reports, conferences, symposia, meetings, journal articles, newspaper articles. Subject coverage of interest to this audience includes: air pollution, environmental design and urban ecology, energy, general environmental topics, renewable and non-renewable resources, oceans and estuaries, waste management, water pollution, and weather modification and geophysical change.

LexisNexis Environmental provides access to the contents of Environment Abstracts plus LexisNexis legal and regulatory information such as environmental codes, case law, regulatory agency decisions, and law reviews.

Environmental Engineering Abstracts: Cambridge Scientific Abstracts (CSA) <http://www.csa.com> (accessed May 18, 2005). Coverage from 1990 to present, with most records from 1997 or later, updated monthly. Types of publications indexed: More than 700 primary journals plus over 2,500 additional sources, including monographs and conference proceedings. A subfile of CSA Engineering Research Database, Environmental Engineering Abstracts, indexes the literature on technological and engineering aspects of air and water quality, environmental safety, and energy production.

Environmental Sciences and Pollution Management: Cambridge Scientific Abstracts (CSA) <http://www.csa.com/> (accessed May 18, 2005). Coverage from 1967 to present, updated monthly. Indexes serials, conference proceedings, reports, monographs, books, and government publications.

Environmental Sciences and Pollution Management includes the following subfiles, searchable separately:

Agricultural and Environmental Biotechnology Abstracts
ASFA – Aquatic Science and Fisheries Abstracts
Aquatic Pollution and Environmental Quality
Ecology Abstracts
EIS Digests of Environmental Impact Statements
Environmental Engineering Abstracts
Health and Safety Sciences Abstracts
Microbiology A – Industrial and Applied Microbiology
Microbiology B – Bacteriology
Pollution Abstracts
Risk Abstracts

Toxicology Abstracts
Water Resources Abstracts

Subject coverage of interest to environmental engineers includes air quality, aquatic pollution, energy resources, environmental biotechnology, environmental engineering, environmental impact statements (U.S.), hazardous waste, industrial hygiene, microbiology related to industrial and environmental issues, pollution: land, air, water, noise, solid waste, radioactive, risk assessment, toxicology and toxic emissions, water pollution, waste management, and water resource issues.

GeoRef (Bibliography and Index of Geology – print counterpart) American Geological Institute, <http://www.agiweb.org> (accessed May 18, 2005) 1785 to present (North America coverage); 1933 to present (Global coverage), updated biweekly. Types of publications indexed: publications of the U.S. Geological Survey, U.S. and Canadian theses and dissertations, journal articles, books, maps, conference papers, and reports. GeoRef provides access to the world's geoscience literature.

Global Mobility Society of Automotive Engineers Inc. (SAE International). Of interest for environmental engineering is the technological information on mobility engineering such as emissions, environment, fuels and lubricants, noise and vibration (see Chapter 20 on Transportation engineering).

ISI Current Contents Connect (Current Contents – print counterpart) Thomson ISI (Institute for Scientific Information) <http://www.isinet.com/products/cap/ccc/> (accessed May 18, 2005). One year rolling file, updated weekly. Types of publications indexed: articles, editorials, meeting abstracts, commentaries, and all other significant items in recently published editions of over 1,120 journals and books. Other features: A subject search will also find related external websites, evaluated by ISI editors.

Current Contents is a weekly alerting service useful for keeping current with the most recent publications. Citations are subsequently added to Web of Science. Options include searching, browsing by journal title, viewing table of contents, and setting up automatic alerts delivered weekly by e-mail. The Current Contents weekly editions of most interest to environmental engineers are *Engineering, Computing, and Technology* edition, and *Agriculture, Biology, and Environmental Sciences* edition.

Oceanic Abstracts: Cambridge Scientific Abstracts (CSA) <http://www.csa.com> (accessed May 18, 2005) (1981–) Updated monthly. Indexes serials. *Oceanic Abstracts* focuses on the worldwide technical literature on marine and brackish-water environment. Subjects covered include marine biology, physical oceanography, fisheries, aquaculture, non-living resources, meteorology and geology, as well as environmental, technological, and legislative topics.

SciFinder (see Chapter 7).

TRIS Online (see Chapter 20 on Transportation engineering).

Web of Science (see Chapter 2).

Databases and data sets

American Water Works Association. Water:\STATS [electronic resource]. Denver, CO: American Water Works Association. The comprehensive source of statistical information on water utilities in the United States, Canada, and internationally offers a wealth of general utility information on topics such as treatment practices, distribution systems, water quality, revenue, and financial data. The data are available on CD-ROM.

Darnay, A.J. (ed.) (1992) *Statistical Record of the Environment*, Detroit: Gale Research. The three editions, published from 1992 to 1995, provide statistics on the status of U.S. environmental conditions.

Handbook of Chemical Risk Assessment Health Hazards to Humans, Plants, and Animals (2000) Boca Raton, FL: Lewis Publishers <http://www.environetbase.com/pdf/enb/L1506/L1506PDFTOC.pdf> (accessed May 18, 2005). *Note:* Access full-text pdf file online via ENVIROnetBASE (Restricted to subscribing libraries) <http://www.environetbase.com/pdf/enb/L1506/L1506PDFTOC.pdf.>

NWISWeb data collected and published by the U.S. Geological Survey <http://waterdata.usgs.gov/nwis> (accessed May 18, 2005). These pages provide access to water-resources data collected at approximately 1.5 million sites in all fifty states, the District of Columbia, and Puerto Rico. Online access to this data is organized around surface water, ground water and water quality. The USGS investigates the occurrence, quantity, quality, distribution, and movement of surface and underground waters and disseminates the data to the public, state and local governments, public and private utilities, and other federal agencies involved with managing our water resources.

National Water Quality Assessment Data Warehouse (NAWQA). The U.S. Geological Survey (USGS) began its NAWQA program in 1991, systematically collecting chemical, biological, and physical water quality data from forty-two study units (basins) across the nation. This database contains chemical concentrations in water, bed sediment, and aquatic organism tissues for about 609 chemical constituents; daily stream flow information for fixed sampling sites; ground water levels and more.

United Nations Environment Programme (1987–1994) *Environmental Data Report*, Oxford; New York: Blackwell. A biennial publication from 1987 through 1994 prepared by the GEMS Monitoring and Assessment Research Centre in cooperation with the World Resources Institute, it provides annotated data on environmental pollution, climate, natural resources, human health and population, energy, wastes, and natural disasters for developed and developing countries. The information was gathered from government agencies, environmental organizations, and published sources. It ceased publication in 1994; records are useful for comparison and historical studies.

Bibliographies and guides to the literature

Printed bibliographies are much less commonly produced now that there is easier access to full-text reports and journal articles on the World Wide Web, websites, and specialized web portals. However, a few are worth noting.

EPA Publications Bibliography, sponsored by Library Systems Branch, U.S. Environmental Protection Agency in cooperation with EPA Offices of Air and Waste

Management. Washington, DC: The Branch; Springfield, VA: National Technical Information Service, distributor. Quarterly, began in October/November 1977 and ceased in 2000. Consolidates all reports published by the U.S. Environmental Protection Agency and covers all aspects of the environment.

Lane, C.N. (2003) *Acid Rain: Overview and Abstracts*, New York: Nova Science Publishers. Covering both wet and dry deposition of acid rain, this work offers both a background article and over 900 abstracts and book citations. Title, author, and subject indexes are provided for easy access.

New Publications of the Geological Survey, Washington, DC: Geological Survey, U.S. Government Printing Office <http://pubs.usgs.gov/publications/index.shtml> (accessed May 18, 2005). As of January 2004, this catalog is available monthly and annually only and only online. This publication has had many variant titles since its inception in 1879. It is a comprehensive listing of new materials published by the Geological Survey in the following series: digital data, professional papers, bulletins, water supply papers, techniques of water resources investigations, fact sheets, water resources investigation reports, open-file reports, thematic maps and charts, and topographic maps.

Van der Leeden, F. (ed.) (1991) *Geraghty & Miller's Groundwater Bibliography* (5th edn), Plainview, NY: Water Information Center. A compilation of more than 5,500 selected references covering ground water environments and systems, ground water contamination, saltwater intrusion, ground water models.

Directories

Print directories are no longer the essential resource that they were a few years ago, due to the World Wide Web. Most professional societies, academic, and research institutions or government organizations have membership directories that are published and distributed to their members in print, are searchable online, or may be available for purchase.

American Academy of Environmental Engineers (Annual) *Who's Who in Environmental Engineering*, Annapolis, MD: The Academy <www.aaee.net> (accessed May 18, 2005). *Who's Who in Environmental Engineering* includes a roster of all the Board Certified Diplomates of Environmental Engineering who have demonstrated their expertise in one or more of seven environmental specialties. This directory provides an alphabetical listing of each Diplomate together with a biographic profile, and is cross-referenced geographically and by specialty.

American Academy of Environmental Engineers (Annual) *Environmental Engineering Selection Guide*, Annapolis, MD: The Academy <http://www.aaee.net/newlook/selecton_guide.htm> (accessed May 18, 2005). A free public service of the American Academy of Environmental Engineers, this directory is a source for consulting firms employing board-certified engineers. It also lists individuals who have earned the title Diplomate Environmental Engineer.

AWWA Sourcebook (Annual) Denver, CO: American Water Works Association <www.awwa.org> (accessed May 18, 2005). The *Sourcebook* is a comprehensive directory of drinking-water products and services. It includes a directory of service providers arranged by company; a guide to suppliers; and a guide to consultants.

Environmental Key Contacts and Information Sources, Rockville, MD: Government Institutes. This directory evolved from two previous publications: the *Environmental Telephone Directory* and the *Directory of Environmental Information Sources.*

Gale Environmental Sourcebook (1992–1994) (2 vols), Detroit: Gale Research Inc. An excellent source of information about organizations and agencies concerned with the environment.

Encyclopedias and dictionaries

Encyclopedias

Bisio, A. and Boots, S. (eds) (1995) *Encyclopedia of Energy Technology and the Environment* (4 vols), New York: Wiley. Not to be confused with Wiley's condensed version of this work, this encyclopedia won the 1995 Association of American Publishers, Inc. Award for Excellence in Professional/Scholarly Publishing, Chemistry. In four volumes, the encyclopedia covers the impact of energy production technologies through discussions of such topics as acid rain, air pollution, aircraft fuel, building systems, coal combustion, computer applications for energy-efficient systems, risk assessment, solar heating, waste management planning, water power, and so on. It contains illustrations, photographs, tables, and a list of environmental organizations.

Cheremisinoff, P.N. (ed.) (1989–) *Encyclopedia of Environmental Control Technology* (9 vols), Houston, TX: Gulf Publishing. Provides in-depth coverage of specialized topics relating to environmental and industrial pollution control technology and state-of-the-art information as well as projections of future trends.

Maroni, M., Seifert, B., and Lindvall, T. (1995) *Indoor Air Quality: A Comprehensive Reference Book*, Amsterdam; New York: Elsevier. International in scope, the intent of this reference source is to compile and integrate information from all disciplines involved in indoor air quality issues. Among these are building design and building sciences, health effects and medical diagnosis, toxicology of indoor air pollutants, and air sampling and analysis.

Meyers, R.A. (ed.) (1998) *Encyclopedia of Environmental Analysis and Remediation* (8 vols), New York: John Wiley. The preface categorizes this eight-volume encyclopedia as a "professional level compendium" of all aspects of the environment. Emphasis is on sampling, analysis, and remediation, but the work also includes discussions of pollution sources, transport, regulations, and health effects. There are about 280 articles, many including tables, figures, equipment procedures, and standards. Information to assist in preparation of parts of environmental impact statements and air permitting documents is also included.

Meyers, R.A. (ed.-in-chief) and Kender, D.D. (ed.) (1999) *Encyclopedia of Environmental Pollution and Cleanup* (2 vols), New York: John Wiley. This is a concise, up-to-date, two-volume encyclopedia that provides all readers with a working knowledge of contemporary issues in environmental pollution and clean-up. It includes over 200 self-contained, cross-referenced articles with coverage of key topics in hazardous waste, air pollution control, biosphere pollution, health effects, nuclear waste, environmental law and regulation, water reclamation, and more. Hundreds of photographs, figures, charts, and tables illustrate major points, explain difficult material, and summarize important data. Bibliographic entries list sources of additional information on selected topics.

Pfafflin, J.R. and Ziegler, E.N. (eds) (1998) *Encyclopedia of Environmental Science and*

Engineering (4th edn, rev. and updated). Amsterdam; New York: Gordon and Breach Science Publishers. This work is written for the practitioner and those with a science background. After brief definitions or introductions, each chapter quickly focuses on technical details. For example, the chapter on aerosols includes a table of physical interpretation for characteristic diameters of various particles, provides defining equations, and methods of aerosol particle size analysis.

Van der Leeden, F., Troise, F.L., and Todd, D.K. (1998) *The Water Encyclopedia* (2nd edn), Chelsea, MI: Lewis Publishers. This one-volume encyclopedia is really a handbook. It is a standard in the field with international coverage on all aspects of water. It contains tables, charts, graphs, and some maps. Subject matter includes climates, hydrology, surface and ground water, water use, water quality, and water management. This encyclopedia contains unique statistical information that is often difficult to find.

Dictionaries

Frick, G.W. and Sullivan, T.F.P. (eds) (1990) *Environmental Regulatory Glossary* (5th edn), Rockville, MD: Government Institutes. This dictionary covers terms pertaining to regulatory issues.

Lee, C.C. (compiler and ed.) (2004) *Environmental Engineering Dictionary* (4th edn), Boca Raton, FL: Government Institutes. This newly updated dictionary provides a comprehensive reference of hundreds of environmental engineering terms in use throughout the field. This edition draws from government documents and legal and regulatory sources, and includes terms relating to pollution control technologies, monitoring, risk assessment, sampling and analysis, quality control and permitting, fuel cell technology, and basic environmental calculations. Users of this dictionary will find exact and official U.S. Environmental Protection Agency definitions for statute and regulation-related terms. The book is available online through ENVIROnetBASE <www.environetbase.com/> (accessed May 18, 2005). An annual subscription is required.

Pankratz, T.M. (2001) *Environmental Engineering Dictionary and Directory*, Boca Raton, FL: Lewis Publishers. This book includes more than 8,000 terms, acronyms, and abbreviations applying to waste-water, potable water, industrial water treatment, seawater desalination, air pollution, incineration, and hazardous waste treatment. Its most unique feature is the inclusion of 3,000 trademarks and brand names. Also available online through ENVIROnetBASE. An annual subscription is required.

Pfafflin, J.R., Baham, P., and Gill, F.S. (1996) *Dictionary of Environmental Science and Engineering*, Amsterdam, the Netherlands: Gordon and Breach Publishers. The dictionary explains specialist environmental terms concisely, provides a guide to acronyms, and describes acts, organizations, and requirements related to the legislation of the environment, particularly those of the United States.

Porteous, A. (2000) *Dictionary of Environmental Science and Technology* (3rd edn), Chichester; New York: John Wiley. With a mix of short and long entries, this work provides basic definitions and data. For example, five pages with diagrams are devoted to anaerobic digestion but other entries are a single line. Some include references. The dictionary covers the broad range of environmental terminology. The publisher describes this work as "over 4,000 in-depth entries on scientific and technical terminology associated with environmental protection and resource management."

Smith, P.G. and Scott, J.S. (2002) *Dictionary of Water and Waste Management* (2nd edn), Oxford; Boston, MA: Butterworth-Heinemann; London: IWA Publications. Reference to U.S./U.K. and European standards, legislation, and spelling ensures that the reader will find the book of global relevance. Illustrations throughout aid the reader's understanding of the explanations. Over 7,000 terms on water quality, engineering, and waste management make this the reference of choice for the professional. Over the past twenty years, areas such as air pollution control, solid waste management, hazardous waste management, pipeline management (leakage control, pipeline and sewer renewal), and environmental management systems have all become increasingly important. To reflect this shift, this completely revised and updated edition now covers water and waste management as well as treatment.

Webster, L.F. (ed.) (2000) *Dictionary of Environmental & Civil Engineering*, New York: Parthenon. Focusing on the mechanical aspects of environmental engineering, this dictionary is more comprehensive than most and includes some unique terms. Definitions are often brief; therefore, it may best be used in conjunction with other resources.

Handbooks, manuals, and properties

General

Bitton, G. (1998) *Formula Handbook for Environmental Engineers and Scientists*, New York: Wiley. This work provides formulas and equations from a range of disciplines and sources. It complements the *CRC Handbook of Physics and Chemistry*. Arranged in alphabetical order, some cross-references are included. Appendices contain conversion tables.

Bregman, J.I. (1999) *Environmental Impact Statements* (2nd edn), Boca Raton, FL: Lewis Publishers. This extensively revised second edition addresses all the requirements for federal, state, and local environmental impact statements (EISs) and provides detailed "how to" information for their preparation. It is available in print or as an online subscription <http://www.environetbase.com/pdf/enb/L1369/L1369%5FPDF%5FTOC.pdf>

Burke, G., Singh, B.R., and Theodore, L. (2000) *Handbook of Environmental Management and Technology*, New York: John Wiley. This thoroughly updated edition offers a historical perspective on pollution problems and solutions along with an introduction to the scientific and technical literature in the field.

CFR's Made Easy (series) Rockville, MD: Government Institutes. These books provide an in-depth, focused look at environmental regulations. Each handbook provides an overview of compliance programs, guidelines for understanding requirements, and detailed explanations and check-lists. *Air CFR's Made Easy* (2nd edn), *Water CFR's Made Easy*, *RCRA CFR's Made Easy* as well as *Environmental Compliance Made Easy* are of particular interest to this audience.

Cheremisinoff, N.P. (2001) *Handbook of Pollution Prevention Practices*, New York: Marcel Dekker. This book focuses on reducing manufacturing and environmental compliance costs by instituting improved operational schemes, recycling and by-product recovery, waste minimization, and energy efficiency policies, and offers project cost accounting tools that assist in evaluating pollution prevention technologies.

Corbitt, R.A. (ed.) (1999) *Standard Handbook of Environmental Engineering* (2nd edn), New York: McGraw-Hill. This second edition serves as a guide to environmental engineering, including information on tools, techniques, and regulations in the field. In addition to covering air, water, and waste treatment and handling methods, there are discussions of project management, environmental legislation and regulations, environmental assessment, and air and water quality standards. There are numerous tables and charts. One small drawback is that references do not appear to be as current as possible for a 1999 copyright date.

Eccleston, C.H. (2001) *Effective Environmental Assessments: How to Manage and Prepare NEPA EAs*, Boca Raton, FL: CRC Press. This is a comprehensive, step-by-step guide on the preparation of defensible Environmental Assessments (EAs).

Environetbase Environmental Resources Online (2000) Boca Raton, FL: CRC Press <http://www.environetbase.com/> (www.environetbase.com/). A subscription is required for online version. This is a collection of over 154 CRC Press electronic handbooks, many of which are listed individually in this section. Restricted to subscribers, it is a useful resource for rapid access to precise information on environmental modeling, systems analysis, risk assessment, health and safety, chemistry and toxicology, environmental engineering, law and compliance, and water science. Visit <http://www.environetbase.com/> for a complete list of titles and subscription information.

Environmental Law Handbook (2003) (17th edn), Washington, DC: Government Institutes, Inc. A comprehensive and up-to-date analytical review of the major environmental, health, and safety laws affecting U.S. businesses and organizations. The authors provide easy-to-read interpretations of major environmental laws.

Ghassemi, A. (ed.) (2000) *Handbook of Pollution Control and Waste Minimization*, New York: Marcel Dekker. This handbook covers the broad spectrum of pollution prevention including process design, lifestyle analysis, risk, and decision-making. The author's aim is to make environmental issues a major design consideration. The book presents the fundamentals of pollution prevention – life-cycle analysis, designs for the environment, and pollution prevention in process design. The various components of pollution prevention are discussed in detail, and current legislative and regulatory policies governing the management of waste in Europe and the U.S.A. are covered.

Gottlieb, D.W. (2003) *Environmental Technology Resources Handbook*, Boca Raton, FL: Lewis Publishers. Intended to guide users to the proper technology for solving environmental problems, this work focuses primarily on Internet resources for control, remediation, assessment, and prevention. Although it risks becoming dated quickly given the constantly changing nature of the Internet, it does guide the user to major resources that the author has carefully studied.

Higgins, T.E. (ed.) (1995) *Pollution Prevention Handbook*, Boca Raton, FL: Lewis Publishers. This fully updated and revised edition shows engineers and managers how to plan and implement an effective waste minimization program. Numerous figures and tables make access to information easy. A cross-section of industries is included.

Khandan, N. (2002) *Modeling Tools for Environmental Engineers and Scientists*, Boca Raton, FL: CRC Press. The author describes some fifty computer models developed with eight different software packages. Intended for nonprogrammers to develop computer-based mathematical models for natural and engineered environmental systems, this book includes a review of mathematical modeling and fundamental concepts such as material balance, reactor configurations, and fate and transport of environmental contaminants (available in print or electronic format, the latter on a subscription basis, <http://www.environetbase.com/pdf/enb/TX69957/TX69957%5FPDF%5FTOC.pdf.>).

Kingston, M.A. (2005) *Environmental Remediation. Cost Data-assemblies* (11th edn), R.S. Means Company; Englewood, CO: Talisman Partners. This book includes pricing for more than seventy standard remediation technologies and related tasks, and includes costs for every kind and size of project.

Kingston, M.A. (2005) *Environmental Remediation. Cost Data-unit Price* (11th edn), R.S. Means Company; Englewood, CO: Talisman Partners. This book provides the detailed line items, component costs, forms, instructions, and guidelines needed to prepare or verify cost estimates for almost any type of environmental remediation project, ranging from simple underground storage tank removals to complex hazardous waste sites.

Lee, C.C. (ed.-in-chief) and Lin, S.D. (assoc. ed.) (2000) *Handbook of Environmental Engineering Calculations*, New York: McGraw-Hill. The purpose of this handbook is to provide fully illustrated, step-by-step calculation procedures for solid waste management; air resources management; water quality assessment and control; surface water; lakes and reservoirs; ground water; public water supply; waste water treatment; and risk assessment/pollution prevention. Co-author C.C. Lee, with experience as an EPA Research Program Manager, integrates regulatory requirements into the discussions.

Liu, D.H.F. and Lipták, B.G. (eds) (1997) *Environmental Engineers' Handbook* (2nd edn), Boca Raton, FL: Lewis Publishers. The many contributors to this work are from industry, consulting firms, and academia and have a wide range of backgrounds from engineering, law, medicine, agriculture, meteorology, biology, and so on. This handbook on the prevention and management of pollution illustrates the technology and techniques through tables, schematic diagrams, and in-depth discussions. For example, some 140 pages are devoted to ground water and surface water pollution; waste-water treatment covers sources, monitoring, sewers, treatment, biological treatment, and sludge disposal in 420 pages.

Mackay, D., Shiu, W.Y., and Ma, K.C. (1992–) *Illustrated Handbook of Physical-chemical Properties and Environmental Fate for Organic Chemicals*, Boca Raton, FL: Lewis Publishers. This is a comprehensive series in five volumes that focuses on environmental fate prediction. These books tackle environmental fate calculations and QSAR (quantitative structure–activity relationship) plots. This shows where the chemicals will go, relative concentrations, persistence, and important intermediate transport processes.

Manly, B.F.J. (2001) *Statistics for Environmental Science and Management*, Boca Raton, FL: Chapman & Hall/CRC Press. This work features a non-mathematical approach to statistical methods used in environmental data analysis. Techniques such as environmental monitoring, impact assessment, assessing site reclamation, censored data, and Monte Carlo risk assessment are discussed (available both in print and electronic format, the latter by subscription, <http://www.environetbase.com/pdf/enb/C0295/C0295%5FPDF%5FTOC.pdf.>).

Moore, G.S. (2000) *Environmental Compliance: A Web-enhanced Resource*, Boca Raton, FL: CRC Press. The text provides URLs to a supporting website updated continually by the author. Current compliance with U.S. environmental regulations at federal, state, and local levels.

Patnaik, P. (1997) *Handbook of Environmental Analysis: Chemical Pollutants in Air,*

Water, Soil, and Solid Wastes, Boca Raton, FL: CRC Press/Lewis Publishers. The handbook provides thorough treatment of the analysis of toxic pollutants in the environment, addressing ambient air, ground water, surface water, industrial waste-water, and soils and sediments.

Reynolds, J.P., Jeris, J.S., and Theodore, L. (2002) *Handbook of Chemical and Environmental Engineering Calculations*, New York: John Wiley. The scientific and mathematical cross-over between chemical and environmental engineering is the key to solving a host of environmental problems. Many problems included in this book demonstrate this cross-over, as well as the integration of engineering with current regulations and environmental media such as air, soil, and water. Solutions to the problems are presented in a programmed instructional format.

Salvato, J.A., Nemerow, N.L., and Agardy, F.J. (eds) (2003) *Environmental Engineering* (5th edn), Hoboken, NJ: John Wiley. Applies the principles of sanitary science and engineering to sanitation and environmental health.

The air environment

Alley, E.R., Stevens, L.B., and Cleland, W.L. (1998) *Air Quality Control Handbook*, New York: McGraw-Hill. Starting with an overview of air pollution control management, this work discusses the sources and effects of air pollution, and then explains the available pollution treatment and control technologies from the perspective of compliance with the 1990 Clean Air Act. Air-moving, vapor, and particulate control equipment are included as well as ambient air, continuous air, and stack monitoring and various forms of pollution testing.

Beim, H.J., Spero, J., and Theodore, L. (1998) *Rapid Guide to Hazardous Air Pollutants*, New York: Van Nostrand Reinhold. This portable, concise guide is a convenient compilation of information on the 189 substances deemed to be hazardous air pollutants (HAPs) under the 1990 Amendments to the Clean Air Act. In addition to chemical substance names and synonyms, each record provides the CAS Registry number, uses, physical and chemical properties, acute and chronic health risks, hazardous risks, and regulatory information.

Brownell, F.W. (1998) *Clean Air Handbook* (3rd edn), Rockville, MD: Government Institutes. The third edition provides an overview of the regulatory requirements of the Clean Air Act and its amendments. Chapters cover such topics as the federal–state partnership, control technology regulation, operating and preconstruction permitting programs, acid deposition control program, and regulation of mobile sources of air pollution.

Cheremisinoff, N.P. (2002) *Handbook of Air Pollution Prevention and Control*, Boston, MA: Butterworth-Heinemann. For process engineers and environmental managers, this work focuses on the industrial sector, providing discussions of hardware, cost accounting, estimation methods for emissions, indoor air quality overview, point source, and regulations.

Cooper, A.R. (2004) *Air CFRs Made Easy* (2nd edn), Rockville, MD: Government Institutes. Addressing the air regulations covered in CFR Title 40, beginning with Part 104, and in Title 33, beginning with Part 320, core elements of each program are explained. Subjects covered include accident prevention regulations, acid rain control, emissions offsets, enforcement provisions, hazard assessments, mobile sources, national ambient air quality standards, and state implementation plans.

Davis, W.T. (ed.) (2000) *Air Pollution Engineering Manual* (2nd edn), Air & Waste Management Association; New York: Wiley. Organized into twenty chapters by industry, this manual discusses the different processes that generate air pollution, equipment used with all types of gases and particulate matter, and emissions control for such industries as graphic arts, chemical processes, metallurgy, and water treatment plants. This updated, second edition reflects recent emission factors and control measures for reducing air pollutants, providing technological and regulatory information for compliance with the air pollution standards. The work contains detailed flow charts and photographs as well as Internet resources.

Hewitt, C.N. and Jackson, A.V. (eds) (2003) *Handbook of Atmospheric Science*, Malden, MA: Blackwell. Divided into two major parts with contributions from twenty-eight authors, this work begins with a discussion of the behavior of the Earth's atmosphere. Part 2 then examines the problems of air pollution and the tools and practices for air quality management. Emissions monitoring and modeling tools are discussed for large-scale applications. Charts, maps, and color plates are used to illustrate the concepts, and extensive references are included for each chapter.

Schifftner, K.C. (2002) *Air Pollution Control Equipment Selection Guide*, Boca Raton, FL: Lewis Publishers. This guide assists with appropriate selection of equipment for air pollution control. Organized by primary technology employed (i.e., quenching, cooling, particulate removal, gas absorption), each section covers basic physical forces used in the technology, common sizes, and common uses. Text includes current photographs or drawings of typical equipment within that device type (available both in print and online, the latter being on a subscription basis, <http://www.environetbase.com/pdf/ENB/TX692/TX692PDFTOC.pdf.>).

Schnelle, K.B. and Brown, C.A. (2002) *Air Pollution Control Technology Handbook*, Boca Raton, FL: CRC Press. The handbook is a resource for commonly used air pollution control technology in stationary sources for the control of gaseous pollutants and particulate matter. Selection, evaluation, and design are covered as well as alternative air pollution control processes. Environmental regulations and costs are also addressed (available both in print and online, the latter being on a subscription basis, <http://www.engnetbase.com/ejournals/books/book%5Fsummary/summary.asp?id=493>).

Spicer, C.W. (2002) *Hazardous Air Pollutant Handbook: Measurements, Properties, and Fate in Ambient Air*, Boca Raton, FL: Lewis Publishers. This work describes the 188 substances designated as hazardous air pollutants (HAPs) under the Clean Air Act Amendments of 1990. Selected chemical and physical properties, measurement methods in ambient air, mean and range of ambient levels reported in the literature and reaction mechanisms are covered for each substance.

Compilation of Air Pollutant Emission Factors, Stationary Point and Area Sources (AP-42). Volume I, Supplements A–F, and Updates 2001–2004. Washington, DC: U.S. Environmental Protection Agency, Office of Air Quality Planning and Standards, Office of Air and Radiation (1995, with continuous updates) <http://www.epa.gov/ttn/chief/ap42/> (www.environetbase.com/). Prepared by the Emission Factor And Inventory Group (EFIG) of the U.S. Environmental Protection Agency's (EPA) Office Of Air Quality Planning And Standards (OAQPS), the AP-42 series is the principal means of documenting stationary point and area sources emission factors. These factors are cited in numerous other EPA publications and electronic data bases but without the process details and supporting reference material provided in AP-42.

As noted in the introduction to this work, emission estimates are important for developing emission control strategies, determining applicability of permitting and control programs, and ascertaining the effects of sources and appropriate mitigation strategies. Therefore, emission factors are frequently the best or only method available for estimating emissions, in spite of their limitations. (EPA-454/F-99-003).

The water environment

RCRA Orientation Manual [microform] developed by the U.S. Environmental Protection Agency, Office of Solid Waste, Communications, Information, and Resources Management Division. Washington, DC: U.S. Environmental Protection Agency, Solid Waste and Emergency Response, [2003]. Also available online at <www.epa.gov> (www.environetbase.com/). This updated manual provides introductory information on the solid and hazardous waste management programs under the Resource Conservation and Recovery Act (RCRA). Designed for EPA and state staff, members of the regulated community, and the general public who wish to better understand the RCRA program.

Standard Methods for the Examination of Water and Wastewater (1998) Prepared and published jointly by the American Public Health Association, American Water Works Association, Water Environment Federation (20th edn). Joint Editorial Board: Lenore S. Clesceri, Arnold E. Greenberg, Andrew D. Eaton; managing editor Mary Ann H. Franson. Washington, DC: American Public Health Association. This book is a standard in the field. The methods presented in this book, as in previous editions, are believed to be the best available and generally accepted procedures for the analysis of water and waste-water. The selection of methods that are included as well as the formal procedure for approval have been reviewed by a broad range of experts in the field.

Boulding, J.R. (2004) *Practical Handbook of Soil, Vadose Zone, and Ground-water Contamination: Assessment, Prevention, and Remediation* (2nd edn), Boca Raton, FL: Lewis Publishers. This handbook provides a comprehensive yet practical guide to soil, vadose zone, and ground water contamination and remediation. It discusses the basics of soils, hydrology, and hydrogeology as well as addressing assessment and monitoring, and prevention.

Burton, G.A. Jr and Pitt, R.E. (2002) *Stormwater Effects Handbook: A Toolbox for Watershed Managers, Scientists, and Engineers*, Boca Raton, FL: Lewis Publishers. The handbook assists in determining when storm water run-off causes adverse effects in receiving waters. It includes case studies, many photographs, and figures that allow easy visualization of methods.

Cheremisinoff, N.P. (2003) *Handbook of Solid Waste Management and Waste Minimization Technologies*, Boston, MA: Butterworth-Heinemann. An essential tool for plant managers, process engineers, environmental consultants, and site remediation specialists that focuses on practices for handling a broad range of industrial solid waste problems and presents information on waste minimization practices. Included in the text are sidebar discussions, questions for consideration and discussion, recommended resources (print and web) for the reader, and a comprehensive glossary.

Cheremisinoff, N.P. (1995) *Handbook of Water and Wastewater Treatment Technology*, New York: Dekker. This handbook serves as a collection of exact and useful information related to the treatment of water and waste-water for municipal, sanitary, and industrial uses.

De Zuane, J. (1997) *Handbook of Drinking Water Quality* (2nd edn), New York: Van Nostrand Reinhold. This handbook can be a quick reference for anyone dealing with water quality issues. Appendices include World Health Organization Guidelines and European Drinking-water Directives.

Freeman, H.M. (ed.) (1998) *Standard Handbook of Hazardous Waste Treatment and Disposal* (2nd edn), New York: McGraw-Hill. A reference of alternative and innovative technologies for managing hazardous waste and cleaning up abandoned disposal sites.

Gallagher, L.M. (2003) *Clean Water Handbook*, Rockville, MD: Government Institutes. This handbook was designed to provide a comprehensive road-map to the requirements, legal theories, and critical issues of water pollution control.

HDR Engineering, Inc. (2001) *Handbook of Public Water Systems*, New York: John Wiley. This book has become a standard reference in this field. The numerous contributors are experts on the subject and they have been involved in the development of several new water treatment processes that have been incorporated into water treatment plant designs.

Letterman, R.D. (technical ed.) (1999) *Water Quality and Treatment: a Handbook of Community Water Supplies*, American Water Works Association; New York: McGraw-Hill. This book consists of eighteen illustrated chapters detailing state-of-the-art technologies and methods. It features updated discussions of all water treatment processes.

Lewis, R.J. Sr (2002) *Hazardous Chemicals Desk Reference* (5th edn), New York: J. Wiley-Interscience. This book is intended to be a quick reference and includes over 6,000 common industrial and laboratory materials. The information is taken from the more comprehensive *Sax's Dangerous Properties of Industrial Materials* by Irving Sax, but it is less technical.

Lin, S.D. and Lee, C.C. (2001) *Water and Wastewater Calculations Manual*, New York: McGraw-Hill. This manual offers streamlined, step-by-step procedures for problems in water and waste-water engineering from the simple to the complex.

Mays, L.W. (ed.-in-chief) (2001) *Stormwater Collection Systems Design Handbook*, New York: McGraw-Hill. This book is a comprehensive reference of state-of-the-art design of storm water collection systems and their component parts. It includes problem examples, discussion of what can go wrong in the design process, references, formulae, and diagrams.

Montgomery, J.H. (2000) *Ground Water Chemicals Desk Reference* (3rd edn), Boca Raton, FL: CRC Press/Lewis Publishers. This comprehensive work deals with hazardous chemicals in ground water. Each entry gives a complete description of the chemical, including physical and chemical properties.

Rizzo, J.A. (ed.) (1998) *Underground Storage Tank Management: A Practical Guide* (5th edn), Rockville, MD: Government Institutes. This guide includes all updates and requirements to comply with the U.S. EPA's federal requirements including soil sampling, analytical guidelines, and the evolution of tank testing strategies.

Tchobanoglous, G. and Kreith, F. (2002) *Handbook of Solid Waste Management*, New York: McGraw-Hill. This handbook offers an integrated approach to the planning, design, and management of economical and environmentally responsible solid waste disposal systems.

White, G.C. (1999) *The Handbook of Chlorination and Alternative Disinfectants*. New York: John Wiley. This book discusses all aspects of water purification using chlorination.

Monographs and textbooks

General

CRC Press, one of the leading publishers of engineering books, has created three valuable resources for environmental engineers. The books are available online at ENGnetBase <http://www.engnetbase.com/, ENVIROnetBase http://www.environetbase.com/>, and STATSnetBASE <http://www.statsnetbase.com/>. Using a simple search screen, you enter the search terms you want either as a single word, phrase, or with Boolean operators. These three databases are only available as an annual subscription. Once you are a subscriber, any new books added during the year are yours to browse for no additional cost.

Berthouex, P.M. and Brown, L.C. (2002) *Statistics for Environmental Engineers* (2nd edn), Boca Raton, FL: Lewis Publishers (electronic resource and print). This edition consists of fifty-four short, stand-alone chapters with each chapter addressing a particular environmental problem or statistical technique.

Davis, L.R. (1999) *Fundamentals of Environmental Discharge Modeling*, Boca Raton, FL: CRC Press. This book focuses on engineering and mathematical models for documenting and approving mechanical and environmental discharges with an emphasis on waste-water and atmospheric discharges. Various diffuser and surface discharge models are discussed, as well as the fundamentals of turbulent jet mixing. Case studies are used to illustrate problems.

Gupta, R.S. (2004) *Introduction to Environmental Engineering and Science* (2nd edn), Rockville, MD: ABS Consulting, Government Institute. An introduction to the fundamental principles common to most environmental problems is followed by major sections on water pollution, hazardous wastes and risk assessment, waste treatment, air pollution, global climate change, and hazardous substances. Includes problems to develop skills learned in the text.

Jørgensen, S.E. (2000) *Principles of Pollution Abatement: Pollution Abatement for the 21st Century*, New York; Amsterdam: Elsevier. This is a revised and expanded version of the 1988 *Principles of Environmental Science* by the same author. Contents include mass conservation, energy conservation, risks and effects, water and waste water problems, solid waste problems, and air pollution problems. The work features new tools such as ecotechnology, cleaner technology, life-cycle analysis, and new environmental management techniques by changes in products and production methods.

Manly, B.F.J. (2001) *Statistics for Environmental Science and Management* (electronic resource and print). Boca Raton, FL: Chapman & Hall/CRC Press. The use of appropriate statistical methods is essential when working with environmental data. This book is intended to introduce environmental scientists and managers to the statistical methods that will be useful in their work. The book is not meant to be a complete introduction to statistics but Appendix A provides a quick refresher.

Mulligan, C.N. (2002) *Environmental Biotreatment: Technologies for Air, Water, Soil, and Waste*, Rockville, MD: Government Institutes. The author summarizes the application of twenty-six bioremediation techniques to cleaning air, soil, water, and wastes. For each method, advantages, disadvantages, costs, and other considerations are discussed, and then each method is compared to more conventional technology. Case studies are presented and extensive references are included.

Nazaroff, W.W. and Alvarez-Cohen, L. (2001) *Environmental Engineering Science*, New York: Wiley. This textbook covers the fundamentals of environmental engineering and applications in water quality, air quality, and hazardous waste management.

Revelle, C. and McGarity, A. (1997) *Design and Operation of Civil and Environmental Engineering Systems*, New York: Wiley. This reference work applies the tools of operations research to the problems of civil and environmental engineering. Contents include discussions of water quality management, management of hazardous waste, air quality management, solid waste management, hazardous waste management, environmental planning for electric utilities, effluent charges, and transferable discharge permits.

Salvato, J.A., Nemerow, N.L., and Agardy, F.J. (eds) (2003) *Environmental Engineering* (5th edn), Hoboken, NJ: John Wiley. This book applies principles of sanitary science and engineering to sanitation and environmental health.

Wiersma, G.B. (ed.) (2004) *Environmental Monitoring*, Boca Raton, FL: CRC Press. The book is a compilation that brings together the activities and complex approaches to monitoring air, water, and land.

Air environment

Baumbach, G. (1996) *Air Quality Control: Formation and Sources, Dispersion, Characteristics and Impact of Air Pollutants – Measuring Methods, Techniques for Reduction of Emissions and Regulations for Air Quality Control*, New York: Springer. A translation from the German, this English edition has been revised to include situations and regulations applicable in the U.S. It received "Outstanding Title!" status and was strongly recommended in the June 1997 CHOICE review.

Devinny, J.S., Deshusses, M.A., and Webster, T.S. (1999) *Biofiltration for Air Pollution Control*, Boca Raton, FL: Lewis Publishers. This is a comprehensive discussion of biofiltration technology (the use of micro-organisms growing on porous media) for air pollution control. Materials, designs, monitoring methods, as well as examples of successful applications are included.

Friedlander, S.K. (2000) *Smoke, Dust, and Haze: Fundamentals of Aerosol Dynamics* (2nd edn), New York: Oxford University Press. Updating a much earlier edition, this work summarizes current aerosol knowledge. It was awarded "Outstanding Title!" status in the December 2000 CHOICE review.

Godish, T. (2004) *Air Quality* (4th edn), Boca Raton, FL: Lewis Publishers. This fourth, revised edition of a classic offers a comprehensive overview of air quality issues, including atmospheric chemistry, analysis of the control of emissions from stationary sources, the effects of pollution on public health and the environment, public policy concerns, and technology and regulatory practices. Among the new sections are toxicological principles and risk assessment. According to the preface, the work is useful as a supplement to engineering curricula for the design and operation of pollution control equipment.

Heinsohn, R.J. and Cimbala, J.M. (2003) *Indoor Air Quality Engineering: Environmental Health and Control of Indoor Pollutants*, New York: Basel. Intended as an upper-level textbook and reference for professionals, this monograph seeks to bridge engineering and industrial hygiene. In addition to the ten chapters covering risk, contaminant concentration, respiratory systems, design criteria, ventilation, particle emission, and emission rates, the authors provide some supplementary information at http://www.mne.psu.edu/cimbala/Heinsohn_

Cimbala_book/index.htm. Here they provide Excel and MathCAD files corresponding to examples in the book and a free program for calculating flow in ducts.

Kowalski, W.J. (2003) *Immune Building Systems Technology*, New York: McGraw-Hill. Advertised as a one-stop guide to building ventilation and air treatment systems design, this book aims for a comprehensive approach to the protection of buildings against biological pathogens. Designing, retrofitting, and building state-of-the-art ventilation and air treatment systems are covered.

Stern, A.C. (ed.) (1968) *Air Pollution* (2nd edn), New York: Academic Press. Despite the publication date, Stern's three-volume second edition is still cited regularly and is available in numerous libraries. Written for professionals and practitioners, the sixty contributors address the cause, effect, transport, measurement, and control of air pollution.

Wang, L.K., Pereira, N.C., and Hung, Y-T. (2004–2005) *Air Pollution Control Engineering* (2 vols), Totowa, NJ: Humana Press. Available either as an electronic book or as hard copy, this work surveys principles and practices underlying control processes, illustrating them with detailed design examples. Chapters include fabric filtration, cyclones, electrostatic precipitation, wet and dry scrubbing, condensation as a basis for intelligent planning of abatement systems, flare processes, thermal oxidation, catalytic oxidation, gas-phase activated carbon adsorption, and gas-phase biofiltration. Best Available Technologies (BAT) for air pollution control are detailed and cost data provided as well as engineering methods for the design, installation, and operation of air pollution process equipment.

Zhang, Y. (2005) *Indoor Air Quality Engineering*, Boca Raton, FL: CRC Press. Based on the author's own course lecture notes, this text covers properties and mechanics of airborne pollutants, measurement and sampling, and indoor air quality control technologies. Since it was written as a textbook, it provides discussion topics, problems, and references to further reading for each topic.

Hazardous waste, solid waste, and water environment

Asano, T. (ed.) and Eckenfelder, W.W., Malina, J.F. Jr, and Patterson, J.W. (library eds) (1998) *Wastewater Reclamation and Reuse*, Lancaster, PA: Technomic Publications. Experts from around the world contributed to this useful and unique text which analyzes and reviews aspects of waste-water reclamation, recycling, and reuse in countries around the world. This is Volume 10 of an eleven-volume series, the Water Quality Management Library, which thoroughly addresses issues in waste-water treatment, sludge, non-point pollution, toxicity reduction, and ground water remediation.

Bagchi, A. (2004) *Design of Landfills and Integrated Solid Waste Management* (3rd edn), Hoboken, NJ: John Wiley. This book combines integrated solid waste management with the traditional coverage of landfills. This new edition offers the first comprehensive guide to managing the entire solid waste cycle, from collection, to recycling, to eventual disposal. Includes new material on source reduction, recycling, composting, contamination soil remediation, incineration, and medical waste management.

Baruth, E.E. (technical ed.) (2005) *Water Treatment Plant Design* (4th edn), New York: McGraw-Hill. Water treatment plant design has become increasingly complex. This book has become the standard reference for engineers involved in this area. It contains articles by over forty international design experts and is a comprehensive reference on modernizing existing water treatment facilities and planning new ones – from initial plans and permits through design, construction, and start-up. The third edition of this book was written as a companion to AWWA's *Water Quality and Treatment: A Handbook of Community Water Supplies.* American Water Works Association; Raymond D. Letterman (technical ed.) (1999).

Grady, C.P.L. Jr, Daigger, G.T., and Lim, H.C. (1999) *Biological Wastewater Treatment* (2nd edn), New York: Marcel Dekker. This book integrates the principles of the biochemical processes with applications to the design, operation and optimization of biochemical operations. It is volume 19 in the Environmental Science and Pollution Control series that includes twenty-seven volumes. This comprehensive series covers a vast array of topics pertinent to environmental engineers, including pollution prevention practices and control, bioremediation of contaminated soils, combustion and control, biosolids treatment and management, ground water contamination, and hazardous waste.

Hammer, D.A. (ed.) (1989) *Constructed Wetlands for Wastewater Treatment: Municipal, Industrial, and Agricultural*, Chelsea, MI: Lewis Publishers. New advancements in this area continue to be covered in the journal literature; however, this book is a standard in the field for its comprehensive coverage of this topic. It covers general principles of wetlands, ecology, hydrology, soil chemistry, vegetation, and wildlife as well as case studies and specific applications.

Kadlec, R.H. and Knight, R.L. (1996) *Treatment Wetlands*, Boca Raton, FL: Lewis Publishers. The most complete guide to using wetlands for water quality management and habitat creation. This book provides scientific and operational information on treatment wetlands technologies for entry-level practitioners. Contents include an introduction to wetlands for treatment, wetland structure and function, effects of wetlands on water quality, wetland project planning and design, wetland treatment system establishment, operation and maintenance, and wetland data case histories.

Metcalf & Eddy, Inc. revised by Tchobanoglous, G., Burton, F., and Stensel, H.D. (2003) *Wastewater Engineering: Treatment and Reuse* (4th edn), Boston, MA: McGraw-Hill. Gives a solid overall perspective on waste-water engineering.

Methods, H. and Durrans, S.R. (eds) (2003) *Stormwater Conveyance Modeling and Design*, Waterbury, CT: Haestad Press. The academic version includes a CD-ROM, and StormCAD Stand-alone, PondPack, CulvertMaster, and FlowMaster software. This book guides you through the design and analysis process using an approach that combines theoretical fundamentals with practical design guidance and hydraulic monitoring techniques.

Mitsch, W.J. and Gosselink, J.G. (2000) *Wetlands* (3rd edn), New York: John Wiley. The third edition, this standard textbook is more international in scope and has been expanded in other areas. It includes information about all types of wetlands, and chapters on wetland law and regulation and wetland delineation.

Pierzynski, G.M., Sims, J.T., and Vance, G.F. (2005) *Soils and Environmental Quality* (3rd edn), Boca Raton, FL: CRC Press. This revised and updated text provides detailed discussions about soils science, hydrology, and the classification of pollutants.

Todd, D.K. and Mays, L.W. (2005) *Groundwater Hydrology* (3rd edn), Hoboken, NJ: John Wiley. All aspects of ground water hydrology are covered; includes chapters on the quality and pollution of ground water and saline water intrusion.

Journals

Unless otherwise noted, all journals are peer reviewed. For further identification, the ISSN (International Standard Serial Number) appears after each title. When a separate ISSN for electronic format is available, it is indicated as the second number.

Advances in Environmental Research: An International Journal of Research in Environmental Science, Engineering and Technology (1093–0191). This publication contains original full-length research papers, case studies, notes, and critical reviews on major advances in protection of the quality of air, water, and land environments; improvements to existing technology; and contributions to the knowledge of transport and fate of pollutants in the environment.

Aerosol Science and Technology: The Journal of the American Association for Aerosol Research (0278–6826). This journal covers theoretical and experimental investigations of aerosol and closely related phenomena, as well as papers on fundamental and applied topics.

Ambio: A Journal of the Human Environment (0044–7447). This journal has been published since 1972 by the Royal Swedish Academy of Sciences, an independent, non-governmental organization that aims to promote research in mathematics and the natural sciences.

Annual Review of Environment and Resources (1056–3466). This review focuses on the emerging scientific and policy issues at the crux of sustainable development. The reviews assess critical scientific, policy, technological, and methodological issues related to the Earth's global life support systems, sectors of human use of environment and resources, and the human dimensions and management of resources and environmental change.

Applied Catalysis. B, Environmental (0926–3373). This journal covers the catalytic chemistry of polluting substances.

Archives of Environmental Contamination and Toxicology (0090–4341). This journal includes significant, full-length articles describing original experimental or theoretical research work pertaining to the scientific aspects of contaminants in the environment.

Atmospheric Environment (1352–2310). Focusing on the consequences of natural and human-induced perturbations on the Earth's atmosphere, this journal covers air pollution research and its applications, air quality and its effects, dispersion and transport, deposition, biospheric–atmospheric exchange, global atmospheric chemistry, radiation, and climate.

Biodegradation (0923–9820). This journal publishes papers on all aspects of science pertaining to the detoxification, recycling, amelioration, or treatment of waste materials and pollutants by naturally occurring microbial strains or recombinant organisms.

Building and Environment (0360–1323). This journal publishes original papers and review articles on building research and its applications, and on the social, cultural, and technological contexts of building research and architectural science.

Bulletin of Environmental Contamination and Toxicology (0007–4861). This journal provides rapid publication of significant advances and discoveries in the fields of air, soil, water, and food contamination and pollution as well as articles on methodology and other disciplines concerned with the introduction, presence, and effects of toxicants in the total environment.

Chemosphere (0045–6535). This international, multi-disciplinary journal disseminates original articles describing new discoveries or developments in fields related to the environment and human health and developing areas of environmental science.

Clean Water Report (0009–8620). This journal provides comprehensive coverage of drinking-water and sewer systems, lakes, rivers and streams, coastal protection, tributaries and bays, as well as the Safe Drinking Water Act, the Clean Water Act, and other major legislative initiatives. It also includes articles on biosolids, pathogens, arsenic, chlorine, dioxin, and other pollutants, problems such as flooding, silting, sedimentation, nutrients, and more.

Climate Research (0936–577X). This journal evaluates, selects, and disseminates important new information about basic and applied research devoted to all aspects of climate – present, past, and future; effects of human societies and organisms on climate; and effects of climate on the ecosphere.

Cold Regions Science and Technology (0165–232X). This journal primarily addresses problems related to the freezing of water, and especially with the many forms of ice, snow, and frozen ground. Emphasis is given to applied science, mainly in the physical sciences, with broad coverage of the physics, chemistry, and mechanics of ice, ice-water systems, and ice-bonded soils.

Critical Reviews in Environmental Science and Technology (1064–3389) (1547–6537). This journal serves as an international forum for the critical review of current knowledge on the broad range of topics in environmental science. It addresses current problems and the scientific basis for new pollution control technologies.

Desalination (0011–9164). *Desalination* covers all desalting fields – distillation, membranes, reverse osmosis, electrodialysis, ion exchange, freezing, water purification, water reuse, and waste-water treatment – and aims to provide a forum for any innovative concept or practice.

Ecological Engineering (0925–8574) (0925–8574). This journal is meant for ecologists who are involved in designing, monitoring, or constructing ecosystems and serves as a bridge between ecologists and engineers.

Environmental Engineering Science (1092–8758). This journal publishes papers on environmental science topics that include the development and application of fundamental principles toward solving problems in air, land, and water media, including environmental applications of the basic science.

Environmental Health Perspectives (EHP) (0091–6765). A peer-reviewed journal of the National Institute of Environmental Health Sciences, this is an important vehicle for the dissemination of environmental health information and research findings.

Environment International (0160–4120). This journal covers all disciplines engaged in the field of environmental research, and seeks to quantify the impact of contaminants in the human environment, and to address human impacts on the natural environment itself. It covers the entire spectrum of sources, pathways, sinks, and interactions between environmental pollutants, whether chemical, biological, or physical.

Environmental Management (0364–152X) (1432–1009). This journal serves to improve cross-disciplinary communication. Contributions are drawn from biology, botany, climatology, ecology, ecological economics, environmental engineering, fisheries, environmental law, management science, forest sciences, geography, geology, information science, law politics, public affairs, and zoology.

Environmental Modelling and Software: With Environment Data News (1364–8152). This journal publishes contributions in the form of research articles, reviews, and short communications as well as software and data news on recent advances in environmental modeling and/or software to improve the capacity to represent, understand, predict, or manage the behavior of environmental systems at all practical scales.

Environmental Pollution (0269–7491). This is an international journal that addresses issues relevant to the nature, distribution, and ecological effects of all types and forms of chemical pollutants in air, soil, and water. It includes articles based on original research, findings from re-examination and interpretation of existing data, and articles on new methods of detection and remediation of environmental pollutants.

Environmental Science and Technology (0013–936X), (1520–5851). This journal is a unique source of information for scientific and technical professionals in a wide range of environmental disciplines. Contributed materials may appear as current research papers, policy analyses, or critical reviews. Also includes a magazine section that provides authoritative news and analysis of the major developments, events, and challenges shaping the field.

Ground Water (0017–467X). This technical publication is strictly for ground water hydrogeologists. Each issue of the journal contains peer-reviewed scientific articles on pertinent ground water subjects.

Ground Water Monitoring and Remediation (069–3629). Since its inception in 1981, this journal has been the leader in the field of ground water monitoring and clean-up. It contains a mixture of original columns authored by industry leaders, industry news, EPA updates, product and equipment news, and peer-reviewed papers.

Hazardous Waste Consultant (0738–0232). This journal has articles about the newest developments relating to hazardous waste assessment, treatment, storage, and disposal, as well as waste minimization technologies. In addition, summaries of federal and state legal cases with a focus on regulatory interpretation and enforcement related to hazardous waste compliance are reported.

Hazardous Waste/Superfund Week (1521–2882). This journal covers federal and state regulations, their interpretations and the effect they have on organizations, as well as contract opportunities and lawsuit coverage. Every other week, there is comprehensive, behind-the-scenes coverage of congressional action, EPA initiatives, Department of Defense clean-ups, Superfund sites, regulatory changes, court cases, enforcement news, contract opportunities, research findings, and business developments.

Indoor and Built Environment: The Journal of the International Society of the Built Environment (1420–326X). This journal publishes reports on any topic pertaining to the quality of indoor or built environment, and how this may affect the health, performance, efficiency, and comfort of persons in these environments.

Journal of the Air and Waste Management Association (1096–2247). One of the oldest continuously published, peer-reviewed, technical environmental journals in the world, this journal serves those occupationally involved in air pollution control and waste management. It covers air pollution, hazardous waste, management, regulations, measuring, modeling, emissions, testing, monitoring, and more.

Journal of Aerosol Science (0021–8502). Covering all aspects of basic and applied aerosol research, the original papers in this journal describe recent theoretical and experimental research relating to the basic physical, chemical, and biological properties of systems of airborne particles of all types; their measurement, formation, transport, deposition and effects; and industrial, medical, and environmental applications.

Journal of the American Water Works Association (0003–150X). Both a professional and a scholarly publication, this journal contains information about water quality, water resources and supply, as well as the management and operation of water utilities.

Journal of Atmospheric Chemistry (0167–7764). This journal includes, in particular, studies of the composition of air and precipitation and the physiochemical processes in the Earth's atmosphere; the role of the atmosphere in biogeochemical cycles; the chemical interaction of the oceans, land surface, and biosphere with the atmosphere; as well as laboratory studies of the mechanics in transformation processes and descriptions of major advances in instrumentation.

Journal of Environmental Engineering (0733–9372). This journal presents broad interdisciplinary information on the practice and status of research in environmental engineering science, systems engineering, and sanitation. Contributors include consultants, practicing engineers, and researchers.

Journal of Environmental Science and Health. Part A, Toxic/Hazardous Substances and Environmental Engineering (1093–4529) (1532–4117). This is a comprehensive journal that provides an international forum for the rapid publication of essential information including the latest engineering innovations, effects of pollutants on health, control systems, laws, and projections pertinent to environmental problems whether in the air, water, or soil.

Journal of Environmental Systems (0047–2433). This journal includes articles that range from case studies of particular environmental/energy/waste problems or technologies, to assessments of overall environmental system (or cost, risk, energy) impacts, to broad discussions of issues of theory, methodology, and policy. The emphasis is on practical environmental problems, such as recycling and waste minimization.

Journal of Hazardous Materials (0304–3894). This journal publishes full-length research papers, reviews, project reports, case studies, and short communications that improve understanding of the hazards and risks which certain materials pose to people and the environment, and with ways of controlling the hazards and associated risks. The journal is published in two parts: Part A: Risk Assessment and Management; and Part B: Environmental Technologies.

Journal of Hydrology (0022–1694). Papers comprise but are not limited to the physical, chemical, biogeochemical, stochastic and systems aspects of surface and ground water hydrology, hydrometeorology and hydrogeology. Relevant topics in related disciplines such as climatology, water resource systems, hydraulics, geomorphology, soil science, instrumentation, remote sensing, and civil and environmental engineering are also included.

Journal of Hydrologic Engineering (1084–0699). This journal disseminates information on the development of new hydrologic methods, theories, and applications to current engineering problems. It publishes papers on analytical, numerical, and experimental methods for the investigation and modeling of hydrological processes.

Journal of Water Resources Planning and Management (0733–9496). This journal reports on all phases of planning and management of water resources. Papers examine social, economic, environmental, and administrative concerns related to the use and conservation of water. Social and environmental objectives in fish and wildlife management, water-based recreation, and wild and scenic river use are assessed.

Noise Control Engineering Journal (0736–2501). A refereed journal published by the

Institute of Noise Control Engineering, it contains information about noise control solutions and manufacturers.

Pollution Engineering (0032–3640). Includes feature articles devoted to practical engineering applications useful for the recognition, measurement, control, and disposal of hazardous solids, air and liquid containments.

Remediation: The Journal of Environmental Cleanup Costs, Technologies and Techniques (1051–5658). This quarterly journal focuses on the practical application of remediation techniques and technologies; how to diagnose problems at hazardous waste disposal sites; and how to select the best, most cost-effective clean-up technology

Resources, Conservation, and Recycling (0921–3449). This journal emphasizes the processes involved in more sustainable production and consumption systems. Emphasis is upon technological, economic, institutional, and policy aspects of specific resource management practices, such as conservation, recycling, and resource substitution, and of strategies, such as restructuring of production and consumption profiles, and the transformation of industry.

Science of the Total Environment (0048–9697). With an emphasis on applied environmental chemistry and environmental health, this international journal publishes research about changes in the natural level and distribution of chemical elements and compounds which may affect man and the natural world.

Spill Science and Technology Bulletin (1353–2561). This journal focuses on the impacts and control of discharges of oil, oil products, and other hazardous substances.

Tellus. Series B, Chemical and Physical Meteorology (0280–6509). The series focuses on air chemistry, surface exchange processes, long-range and global transport, aerosol science, and cloud physics including related radiation transfer.

Waste Management (0956–053X). This is the official journal of the International Waste Working Group (IWWG). It was started to create an intellectual forum to encourage and support economical and ecological (integrated and sustainable) waste management worldwide, and to promote scientific advancement in the field. This aim will be accomplished by learning from the past, analyzing the present for developing new ideas, and visions for the future.

Water, Air, and Soil Pollution (0049–6979). This international, interdisciplinary journal covers all aspects of pollution and solutions to pollution in the biosphere. This includes chemical, physical, and biological processes affecting flora, fauna, water, air, and soil in relation to environmental pollution.

Water Environment and Technology (1044–9493). This journal covers issues such as expansions and upgrades, nutrient removal, biosolids management, new technology, safety, permitting and regulations, collection systems, disinfection, pumps, odor control, watershed management, storm water, and ground water clean-up.

Water Environment Research: A Research Publication of the Water Environment Federation (1061–4303). An environmental journal for the dissemination of fundamental and applied research in all scientific and technical areas related to water quality and pollution control. Topics of interest include hazardous wastes, ground water and surface water, drinking-water, source water protection, remediation and treatment systems, reuse, and environmental risk and health.

Water Research (0043–1354). Covering all aspects of the science and technology of water quality and its management worldwide, the journal's scope includes treatment processes for water and waste-water; water quality standards and monitoring; studies on inland, tidal, or coastal waters and urban waters; limnology of lakes, impoundments, and rivers; solid and hazardous waste management; soil and ground water remediation; analysis of the interfaces between sediments and water, and water/atmosphere interactions; modeling techniques and public health and risk assessment.

Water Resources Research (0043–1397). This is an interdisciplinary journal that contains original articles about hydrology; in the physical, chemical, and biological sciences; and in the social and policy sciences, including economics, systems analysis, sociology, and law.

Water Science and Technology: A Journal of the International Association on Water Pollution Research (0273–1223). This journal selects the best papers from biennial, regional, and specialized conferences sponsored by the International Water Association (IWA) encompassing important developments in all aspects of water quality management and pollution control. It is a useful resource for those unable to attend the conferences.

Wetlands Ecology and Management (0923–4861). An international journal that publishes original articles in the field of wetlands ecology, the science of the structure and functioning of wetlands for their transformation, utilization, preservation, and management on a sustainable basis. The journal covers pure and applied science dealing with biological, physical, and chemical aspects of freshwater, brackish, and marine coastal wetlands.

Standards, test methods, guidance, and criteria documents

Air

Federal and state government resources

California Air Resources Board Ambient Air Quality Standards, <http://www.arb.ca.gov/research/aaqs/aaqs.htm.> (accessed May 18, 2005). Both the Air Resources Board (ARB) and the U.S. Environmental Protection Agency (USEPA) are authorized to set ambient air quality standards. This website includes both California and U.S. federal ambient air quality standards.

U.S. Department of Labor, Occupational Health and Safety Administration (OSHA) <http://www.osha.gov/SLTC/indoorairquality/standards.html> (accessed May 18, 2005). To quote from this federal website: "Indoor Air Quality (IAQ) hazards are addressed in specific standards for general and construction industries. This page provides links to those standards as well as references related to OSHA enforcement policy such as interpretation letters. . . . Note: Some states have OSHA-approved State Plans and have adopted their own standards and enforcement policies."

U.S. Environmental Protection Agency, Air and Radiation, National Ambient Air Quality Standards (NAAQS) <http://www.epa.gov/air/criteria.html> (accessed May 18, 2005). The Clean Air Act requires EPA to set National Ambient Air Quality Standards for pollutants considered harmful to public health and the environment. The EPA Office of Air Quality Planning and Standards (OAQPS) <http://www.epa.gov/air/oaqps/> (accessed May 18, 2005) has set National Ambient Air Quality Standards for six principal pollutants (ozone, lead, carbon monoxide, sulfur dioxide, nitrogen dioxide, and respirable particulate matter), which are called "criteria" pollutants.

Associations, societies, organizations, and commercial resources

Acoustical Society of America (ASA) <http://www.acoustics.org/index.html> (accessed May 18, 2005). The ASA maintains four standards committees accredited by the American National Standards Institute (ANSI): S1 on Acoustics, S2 on Mechanical Vibration and Shock, S3 on Bioacoustics, and S12 on Noise. These four accredited standards committees also provide the U.S. input to several international committees.

American Society of Heating, Refrigeration and Air-conditioning Engineers, Inc. (ASHRAE) <http://www.ashrae.org/template/Index> (accessed May 18, 2005). The ASHRAE develops standards for both its members and others professionally concerned with refrigeration processes and the design and maintenance of indoor environments; for example, "Standard 62-1989: Ventilation for Acceptable Indoor Air Quality." This is a voluntary standard for "minimum ventilation rates and indoor air quality acceptable to human occupants and intended to avoid adverse health effects. See list of standards at <http://xp20.ashrae.org/STANDARDS/standa.htm> (accessed May 18, 2005).

Society of Automotive Engineers, SAE International, Technical Standards Development <http://www.sae.org/standardsdev/> (accessed May 18, 2005). SAE International maintains over 8,300 technical standards and related documents, which may be purchased through their website. Selected standards related to fuels, emissions, and the environment are listed in the ground vehicle, and aerospace sections may be of interest to this audience.

Water

American Water Works Association (1990–) *AWWA Standards* (loose-leaf), Denver, CO: AWWA. AWWA standards are approved as American National Standards and are used by water supply professionals throughout North American and in other parts of the world. AWWA develops and maintains standards to improve drinking-water quality and supply for the public.

EPA Test Methods and Guidelines <http://www.epa.gov/epahome/Standards.html> (accessed May 18, 2005). This resource provides a series of links to related EPA websites containing methods and standards for air, drinking-water, and wastewater, solid and hazardous waste including underground storage tanks, pesticides and toxic substances, and microbial methods.

Franson, M.A.H. (managing ed.), Clesceri, A.E., Greenberg, A.E., and Eaton, A.D. (Joint Editorial Board) (1998) *Standard Methods for the Examination of Water and Wastewater* (20th edn), Washington, DC: American Public Health Association, American Water Works Association, Water Environment Federation, prepared and published jointly. This standard work details comprehensive tests for all major pollutants, giving precise instructions for procedures, apparatus set-up, calibrations, and current reference data. Subscription information at <http://www.standardmethods.org/>.

Index to EPA Test Methods <http://www.epa.gov/region01/oarm/testmeth.pdf> (accessed May 18, 2005). Lists over 700 air, water, and waste methods. EPA test methods are approved procedures for measuring the presence and properties of chemical substances or measuring the effects of substances under various conditions.

Keith, L.H. (1996) *Compilation of EPA's Sampling and Analysis Methods* (electronic resource) (2nd edn), Boca Raton, FL: CRC Press/Lewis Publishers. This is a printed edition of the EPA database. The book and the database are intended to help people select the most appropriate methods of sampling and analysis for a particular situation.

Kopp, J.F. and McKee, G.D. (1983) *Methods for Chemical Analysis of Water and Wastes*, Cincinnati, OH: Environmental Monitoring and Support Laboratory, Office of Research and Development, U.S. Environmental Protection Agency, 1983. Contains chemical analytical procedures used in EPA laboratories for the examination of ground and surface waters, domestic and industrial waste effluents, and treatment of process samples. These methods are also included on Selected Office of Water methods and guidance (electronic resource)/United States Environmental Protection Agency, Office of Water. Version 4. Washington, DC: U.S. Environmental Protection Agency, Office of Water (2002). This series is being frequently updated and includes recently developed and requested analytical methods and related guidance documents.

National Ambient Air Quality Standards for Criteria Pollutants (NAAQS) <http://www.epa.gov/air/criteria.html> (accessed May 18, 2005). The Clean Air Act, which was last amended in 1990, requires EPA to set National Ambient Air Quality Standards for pollutants considered harmful to public health and the environment. The Clean Air Act established two types of national air quality standards. Primary standards set limits to protect public health, including the health of "sensitive" populations, and Secondary standards that set limits to protect public welfare, including protection against decreased visibility, damage to animals, crops, vegetation, and buildings.

National Primary Drinking Water Standards and National Secondary Drinking Water Standards. Visit the EPA website <http://www.epa.gov/safewater/standards.html#rules> (accessed May 18, 2005).

Telliard, W.A. (prepared under the direction of) (2004) *Selected Office of Water Methods and Guidance* (electronic resource), Version 5, Washington, DC: U.S. Environmental Protection Agency, Office of Water. This CD-ROM includes more than 500 recently developed and frequently requested analytical methods and related guidance documents for the detection of trace metals, microbiological contaminants, and organics in waste-water and drinking-water; chemical and biological methods for biosolids; guidance documents on the Office of Water's revised approach to method approval, whole effluent toxicity testing, and oil and grease analysis. Version 5 contains all the methods and guidance contained on previous versions. CD-ROM available from NTIS.

Test Methods for Evaluating Solid Waste: Physical/Chemical Methods (1986) Washington, DC: U.S. Environmental Protection Agency, Office of Solid Waste and Emergency Response (Supt. of Docs., U.S. G.P.O., distributor). "SW-846." A four-volume basic manual, currently in its third edition. Publication SW-846 is the Office of Solid Waste's official compendium of analytical and sampling methods that have been evaluated and approved for use in complying with the Resource Conservation and Recovery Act (RCRA) regulations. SW-846 functions primarily as a guidance document setting forth acceptable, although not required, methods to use in responding to RCRA-related sampling and analysis requirements. SW-846 is now available on CD-ROM and online <http://www.epa.gov/epaoswer/hazwaste/test/sw846.htm> (accessed May 18, 2005).

Criteria documents

Criteria documents accurately reflect the latest scientific knowledge and are based solely on data and scientific judgments on the relationship between pollutant concentrations and environmental and human health effects. Criteria provide guidance to EPA when promulgating federal regulations and developing standards.

National Recommended Water Quality Criteria (electronic resource) (2002) Washington, DC: United States Environmental Protection Agency, Office of Water, Office of Science and Technology <http://www.epa.gov/OST/standards/wqcriteria.html> (accessed May 18, 2005). This is a compilation of recommended water quality criteria for 157 pollutants and is an update of *Quality Criteria for Water*, Washington, DC: U.S. Environmental Protection Agency, Office of Water Regulations and Standards. For sale by the Supt. of Docs., U.S. G.P.O., 1986, and 1995 update. This book was also known as the Gold Book.

1995 updates: water quality criteria documents for the protection of aquatic life in ambient water/United States Environmental Protection Agency, Office of Water. Washington, DC: U.S. Environmental Protection Agency, Office of Water; Springfield, VA: U.S. Department of Commerce, National Technical Information Service (distributor) (1996). A Clean Water Act Section 304(a) water quality criterion is a qualitative or quantitative estimate of the concentration of a contaminant or pollutant in ambient waters, which, when not exceeded, will ensure water quality is sufficient to protect a specified water use.

Regulations

Laws and regulations are a major tool in protecting the environment. Congress passes the laws that govern in the United States. To put those laws to work, Congress authorizes certain government agencies to create and enforce regulations. Regulations state specific details about what is required by various communities to comply with the law and regulations set specific levels of pollutants, and so on. Once an authorized agency (such as EPA) decides that a regulation is needed, it lists the proposed regulation in the *Federal Register* (FR). The public is allowed to comment on the impact of the regulation. The agency considers the comments, revises the regulation, and issues a final rule. At each stage in the process, the agency publishes a notice in the FR. These notices include the original proposal, requests for public comment, notices about meetings that are open to the public where the proposal will be discussed, and the text of the final regulation. The FR is published every work day of the federal government.

Twice a year, each agency publishes a comprehensive report that describes all the regulations it is working on or has recently finished. These are published in the *Federal Register*, usually in April or October, as the Unified Agenda of Federal and Regulatory and Deregulatory Actions.

Once a regulation is completed and has been printed in the FR as a final rule, it is codified and published in the *Code of Federal Regulations* (CFR). The

CFR is the official record of all regulations created by the federal government. It is divided into fifty volumes, called titles, each with a specific focus. Almost all environmental regulations appear in Title 40 CFR. The CFR is revised yearly with a quarter of the volumes updated every three months. Title 40 is revised every July 1 (source: U.S. EPA website).

Bureau of National Affairs (BNA) <http://www.bna.com> (accessed May 18, 2005). BNA is the foremost publisher of print and electronic news, analysis, and reference products, providing intensive coverage of legal and regulatory developments for professionals in business and government. BNA's practical and thorough professional tools make it easier to track international, federal, and state requirements, and help you assess the impact of developments on your company or clients. Review the wealth of materials BNA has in its environment, health and safety library at <http://www.bna.com/products/ens/index.html> (accessed May 18, 2005).

Code of Federal Regulations (CFR) Washington, DC: General Services Administration, National Archives and Records Service, Office of the Federal Register. Also online at <http://www.gpoaccess.gov/cfr/index.html> (accessed May 18, 2005).

Government Printing Office (GPO) Access <http://www.gpoaccess.gov/> (accessed May 18, 2005). Service of the U.S. Government Printing Office that provides free electronic access to information products produced by the federal government. Included are congressional hearings, committee print and directory information, federal regulations, public laws, the Congressional Record, and many more titles.

Federal Register – Government Printing Office, Washington, DC: Office of the Federal Register, National Archives and Records Service, General Services Administration: Supt. of Docs., U.S. G.P.O., distributor. Available online at <http://www.gpoaccess.gov/fr/> (accessed May 18, 2005). Published each federal work day. Includes proposed and final regulations by federal regulatory agencies.

Search engines, important websites, portals, discussion lists, weblogs

E-Print Network – Research Communications for Scientists and Engineers. Office of Scientific and Technical Information, Department of Energy, 2004 <http://www.osti.gov/eprints/> (accessed May 18, 2005). Provides a searchable gateway to thousands of science and technology websites and databases worldwide and is a unique Deep Web search through scientific and technical e-prints and preprints generated by researchers. Subject areas of interest to this audience include Biology and medicine; Chemistry; Energy storage, conversion and utilization; Engineering; Environmental management and restoration technologies; Environmental sciences and ecology; Fossil fuels; Geosciences; Power transmission, distribution and plants; and Renewable energy. It is possible to browse or search by these subject headings.

National Environmental Methods Index (NEMI) by the U.S. Geological Survey is a free, web-based online clearinghouse of environmental monitoring methods <http://www.nemi.gov> (accessed May 18, 2005). The NEMI database contains chemical, microbiological, and radiochemical method summaries of laboratory and field protocols for regulatory and non-regulatory water quality analyses. In future, NEMI will be expanded to meet the needs of the monitoring community.

The tool also allows monitoring data to be shared among different agencies and organizations that use different methods at different times. This database was developed in conjunction with the U.S. Environmental Protection Agency (USEPA), and other partners in the federal, state, and private sectors.

SciTechResources.gov <http://www.scitech.gov/> – Environment and Environmental Quality. A companion website to Science.gov.

Air environment

Global Change Research Information Office (GCRIO) <http://www.gcrio.org/> The U.S. GCRIO provides access to data and information on climate change research, adaptation/mitigation strategies and technologies, and global change-related educational resources on behalf of the various U.S. federal agencies that are involved in the U.S. Global Change Research Program (USGCRP).

Water environment

StormwaterAuthority.org <www.stormwaterauthority.org> (accessed May 18, 2005). This website is the first comprehensive Internet resource for all aspects of the storm water industry – covering news, events, state regulations, education, and more. The mission of StormwaterAuthority.org is to assist professionals in making educated and environmentally sound storm water decisions.

WaterWiser – The Water Efficiency Clearinghouse – American Water Works Association <http://www.awwa.org/waterwiser/> (accessed May 18, 2005). WaterWiser is the premier water efficiency and water conservation online information resource.

Government agencies

There are several federal agencies that regulate different aspects of the environment and are essential to the field of environmental engineering. These agencies are:

U.S. Army Corps of Engineers <http://www.usace.army.mil/>
U.S. Department of Energy <http://www.doe.gov/>
U.S. Department of the Interior <http://www.doi.gov/>
U.S. Bureau of Reclamation <http://www.usbr.gov/>
U.S. Fish and Wildlife Service <http://www.fws.gov/>
U.S. Geological Survey <http://www.usgs.gov>
U.S. Environmental Protection Agency <http://www.epa.gov>
U.S. National Oceanic and Atmospheric Administration <http://www.noaa.gov/> (all accessed May 18, 2005)

The websites for these key agencies contain a wealth of information. However, it is important to point out some essential elements of the individual websites and specific work of the agencies.

U.S. Army Corps of Engineers (USACE)

The United States Army Corps of Engineers <http://www.usace.army.mil/> is made up of approximately 34,600 civilian and 650 military men and women and is headquartered in Washington, DC. The Corps staff provides engineering services that include planning, designing, building, and operating water resources and other civil works projects (e.g., navigation, flood control, environmental protection). They also design and manage the construction of military facilities for the Army and Air Force, and provide design and construction management support for other defense and federal agencies.

The Corps has eight regional divisions in the United States and forty-one subordinate districts throughout the U.S., Asia, and Europe. The Engineer Research and Development Center (ERDC) <http://www.erdc.usace.army.mil/> consists of eight laboratories with each conducting specialized research. These laboratories include the Waterways Experiment Station (WES), Coastal and Hydraulics, Cold Regions Research and Engineering, Construction Engineering, Geotechnical and Structures, Environmental, Information Technology, and Topographical Engineering.

Another organization of interest to environmental engineers within the Corps is the Institute for Water Resources (IWR) <http://www.iwr.usace.army.mil/>. The IWR develops and applies new planning evaluation methods, policies, and data in anticipation of changing water resources management conditions.

For publications search <http://www.usace.army.mil/search.html> (all accessed May 18, 2005).

U.S. Department of Energy (DOE)

The DOE was established and effective from October 1, 1977 and consolidated the federal energy functions into a cabinet-level department. The DOE's mission is to advance the national, economic, and energy security of the United States, to promote scientific and technological innovation, and to ensure the environmental clean-up of the national nuclear weapons complex. Several programs that are part of the DOE are the Energy Information Administration, National Nuclear Security Administration, National Laboratories and Technology Centers, Power Marketing Administration, and Operations Offices and Field Organizations.

U.S. Department of the Interior (DOI)

The Department of the Interior manages public lands and mineral resources, national parks, national wildlife refuges, Western water resources, surface-mined lands, and upholds federal trust responsibilities to Indian tribes. The DOI is a large agency with extensive responsibilities essential to environmental engineers. Of particular importance are the Bureau of Reclamation, U.S. Fish and Wildlife Service, and the US. Geological Survey.

U.S. Bureau of Reclamation (USBR)

The Reclamation Service was established in 1902 to manage, develop, and protect water and related resources in an environmentally and economically sound manner. It is best known for the dams, power plants, and canals it constructed in seventeen Western states to secure a year-round water supply for irrigation. These projects led to homesteading and promoted the economic development of the West.

The Reclamation Service was initially created within the US. Geological Survey, but in 1907 the Reclamation Service was separated from USGS and in 1923 was renamed the Bureau of Reclamation. The Bureau is made up of five regions with project facilities that include over 600 reservoirs and dams, thousands of miles of canals and other distribution facilities, and fifty-two hydroelectric plants.

For publications search <http://www.usbr.gov/library/>.

U.S. Fish and Wildlife Service (FWS)

The FWS employs approximately 7,500 people at facilities across the U.S. It is a decentralized organization with a headquarters office in Washington, DC, seven geographic regional offices, and nearly 700 field units. The Service manages the ninety-three-million-acre National Wildlife Refuge System and thousands of small wetlands and other special management areas. It also operates sixty-six National Fish Hatcheries, sixty-four fishery resource offices and seventy-eight ecological services field stations.

The FWS enforces federal wildlife laws, protects and manages migratory birds, restores nationally significant fisheries, conserves and restores wildlife habitat such as wetlands, and helps foreign governments.

For publications search <http://library.fws.gov/>.

U.S. Geological Survey (USGS)

The USGS was established in 1879. The USGS serves the Nation by providing reliable scientific information to describe and understand the Earth; minimize loss of life and property from natural disasters; manage water, biological, energy, and mineral resources; and enhance and protect our quality of life. To attain these objectives, the USGS prepares maps, collects and interprets data on energy and mineral resources; conducts nationwide assessments of quality, quantity, and use of the nation's water resources; performs fundamental and applied research in the sciences and techniques involved; and publishes their investigations through maps and a variety of technical publications.

The USGS employs 10,000 staff and maintains an extensive network of regional offices and research centers. The Water Resources Division <http://water.usgs.gov> is particularly relevant to environmental engineers. The National Water Information System (NWIS) web page contains historical and real-time data on surface water, ground water, and water quality. The USGS manages water information at offices located throughout the U.S. Although all offices are tied together through a national network, each office collects data and conducts studies in a particular area. Local information is best found at the following sites: Cooperative Water Program, National Streamflow Information Program (NSIP), National Water Quality Assessment Program (NAWQA), Toxic Substances Hydrology (TOXICS) Program, Ground Water Resources Program, Hydrologic Research and Development, State Water Resources Research Institute Programs. The agency also conducts international research.

For publications search <http://infotrek.er.usgs.gov/pubs/>.

U.S. Environmental Protection Agency (EPA)

The mission of the EPA is to protect human health and the environment. The EPA employs 18,000 people across the country at its headquarters offices in Washington, DC, ten regional offices, and more than a dozen laboratories.

There are more than a dozen statutes and laws that form the legal basis for the programs that the EPA administers. A complete list may be found on the website. The following list contains the laws that are most applicable to the environmental engineering community:

Clean Air Act (CAA)
Clean Water Act (CWA)
Comprehensive Environmental Response, Compensation, and Liability Act (CERCLA or Superfund)
Emergency Planning and Community Right-to-know Act (EPCRA)
Endangered Species Act (ESA)
Federal Insecticide, Fungicide, and Rodenticide Act (FIFRA)
National Environmental Policy Act of 1969 (NEPA)
Oil Pollution Act of 1990 (OPA)
Pollution Prevention Act (PPA)
Resource Conservation and Recovery Act (RCRA)
Safe Drinking Water Act (SDWA)
Solid Waste Disposal Act
Superfund Amendments and Reauthorization Act (SARA)
Toxic Substances Control Act (TSCA)

The EPA operates extensive information networks. Some of the most important are listed below:

Hotlines and clearinghouses
Technology transfer networks
Dockets
Libraries

Publications
Databases and software

Visit the Web pages of the following EPA program offices:

Air and Radiation
Compliance and Enforcement
Emergency Response
Hazardous Waste Program
Indoor Air – Radon
National Center for Environmental Assessment (NCEA)
National Center for Environmental Research (NCER)
National Estuary Program
Persistent Bioaccumulative and Toxic (PBT) Chemical Program
Research and Development
Science Advisory Board
Solid Waste and Emergency Response
Superfund Information
Underground Injection Control (UIC) Program
Underground Storage Tanks (UST)
Water: aquatic ecosystems, drinking water, ground water, storm water, surface water, waste-water, water pollution, water quality monitoring
Watersheds
Wetlands Program

The US Environmental Protection Agency's website is an essential resource for anyone dealing with environmental issues. Given its scope and size, it can be challenging to identify all the pertinent information on this site. In early 2005, the EPA initiated the Web Ambassadors Program <http://www.epa.gov/ambassad/about_web_ambassadors.htm> (accessed May 18, 2005) to address the challenge. Its first target audience is librarians, for whom it has developed a tool kit, training sessions, and other support. See <http://www.epa.gov/ambassad/> (accessed May 18, 2005) for more information and watch for the addition of similar materials for other target audiences.

U.S. National Oceanic and Atmospheric Administration (NOAA)

Although NOAA began in 1970, it was formed from agencies that are among the oldest in the federal government. The agencies included the United States Coast and Geodetic Survey formed in 1807, the Weather Bureau formed in 1870, and the Bureau of Commercial Fisheries formed in 1871. Individually, these organizations were America's first physical science agency, America's first agency dedicated specifically to the atmospheric sciences, and America's first conservation agency. Much of America's scientific heritage resides in these agencies. They brought their cultures of scientific accuracy and precision, stewardship of resources, and protection of life and property to the newly formed agency.

NOAA researches and gathers data on weather, oceans, satellites, fisheries, climate, coasts, and charting and navigation. NOAA is made up of six divisions: National Weather Service (NWS), National Environmental Satellite, Data, and Information Service (NESDIS), National Marine Fisheries Service (NMFS), National Ocean Service (NOS), Office of Oceanic and Atmospheric (OAR), and Marine and Aviation Operations (NMAO).

For publications search <http://www.lib.noaa.gov/>.

Associations, organizations, societies, and conferences

Most of the entities listed here alphabetically offer annual meetings, education opportunities, publications, buyer's guides, job listings, and membership directories. More recently, some have begun to offer online discussion groups, electronic mailing lists, and virtual seminars.

Air and Waste Management Association (A&WMA) <http://www.awma. org> (accessed May 18, 2005). An environmental, educational, and technical organization, A&WMA seeks to provide a neutral forum for the exchange of technical information on a wide variety of environmental topics. Committees and divisions include Air; Environmental management, and Waste. Formerly Smoke Prevention Association of America; (1987) Air Pollution Control Association (APCA) (1989).

Special events or services: Annual conference and exhibition, with the International Urban Air Quality Forum as a satellite event.

American Academy of Environmental Engineers (AAEE) <http://www.aaee.net> (accessed May 18, 2005). Environmentally oriented registered professional engineers are certified by examination as Diplomats of the Academy. Purposes are: to improve the standards of environmental engineering; to certify those with special knowledge of environmental engineering; and to furnish lists of those certified to the public. The AAEE works with other professional organizations on environmentally oriented activities.

Special events or services: Certification program.

American Association for Aerosol Research (AAAR) <http://www.aaar.org/> (accessed May 18, 2005). AAAR promotes and communicates technical advances in the field of aerosol research.

Special events or services: Annual conference.

American Chemical Society (ACS) <http://www.chemistry.org/portal/a/c/s/1/home.html> (accessed May 18, 2005). ACS is a self-governed individual membership organization that consists of more than 159,000 members at all degree levels and in all fields of chemistry. The Division of Environmental Chemistry sponsors an annual specialty meeting on Green Chemistry and Engineering.

Special events or services: Annual national meeting and regional meetings.

American Conference of Governmental Industrial Hygienists (ACGIH) <http://www.acgih.org/> (accessed May 18, 2005). The best known of ACGIH's activities is the Threshold Limit Values for Chemical Substances (TLV®-CS). Today, ACGIH has eleven committees focusing on a range of topics that include air sampling instruments, bioaerosols, exposure indices, and industrial ventilation.

Special events or services: Partners with Society of Toxicology, AIHA, and AIHce in their conferences.

American Industrial Hygiene Association (AIHA) – <http://www.aiha.org/> (accessed May 18, 2005). Serving the needs of occupational and environmental health professionals practicing industrial hygiene, AIHA's more than thirty technical committees deal with such concerns as exposure and risk assessment strategies, indoor environmental quality, workplace environmental exposure levels, noise hazards, and respiratory protection.

Special events or services: AIHce annual conference, co-sponsored by AIHA and ACGIH. Website offers an electronic discussion list.

American Membrane Technology Association (AMTA) <http://www.membranes-amta.org> (accessed May 18, 2005). Water supply agencies, manufacturers of desalting equipment, design and construction companies, advanced water sciences and technologies consultants, individuals, academicians, and librarians form the membership. The AMTA advances research and development programs in desalination, waste-water reclamation, and other water sciences; promotes programs of water supply and urban environment improvement; and sponsors training of water treatment plant operators. Formerly National Water Supply Improvement Association (1993); American Desalting Association (2003).

Special events or services: Various technology conferences; website hosts a message board.

American Water Resources Association (AWRA) <http://www.awra.org> (accessed May 18, 2005). Membership encompasses engineers; natural, physical, and social scientists; other persons engaged in any aspect of the field of water resources; business concerns and other organizations. AWRA seeks to advance water resources research, planning, development, and management.

Special events or services: Annual conference. Website offers a virtual mentoring program.

American Water Works Association (AWWA) <http://www.awwa.org> (accessed May 18, 2005). Membership is drawn from water utility managers, superintendents, engineers, chemists, bacteriologists, and other individuals interested in public water supply; municipal- and investor-owned water departments; boards of health; manufacturers of waterworks equipment; government officials and consultants. AWWA develops standards and supports research programs in waterworks design, construction, operation, and management, conducts in-service training schools and prepares manuals for waterworks personnel. The database, Waternet, is available on Dialog as file 245 and on CD-ROM. Affiliated with the Water Environment Federation.

Special events or services: Annual conference and exposition as well as numerous regional conferences and workshops. Electronic discussion forums available through website.

Association of Engineering Geologists (AEG) <http://www.aegweb.org/indexf.html> (accessed May 18, 2005). Serving professionals in ground water, environmental, and engineering geology, the mission of AEG is to provide leadership in the development and application of geologic principles and knowledge to serve engineering, environmental, and public needs.

Special events or services: Annual meeting.

Association of Environmental Engineering and Science Professors (AEESP) <http://www.aeesp.org> (accessed May 18, 2005). The Association works to improve education and research programs in the science and technology of environmental protection. It provides information to government agencies and the public, encourages graduate education by supporting research and training for students, and maintains a speaker's bureau. Formerly Association of Environmental Engineering Professors (1999).

Special events or services: Annual conference.

Association of Ground Water Scientists and Engineers (AGWSE), a division of the National Groundwater Association <http://www.ngwa.org/membership/agbenft.html> (accessed May 18, 2005). A technical division of the National Ground Water Association, members are hydrogeologists, geologists, hydrologists, civil and environmental engineers, geochemists, biologists, and scientists in related fields. The Association seeks to provide leadership and guidance for scientific, economical, and beneficial ground water development and to promote the use, protection, and management of the world's ground water resources. Formerly known as the Ground Water Technology Division of the National Water Well Association prior to 1985.

Special events or services: Specialty conferences and expositions.

Association of State and Interstate Water Pollution Control Administrators (ASIWPCA) <http://www.asiwpca.org> (accessed May 18, 2005). With a membership of administrators of state and interstate governmental agencies legally responsible for the prevention, abatement, and control of water pollution, the Association promotes coordination among state agency programs and those of the Environmental Protection Agency, Congress, and other federal agencies.

Special events or services: Annual and mid-year meetings.

Association of State Drinking Water Administrators (ASDWA) <http://www.asdwa.org> (accessed May 18, 2005). Comprises managers of state and territorial drinking-water programs and state regulatory personnel. The ASDWA works to meet communication and coordination needs of state drinking-water program managers; facilitates the exchange of information and experience among state drinking-water agents; acts as a collective voice for the protection of public health through assurance of high-quality drinking-water; and oversees the implementation of the Safe Drinking Water Act.

Special events or services: Annual conference.

Environmental and Engineering Geophysical Society (EEGS) <http://www.eegs.org/> (accessed May 18, 2005). This applied scientific organization's mission is to promote the science of geophysics, especially as it is applied to environmental and engineering problems and to foster common scientific interests of geophysicists and their colleagues in other related sciences and engineering.

Special events or services: Annual meeting.

Institute of Environmental Science and Technology (IEST) <http://www.iest.org> (accessed May 18, 2005). IEST, an ANSI-accredited developer of American National Standards, focuses on education and the development of recommended practices and standards. Absorbed the American Association for Contamination Control (1973). Formed by merger of the Institute of Environmental Engineers and Society of Environmental Engineers (1994). Formerly Institute of Environmental Sciences (1999).

Special events or services: ESTECH, the IEST annual technical meeting and exposition. Website offers a conferencing and chat service.

Institute of Noise Control Engineering of the USA (INCE-USA) <http://www.inceusa.org/> (accessed May 18, 2005). Promoting engineering solutions to environmental noise problems, INCE/USA is a member society of the International Institute of Noise Control Engineering, an international consortium of organizations with interests in acoustics and noise control.

Special events or services: National conference, which often includes the Meeting of the Acoustical Society of America.

International Environmental Modelling and Software Society (iEMSs) <http://www.iemss.org/> (accessed May 18, 2005). Dealing with environmental modeling, software, and related topics, the aims of the iEMSs include the development and use of environmental modeling and software tools to advance the science and improve decision-making with respect to resource and environmental issues. iEMSs was founded in 2000.

Special events or services: Biennial general meeting usually held on even years.

International Institute of Noise Control Engineering (I-INCE) <http://www.i-ince.org/> (accessed May 18, 2005). This is a worldwide consortium of organizations concerned with noise control, acoustics, and vibration. The primary focus of the Institute is on unwanted sounds and on vibrations producing such sounds when transduced.

Special events or services: Annual congress.

International Society of Exposure Analysis (ISEA) <http://www.iseaweb.org/> (accessed May 18, 2005). Established in 1989 to foster and advance the science of exposure analysis related to environmental contaminants, both for human populations and ecosystems, the membership promotes communication among all disciplines involved in exposure analysis, recommends exposure analysis approaches to address substantive or methodological concerns, and works to strengthen the impact of exposure assessment on environmental policy.

Special events or services: Annual conference.

International Society of Indoor Air Quality and Climate (ISIAQ) <www.ie.dtu.dk/isiaq/default.asp> (accessed May 18, 2005). An international scientific organization, the purpose of ISIAQ is to support the creation of healthy, comfortable, and productive indoor environments by advancing the science and technology of indoor air quality and climate as it relates to indoor environment design, construction, operation and maintenance, air quality measurement, and health sciences.

Special events or services: Annual conference.

International Water Association (IWA) <http://www.iawq.org.uk/template.cfm?name=home> (accessed May 18, 2005). Integrating the leading edge of professional thought on research and practice across the drinking-water, waste-water and storm water disciplines, the IWA was founded in September 1999 by the merger of the International Association of Water Quality (IAWQ) and the International Water Supply Association (IWSA).

Special events or services: Biennial World Water Congress plus regional and specialty conferences. Website offers a members-only Extranet.

National Association of Environmental Professionals (NAEP) <http://www.naep.org/> (accessed May 18, 2005). NAEP is a multi-disciplinary, professional association dedicated to the promotion of ethical practices, technical competency, and professional standards in the environmental fields. Members have access to the most recent developments in environmental practices, research, technology, law, and policy.

Special events or services: Annual conference. Professional certification program.

National Ground Water Association (NGWA) <http://www.ngwa.org/> and <http://www.wellowner.org> (accessed May 18, 2005). With a membership of ground water geologists and hydrologists, engineers, ground water contractors, manufacturers, and suppliers of ground water-related products and services, the Association's purpose is to provide guidance to members, government representatives, and the public for sound scientific, economic, and beneficial development, protection, and management of the world's ground water resources. (Formerly the National Water Well Association.)

Special events or services: Annual exposition. Professional certification programs.

Society of Environmental Toxicology and Chemistry (SETAC) <http://www.setac.org/> (accessed May 18, 2005). A non-profit, worldwide professional society, SETAC promotes the advancement and application of scientific research related to contaminants and other stressors in the environment, education in the environmental sciences, and the use of science in environmental policy and decision-making.

Special events or services: SETAC North America Annual Meeting, an annual World Congress, other international meetings.

State and Territorial Air Pollution Program Administrators (STAPPA)/Association of Local Air Pollution Control Officials (ALAPCO) <http://www.cleanairworld.org/> (accessed May 18, 2005). STAPPA and ALAPCO are the two national associations representing air pollution control agencies in fifty-four states and territories and over 165 major metropolitan areas across the United States. Formed over thirty years ago, their purpose is to improve their effectiveness as managers of air quality programs through the exchange of information among air pollution control officials, communication and cooperation among federal, state and local regulatory agencies, and management of air resources.

Water Environment Federation (WEF) <http://www.wef.org> (accessed May 18, 2005). The WEF seeks to advance fundamental and practical knowledge concerning the nature, collection, treatment, and disposal of domestic and industrial waste-waters, and the design, construction, operation, and management of facilities for these purposes. Formerly Federation of Sewage Works Associations (1949); Federation of Sewage and Industrial Wastes Associations (1959); Water Pollution Control Federation (1991).

Special events or services: WEFTEC, the annual technical exhibition and conference.

Conclusion

Environmental engineering is a diverse and challenging discipline. Environmental engineers are learning more about our environment every day and they are finding solutions to many complex problems. However, as test methods and instrumentation become more sophisticated, environmental engineers are discovering more pollutants and realizing that the number of unanswered questions continues to grow.

The information resources recommended in this chapter are starting points to help penetrate the increasing volume and complexity of knowledge about environmental engineering as a means of managing human impact on the environment. With its broad range of specialties and subspecialties – air pollution control, industrial hygiene, radiation protection, hazardous waste management, toxic materials control, water supply, waste-water management, storm water management, solid waste disposal, public health, and land management – keeping up-to-date in one's area of expertise and aware of the bigger picture is becoming increasingly difficult.

We encourage those teaching, studying, and working in environmental engineering to further the scientific knowledge in support of good stewardship of the planet.

Acknowledgments

I would like to acknowledge my co-author, Lois Widmer, for her assistance and encouragement in writing this chapter. I would also like to acknowledge my colleagues on the UC Berkeley campus, and Deborra Samuels at the U.S. Environmental Protection Agency Library in San Francisco for her support. We would like to thank all the ChoiceReviews reviewers. Their comments certainly helped in our decision-making (LV).

In addition to thanking Linda Vida for the opportunity to work on this project, I want to thank colleagues, known and unknown, for the constant exchange of valuable information on many listservs and at workshops and conferences. Serendipity remains a valid part of information gathering (LW).

References

Balachandran, S. (ed.) (1993) *Encyclopedia of Environmental Information Sources: A Subject Guide to about 34,000 Print and Other Sources of Information on All Aspects of the Environment*, Detroit: Gale Research.

Lord, C.R. (2000) *Guide to Information Resources in Engineering*, Englewood, CO: Libraries Unlimited.

Nazaroff, W.W. and Alvarez-Cohen, L. (2001) *Environmental Engineering Science*, New York: John Wiley.

ChoiceReviews online, American Library Association <http://www.choiceonsite.org>.

13 History of engineering

Nestor L. Osorio and Mary A. Osorio

Introduction, history, and scope of discipline

According to the *Oxford English Dictionary* (1989) one of the first usages of the word "engineering" in the English language dated back to 1681; it also known that in 1729 Bernard Fores de Belidor published one of the first books with the word "engineering" in the title: *La Science de Ingenieures*. These are two literary indications that during the Industrial Revolution engineering began to solidify as a distinct profession. There have been continuing discussions among experts about the differences and the definitions of science, technology, and engineering. The many sources presented in this chapter may help those readers interested in these types of arguments but the focus of this chapter is far from presenting a critical review of their definitions and relationships. Nevertheless, it is important to raise some of the basic concepts related to the history of engineering: it is appropriate to bring up the connection between applied sciences and engineering; to make a distinction about the meaning of technology and engineering; and to present a brief chronology of engineering in modern times.

To better understand the state of engineering history today it is necessary to look back at its origins. If one were to choose one magnificent human creation that signified the ingenuity of early engineering wonders it would be the pyramids of Egypt. The Golden Age of Egyptian pyramid-building occurred between 2700 and 2550 BC. At this time, and without the use of the wheel, about eleven million cubic meters of stone were transported (Garrison, 1999). Early civilizations in Mesopotamia, Greece, Rome, India, and China – to mention a few – show highly sophisticated activities that used resources for the production of goods. Progress continued through the years but a new, invigorated era of technological advances and scientific knowledge began during the Renaissance. Domenico Fontana is an example of an engineer of this era. He: "1. Mastered difficult technical problems. 2. Designed road and hydraulic work. 3. Worked at practical building construction. 4. Had knowledge of mathematics, geometry, and the deliberate use of calculation" (Garrison, 1999). Some of the remarkable things that happened prior to the development of steam power include new techniques for working with raw materials, the use of new agriculture techniques, the extraction of chemical products, the beginning of the textile industry, the first stage of industrial mechanization, the mastering of clocks construction, and the improvement of land and sea transportation (Daumas, 1969–1979).

The relationship between engineering and applied science is readily noted by observing the number of colleges that include these two disciplines in their name. Several engineering specialties have evolved from the sciences; for example, chemical engineering, from chemistry, metallurgical engineering from geology, chemistry and physics and electrical engineering from mathematics and physics. Scientific knowledge can attain practical applications that may be used by engineers in the development of goods and services that produce usable and economical benefits to society. Because of this strong connection between applied sciences and engineering, the study of the history of science overlaps with the study of the history of technology. In this chapter science resources that are part of this overlap have been presented.

As in the case of applied sciences there is not a real clear line of demarcation between engineering and technology. According to the following definitions, engineering deals with the use of resources and the production of goods and services; while technology deals with the tools and techniques used to achieve those goals. This is how the *McGraw-Hill Encyclopedia of Science and Technology* (2002) defines engineering:

> Most simply, the art of directing the great sources of power in nature for the use and the convenience of humans. In its modern form engineering involves people, money, materials, machines, and energy. It is differentiated from science because it is primarily concerned with how to direct to useful and economical ends the natural phenomena that scientists discover and formulate into acceptable theories. Engineering therefore requires above all the creative imagination to innovate useful applications of natural phenomena. It is always dissatisfied with present methods and equipment. It seeks newer, cheaper, better means of using natural sources of energy and materials to improve the standard of living and to diminish toil.

The same source defines technology as follows:

> Systematic knowledge and action, usually of industrial processes but applicable to any recurrent activity. Technology is closely related to science and to engineering. Science deals with humans' understanding of the real world about them – the inherent properties of space, matter, energy, and their interactions. Engineering is the application of objective knowledge to the creation of plans, designs, and means for achieving desired objectives. Technology deals with the tools and techniques for carrying out the plans.

The terms "engineering" and "technology" are in many cases inclusive, interchangeable, and sometimes misused. That is the reason why a good number of the sources listed in this chapter include the term "technology." An effort has been made to include only those sources that have an engineering content.

According to Auyang (2004) there are three overlapping phases in the history of modern engineering: the Industrial Revolution, the second Industrial Revolution, and the Information Revolution.

The Industrial Revolution, which includes the eighteenth and early nineteenth centuries, is characterized by the changes that occurred in civil, mechanical, mining, and metallurgical engineering, when they went from being practical arts to becoming scientific professions. This is also the time when the first engineering schools are founded. Some of the most significant engineering advances of this era are the steam-engines, the improvements made to water power, the beginning of the study of the strength of materials, the development of the iron industry, and the use of cement as a building material (Kirby *et al.*, 1956).

The second Industrial Revolution occurred the century before World War II (Auyang, 2004); it is characterized by the improvement of steel production, the manufacturing of the rapid-fire gun, the shift from steam power to the internal combustion engine, the further development of metallurgy, the generation of electric power, the filament lamp, the transmission of sound over long distances, and the beginning of radio communication, radar, and television. Later in this era the aircraft industry, the system of mass production, and the automobile were also created. (Engineering and Technology, 2005).

World War II serves as a catalyst to many innovations that highlight the Information Revolution. It may be summarized by listing some of the major engineering accomplishments: space travel, atomic power, electronic computers, and the advent of the microchip. In this last era, the capacity to create, store, and exchange data in its multiple forms has propelled industry, commerce, the production of services and products, and is present today in all aspects of human endeavor.

The future of engineering as a profession is assured, since it will continue to have a significant impact on the transformation of society. In writing about the future of engineering, Bell and Dooling (2000) have stated: "Some of the challenges will require a purely technical approach. Others will be primarily societal, requiring a balancing of material priorities with sociological values – with the outcome expressed by funding availability and regulations. All involve engineering techniques and expertise. And never before have humans been so technically well-prepared." The history of engineering is the story of how engineers and their organizations have brought about changes throughout society through the ages.

Searching the library catalog, keywords, LC Subject Headings, LC call numbers

Searching the online catalog is an effective way of finding documents about the history of engineering. Online catalogs of libraries in the United States, in Europe, in Japan, and in most countries of the world do not just list books. Their databases include data about libraries' holdings in many other formats such as journals, films, audiovisuals, federal and state government publications, reports, and dissertations. To find non-cataloged documents held by special collections in libraries and museums the best option is to get direct information from their curators. There are several ways of searching for documents in an online catalog, as will be shown through the following examples.

This is a list of Subject Headings used by the Library of Congress (LC) of the United States. The first group includes some of the most commonly used terms in finding materials in major general subject areas relating to the history of engineering:

Engineering – History
Technology – History
Science – History
Inventions – History
Technology and civilization

This second group of terms shows examples of how to find information in more specific areas in the history of engineering:

Chemical engineering – History
Civil engineering – History
Electric engineering – History
Human engineering – History
Industrial engineering – History
Mechanical engineering – History
Petroleum engineering – History
Women in technology – History

Specific areas of engineering may also be limited to geographical locations, as indicated in the following examples:

Aerospace engineering – United States – History
Electric industry workers – Great Britain – History
Electronics – United States – History
Engineering – Great Britain – History
Industrial management – United States – History
Mass production – United States – History
Technology – Social aspects – United States – History
Telecommunication – United States – History

The names of relevant people in engineering may be searched in the subject field by using their last name followed by the first name. For example:

Alexanderson, Ernst Fredrik Werner, 1878–1975

Jervis, John B. (John Bloomfield), 1795–1885
Coulomb, Charles Augustin de

The name of corporations or institutions may also be searched in the subject field, for example:

H.H. Franklin Manufacturing Company – History
American Telephone and Telegraph Company – History

The following subject terms show how to find biographical documents about a person or a group of persons:

Civil engineers – United States – Biography
Electric engineers – United States – Biography
Electric engineers
Inventors – Biography – Dictionaries

Finally, readers could also limit their search to a specific period of time as is indicated below:

Civil engineering – Early works to 1850
Engineering – Europe – History – To 1500
Statics – Early works to 1800
Technology – History – 20th century
Technology – United States – History – 19th century

Browsing the shelves is another way of finding documents. These are some of the areas in the LC classification system where readers may find materials:

T14.5 Technology – Social aspects – History
T15–T40 Technology – History. It includes: general works, ancient, medieval, modern, nineteenth to twentieth centuries, special counties, United States, Canada, Latin America, South America, Europe, Asia, Africa, Australia, Pacific Islands, Arctic regions, lost arts, historical atlases, industrial archaeology, and biographies.
T144–145 Technology – General works, prior to nineteenth century, nineteenth century, and later

Similar coverage in the 15–145 range exists for the following engineering areas:

TA15–T145 General engineering and civil engineering
TC15–T145 Hydraulic engineering
TD15–T145 Environmental technology. Sanitary engineering
TE15–T145 Highway engineering. Roads and pavements
TF15–T145 Railroad engineering and operation

TG15–T145 Bridge engineering
TH15–T145 Building construction
TJ15–T145 Mechanical engineering and machinery
TK15–T145 Electrical engineering. Electronics. Nuclear engineering
TL15–T145 Motor vehicles. Aeronautics. Astronautics
TN15–T145 Mining engineering. Metallurgy
TP15–T145 Chemical technology
TS15–T145 Manufactures

In addition, materials on the history of engineering may be found throughout the entire T subject classification of the LC system. For example: history of automation control in TJ213; history of petroleum engineering in TN871; and history of mechanical drawing in T353.

Furthermore, since there is a close relationship between the applied sciences and engineering, other works on the history of engineering may be found in these LC sections: mathematics (QA); physics (QC); chemistry (QD); geology (QE); as well as in psychology (BF); and industrial management (HF). In this case, and to identify specific areas, it is better first to search the online catalog by using keywords or subject terms.

The Library of Congress classification system is the most widely used system in North America. It is advisable that readers in other countries follow the classification and Subject Heading guidelines of their native system.

Indexes and abstracts

Indexing and abstracting databases are excellent sources for finding contemporary publications on the history of engineering. They index articles in journals and in conference proceedings which represent a major portion of the publications on this subject. Indexes and abstracts also list reviews of books, conference proceedings, dissertations, and reports. To search for older materials it is advisable to consult a librarian.

America: History and Life (1964–) Santa Barbara, CA: ABC-CLIO. This is a basic but comprehensive bibliographic resource for the study of U.S. and Canadian history. There are two parts to this work: Part A, Article abstracts and citations, and Part B, Index to book reviews and Doctoral dissertations. Entries are listed alphabetically by author; entries without authors are listed alphabetically by title at the end of each classification section.

Bibliografia Italiana di Storia della Scienza (1982–1998) Florence, Italy: L.S. Olschki. Produced by the staff of the Istituto e Museo di Storia della Scienza (IMSS), Florence, Italy. This bibliography includes mainly books, articles in periodicals, book chapters, and book reviews primarily in Italian. Paper issues of this publication were last published in 1998. Since 1999 the bibliography is available on the Web and searchable from the IMSS library catalog search engine (http://biblioteca.imss.fi.it/LV2_1bin/LibriVision) (cited December 16, 2004). It has been incorporated into the *History of Science, Technology, and Medicine* database.

British Humanities Index (1962–) London: Library Association Publishing. This is an abstract and index database that covers around 400 British journals and newspapers. It includes citations on the history of technology within the coverage of philosophy, economics, politics, history, and society. Major obituaries of prominent figures are noted. It is arranged according to abstract number, followed by the subject index term and title.

Current Bibliography in the History of Technology (1964–2002) Baltimore, MD: The Johns Hopkins University Press. This publication is a supplement of the journal *Technology and Culture* of the Society for the History of Technology. The paper edition ceased in 2002. Since then it is published only in electronic format as part of the *History of Science, Technology, and Medicine* database. It is divided into seventeen subject classifications for each of the five defined historical time periods. It provides extensive coverage in all formats of the literature on the history of technology.

Francis (1991–) Paris, France: Institut de L'Information Scientifique et Technique. Since 1994 *Francis* is published in electronic format only. It covers articles in journals, conference proceedings, books, research reports, and French dissertations from 1972 to the present. It is an index for the humanities and social sciences including the history of science and technology. Prior to 1994 it was published in paper. The sections on the history of technology are known as: *Francis Bulletin Signalétique. 522, Histoire des Sciences et des Techniques*, 1991–1994; *Bulletin Signalétique. 522, Histoire des Sciences et des Techniques*, 1969–1990; and *Bulletin Signalétique. 22, Histoire des Sciences et des Techniques*, 1961–1968.

Historical Abstracts (1955–) Santa Barbara, CA: ABC-CLIO. This is considered to be the most comprehensive index of abstracts for articles, book reviews, essays, papers, conference proceedings, and dissertations in world history and the social sciences. The time frame of the topics in history abstracts starts from the 1450s and continues into the present. It covers 2,100 journals, book reviews, and dissertations.

History of Science, Technology, and Medicine (1976–) Washington, DC: Research Libraries Group. Available through RLIN, this database contains four bibliographies: *Isis Current Bibliography of the History of Science and its Cultural Influences* (1975–), *Current Bibliography in the History of Technology* (1987–), *Bibliografia Italiana di Storia della Scienza* (1982–), and the *Wellcome Bibliography for the History of Medicine* (1991–2004). This agglomeration makes this database the most comprehensive index to the literature on the history of engineering, technology, and related fields. It covers all formats.

Isis Current Bibliography of the History of Science and its Cultural Influences (1975–) Chicago, IL: University of Chicago Press. This is published as a supplement of the History of Science Society journal *Isis*. It indexes mainly books, articles in journals and symposia, dissertations, book chapters, and book reviews from all areas of the history of science. These data are being incorporated into the *History of Science, Technology, and Medicine* database. It has a journal list and is divided into twenty-four subject sections.

Private and institutional databases

The previous section has a selected list of indexing services that include mainly contemporary materials on this subject. In this section some examples of sources have been included where older materials can be found. In addition, numerous institutions such as museums and historical archives have their own databases. To localize materials from collections in most archives and museums it is necessary to search their own databases. Another option is to contact or visit them.

Bibliografía Española de Historia de la Ciencia y de la Técnica <http://161.111.141.93/hcien/> (accessed May 17, 2005). This database is maintained by the Unidad de Historia de la Ciencia del Instituto de Historia de la Ciencia y Documentación "López Piñero," Universidad de Valencia-CSIC. It covers all works about the history of science and technology in Spain since 1989.

The British Library – Manuscripts Catalogue <http://www.bl.uk/catalogues/manuscripts/> (accessed May 17, 2005). "Use this website to search the main catalogues of the British Library's collection of Western manuscripts, covering handwritten documents of all kinds from pre-Christian, Classical, medieval and modern times" (source: BL Web page).

National Cataloguing Unit for the Archives of Contemporary Scientists (NCUACS) <http://www.bath.ac.uk/ncuacs/> (accessed May 17, 2005). "The (NCUACS) was established in 1973 to locate, sort, index and catalogue the manuscript papers of distinguished contemporary British scientists and engineers. They include correspondence of all kinds, professional or technical documents such as laboratory notebooks, experimental drawings and calculations, lecture notes, engagement diaries and journals" (source: NCUACS Web page).

The History Journals Guide. Periodicals Directory <http://www.history-journals.de/journals/index.html>. This database is maintained by Stefan Blaschke. It includes nearly 6,000 serials titles from all periods, all regions, and all fields of history.

The Royal Historical Society Bibliography <http://www.rhs.ac.uk/bibwel.asp> (accessed May 17, 2005). "The Royal Historical Society bibliography, which is hosted by the Institute of Historical Research, is an authoritative guide to writing on British and Irish history from the Roman period to the present day. It contains over 350,000 entries including articles in journals and collective volumes, and including data from London's Past Online. You can search by author, by publication details, by subject or by period covered" (source: RHS Web page).

Bibliographies and guides to the literature

Ferguson's bibliography on the history of technology listed in this section is considered one of the foremost printed bibliographies in the field; it includes a good number of valuable sources for researchers in the history of engineering. Channell's works (also listed here) is one of the first bibliographies in the English language on the history of engineering, and both titles are highly recommended. Included are more recent bibliographies on contemporary topics such as gender, human factors, and computing. Two general guides on the history of engineering are included.

Bibliography for History of Engineering <http://www.iit.edu/~misa/biblios/hist_engineering.html> (accessed May 17, 2005). This Web bibliography is maintained by T.J. Misa of the Department of Humanities, Illinois Institute of Technology, Chicago, IL. Last updated in December of 2003, it contains a listing of books without annotations and is organized into two sections: Chronology and Topics; and Branches of Engineering (cited February 1, 2005).

Bindocci, C.G. (1993) *Women and Technology: an Annotated Bibliography*, New York: Garland. The purpose of this bibliography is to identify scholarly research about women and technology, an area of research that has been greatly overlooked through the years. The bibliography is meant to be the starting point for research by summarizing what others have already accomplished in this field. This work includes only secondary works in the English language published from 1979 to 1991 in the forms of articles, books, published conference proceedings, and dissertations.

Channell, D.F. (1989) *The History of Engineering Science: An Annotated Bibliography*, New York: Garland. The author of this work intended it to be a guide for students, scholars, and researchers into the history of engineering science, a field that can encompass both the history of science and the history of technology. Numerous entries may be overlooked by both groups of researchers when it comes to finding scholarly work in the history of engineering science. This comprehensive bibliography fills the void. Both primary and secondary materials are found.

Cortada, J.W. (1996) *Second Bibliographic Guide to the History of Computing, Computers, and the Information Processing Industry*, Westport, CT: Greenwood Press. This volume primarily emphasizes historically consequential published materials since the late 1980s. The titles are annotated and include both an author and subject index. Each citation is therefore numbered; all references are to item numbers instead of pages. It favors documenting hardware and industry literature. This bibliography is meant to be a reputable source for the student of the history of computing, computers, and information processing, and for all those who wish to gain valuable insights into the world of computer technology.

Dekosky, R.K. (1995) "Science, Technology, and Medicine," in: *The American Historical Association's Guide to Historical Literature* (3rd edn), ed. M.B. Norton and P. Gerardi, New York: Oxford University Press. This guide presents the highest quality and the most insightful books and articles from every field of historical scholarship. By providing a critical overview of the best historical scholarship it helps to regain the unifying or integral vision that can be lost in a time of many specialists and specialties. Section 4 in Volume 1 delves into the areas of science, technology, and medicine. This work contains a list of journals, an author index, and a subject index.

Ferguson, E.S. (1968) *Bibliography of the History of Technology*, Cambridge, MA: Society for the History of Technology. The purpose of this bibliography is to gather all relevant information for a comprehensive introduction to primary and secondary sources in the history of technology. It is geared to the student, sending him to the tools and resources of the scholar. However, all those who have an interest in the subject profit greatly through its use. Not only is the what and how of technology probed but the why of technology is also studied, since it is essential to the understanding of the culture and its implications for technology.

Fritze, R.H., Coutts, B.E., and Vyhnanek, L. (2004) *Reference Sources in History: An Introductory Guide* (2nd edn), Santa Barbara, CA: ABC-CLIO. This is a general reference book that consists of 930 reference works listed in numbered bibliographic entries with annotations. It is an introduction to major historical works for all time periods of history and for all geographical areas. Chapter 10 is devoted to biographical sources that will aid in the search for persons in the field of engineering or technology, while Chapter 13 is a resource for information from archives, manuscripts, special collections, and digital sites from around the world.

Green, R.J., Self, H.C., and Ellifritt, T.S. (eds) (1995) *50 Years of Human Engineering: History and Cumulative Bibliography of the Fitts Human Engineering Division*, Wright-Patterson Air Force Base, OH: Crew Systems Directorate, Armstrong Laboratory, Air Force Materiel Command. This book is a presentation of the unclassified publications of the Human Engineering Division of the U.S. Federal Government. Technical reports are the most common form of publication but some of the research and development work is published in journals and proceedings of scientific and technical societies, or as chapters in books, handbooks, military specifications, and standards or special reports.

National Museum of History and Technology (1978) *Guide to Manuscript Collections in the National Museum of History and Technology*, Washington, DC: Smithsonian Institution Press. Volume 3 of the series Archives and Special Collections of the Smithsonian Institution. The Museum of History and Technology holds important archival collections. One can find in their collections both fundamental source materials and illustrative material not found elsewhere. Most of the entries in this guide concern personal papers, business records, or document files; however, one can also find graphic material, trade literature, and information and reference files.

Niskern, D. (2001) *Library of Congress. Science and Technology Division. Reference Section. The History of Technology*, Washington, DC: Science Reference Section, Science, Technology and Business Division, Library of Congress. This tracer bullet published by the Library of Congress lists sources for the history of science, invention, medicine, and technology in colonial America. It provides a variety of materials in the collections of the Library of Congress helpful in researching the science and technology of eighteenth-century America. It includes journals, government publications, conference proceedings, dissertations, bibliographies, biographies, and books.

Rider, K.J. (1970) *History of Science and Technology: A Select Bibliography for Students* (2nd edn), London: Library Association. Although it is more than thirty years since publication by the Library Association of England, this selective bibliography is divided into two sections, the first being the history of science and the second being the history of technology. The time periods covered are Ancient, Mediaeval and Renaissance, and Modern (anything after 1600 to the present). Subjects covered are agriculture, building, chemical industries, clocks, engineering (civil, electrical and mechanical), firearms, machinery, metallurgy, papermaking, photography, printing, textiles, transportation, and woodworking.

Sterling, C.H. and Shiers, G. (2000) *History of Telecommunications Technology: An Annotated Bibliography*, Lanham, MD: Scarecrow Press. A selective guide to historical material of import, it records the history of major telecommunication technologies going back almost two hundred years. It contains more than 2,500 annotated bibliographic entries that are restricted to items in English. Included are general reference works; serial publications; general surveys; institutional and company histories; biographies; plus name and title indexes.

Turner, R. (ed.) (2002) *Guide to the History of Science* (9th edn), Chicago, IL: History of Science Society. This guide was written by members of the History of Science Society. It is the world's oldest and largest society dedicated to understanding the historical context of science, technology, medicine, and society. It includes a membership directory; graduate programs; research centers; libraries; museums; societies and organizations; journals and newsletters; inactive publications, and an index. It is also available online at <http://www.hssonline.org> (cited January 25, 2005).

Encyclopedias and dictionaries

In Eugene S. Ferguson's *Bibliography of the History of Technology* which has been annotated in this chapter, there is a good section on older encyclopedias and dictionaries from the 1700s up to the early 1900s offering excellent sources of information pertaining to the early developments of modern technology. Encyclopedias listed in Sarton (1952) complement the list by Ferguson. The focus of this section is to present some more contemporary works.

A Pictorial History of Science and Engineering; The Story of Man's Technological and Scientific Progress from the Dawn of History to the Present, Told in 1,000 Pictures and 75,000 Words (1957) New York: Year, Inc. This book is a general history of science and technology. Since no one can be a subject specialist in every field this work gives us the outline of scientific and technical progress through the ages. It begins in ancient times and proceeds to the twentieth century. There are 1,000 pictures collected from numerous museums and prestigious organizations.

Day, L. and McNeil, I. (eds) (1996) *Biographical Dictionary of the History of Technology*, New York: Routledge. This dictionary includes about 1,500 of the most influential people who have made significant contributions to new technological advances through history. The coverage is worldwide and from ancient times to the twentieth century.

Jones, W.R. (1996) *Dictionary of Industrial Archaeology*, Phoenix Mill, UK: Sutton Publishing. This dictionary contains over 2,600 terms with a strong focus on the industrial history of Great Britain. Its major emphasis is on the developments that occurred between 1750 and 1850.

Knight, E.H. (1876) *Knight's American Mechanical Dictionary: Being a Description of Tools, Instruments, Machines, Processes, and Engineering: History of Inventions: General Technological Vocabulary: and Digest of Mechanical Appliances in Science and the Arts* (3 vols), Boston, MA: Houghton, Mifflin. Normally, a dictionary describes topics in the alphabetical order of their names and its general scope is the history of inventions until about 1876. The goal was to place the entries in the most systemic order so that any detail could be readily reached at a moment's notice. This informative work boasts upwards of 7,000 engravings. Subject matter indexes are found throughout the three volumes and a list of the principal ones follows the preface.

Langmead, D. and Garnaut, C. (2001) *Encyclopedia of Architectural and Engineering Feats*, Santa Barbara, CA: ABC-CLIO. This work gives an overview of the architectural and engineering marvels and innovations that helped improve community living through each successive generation. The many differing factors that go into the progress of a community are mentioned in this book. The feats in question are arranged alphabetically and a glossary is provided for the definitions of terms used. There are citations at the end of each article for further reading.

McNeil, I. (ed.) (1989) *An Encyclopaedia of the History of Technology*, New York: Routledge. Put together by a staff of scholars from England, this book tries to simplify and create interest in the study of technology by telling the story of inventions from "stone age axe to spacecraft" in one volume containing twenty-two chapters. A scholar who is lauded by the academic community for their expertise writes each chapter. At the end of each chapter other publications are listed for further reading.

Selin, H. (ed.) (1997) *Encyclopaedia of the History of Science, Technology, and Medicine in Non-Western Cultures*, Boston, MA: Kluwer Academic. The *Encyclopedia* contributes to the cultural diversity of the sciences by legitimizing the study of other cultures' science and technology. The goal is to engage in a mutual exchange of ideas that enlightens both Western culture and non-Western culture alike. At the beginning of this sizable volume is a list of entries. In the following section the entries are alphabetically arranged. There are references at the end of each entry for further study, a list of authors, and an index at the end of the book.

Trinder, B. (ed.) (1992) *The Blackwell Encyclopedia of Industrial Archaeology*, Oxford: Blackwell. This encyclopedia contains sizable information about the industrial societies that emerged in the West from the mid-eighteenth century to the twentieth century. Scholars from thirty-one countries have contributed articles that also consider the pre-industrial period of each country. It is a guide to the monuments, settlements, and museums that house artifacts from the mid-eighteenth century. It includes pictures, maps, bibliography, index, and an appendix.

Handbooks, manuals, and guides

Handbooks, manuals, and guides are good reference sources because they provide information about how things work, and often include diagrams, pictures, graphics, and tables to help visualize the topic or object being studied. In this section some titles listed are highly recognized such as the work of Singer; others are basic in scope and clearly written for the general public. Nevertheless, these types of materials, whether scholarly or popular in scope, are excellent sources in which to demonstrate the marvels of engineering creativity.

Berlow, L.H. (1998) *The Reference Guide to Famous Engineering Landmarks of the World: Bridges, Tunnels, Dams, Roads, and other Structures*, Phoenix, AZ: Oryx Press. Gives brief but informative entries for over 600 of the most well-known engineering achievements of the last 5,000 years. Entries are cited for additional information and many are illustrated. Appendix A demonstrates bridge and truss designs; Appendix B illustrates a portfolio of Ohio covered bridges; Appendix C cites the tallest, longest, and highest for 1997; there is a section for biographies; a section for chronology; a bibliography; a geographical index, and a subject index.

Bunch, B.H. and Hellemans, A. (1993) *The Timetables of Technology: A Chronology of the Most Important People and Events in the History of Technology*, New York: Simon & Schuster. This work devoted to the history of technology is divided into seven time frames: the Stone Ages (24000000–4000 BC); the Metal Ages (4000 BC–1000 AD; the Age of Water and Wind (1000–1732); the Industrial Revolution (1733–1878); the Electric Age (1879–1946); the Electronic Age (1947–1972); and the Information Age (1973–1993). It includes a name index and a subject index.

Daumas, M. (ed.) and Hennessy, E.B. (trans.) (1969–1979) *A History of Technology and Invention: Progress Through the Ages* (3 vols), New York: Crown Publishing. This encyclopedic handbook is the English translation of *Histoire Générale des Techniques*. It is an extensive presentation of "the history of the methods that man has discovered and utilized to improve the conditions of his existence." It covers prehistoric times to 1860. It includes detailed diagrams and pictures of hundreds of technological methods.

National Geographic Society (1992) *The Builders: Marvels of Engineering*, Washington, DC: National Geographic Society, Book Division. This is a popular chronicle of the engineering wonders of the world; it includes famous roads, canals, bridges, railroads, pipelines, towers, tunnels, skyscrapers, sport arenas, exposition halls, domes, cathedrals, wind, solar, and electric power stations. It has over 400 photographs, detailed diagrams, and drawings.

Singer, C.J., Holmyard, E.J., and Hall, A.R. (eds) (1954–1978) *A History of Technology* (7 vols), London: Oxford University Press. Written in the form of an encyclopedic handbook, this work is an extensive treatise of the history of technology; it includes thousands of text figures, and hundreds of plates. Each chapter of its seven volumes is written by an expert, and covers early society to modern times. It represents one of the contemporary major accounts of the history of technology. Its supplemental volume includes names, places and names, and a subject index.

Wikander, O. (ed.) (2000) *Handbook of Ancient Water Technology*, Boston, MA: Brill. This volume presents water management and hydrotechnology from the Mesopotamia, Iran, the Indus Valley to the Atlantic Ocean from prehistoric times to the sixth or even seventh century AD. Topics covered are: water-supply; urban use; irrigation and rural drainage; larger hydraulic infringements on nature; water power; water as an aesthetic and recreational element; water legislation in the ancient world; and historical context.

Monographs and textbooks

Monographic publications are a very important form in the literature of the history of engineering. Some books listed in this section are about a specific area of engineering such as civil or chemical engineering. Other books deal with social aspects of engineering and its contributions to society. This list is intended to be representative of the topics being published as books in this field. Readers can find a more extensive list of monographs in the bibliographies listed above, searching the indexes and databases listed in this chapter, and by searching library online catalogs.

Armytage, W.H.G. (1976) *A Social History of Engineering* (4th edn), Boulder, CO: Westview Press. The goals of this book are to timeline technological developments as they refer to Britain, and to shed light on how these developments impacted upon the social life of the era and on how certain needs of a community at a certain time gave impetus to certain technological developments; further to discover the origins of innovations and institutions. It begins at the Stone Age and carries through into the twentieth century. Selected bibliographic references are given for each chapter.

Auyang, S.Y. (2004) *Engineering: An Endless Frontier*, Cambridge, MA: Harvard University Press. The author's purpose in this book is to explain how engineers create technology by presenting a broad picture of modern engineering in both its physical and human dimensions. Titles for topics of study are: technology takes off; engineering and information; engineers in society; innovation by design; science of useful systems; and leaders who are engineers. Each chapter includes an extended bibliography. Also included are two appendixes: Statistical Profiles of Engineers, and U.S. Research and Development.

Belanger, D.O. (1998) *Enabling American Innovation: Engineering and the National*

Science Foundation, West Lafayette, IN: Purdue University Press. This work in the History of Technology series was assembled by a history of technology scholar. It contains a large amount of research: document files, a comprehensive bibliography, synopses of oral history interviews, summaries, analyses of various topics, charts and graphs, and formatted endnotes. Sponsored by the Engineering and the National Science Foundation, the goal of this book is to provide a historical analysis of the Foundation's policies and its support of certain disciplines showing both the success and failure of particular funded programs and projects.

Burstall, A.F. (1965) *A History of Mechanical Engineering*, Cambridge, MA: MIT Press. This book was intended for mechanical engineering students. The author presents the most important events in the science of mechanical engineering from prehistoric times until the twentieth century. This work is divided into nine chapters each for a different period in history. Every chapter includes materials available to the mechanical engineer; tools; mechanisms and machines for the mechanical transmission of power and motion; fluid machinery; and heat engines; the final chapter gives a revealing review of the progress of these achievements.

Carter, D.V. (1961) *History of Petroleum Engineering*, Dallas, TX: Boyd Printing Company. The American Petroleum Institute assembled a board of scholars to provide this history of the birth and growth of petroleum engineering. Topics such as the percussion-drilling system; the hydraulic rotary-drilling system; cementing; logging, sampling, and testing; completion methods; production equipment; production techniques and control; reservoir engineering; fluid injection; handling oil and gas in the field; evaluation; research; conservation; unitization; and standardization of oil-field equipment are found. An index is included.

Cutcliffe, S.H. and Reynolds, T.S. (1997) *Technology and American History. A Historical Anthology from Technology and Culture*, Chicago, IL: University of Chicago Press. This is a collection of essays published in *Technology and Culture*. The fifteen articles presented are an attempt to cover the history of American technology from the late 1700s to the end of the twentieth century.

Furter, W.F. (ed.) (1980) *History of Chemical Engineering: Based on a Symposium Cosponsored by the ACS Divisions of History of Chemistry and Industrial and Engineering Chemistry at the ACS/CSJ Chemical Congress, Honolulu, Hawaii, April 2–6, 1979*, Washington, DC: American Chemical Society. "The central theme of the book is the historical identification and development of chemical engineering as a profession in its own right, distinct not only from all other forms of engineering, but particularly from all forms of chemistry including applied chemistry and industrial chemistry." The history of chemical engineering in industry and in education is cited, as well as the origins in Canada, Britain, Germany, Japan, Italy, India and the United States.

Garrison, E.G. (1999) *A History of Engineering and Technology: Artful Methods* (2nd edn), Boca Raton, FL: CRC Press. The purpose of this book is to aid us in discovering the persons, concepts, and events that helped make engineering what it is today. This book covers the history of engineering from the time of early stone tools studied by archaeologists to the present, with additional insights into what may lie ahead. It includes bibliographical references and an index, and contains information from authentic and highly regarded sources.

Hill, D. (1984) *A History of Engineering in Classical and Medieval Times*, LaSalle, IL: Open Court Publishing Company. This work records the major engineering achievements of the peoples of Europe and Western Asia in the period from 600 BC to 1450. It is concerned with origins of these achievements and their dispersion, paying attention to descriptions of techniques and machines. Research is brought to bear on irrigation and water supplies; dams, bridges, and roads; building construction; surveying; water-raising machines, power from water and wind; instruments; automata; and clocks. A bibliography and index are included at the end of the volume.

Hughes, A.C. and Hughes, T.P. (eds) (2000) *Systems, Experts, and Computers: The Systems Approach in Management and Engineering, World War II and After*, Cambridge, MA: MIT Press. A systems approach to solving complex problems and managing complex systems came into being after World War II. Thus, the "systems engineer" was born and so too was the field of "systems engineering." This book studies the origins of the systems approach; the organizations and individuals involved; the different systems and projects; the experts and their expertise; computers; and the applications of all these innovations in the United States and the world over. Bibliographies are included at the end of each chapter. Notes on the contributors are given at the end of the book.

Kirby, R.S., Withington, S., Darling, A.B., and Kilgour, F.G. (1956) *Engineering in History*, New York: McGraw-Hill. Both engineers and historians working together have written this book on the history of engineering. It is intended to present the development of engineering in Western civilization from its origins into the twentieth century; it is a general introduction to this pursuit rather than a definitive history. The interaction between the activities of the engineers and the activities of the people of the communities in which they lived are also discussed because progress in this science cannot come in any other manner. A bibliography is included at the end of the book, and selected bibliographies at the end of most chapters are provided.

Neuburger, A. (2003) *The Technical Arts and Sciences of the Ancients*, New York: Columbia University Press. This is a comprehensive and well-regarded book on ancient technical science concerning daily life up until the end of the fifth century. Topics covered are: mining, metal extractions and metalworking; woodworking; the treatment and preparation of leather; agriculture; fermentation; the production of oils and perfumes; preservation and mummification; ceramic arts, glass, yarns and textiles; dyes, painting, mechanics, lighting and heating; town planning; fortification; houses, building methods and materials; water supply and drains; and roads, bridges, and shipbuilding.

Rae, J. and Volti, R. (2001) *The Engineer in History* (rev. edn), New York: Peter Lang. This is a recollection of details that exemplify engineers in their lives such as their education and working methods; their influence and relationship with their employers; their perceived value in society, and their power in the sociopolitical arena. The historical coverage of this book is from antiquity to the end of the Industrial Revolution.

Straub, H. and Rockwell, E. (trans.) (1964) *A History of Civil Engineering: An Outline from Ancient to Modern Times*, Cambridge, MA: MIT Press. The author's keen interest in the history of his own profession prompted him to share just how modern construction technique and present-day engineering have gradually developed from two distinct sources – first, the science of mechanics and second, the creative craft of building. The book is divided into time frames starting with the ancient world and going forward into the twentieth century.

Journals and monographic series

Since the publication of the first scientific journal *Philosophical Transactions* of the Royal Society of London in 1665, journals have played a very important role in the archiving and dissemination of technical and scientific information. From the historical point of view it is also possible to say that the archives of these journals are valuable sources for the history of engineering. In this section we are more concerned about journals in the history of engineering; most of these journals started to publish in the 1900s; before this time works were published in subject-related engineering or technology journals. This is a representative list. Sources with an extensive list of history journals are: The *History Journals Guide Periodicals Directory*, a website maintained by Stefan Blaschke; the *Bibliography of the History of Technology* by Eugene S. Ferguson; the list of journal titles indexed in the *History of Science, Technology, and Medicine* database (http://www.rlg.org/en/page.php?Page_ID=194) (cited December 14, 2004); and in the *2002 Guide to the History of Science* (9th edn), edited by Roger Turner. All these sources are annotated in this chapter.

American Heritage of Invention & Technology (1985–) (8756–7296) New York: American Heritage.

Berichte zur Wissenschaftsgeschichte: Organ der Gesellschaft fuer Wissenschaftsgeschichte (1978–) (0170–6233) Weinheim, Germany: Wiley – VCH Verlag GmbH & Co. KgaA.

Blaetter fuer Technikgeschichte (1932–2003) (0067–9127) New York: Springer-Verlag. Originally: *Blaetter für Geschichte der Technik* (1932–1938).

British Journal for the History of Science (1962–) (0007–0874) Cambridge: Cambridge University Press.

Historia Scientiarum (1980–) (0285–4821) Tokyo: Nihon Kagakushi Gakkai (History of Science Society of Japan). Originally: *Japanese Studies in the History of Science* (1962–1980).

History and Technology: An International Journal (1983–) (0734–1512) London: Routledge.

History of Science: Review of Literature and Research (1962–) (0073–2753) Cambridge: Science History Publications.

History of Technology (1976–) (0307–5451) London: Mansell Publishing.

IA: The Journal of the Society for Industrial Archeology (1976–) (0160–1040) Houghton, MI: Society for Industrial Archeology.

Icon: Journal of the International Committee for the History of Technology (1996–) (1361–8113) Hamburg, Germany: International Committee for the History of Technology.

IEEE Annals of the History of Computing (1979–) (1058–6180) Piscataway, NJ: Institute of Electrical and Electronics Engineers.

Indian Journal of History of Science (1966–) (0019–5235) New Delhi, India: Indian National Science Academy.

Isis: International Review Devoted to the History of Science and its Cultural Influences (1912–) (0021–1753) Chicago, IL: University of Chicago Press, Journals Division.

Jahrbuch fuer Wirtschaftsgeschichte (1960–) (0075–2800) Berlin, Germany: Akademie Verlag GmbH.

Journal of the Association for History and Computing (1998–) (not supplied) Forest Grove, OR: American Association for History and Computing.

Journal of the Society for Army Historical Research (1921–) (0037–9700) Cambridge: Society for Army Historical Research.

Journal of Transport History (1953–) (0022–5266) Manchester:, UK: Manchester University Press.

Kagakushi Kenkyu (Journal of History of Science, Japan) (1941–) (0022–7692) Tokyo: Nihon Kagakushi Gakkai (History of Science Society of Japan).

Kultur und Technik (1977–) (0344–5690) Munich, Germany: Verlag C.H. Beck oHG.

Kwartalnik Historii Nauki i Techniki (1956–) (0023–589X) Warsaw, Poland: Polska Akademia Nauk, Instytut Historii Nauki.

Notes and Records of the Royal Society of London (1938–) (0035–9149) London: Royal Society of London.

Osiris: A Research Journal Devoted to the History of Science and its Cultural Influences (1936–) (0369–7827) Chicago, IL: University of Chicago Press, Journals Division.

Revista da Sociedade Brasileira de Historia da Ciencia (1985–) (0103–7188) Campinas, SP, Brazil: Universidade Estadual de Campinas (UNICAMP), Grupo de Historia e Teoria da Ciencia.

Revue D'histoire de la Culture Matérielle (1979–) (1183–1073) Ottawa, Ontario, Canada: Musée des sciences et de la technologie du Canada.

Revue D'histoire des Sciences (1972–) (0151–4105) Paris, France: Presses Universitaires de France.

Studies in History and Philosophy of Science Part A (1970–) (0039–3681) Oxford: Pergamon Press.

Technikgeschichte (1909–) (0040–117X) Berlin, Germany: Kiepert GmbH und Co. KG.

Technology and Culture (1960–) (0040–165X) Baltimore, MD: The Johns Hopkins University Press, Journals Publishing Division.

Transactions of the Newcomen Society (1922–) (0372–0187) London: Newcomen Society for the Study of the History of Engineering and Technology.

Voprosy Istorii Estestvoznaniya i Tekhniki (1956–) (0205–9606) Moscow, Russian Federation: Izdatel'stvo Nauka.

Zhongguo Keji Shiliao (China Historical Materials of Science and Technology) (1991–) (1000–0798) Beijing, China: Zhongguo Kexue Jishu Xuehui (Chinese Association of Science and Technology).

Conferences and symposia

Conferences, symposia, and seminars are very common places for scientists, engineers, and historians interested in the history of engineering to present the results of their research. Some of these gatherings are sponsored by professional societies on the history of science, technology, or engineering, while others are sponsored by subject-specific societies such as the American Chemical Society, and many other programs are sponsored by academic departments, research centers, and museums. The sources listed in this chapter under Article indexes are recommended to locate the proceedings of these meetings.

Museums and special collections

Museums and special collections represent sources of importance for researchers. Special collections held at universities, public libraries, and at other institutions provide works published sometimes several centuries back or their reproductions, and in many instances also collections of graphic works and objects. Most museums have the purpose of preserving large collections of objects, graphics, and other materials; they also provide educational programs for the public. Museums and special collection departments are also involved in research programs. In this section special collections are listed first followed by a list of museums.

Libraries

Case Western Reserve University. Special Collections <http://www.cwru.edu/UL/SpecColl/histtech> (accessed May 17, 2005). "In the History of Science collection one finds early editions of major works such as De Fabrica by Versalius and Opticks by Newtons. Materials on the History of Technology can be found in the technical papers of the Cleveland firms of Warner and Swasey and Charles F. Brush. In addition, the collections contain important early German, French, English and American journals. The Natural History collection includes 220 plates of Audubon's Birds of America, Catesby's The Natural History of Carolina, Florida and the Bahama Islands, and Travels of Lewis and Clark" (source: CWR Web page).

Cornell University Library. History of Science Collections, Division of Rare Books and Manuscript Collections <http://rmc.library.cornell.edu> (cited January 30, 2005). "In the history of technology, the collection's strengths include the Hollister Collection (300 vols.), which focuses on civil engineering, and the Cooper Collection on American railroad bridges, which includes the original blueprints for many structures that no longer survive. Notable among the technology holdings is Gustave Eiffel's La tour de trois cents mêtres, a scarce work which includes extraordinarily detailed plans of the Eiffel Tower, photographs of the construction process, and facsimile signatures of the dignitaries who were first to ascend what was then the world's tallest manmade structure" (source: Cornell Web page).

The Dibner Library of the History of Science <http://www.sil.si.edu/libraries/Dibner/> (cited February 9, 2005). The Dibner Library is the Smithsonian's collection of rare books and manuscripts relating to the history of science and technology. "Contained in this world-class collection of 25,000 rare books and 2,000 manuscript groups are many of the most important works dating from the fifteenth to the nineteenth centuries in the history of science and technology including engineering, transportation, chemistry, mathematics, physics, electricity, and astronomy" (source: DL Web page).

The Hagley Museum and Library. Research – Manuscripts and Archives <http://www.hagley.lib.de.us/> (cited January 30, 2005). "Hagley houses an important collection of manuscripts, photographs, books, and pamphlets documenting the history of American business and Technology. Hagley's main strength is in the Middle Atlantic region, but the scope of collecting includes business organizations and companies with national and international impact" (source: HM&L Web page).

The Huntington Library, Art Collections, and Botanical Gardens <http://www.huntington.org> (cited January 30, 2005). The library's rare books and manuscripts has collections in electricity and/or from the works of Franklin Taisnier, Gilbert, Musschenbroek, Beccaria, Galvani, Volta, Ampere, Oersted, Ohm, and Faraday. The civil engineering collections include works on early printed books; early English and American surveying books; eighteenth- to nineteenth-century American and English engineering materials; maritime technology and engineering; nineteenth- to twentieth-century railroad books and records; historical photograph collection; California and the American West: mining, geology, architecture, and civil engineering papers. In transportation the holdings include printed and manuscript materials on British and American railroads, sea transportation, and pre-Wright brothers aeronautics.

Iowa State University. Special Collections – Archives and Manuscripts <http://www.lib.iastate.edu/spcl/index.html> (cited January 30, 2005). "The Special Collections Department identifies, selects, preserves, creates access to, provides reference assistance for, and promotes the use of rare and unique research materials in the areas of Agricultural Engineering, Civil Engineering and Transportation, Energy and Electrical Power Management (i.e. energy management, consumption, efficiency, production, conversion and transmission, primarily of a non-nuclear nature). They also maintain an Archives of Women in Science and Engineering." (source: ISU Web page).

Library of Congress. Science, Technology and Business Division <http://www.loc.gov/rr/scitech/> (cited January 30, 2005). The highlights of the special collections holdings of this division are: Thomas Jefferson's personal library; 40,000 volumes from the original collection of the Smithsonian Institute; a large collection of manuscripts of major American scientists, inventors, engineers, explorers, and business pioneers; and several special collections in the history of aeronautics. In addition, a significant rare book collection in the history of computers and data processing, and several special collections of technical reports, standards, and gray literature in the sciences, technology, and engineering.

Linda Hall Library of Science, Engineering and Technology <http://www.lindahall.org/> (cited January 30, 2005). "The Library's major emphasis in engineering is on collecting important current research material. In addition to civil engineering, applied mechanics and related disciplines are the most extensively covered subjects in the Library's collection. The collection includes meeting papers from many engineering societies, an extensive collection of individual papers collected by AIAA as well as NASA papers, a variety of Government Documents and Technical Reports; the Library owns an impressive collection of engineering standards and specifications issued by industry and government. The collection contains not only current standards and specifications, but a large collection of historical documents, as well" (source: LHL Web page).

Rutgers, The State University of New Jersey. Thomas A. Edison Papers <http://edison.rutgers.edu/> (cited January 30, 2005). "The Thomas A. Edison Papers is a documentary editing project sponsored by Rutgers, The State University of New Jersey, the National Park Service, the Smithsonian Institution, and the New Jersey Historical Commission. Here one can view 180,000 document images, search a database of 121,000 document records and 19,250 names or Keyword search 4,000 volume-and-folder descriptions" (source: Rutgers Web page).

Stanford University Libraries. History of Science and Technology Collections <http://www-sul.stanford.edu/depts/hasrg/histsci/scihome.html> (cited January 30, 2005). "The History of Science and Technology Collections in the Stanford University Libraries support research and instruction in history of science and technology and related fields. The History of Science and Technology Collection covers the History of Computing; Rare Books in the History of Science; Civil and Electrical Engineering, Classical and High-Energy Physics, Computer Science, Industrial Revolution & Heavy Industry, Military Science, Photography, and Telecommunication" (source: SUL Web page).

Wright State University. Fordham Health Sciences Library, Special Collections and Archives <http://www.libraries.wright.edu/special> (cited January 30, 2005). "Dunbar Library's emphasis on aviation history and Fordham Library's emphasis on aerospace medicine and human factors engineering combine to make Wright State a nationally known repository for the documentation of some of the twentieth century's most dramatic technologies. The other major focus for both libraries is the local and regional history of the Miami Valley area of Ohio. Together the libraries offer a comprehensive historical perspective on the region, aviation, and aerospace medicine." (source: WSU Web page).

Museums

The Caltech Institute Archives California Institute of Technology <archives.catech.edu> (accessed May 9, 2006) "The Institute Archives serves as the collective memory of Caltech by preserving the papers, documents, artifacts and pictorial materials that tell the school's history, from 1891 to the present. Researchers will also find here a wealth of sources for the history of science and technology worldwide, stretching from the time of Copernicus to today." (source: archives home page).

Canada Science and Technology Museum <http://www.sciencetech.technomuses.ca/english/index.cfm> (cited January 30, 2005). "In the library of the Canada Science and Technology Museum, you will find a wealth of material documenting the history and development of science and technology, with special reference to Canada, and the Museum's collection. The library collection focuses on the Museum's special subject areas: agricultural technology, communications, energy and mining, forestry, graphic arts, land and marine transportation, industrial technology, physical sciences, space and museology" (source: CSTM Web page).

Deutsches Museum <http://www.deutsches-museum.de/> (cited January 25, 2005). "The archives of the Deutsches Museum are the leading collections for the history of science and technology in Europe. They hold currently 4.3 kilometers of archival material, sources and documents. The archival collections focus on transportation, aeronautics and astronautics, the history of physics and chemistry. The library offers literature that primarily relates to the history of the natural sciences and technology, ranging from non-fiction titles to scientific manuals or specialist essays and papers from the invention of the letterpress to the present day." (source: DM Web page).

Institute and Museum of History of Science – Florence <http://www.imss.fi.it> (cited January 25, 2005). Istituto e Museo di Storia della Scienza (IMSS) <http://galileo.imss.firenze.it/indice.html>. The scientific instrument collection of the Istituto e Museo di Storia della Scienza includes, several thousand instruments dating from the sixteenth to the early twentieth century. The archival resources include the Archive of the Royal Museum of Physics and Natural History (ARMU), and the Congresses of Italian Scientists Archive (IRIS). This institute is also listed in this chapter under Societies and research institutes.

Museo Nacional de Ciencia y Tecnología <http://mnct.mcyt.es/presentacion.htm> (cited February 6, 2005). This museum has a collection of 12,500 objects representing the scientific and technological heritage of Spain.

National Museum of American History, Behring Center, Smithsonian Institution <http://americanhistory.si.edu/> (cited January 28, 2005). "The Museum is interested in how objects are made, how they are used, how they express human needs and values, and how they influence society and the lives of individuals. NMAH's natural focus is on the history of the United States of America, including its roots and connections with other cultures. As sources for research, the Museum not only offers the historical object but also significant collections of oral histories, prints, photographs, business Americana and trade literature, and engineering drawings "(source: NMAH Web page).

National Technical Museum in Prague. History of Science and Technology <http://www.ntm.cz> (cited January 25, 2005). "The collections of the National Technical Museum comprise approximately 60.000 filing items and contain many-times higher number of individual objects. Nine percent of them are displayed in permanent exhibitions and exhibitions. The subjects covered are: Chemistry and Biotechnology, Food Industry, Mining, Metallurgy, Electrical Engineering, Acoustics, Mechanical Engineering, Transportation, Photography-Cinematography, Consumer Industry, Architecture, Building Industry, and Industrial Design" (source: NTM Web page).

The Franklin Institute Science Museum <http://www.fi.edu> (cited January 28, 2005). "Its Institute features the largest collection of Franklin materials – original works of art, documents, and artifacts – ever assembled, as well as interactive, multi-media installations. The Franklin Institute Award Case Files are a unique repository in the history of science. They are filled with stories of scientific enterprise and social circumstances. In addition, the Franklin Institute Science Museum has the largest collection of artifacts from the Wright brothers' workshop" (source: FI Web page).

The Jerome and Dorothy Lemelson Center for the Study of Invention and Innovation, Smithsonian Institution <http://www.si.edu/lemelson (cited January 28, 2005). "The Lemelson Center's oral and video history projects add historical documentation on invention and innovation in the United States. The National Museum of American History Archives Center has vast collections relating to technology, invention, and innovation in the nineteenth and twentieth centuries" (source: DLC Web page).

Archives and Special Collections of the Smithsonian Institution <http://www.si.edu/> (cited January 28, 2005). "National Archives, and Archives II located just west of the University of Maryland College Park campus, contains records of many federal agencies involved in science and technology, including: Atomic Energy Commission, Coast and Geodetic Survey, Geological Survey, Manhattan Project, National Aeronautics and Space Administration, National Bureau of Standards, National Science Foundation, Office of Naval Research, Office of Scientific Research and Development [World War I], and Weather Bureau" (source: SI Web page).

Museum of the History of Science, Oxford <http//mhs.ox.ac.uk/> (cited January 30, 2005). "Covers almost all aspects of the history of science, from antiquity to the early twentieth century. Particular strengths include the collections of astrolabes, sundials, quadrants, early mathematical instruments generally (including those used for surveying, drawing, calculating, astronomy and navigation) and optical instruments (including microscopes, telescopes and cameras), together with apparatus associated with chemistry, natural philosophy and medicine. In addition, the Museum possesses a unique reference library for the study of the history of scientific instruments that includes manuscripts, incunabula, prints, printed ephemera and early photographic material" (source: MHS Web page).

Société du Musée des Sciences et de la Technologie du Canada <http://technomuses.ca/index_f.asp> (cited January 30, 2005). "The collection of the Canada Science and Technology Museum encompasses a broad cross-section of Canadian scientific and technological heritage. National in scope, this unique collection consists of artifacts, photographs, technical drawings, trade literature, and rare books, all of which are complemented and supported by library holdings of monographs and serials" (source: SMSTC Web page).

Internet resources

This section lists important websites that contain information about the history of engineering. Other resources not listed here but which may be useful for the readers are specialized search engines, discussion lists, newsletters, and weblogs. Although many popular sites may be found by using a search engine such as Google or Yahoo, most Internet resources of scholarly credibility are supported by professional societies; private corporations; government agencies and universities; therefore, it is advisable to explore Web resources offered by these types of institutions. The resources mentioned in the section on Societies and research institutions, and on Museums and special collections, may be helpful in identifying other Web resources.

ASCE History and Heritage of Civil Engineering. Resource Guide <http://www.asce.org/history/hp_resguide3.html> (accessed May 17, 2005). This guide, hosted by the American Society of Civil Engineers, lists resources pertinent to the history of civil engineering.

Carnegie Mellon University Library's History of Technology and Science <http://www.library.cmu.edu/Research/Humanities/History/hots.html> (accessed May 17, 2005). Listed on this Web reference page are archives, libraries, and museums for the history of technology and science. There are also associations, organizations, and institutes and other resources.

ECHO: Exploring and Collecting History Online – Science, Technology and Industry <http://echo.gmu.edu/index.php> (accessed May 17, 2005). "Echo's research center catalogues, annotates, and reviews sites on the history of science, technology, and industry. You can browse our database of over 5,000 sites by topic, time period, publisher or content" (source: ECHO Web page).

Greatest Engineering Achievements of the 20th Century <http://www.greatachievements.org/> (accessed May 17, 2005). "This site is sponsored by the National Academy of Engineering. This project is a collaboration of the American Association of Engineering Societies, National Engineers Week, with 27 other professional engineering societies. The overarching criterion used was that those advancements had made the greatest contribution to the quality of life in the past 100 years. Even though some of the achievements, such as the telephone and the automobile, were invented in the 1800s, they were included because their impact on society was felt on the 20th century" (source: GEA Web page).

History of Engineering at Yale University <http://www.eng.yale.edu/history/> (accessed May 17, 2005). "This website exists to collect historical resources relating to the story of Engineering at Yale. Resources Available: History of Sheffield Scientific School; The Palmer Models – Exquisite models of 19th century railroad engines and a naval cruiser were crafted by Charles L. Palmer and bequeathed to the Sheffield School in 1908. They are available for private viewing on request; Bibliography – Partial list of significant books and articles about our history"(source: Yale Web page).

History of Science Links <http://web.clas.ufl.edu/users/rhatch/pages/10-HisSci/links/> (accessed May 17, 2005). This site is maintained by Robert A. Hatch, University of Florida; it has numerous interesting links. It includes a directory on List serves and chat pages.

History of Science Society Guide to US Graduate Study in History of Science <http://www.depts.washington.edu/hssexec/hss_gradproglist.html> (accessed May 17, 2005). This website contains links to every graduate program in the history of science and technology offered in the United States.

HOST <http://www.kcl.ac.uk/depsta/iss/library/speccoll/host/about.html> (accessed May 17, 2005). This website offers information about the HOST program of the UK; a collaborative retrospective conversion and conservation program for materials on the history of science and technology, 1801–1914.

I.A.Recordings <http://www.iarecordings.org/> (accessed May 17, 2005). "Industrial Archaeology Recordings is dedicated to recording past and present industry on film and video. Industrial Archaeology usually deals with disused buildings and machines, but it is also vital to record the processes of an industry while it is still working. This site provides over 560 links to other Industrial Archaeology Pages" (source: I.A. Web page).

Internet Resources for History of Science and Technology <http://www2.lib.udel.edu/subj/hsci/internet.htm> (accessed May 17, 2005). This Web page from the University of Delaware library lists: bibliographies and research guides, databases and information sources, museums, libraries and archives, exhibits, societies, associations, and other organizations, university departments, programs, and centers, and regional resources.

La Storia e la Filosofia della Scienza, della Tecnologia e della Medicina. The History and Philosophy of Science, Technology and Medicine <http://galileo.imss.firenze.it/~tsettle/> (accessed May 17, 2005). A selection of websites and other sources maintained by Thomas B. Settle, of the Istituto e Museo di Storia della Scienza (IMSS) and the Polytechnic University, Florence, Italy.

Societies and research institutions

In 1920 the first learned society devoted to the history of engineering and technology, the Newcomen Society, was founded in London, U.K. In the United States, the Society for the History of Technology was created in 1956. Both of these societies have had a very positive impact on the development of the field of the history of engineering. Many other societies that initially began studying the history of sciences are also valuable for the scholar of the history of engineering. In this section a number of research institutions that have provided significant support in this field are also included. This is not a comprehensive list and other sources found in this chapter will help to identify others.

Societies, associations, and organizations

American Association for History and Computing (http://www.theaahc.org/) (cited February 5, 2005). "The American Association for History and Computing (AAHC) is dedicated to the reasonable and productive marriage of history and computer technology. To support and promote these goals, the AAHC sponsors a number of activities, including an annual meeting, annual prizes, an electronic journal – the Journal of the American Association for History and Computing (JAHC), a continuing publication series, and a variety of summer workshops" (source: AAHC Web page).

American Philosophical Society <http://www.amphilsoc.org/> (accessed May 17, 2005). "The American Philosophical Society promotes useful knowledge in the sciences and humanities through excellence in scholarly research, professional meetings, publications, library resources, and community outreach. The American Philosophical Society, this country's first learned society, has played an important role in American cultural and intellectual life for over 250 years" (source: AMP Web page).

British Society for the History of Science <http://www.bshs.org.uk/> (accessed May 17, 2005). "BSHS is the largest UK body dealing with all aspects of the history of science, technology and medicine." The society publishes the British Journal for the History of Science (BJHS) (source: BSHS Web page).

History of Science Society (HSS) <http://www.hssonline.org/> (accessed May 17, 2005). "The History of Science Society is the world's largest society dedicated to understanding science, technology, medicine, and their interactions with society in historical context. Over 3,000 individual and institutional members across the world support the Society's mission to foster interest in the history of science and its social and cultural relations" (source: HSS Web page).

International Committee for the History of Technology <http://www.icohtec.org/> (accessed May 17, 2005). "The International Committee for the History of Technology (ICOHTEC) was founded as the Cold War was being waged with particular bitterness between the nations of the Eastern and Western Worlds. The intent was to provide a forum where scholars of both sides might meet and communicate about matters of mutual interest in the history of technology. It provides a new international journal that is essential reading for scholars and researchers in the field of the history of technology and they also sponsor world congresses" (source: ICOHTEC Web page).

Istituto e Museo di Storia della Scienza (IMSS) <http://galileo.imss.firenze.it/indice.html> (accessed May 17, 2005). "The Istituto e Museo di Storia della Scienza [IMSS] is one of the foremost international institutions in the History of Science, combining a noted museum of scientific instruments and an institute dedicated to the research, documentation and dissemination of the history of science in the broadest senses" (source: IMSS Web page).

The Newcomen Society for the Study of the History of Engineering and Technology <http://www.newcomen.com/> (accessed May 17, 2005). "The Newcomen Society is the world's oldest learned society devoted to the study of the history of engineering and technology. The Society is based in London and is concerned with all branches of engineering: civil, mechanical, electrical, structural, aeronautical, marine, chemical and manufacturing" (source: Newcomen Web page).

Scientific Instrument Society <http://www.sis.org.uk/> (accessed May 17, 2005). "The Scientific Instrument Society (SIS) was formed in April 1983 to bring together people with a specialist interest in scientific instruments, ranging from precious antiques to electronic devices only recently out of production. Collectors, the antiques trade, museum staff, professional historians and other enthusiasts will find the varied activities of SIS suited to their tastes. The Society has an international membership" (source: SIS Web page).

Society for the History of Technology (SHOT) <http://www.shot.jhu.edu> (accessed May 17, 2005). "The Society for the History of Technology was formed in 1958 to encourage the study of the development of technology and its relations with society and culture. An interdisciplinary organization, SHOT is concerned not only with the history of technological devices and processes, but also with the relations of technology to science, politics, social change, the arts and humanities, and economics" (source: SHOT Web page).

Research institutions

American Chemical Society, Division of History of Chemistry (DHC) <http://www.scs.uiuc.edu/~mainzv/HIST/> (accessed May 17, 2005). "The objects of the Division shall be to stimulate study and research in the history of chemistry, to offer an opportunity for presentation of the results of such specialized study and research, and to encourage a wide general interest among all chemists in the historical phases of their science" (source: DHC Web page).

American Institute of Physics (AIP). Center for History of Physics <http://www.aip.org/history/> (accessed May 17, 2005). "A division of the American Institute of Physics, The Center for History of Physics is the oldest and best-known institution dedicated to the history of a scientific discipline.... Our mission is to preserve and make known the history of modern physics and allied fields including astronomy, geophysics, optics, and the like" (source: AIP Web page).

ASCE History and Heritage of Civil Engineering (HHCE) <http://www.asce.org/history/hp_main.html> (accessed May 17, 2005). "HHC's mission is to enhance the knowledge and appreciation of our history and heritage.... Nearly 200 National and International Historic Civil Engineering Landmarks are featured within this website, along with biographies of 43 notable civil engineers" (source: HHCE Web page).

Australian Science and Technology Heritage Centre of the University of Melbourne (AUSTEHC) <http://www.austehc.unimelb.edu.au/> (accessed May 17, 2005). "The purpose of the Centre is the development of projects relating to the history and heritage of science, technology and medicine in Australia and to provide access to information resources in this area such as the former Australian Science Archives Project" (source: AUSTEHC Web page).

Centre for the History of Science, Technology and Medicine, University of Manchester (CHSTM) <http://www.chstm.man.ac.uk/> (accessed May 17, 2005). "It is located in Manchester, England. The Centre for the History of Science, Technology and Medicine (CHSTM) was founded in 1986 to bring together the University's interest in the history of science and medicine. It also includes the National Archive for the History of Computing, a major resource for research in the history and culture of informatics" (source: CHSTM Web page).

Dibner Institute for History of Science and Technology <http://dibinst.mit.edu> (accessed May 17, 2005). "The Dibner Institute is an international center for advanced research in the history of science and technology and located on the campus of MIT. Each year the Institute hosts senior, post-doctoral and graduate student fellows, as well as symposia, conferences, lectures and workshops" (source: Dibner Web page).

IEEE History Center <http://www.ieee.org/organizations/history_center/> (accessed May 17, 2005). "IEEE is the Institute of Electrical and Electronics Engineers, Inc. The IEEE established the IEEE History Center in 1980 and in 1990, the Center moved to the campus of Rutgers University, which became a cosponsor. The Center maintains many useful resources for the engineer, for the historian of technology, and for anyone interested in the development of electrical and computer engineering and their role in modern society" (source: IEEE Web page).

Imperial College London – Centre for the History of Science and Technology <http://www.imperial.ac.uk/historyofscience/> (accessed May 17, 2005). Recognized as one of the top research institutes in the UK, the Centre's areas of research include: science and technology in ancient Greece and Rome, early modern mathematics and mechanics, historiography of technology, science policy, war and technology, and the British State.

Landmarks Roster, Volunteer Center, History Resources. ASME History, History Links <http://www.asme.org/history/index.html> (accessed May 17, 2005). "Through its History and Heritage program, ASME encourages public understanding of mechanical engineering, fosters the preservation of this heritage, and helps engineers become more involved in all aspects of history" (source: ASME Web page).

The Sidney M. Edelstein Center for the History and Philosophy of Science, Technology and Medicine (Research Center) Hebrew University of Jerusalem (HUJ) <http://sites.huji.ac.il/edelstein/> (accessed May 17, 2005). "The Sidney M. Edelstein Center at The Hebrew University of Jerusalem was established in 1980 to encourage advanced research in the history and philosophy of science, technology, and medicine. In particular, the Center fosters research based on the resources of the Einstein Archive, the Quantum Archive, the Edelstein collections on the history of dyeing and chemical technology, and the Yahuda Theological Collection of Isaac Newton" (source: HUJ Web page).

Conclusion

The history of engineering has become an important field of study, originally included with the history of science; the history of engineering has evolved into a distinct subject. The history of engineering is part of the history of technology; most publications and professional organizations either use both terms or only use the term "technology." The number of publications produced is substantial and there are a number of well-designed bibliographic databases that cover the field. Since engineering has an important place in the development of human society it is expected that, as the different areas of engineering continue making contributions to mankind, the history of engineering will continue to be a healthy and intriguing field of study.

References

Auyang, S.Y. (2004) *Engineering: An Endless Frontier*, Cambridge, MA: Harvard University Press.

Bell, T.E. and Dooling, D. (2000) *Engineering Tomorrow: Today's Technology Experts Envision the Next Century*, Piscataway, NJ: IEEE Press.

Daumas, M. (ed.) and Hennessy, E.B. (trans.) (1969–1979) *A History of Technology and Invention: Progress Through the Ages* (3 vols), New York: Crown Publishing.

"Engineering and Technology," *Access Science. McGraw-Hill Encyclopedia of Science and Technology Online* http://www.accessscience.com/server-java/Arknoid/science/AS/Reviews/helicon/TE (cited February 8, 2005).

Garrison, E.G. (1999) *A History of Engineering and Technology: Artful Methods* (2nd edn), Boca Raton, FL: CRC Press.

Kirby, R.S., Withington, S., Darling, A.B., and Kilgour, F.G. (1956) *Engineering in History*, New York: McGraw-Hill.

McGraw-Hill Encyclopedia of Science and Technology (2002) (9th edn), New York: McGraw-Hill.

Sarton, G. (1952) *A Guide to the History of Science: A First Guide for the Study of the History of Science, with Introductory Essays on Science and Tradition*, New York: Ronald Press.

Simpson, J.A. and Weiner, E.S.C. (1989) *The Oxford English Dictionary* (2nd edn), Oxford: Oxford University Press.

14 Industrial and manufacturing engineering

Nestor L. Osorio and Andrew W. Otieno

Introduction

Industrial engineering (IE) is concerned with the efficiency of the functioning of operating systems such as manufacturing, supply, service, and transport. The American Institution of Industrial Engineers defines IE as an area of engineering

> concerned with the design, improvement and installation of integrated systems of men, materials equipment and energy. It draws upon specialized knowledge and skill in the mathematical, physical and social sciences together with principles and methods of engineering analysis and design to specify, predict and evaluate the results to be obtained from such systems.
>
> (Zandin, 2001: 1.11)

In the early 1960s, IE was focused primarily on work simplification and improvement, using quantitative, science-based tools, and techniques. Over the years, IE has evolved to include both design and integration of operating systems. The history of IE dates back to the Industrial Revolution of the mid-1700s when the steam-engine and the initial machine tool were developed. The concepts of specialization of labor by Adam Smith, and Eli Whitney's development of interchangeability of machine parts (Zandin, 2001: 1.4–1.5) led to the progression of IE as it is today. Notable pioneers also include Frederic Taylor, whose experiments culminated in the standardization of work; and Frank and Lillian Gilbreth, whose studies of human motions formed the basis of work study, human factors, and ergonomics. In 1912, at the annual meeting of the American Society of Mechanical Engineers (ASME), educators, consultants, engineers, and other pioneers, including Taylor and Gilbreth, formed the Management Division of the ASME, from which a unique new profession of industrial engineering was created.

During World Wars I and II the need for more rigorous techniques in the planning of complex systems in military operations led to the development of a set of techniques now called operations research (OR). By 1943, the functions of IE had been expanded to include manufacturing engineering, cost control, budgeting, and wages and salary administration. The ASME Management Division continued to expand, and in 1948 the American Institution of Industrial Engineers was formed.

The IE discipline as it has evolved today is sometimes referred to as industrial and systems engineering (Salvendy, 2001: 4–10). The discipline is based on a core engineering curriculum, other core courses being mathematics, statistics, engineering economics, and psychology. It comprises four key areas: operations management (OM), human factors engineering, management systems, and manufacturing engineering.

OM is primarily concerned with techniques that focus on the complexity of managing large organizations requiring the effective use of money, materials, equipment, and people. It involves the application of analytical methods from mathematics, science, and engineering that provide an organization with alternatives for the best allocation of resources using OR tools. OM functions include planning, forecasting, resource allocation, performance measurement, scheduling, the design of production facilities and systems, supply chain management, pricing, transportation and distribution, and the analysis of large databases. Human factors engineering, on the other hand, focuses on how efficiently human resources may be utilized. A generic term used interchangeably with this is ergonomics. Ergonomics is concerned with the relationship between human factors and how these influence workplace design (ANSI, 1989). The Ergonomics Society defines ergonomics as "the application of scientific information concerning humans to the design of objects, systems and environment for human use" (Ergonomics Society, 2004). Human factors, on the other hand, involve the coordination and application of physical movements, and cognitive components such as information processing and decision-making to perform a given task (Marmaras and Kontoginis, 2001). Areas also covered include safety, work measurement and simplification, methods study, and the use of anthropometric data in work systems design. Work measurements and methods study aid in determining optimum cycle times and most efficient methods of performing work-related activities. Anthropometry covers the study of human beings in terms of their physical dimensions, with emphasis on how this may be used for work methods and work place design (ANSI, 1989). Current research in ergonomics focuses more on designs that promote the safety of workers while research in work methods and work simplification continues to focus on better time standards and work methods.

The management function of the IE profession is primarily concerned with aspects of cost control, budgeting, and engineering economics. The manufacturing systems engineering function of IE include facilities planning and design, layout planning of physical facilities, simulation, quality control, inventory management, and manufacturing technologies. Plant location and layout problems have evolved traditionally from being implemented by analytical methods to the use of more powerful simulation models. As the world's economies continue to evolve, the production of goods has become more competitive; hence there is more emphasis on quality and liability issues. This has led to the development of more stringent quality standards, changing the face of manufacturing into an integrated product design and development activity. Future trends in management suggest that organizations will lay more emphasis on continuous quality improvement. Niche techniques have since evolved into IE such as lean manufacturing; design for assembly and manufacture; and green manufacturing practices. To support the manufacturing systems, inventory management has moved into a platform of the electronic world, now undertaken by a more rigorous approach of supply chain management. The role of the industrial engineer also includes designing manufacturing processes, equipment, automation and robotics, and materials handling systems.

The term "manufacturing engineering" is usually used interchangeably with others such as "production engineering" and "manufacturing technology." The term "manufacturing" is used to describe a set of activities that change a raw material into a more useful product. The Society of Manufacturing Engineers (SME), founded in 1932, defines manufacturing engineering as "that specialty of professional engineering which requires such education and experience as is necessary to understand, apply, and control engineering procedures in manufacturing processes and practices of manufacturing, to research and develop tools, processes, machines, and equipment, and to integrate facilities and systems for producing quality products with optimal expenditures" (Soska, 1984: 223). A manufacturing engineer performs duties that include design of the product, selecting appropriate materials for the product, and selection and defining the progression of processes and resources needed for the manufacture of the product (Kalpakjian and Schmid, 2001: 2).

The history of manufacturing engineering as a discipline is closely linked to how manufacturing has evolved over the years. Organized manufacturing may be traced back 6,000 years when manual forming processes began with the production of various articles made of wood, ceramic, stone, and metal. The history may be examined from two perspectives: how materials were discovered, developed, and used to make things; and how systems of production were developed (Groover, 2002: 3).

The first materials used for making household utensils and ornamental objects included metals such as gold, copper, and iron. Whereas gold was found in its pure state by early man, copper was most likely the first metal to be extracted successfully. The production of steel is said to have started about AD 600 to 800. The periods that followed saw the development of a wide variety of ferrous and non-ferrous metals. To date, there have been significant developments in engineering materials. Common engineering materials are now classified to include engineered materials such as ceramics and reinforced composite materials, alloys of various types, and nanomaterials.

The advent of modern manufacturing has its significance in the Industrial Revolution, which began in England in the 1750s and lasted until the 1830s. The economies of the world were heavily dependent upon agricultural produce. With the invention of Watt's steam-engine between 1763 and 1765 and steam power replacing wind and other forms of energy, together with Wilkinson's invention of the water-powered boring machine, modern mechanization began in England and spread quickly to other countries. In the mid-1800s to early-1900s there was expansion in transportation industry, especially the railroad system, necessitating an increase in the need for steel. It is during this period that many consumer products were also developed. The need for more efficient production methods became a very important issue in manufacturing, hence the second Industrial Revolution. This was characterized by mass production, scientific management techniques in manufacturing, electrification (first electric power generating plant in 1881), and assembly lines (Henry Ford's assembly line in 1913) (Groover, 2002: 4).

The two world wars separately played significant roles in continuing to improve manufacturing techniques. During these wars, an increased need for transport, battleships, and guns resulted in the improvements in fabrication techniques and the development of the arc welding process which is the most commonly used process in welding today (Lincoln Electric Company, 1994: 1.1–5). After World War II the Cold War that ensued created an enormous arms race. More sophisticated aircraft had to be developed and these required more complex parts to be machined or formed. Joint efforts by the U.S. Air Force, Massachusetts Institute of Technology, and John Parsons resulted in the first numerical control systems being developed (Chang *et al.*, 1998: 316). Since then, and with the development of microprocessors and increased use of computers, major milestones have been established in all aspects of manufacturing.

The evolution of manufacturing engineering over the past decade has made it an integrated concurrent activity where manufacturing engineers consider a product from its design stage, simultaneously with the manufacturing stage. Manufacturing today is characterized by highly automated systems utilizing robotics and integrated enterprise control. Not only is product design very automated through computer-aided design (CAD) systems, but the control of process is equally automated through computer-aided manufacturing (CAM) systems (Benhabib, 2003: 9–13). *The Handbook of Manufacturing Engineering* (Walker, 1996) identifies four major areas of the discipline: product and factory development, factory operations, materials, and assembly process. The current trends in product design emphasize simultaneous engineering and multi-disciplinary approach in the discipline. Niche technologies that replace traditional lengthy product development cycles and prototyping, such as rapid prototyping, are now used to quickly convert designs from computer representations directly into solid objects. As the optimal design is achieved, the manufacturing engineer is faced with the challenge of designing the production facilities, and, like the industrial engineer, today's facilities designs are achieved through analytical techniques and simulation.

Factory operations functions include manufacturing cost control and estimation, process planning and control, work methods, and quality control. With the evolution of computers in manufacturing, techniques used in today's factory operations include computer-aided process planning and computer-aided testing.

Manufacturing engineering must consider materials used in manufacturing and methods of manufacturing parts. Over the past decades there have been major developments in engineering materials with a major emphasis on producing lighter but stronger materials, materials with better corrosion resistance, and with better manufacturing properties. Traditional manufacturing methods that existed were best used for the more conventional metals and alloys, plastics, and ceramics. Newer, non-traditional techniques have been developed for the more difficult-to-form or difficult-to-machine alloys, ceramics, and composite materials. Forming and fabrication techniques have become more automated, and machining has been integrated with CAD and CAM systems into computer numerical control (CNC) machining.

The final function of manufacturing engineering focuses on analyzing and defining processes and machines, and the entities of assembly automation. This area also includes selection of materials-handling systems, robotics, assembly lines, and all other components of manufacturing automation.

Both functions of industrial and manufacturing engineering must address the issue of quality. Quality control ensures conformance to specifications and meeting customer needs. It dates back to the history of manufacturing. With the advent of mechanization in manufacturing, competition became a reality (NIST, 2005). Because of this, manufacturers began to ensure that their products were of high quality in order to be competitive. In the Middle Ages, apprentices had to go through long periods of training to ensure that they had adequate skills to produce high-quality competitive products. This transitioned into the formation of statutory and professional organizations to ensure that products were of high quality and consumers were protected.

Current trends in quality control began with the development of statistical sampling techniques in the 1920s. Statisticians at the Bell Telephone laboratories were pioneers in the development of statistical quality control (SQC), as it has evolved today. The practices of quality control were, until the mid-1980s, closely guarded by individual manufacturers. Stiff competition from countries such as Japan, especially in consumer electronics, forced the rest of the world to reassess quality control practices beyond just statistical quality control. The concept of total quality management (TQM) soon became the norm in the circles of quality control (Hradesky, 1995). Thus, TQM now focuses on customer satisfaction by including SQC, employee involvement, quality function deployment, and continuous improvement. To ensure competitiveness, many companies have become ISO 9000 certified. Because of the current nature of competition, and liability issues, and rapid development of automation techniques such as machine vision, quality now tends to include 100 percent inspection rather than the traditional sampling processes in SQC.

Industrial and manufacturing engineering are very closely related areas with interlinked skills and functions. Both fields are evolving very rapidly today with a shift from macro to micro and nanoscale manufacturing. Researchers must therefore continue to focus on manufacturing practices that are cost-effective and efficient, at a macro level, and on new and innovative methods used to produce micro and nano products. In this chapter the critical resources for industrial and manufacturing engineering will be outlined.

Searching the library catalog

Keywords

Searching the online catalog by using keywords is an effective way to find library materials. The following is a list of keywords that may be used: assembly automation; assembly processes; automated production; automation and robotics; CAD/CAM; computer simulation; control of production; controllers; cost estimating; ergonomics; fabrication process; factory design; factory requirements; finishing; human factors; industrial safety; inventory management; maintenance management; manufacturing process; materials characteristics; motion systems; optimization; organization and work design; probabilistic models and statistics; process planning; product design; product development; quality control; reliability; risk analysis; safety; scheduling; statistical process control; storage and warehousing; tooling and equipment; and work measurement. This is not a comprehensive list, but some of the major areas related to manufacturing and industrial engineering are presented.

LC Subject headings

The Library of Congress Subject Headings are the official terms used in libraries to catalog materials in a specific field. These subject terms may be used in the subject field of a search engine. Examples of Library of Congress Subject Headings are as follows:

Assembly-line Methods
Automation
Computer Integrated Manufacturing Systems
Engineering Economics
Flexible Manufacturing Systems
Human Engineering
Industrial Engineering
Industrial Engineering – Statistical Methods
Industrial Project Management
Industrial Safety
Manufacturing Processes

Manufacturing Processes – Automation
Manufacturing Resource Planning
Methods Engineering
Operations Research
Plant Layout
Production Control
Production Engineering
Psychology, Industrial
Quality Control
Robots, Industrial
Standardization
Systems Engineering
Work Environment
Work Measurement

LC call numbers

Library materials cataloged under the Library of Congress systems may be found under several call number sections. Most materials are in the T, TA, TJ, and TS sections. Other sections with additional sources are RC and HF. The following is a list of some major call numbers:

T55–55.3 Industrial safety
T57.35–57.5 Quantitative methods
T57.6–57.77 Operations research. Systems analysis
T60–t60.8 Work measurement. Methods engineering
TA177.4–185 Engineering economy
TA329–348 Engineering mathematics
TJ212–225 Control engineering. Automatic control systems
TS155–155.8 Production management. Operations management
TS156–170 Control of production systems
TS171–176 Product design
TS177–182 Manufacturing engineering. Process engineering
TS183 Manufacturing processes
TS184–193 Plant engineering
TS195–199 Packaging
TS200–2301 Specific industries (e.g., metals, textiles)
RC963–967 Industrial hygiene
HF5548.8 Industrial psychology

Due to the interdisciplinary nature of manufacturing engineering the reader will find materials in other sections depending on the subjects included in the research.

Abstracts and indexes

The literature of industrial and manufacturing engineering is similar to other technical areas. Articles from journals, papers published in conference proceedings, monographs, technical reports, and other documents, are important types of publications. Abstracting and indexing services adopt the role of collecting, organizing, and summarizing thousands of these documents that are published every year. Possibly one of the most important indexes in this field is *Compendex* and it should be the starting point for searching information about a topic. Moreover, due to the interdisciplinary nature of industrial and manufacturing engineering, it is advisable to search other databases when looking for additional information, such as technical reports in *NTIS*; theses in *Dissertation Abstracts*, and patents in the US Patents and Trademark Office's Website. In addition, other databases related to this field are: *Inspec*, *Ingenta*, *Computer and Information Systems Abstracts*, *Electronics and Communications Abstracts*, *PsycInfo*, and *ABI/Inform*. All the databases mentioned in this section are described in Chapter 2 of this work. Finally, *Ergonomics Abstracts* and *International Abstracts in Operations Research*, not included in other chapters, are also important.

Ergonomics Abstracts (1969–) London: Taylor & Francis. Sponsored by the Ergonomics Information Analysis Centre at the University of Birmingham, UK, this publication covers the international literature of ergonomics, human factors, human machine systems, including physical environmental influences in the workplace, work design, as well as psychological, physiological, and biomechanical aspects of work.

International Abstracts in Operations Research (1961–) Basingstoke, Hants: Palgrave Macmillan. This index covers about 180 journals in operations research, industrial management, decision sciences, information systems, industrial engineering, and other related fields. This publication is sponsored by the International Federation of Operational Research Societies.

Bibliographies and guides to the literature

Bibliographies and guides to the literature of a subject are produced in several different ways. Books are a common format in which to publish these types of documents. The readers will also find them in websites and as electronic publications. Another way to find bibliographies is by searching in engineering databases such as *Compendex*, *Inspec*, or *Ergonomics Abstracts*. These bibliographies are published as articles in journals. In addition, articles about the review of a subject or as the survey of a subject usually include extensive bibliographies at the end. Finally, some bibliographies are written as technical reports. The following are some of the most relevant in industrial and manufacturing engineering.

Cleland, D.I., Rafe, G., and Mosher, J. (1998) *Annotated Bibliography of Project and Team Management*, Newtown Square, PA: Project Management Institute. This is a comprehensive bibliography covering key topics in project management that spans 1956 to 1998. It includes annotations from articles and books, and there is also a list of dissertations and periodicals. The section on applications covers publications about several industries such as chemical, aerospace, and nuclear power. This is still a valuable resource in the areas of project management, team work, and industrial management.

Hanel, N.L.N. (1997) *Selective Guide to Literature on Statistical Information for Engineers*, Washington, DC: American Society for Engineering Education. Although it was published in 1997 this bibliography is still relevant for its coverage. Statistical applications in manufacturing and industrial engineering are numerous. This guide gives a list of basic resources for engineers. It covers indexes and abstracts; bibliographies; Internet resources; encyclopedias and dictionaries; handbooks and manuals; books; proceedings; and journals.

Jürgensen, A., Khan, N., and Vanderburg, W.H. (2001) *Sustainable Production: An Annotated Bibliography*, Lanham, MD; London: Scarecrow Press. This is a by-product of a project at the Center for Technology and Social Development at the University of Toronto on sustainable engineering. It includes over 250 annotations from books and articles.

McCullough, H. (2001) *Industrial Engineering Resource Guide*, Madison, WS: Kurt F. Wendt Library <http://www.wisc.edu/wendt/help/ieguide.html> (accessed May 17, 2005). This is a Web page with information about several databases, websites, patents, standards, and technical reports sources. It also lists some dictionaries, encyclopedias, handbooks, and manuals.

Osorio, N.L. (2002) *Selective Guide to Literature on Industrial Ergonomics* Washington, DC: American Society for Engineering Education. This bibliography focuses on industrial ergonomics, and has thirty-six pages of resources such as indexes and abstracts; literature guides; Internet resources; encyclopedias and dictionaries; handbooks and tables; periodicals; proceedings; books and industrial standards <http://eld.lib.ucdavis.edu/fulltext/Ergonomics.pdf> (accessed May 17, 2005).

Papadopoulos, H.T. and Heavey, C. (1996) "Queueing theory in manufacturing systems analysis and design: A classification of models for production and transfer lines," *European Journal of Operational Research* 92 (1): 1–27. Written with the purpose of compiling a bibliography, this review article is a guide to the modeling of production and transfer lines using queuing networks. The article has 257 citations on the subject and also categorizes queuing network models as they are applied to manufacturing systems. Although published in 1996, this bibliography is highly recommended. It presents a comprehensive treatment of the subject.

Raafat, F. (2002) "A comprehensive bibliography on justification of advanced manufacturing systems," *International Journal of Production Economics* 79 (3): 197–208. This bibliography covers the literature of techniques on the selection of advanced manufacturing systems from 1990 to 2001. This article includes 231 citations.

Recent Advances in Manufacturing (RAM) <http://www2.shu.ac.uk/services/lc/infoservices/eris/info.cfm?db_id=61> (accessed May 17, 2005). "Recent Advances in Manufacturing (RAM) is a database of bibliographic information containing over 30,000 references from over 500 journals. The manufacturing and management related areas covered include: manufacturing industry and management, environmental and quality management, product development, manufacturing systems and processes, applied technologies, supply-chain and e-business issues, production planning and control, maintenance, monitoring and inspection, and education and training" (source: RAM website).

Wong, B.K. and Monaco, J.A. (1995) "Bibliography of expert system applications for business (1984–1992)," *European Journal of Operational Research* 85 (2): 416–432.

Expert systems are used in industry for a variety of applications including in areas of production and operations. This bibliography contains 214 citations classified by application areas. The article also includes an extensive review of the subject. Published in 1995, this bibliography is a valuable resource for offering the foundations of expert systems in business applications.

Directories

Directories provide valuable information to practitioners and students in the form of practical leads to information about industries and manufacturers. For example, they give details about how to contact industrial suppliers of components and materials. Systematically arranged, they usually include listings of products and services, plants and equipment, and instrumentation and testing equipment.

There are thousands of industrial directories in the marketplace. While most are still published in paper form, these publications also have Web or CD-ROM versions. *Directories in Print* (*DiP*), a comprehensive compilation of directories for all areas, is listed in Chapter 2 of this book. *DiP* can provide the user with a list of directories on a very specific industry such as the chemical industry. The directories listed in this section are about manufacturing and industrial engineering, but in a more general sense. Lists of manufacturers from specific states are not included.

CatalogXpress (1900s–) Englewood, CO: Information Handling Services. A subsection of the IHS online engineering resource center, it is also available in CD-ROM. *CatalogXpress* is an index to product and component information from over 16,000 manufacturers' catalogs and other sources. It also provides general information about an additional half million manufacturers. Available to readers in subscribing institutions at: <http://www.ihs.com/procurement/catalogxpress-vendors/index.html> (accessed May 17, 2005).

Electronics Weekly Buyers' Guide <http://http://www.ewbuyersguide.com/> (accessed May 17, 2005). *Electronics Weekly Buyers' Guide*, formerly *DIAL Engineering Directories*, is the fusion of *DIAL Electronics* and *Electronics Weekly*, the two major suppliers in the UK. It is available free on the Internet, and is searchable by product/service, company, and trade name. Users need to subscribe to gain full access to details about UK engineering companies.

FirstIndex <http://www.firstindex.com> (accessed May 17, 2005). *FirstIndex* is a marketplace for industrial custom-manufactured parts and assemblies. It is a good platform to identify suppliers for machining, sheet metal, stamping, forging, casting, electronics, injection molding, or any other manufacturing process. Free registration is required.

Bittence, J.C. (ed.) (1994) *Guide to Engineering Materials Producers*, Materials Park, OH: ASM International. This is a condensed source for finding manufacturers of metallic and non-metallic materials. It includes a list of over 900 manufacturers. The second section of this book includes nearly seventy-five categories of materials.

Industry Search <http://industrysearch.com> (accessed May 17, 2005). From a list of 200 product lines the searchers in this industry portal have access to over 120,000 manufacturing and supplier companies.

International Directory of Testing Laboratories (1992–) Philadelphia, PA: American Society for Testing and Materials. Formerly the *Directory of Testing Laboratories*, this provides information for testing capabilities, type of tests, and materials and products being analyzed.

ISA Directory of Instrumentation (1981–) Research Triangle Park, NC: Instrument Society of America. This is a guide to measurement and control products; it includes manufacturers, services and representatives. Also available online at <http://www.isadirectory.org/> (cited October 25, 2004).

Manufacturing.Net <http://www.manufacturing.net/> (accessed May 17, 2005).

Manufacturing.Net includes a comprehensive database of manufacturers, suppliers, and products. In addition, manufacturing professionals can find in this site a wealth of information for their enterprises. Free registration is required.

Member Products Directory: Machine Tools, Manufacturing Machinery and Related Products Built by Members of AMT, the Association for Manufacturing Technology (1993–) McLean, VA: The Association for Manufacturing Technology. This directory provides eighteen groups of products from about 700 companies. Browsing may also be done by company name or by geographical location. Available also online via the World Wide Web as: Member Product Directory <http://www.amtonline.org/directoryhome.cfm> (accessed May 17, 2005).

Plant Engineering – Update (1947–) Middletown, OH: Reed Business Information. The *Update* section of this journal reviews products and software packages related to plant engineering and lists manufacturers and services providers. Recent updates may be found on: electrical products; compressed air; tools and welding; instrumentation and control; software and components; heating; safety; lubrication; power transmission; and environmental products.

Skelly, K.J. and Skelly, E.M.T. (1997) *Directory of Safety Standards, Literature and Services*, New York: Van Nostrand Reinhold. This is a reference tool that provides sources for available safety standards, guidelines, and other information. It contains eight sections: emergency safety contacts; federal agencies; federal safety standards; societies, organizations and societies; industry safety standards; safety periodicals and directories; safety literature, state agencies and safety standards.

Society of Manufacturing Engineers. *SME Buyers Guide* <http://www.sme.org/cgi-bin/category-list.pl?&&PUB&&SME&> (cited October 25, 2004). Discontinued as a paper publication in 2002 this online site contains a twenty-product list that guides the user to a list of companies. Keyword searching for products and companies is also possible.

Thomas Register of American Manufacturers (1905–) New York: Thomas Publishing. This is a primary source for finding information about U.S. manufacturers. The paper edition is a multi-volume set arranged by product. Available also online at: <http://www.thomasregister.com/> (accessed May 17, 2005). In the online version users may place orders; view and download CAD drawings; and view thousands of full-text company catalogs. Free registration is required.

Thomas Register of European Manufacturers (1990–) Antwerpen, Belgium: Thomas International Publishing. This is a CD-ROM publication serving as a European industrial buying guide. It includes over 10,500 product and services categories; and more than 200,000 industrial suppliers from seventeen European countries.

Top 400 Contractors Sourcebook (1998–) New York: McGraw-Hill, Engineering News-Record, annual. Provides rankings and overviews of the top 400 contractors for eight major industry sectors: building, transportation, manufacturing, industrial processes, petroleum, power, environmental, and telecommunications. Database version available from ENR.

The Top 500 Design Firms Sourcebook (1998–) New York: McGraw-Hill, annual. This lists the major engineering firms from these sectors: building, transportation, manufacturing, industrial processes, petroleum, power, environmental, and telecommunications. Database version available from ENR; it is also a supplement to *Engineering News Record* (*ENR*).

Dictionaries and encyclopedias

Technical dictionaries and glossaries are lists of the acceptable terminology in a subject. It includes term definitions, and in some cases abbreviations, synonyms, symbols, and cross-references. The definitions may also have an explanation of the meaning of the term and usage. An encyclopedia is usually a more extensive treaty in the form of articles or short entries; the purpose of an encyclopedia is to systematically summarize the most important aspects of a field. Encyclopedias define and interpret what is covered and include valuable bibliographies. Dictionaries and encyclopedias are reference sources to give students and practitioners background information about a topic.

Brody, A.L. and Marsh, K.S. (eds) (1997) *The Wiley Encyclopedia of Packaging Technology* (2nd edn), New York: Wiley-Interscience. This encyclopedia has 250 articles covering modern packaging techniques and materials. It offers the user details on all packaging machinery and equipment, and current information about changes in materials, processes, technologies, and regulations that have occurred in recent years.

Confer, R.G. and Confer, T.R. (1999) *Occupational Health and Safety: Terms, Definitions, and Abbreviations* (2nd edn), Boca Raton, FL: Lewis Publishers. This covers about 5,000 terms in the areas of industrial hygiene, safety, occupational medicine and other related areas in acoustics, chemistry, physics, and biology. Terms on bacteriology, environmental health, epidemiology, illumination, mathematics, and microscopy are also included.

Gass, S.I. and Harris, C.M. (eds) (2001) *Encyclopedia of Operations Research and Management Science* (2nd edn), Boston, MA: Kluwer Academic. This encyclopedia provides a comprehensive review of the theoretical and applied areas of operations research and management science. There are 893 major topics covered, and each entry has a list of reference sources. Over 200 experts contributed to this work, which is useful for the professional and students of these fields.

Industrial Engineering Terminology: A Revision of ANSI Z94.0-1989: An American National Standard, Approved 1998 (1998) (rev. edn), Norcross, GA: Industrial Engineering and Management Press. This book contains more than 15,000 terms, acronyms, and abbreviations. It is divided into seventeen sections related to industrial engineering. Definitions are listed alphabetically within each section. Each definition is an official standard of the American National Standards Institute.

Lee, J. (ed.) (2002) *Dictionary of Industrial Administration* (3 vols), Bristol, UK: Thoemmes. This encyclopedic dictionary of industrial management has been published with the contributions of over 100 experts in the field. It is a basic source for quick information on all managerial topics; most entries are from half to two pages long.

McKenna, T. and Oliverson, R. (1997) *Glossary of Reliability and Maintenance Terms*, Houston, TX: Gulf Publishing. This glossary's goal is creating a common terminology among practitioners; it covers the areas of reliability, process engineering, plant operations, repair, and maintenance technologies. More than 1,000 terms are defined.

Philippsborn, H.E. (1994) *Elsevier's Dictionary of Industrial Technology: In English, German, and Portuguese*, Amsterdam; New York: Elsevier. This multilingual dictionary covers an extensive vocabulary of industrial activity, from agricultural machines to wood products. It is one example of multilingual dictionaries usable in industrial engineering and manufacturing.

Soroka, W.G. and Zepf, P.J. (eds) (1998) *The IoPP Glossary of Packaging Terminology*, Herndon, VA: Institute of Packaging Professionals. This glossary includes nearly 8,000 terms of current acceptance in the packaging industry. The five appendices cover information on: weights and measures; container dimensions and tests; package components; terminology; notes on plastics; wooden pallets and pallet bins. They are of great value to practitioners.

South, D.W. (1994) *Encyclopedic Dictionary of Industrial Automation and Computer Control*, Englewood Cliffs, NJ: Prentice Hall PTR. This is a comprehensive reference source for production automation. Some topics covered are flow line production, numerical control, industrial robotics, material handling, group technology, flexible manufacturing systems (FMS), automated inspection, process control, and computer integrated manufacturing (CIM). Most definitions of terms include applications.

Stellman, J.M. (ed.-in-chief) (1998) *Encyclopedia of Occupational Health and Safety* (4th edn) (4 vols), Geneva, Switzerland: International Labour Office. The set is divided into four volumes covering many important areas of occupational safety and health. Major sections in each volume are as follows: Volume 1: The body and health care; prevention, management and policy; tools and approaches. Volume 2: Hazards. Volume 3: Chemicals; and industries and occupations. Volume 4: Indexes and guides.

Swamidass, P.M. (ed.) (2000) *Encyclopedia of Production and Manufacturing Management*, Boston, MA: Kluwer Academic. This covers the field of production and manufacturing management by presenting about 100 articles and more than 1,000 short entries on the most recent technical and strategic innovations in the field. The strategic and technological perspectives of this tool are representative of the competitive nature of today's manufacturing.

Ward, J.L. (2000) *Project Management Terms: a Working Glossary* (2nd edn), Arlington, VA: ESI International. This glossary summarizes the technical and nontechnical terminology used in the field of project management. It contains nearly 2,000 terms, phrases, and acronyms. It is arranged alphabetically and contains multiple cross-references.

Handbooks and manuals

There are specific handbooks for industrial engineering and also for manufacturing engineering, but there is also a wide variety of other handbooks and manuals that complement these. The following is a list of some of the most relevant handbooks and manuals.

Bralla, J.G. (1998) *Design for Manufacturability Handbook* (2nd edn), New York: McGraw-Hill. A comprehensive guide to the principles and procedures of design for manufacturability (DFM). Sections include general principles and historical perspectives of DFM; materials; formed metal components; machined components; castings; non-metallic parts; assemblies; finishes; finishes and new developments.

Charlton, S.G. and O'Brien, T.G. (eds) (2002) *Handbook of Human Factors Testing and Evaluation* (2nd edn), Mahwah, NJ: Lawrence Erlbaum Associates. Useful resource for obtaining information on human factors research and engineering, ergonomics, and experimental psychology, general principles, techniques, and specific applications of human factors tests and evaluation.

Christopher, W.F. and Thor, C.G. (eds) (1993) *Handbook for Productivity Measurement and Improvement*, Cambridge, MA: Productivity Press. Provides managers and engineers with some of the most recent literature on topics related to productivity improvement. It covers topics on the evolving theory and specific practices of world-class organizations, quality, and productivity. It also provides thorough coverage on the most advanced methods for the measurement and improvement of quality and productivity.

Cleland, D.I. and Bidanda, B. (eds) (1990) *The Automated Factory Handbook: Technology and Management*, Blue Ridge Summit, PA: TAB Professional and Reference Books. A general reference on automated manufacturing and management, it includes chapters on design; planning and control of manufacturing processes; implementation of automation; computer integrated manufacturing; and personnel issues.

Dorf, R.C. and Kusiak, A. (eds) (1994) *Handbook of Design, Manufacturing and Automation*, New York: Wiley-Interscience. Provides wide-ranging information on the theory and applications of computer-integrated manufacturing technologies. Coverage includes design and operation of manufacturing processes, fixtures in automation, packaging, costs, robotics, inspection, and manufacturing control.

Gattorna, J. (ed.) (2003) *Gower Handbook of Supply Chain Management* (5th edn), Aldershot, Hants: Gower Publishing. Chapters are written by experts in supply chain management with case studies in Asia, Europe, and North America. They include supply chains in the context of customers and strategy, operational excellence, supply chain integration and collaboration, virtual supply chains, regional and global supply chains, and other practical considerations in supply chains.

Gilleo, K. (2002) *Area Array Packaging Handbook: Manufacturing and Assembly*, New York: McGraw-Hill. This handbook focuses on the rapidly expanding field of electronics and microelectronics packaging. It presents the basics of array packaging. Specifically it describes the use of Ball Grid Array, Chip Scale Package, and Flip Chip technologies. It also shows the applicability of each technology with varying applications.

Greene, J.H. (ed.) (1997) *Production and Inventory Control Handbook* (3rd edn). New

York: McGraw-Hill. A reference book written in conjunction with the American Production and Inventory Control Society (APICS), this is an essential reference for production and inventory control personnel involved in APICS certification.

Harper, C.A. (ed.) (2000) *Electronic Packaging and Interconnection Handbook* (3rd edn), New York: McGraw-Hill. This handbook gives comprehensive coverage of electronic packaging from development and design, to manufacturing, facilities, and testing. It is one of the most commonly used references in electronics packaging. There is also broad coverage on manufacturing for all the major types of electronic products.

Harper, C.A. (ed.) (2001) *Handbook of Materials for Product Design* (3rd edn), New York: McGraw-Hill. This reference provides materials, data, information, and guidelines for designers, manufacturers, and users of electromechanical products, as well as those who develop and market materials useful for these products. Contains an extensive list of property and performance data.

Higgins, L.R., Mobley, R.K., and Smith, R. (eds) (2002) *Maintenance Engineering Handbook* (6th edn), New York: McGraw-Hill. This useful reference tool for engineers and technicians for performing maintenance covers organization of maintenance, computer applications in maintenance, costs and control, plant, mechanical equipment and electrical equipment maintenance, instrumentation and reliability, lubrication, welding, and corrosion.

Humphreys, K.K. and English, L.M. (eds) (2004) *Project and Cost Engineers' Handbook* (3rd edn), New York: Marcel Dekker. A ready reference for cost engineering covering all the basics, cost accounting and estimating, profitability, cost control, project management, operations research, and computer applications in cost engineering.

Ireson, W.G., Coombs, C.F. Jr, and Moss, R.Y. (eds) (1996) *Handbook of Reliability Engineering and Management* (2nd edn), New York: McGraw-Hill. A resourceful reference for engineers on all aspects of product reliability. Chapters include management roles in reliability, design for reliability, failure, fault-tree analysis, reliability specification, concurrent engineering, data collection and analysis, testing, failure, maintainability and reliability, quality assurance, and mathematical and statistical methods. It also contains data, tables, and charts.

Juran, J.M. and Godfrey, A.B. (1998) *Juran's Quality Handbook* (5th edn), New York: McGraw-Hill. One of the most essential and widely used reference tools in quality management and engineering. The chapters include quality planning process, quality control process, quality improvement process, process management, costs, measurement and decision-making, computer applications to quality systems, ISO 9000, benchmarking, strategic deployment, total quality management, human resources and training, product development, inspection and testing.

Kjell, B. and Zandin, K.B. (eds) (2001) *Maynard's Industrial Engineering Handbook* (5th edn), New York: McGraw-Hill. A ready and very exhaustive reference for industrial engineers and managers. The handbook is organized into several sections with chapters. The scope of each section provides a comprehensive overview of the industrial engineering profession. Many illustrations, tables, and references are included.

Kutz, M. (ed.) (2002) *Handbook of Materials Selection*, New York: John Wiley & Sons. A general reference covering materials properties and application including quantitative methods for selection, major materials, corrosion, material information management, testing and inspection, failure analysis, and manufacturing.

McMillan, G.K. and Considine, D.M. (eds) (1999) *Process/Industrial Instruments and Controls Handbook* (5th edn), New York: McGraw-Hill. This reference contains the latest methods for increasing process efficiency, production rate, and quality.

Mobley, K. (ed.) (2001) *Plant Engineer's Handbook*, Boston, MA: Butterworth-Heinemann. This reference is concerned with industrial operations or maintenance. Coverage includes the basics of plant engineering, layout and location, contracts and specifications, energy and water supply, HVAC, safety and health, maintenance, mechanical and electrical equipment, and statistical applications.

Mulcahy, D.E. (1999) *Materials Handling Handbook*, New York: McGraw-Hill. Reference for latest technologies for design, operation and maintenance of materials handling systems. Coverage includes basics of product movement, facilities layout, vertical and horizontal systems, unit loads, transportation, and economics.

Nof, S. (ed.) (1999) *Handbook of Industrial Robotics* (2nd edn), New York: John Wiley & Sons. Useful as a resource for students, engineers, and managers working in the robotics area. Chapters include historical perspectives of industrial robots, machine intelligence, mechanical design, nanorobotics, robot control, programming, economic aspects, robotics in CIM, ergonomics, design and integration, terminology, and applications.

Peach, R.W. (ed.) (2003) *The ISO 9000 Handbook* (4th edn), New York: McGraw-Hill. A ready reference for implementing ISO 9001:2000 standards; also contains CD-ROM with text of ISO 9001:2000, ISO 9000:2000, and ISO 9004:2000 standards.

Pham, H. (ed.) (2003) *Handbook of Reliability Engineering*, New York: Springer. The text provides fundamental and applied work in systems reliability engineering, including methodologies for quality, maintainability, and dependability. It also focuses on methods to find creative reliability solutions, and to improve processes.

Rosaler, R.C. (ed.) (2002) *Standard Handbook of Plant Engineering* (3rd edn), New York: McGraw-Hill. Chapters include maintenance management; facilities management; plant operations; power generation. Includes metric conversion tables.

Rushton, A., Oxley, J., and Croucher, P. (2000) *The Handbook of Logistics and Distribution Management* (2nd edn), London: Kogan Page. Chapters include concepts of logistics and distribution, planning for logistics, procurement and inventory decisions, warehousing and storage, freight transport, information and supply chain management, outsourcing, security and safety, environmental aspects, and new concepts in logistics.

Sage, A.P. and Rouse, W.B. (eds) (1999) *Handbook of Systems Engineering and Management*, New York: John Wiley. Chapters include basics of system engineering, life cycles, tests and evaluation, planning and marketing, systems engineering management, risk management, configuration management, cost management, total quality management, concurrent engineering project management, systems design, human interaction with complex systems, and operations research.

Salvendy, G. (ed.) (1997) *Handbook of Human Factors and Ergonomics* (2nd edn), New York: John Wiley. Necessary handbook for industrial engineering practitioners. Sections and chapters cover human factors, system design, motion analysis, job design, workplace and environmental design including the biomechanical aspects, health and safety, human-computer interaction, and selected applications.

Salvendy, G. (ed.) (2001) *Handbook of Industrial Engineering: Technology and Operations Management* (3rd edn), New York: John Wiley. This is an essential reference for industrial engineers. Organized into parts and chapters, its coverage is very broad and thorough, and includes all major areas in industrial engineering. Includes numerous illustrations, tables, graphs and references.

Walker, J.M. (ed.) (1996) *Handbook of Manufacturing Engineering*, New York: Marcel Dekker. This handbook provides single-source coverage on the full range of activities that comprise the manufacturing engineering process. This includes management, product and process design, tooling, equipment selection, facility planning and layout, plant construction, materials handling and storage, method analysis, time standards, and production control.

Wu, B. (2002) *Handbook of Manufacturing and Supply Systems Design: From Strategy Formulation to System Operation*, New York: Taylor & Francis. This handbook provides an organizational framework with guidelines and worksheets to assist engineers who design and manage manufacturing and supply systems.

Monographs and textbooks

The production of books in manufacturing engineering is an active part of this professional field. Books summarize the accepted knowledge in a field, providing valuable information to the researcher and practitioner. This list offers a selective number of textbooks and treatises covering both research-oriented and practical treatments. It is recommended to use a library online catalog to find more titles, searching a database (e.g., *Books in Print*), searching a bookstore website (e.g., Amazon.com), and searching a publisher online catalog. In addition, many professional journals have book review sections where readers can find out about new books in their areas.

Babu, B.V. (2004) *Process Plant Simulation*, Oxford: Oxford University Press, 2004. Written for undergraduate students, this book covers all relevant topics on the subject. Particularly interesting are the presentations of optimization techniques, mathematical methods, and CAD examples used throughout the book. It is also recommended for practitioners in the field of plant processes.

Benhabib, B. (2003) *Manufacturing: Design, Production, Automation and Integration*, New York: Marcel Dekker. This book is divided into three sections: engineering design; discrete parts manufacturing; and automatic control in manufacturing. The first part covers from concept to prototyping; the second presents examples of fabrication processes. In the third section instrumentation, production control, systems control, and quality control are presented.

Bussmann, S., Jennings, N.R., and Wooldridge, M. (2004) *Multiagent Systems for Manufacturing Control: A Design Methodology*, Berlin; New York: Springer. This book presents the Designing Agent-based Control Systems (DACS) methodology which is a new concept based on the use of semi-autonomous decision-makers in the production of products. It gives an overview of agent technologies, and provides case studies to illustrate the DACS methodology. This book is of interest to researchers and practitioners.

Cooper, C.L. and Locke, E.A. (eds) (2000) *Industrial and Organizational Psychology Linking Theory with Practice*, Oxford; Malden, MA: Blackwell Business. This book contains a set of contributing papers written by experts in the field of industrial and organizational psychology. The theoretical and practical approaches presented allow for a blend of practical-based and research-oriented discussions that will benefit students and professionals. It covers most of the relevant topics in the field.

Gershwin, S.B., Dallery, Y., Papadopoulos, C.T., and MacGregor-Smith, J. (eds) (2003) *Analysis and Modeling of Manufacturing Systems*, Boston, MA: Kluwer Academic. This collection of papers represents current research on mathematical and computational techniques applied to manufacturing systems. The prediction of factory performance, the stochastic analysis of failure, demands, and other typical events in the manufacturing process are discussed.

Groover, M.P. (2001) *Automation, Production Systems, and Computer-integrated Manufacturing* (2nd edn), Upper Saddle River, NJ: Prentice Hall. This book provides basic and fundamental coverage of topics in manufacturing systems such as flow line production, numerical control, industrial robotics, materials handling, flexible manufacturing systems, automated inspection, and process control. It is a textbook for advanced engineering students and includes many examples and exercises.

Groover, M.P. (2002) *Fundamentals of Modern Manufacturing: Materials, Processes, and Systems* (2nd edn), New York: Wiley. Text covers the three building blocks of manufacturing with major coverage of manufacturing processes and secondary coverage of engineering materials and production systems. Materials included are: metals, ceramics, polymers, composites, and silicon. It is an introductory treatment of the subject.

Lee, E.S. and Shih, H. (2001) *Fuzzy and Multi-level Decision Making: an Interactive Computational Approach*, London; New York: Springer. This is a theoretical approach to decision-making processes found in hierarchical organizations. The first part of the book covers multi-level programming algorithms, and the second part discusses knowledge representation and fuzzy decision-making.

Lee, M.H. and Rowland, J.J. (eds) (1995) *Intelligent Assembly Systems*, Singapore; Hackensack, NJ: World Scientific. Written for robotics, mechanical, and systems engineers, this book covers topics such as: software tools and administration, diagnosis systems and error handling, and sensor–actuator integration methods all essential for the development of flexible automation systems and robotics.

Monplaisir, L. and Singh, N. (2002) *Collaborative Engineering for Product Design and Development*, Stevenson Ranch, CA: American Scientific Publishers. This book presents the use of collaborative technologies as a tool to support product design and the development process. It covers basic concepts, software, hardware, and network systems necessary to implement collaborative product design.

Rehg, J.A. (2003) *Introduction to Robotics in CIM Systems* (5th edn), Upper Saddle River, NJ: Prentice Hall. Written for students and industrial practitioners, this book covers the essentials of automated manufacturing including design, development, implementation, and support for automated production systems. It also includes information about hardware specifications and about software packages that are needed for the implementation of these systems.

Ridley, J. and Channing, J. (eds) (2003) *Safety at Work* (6th edn), Amsterdam; Boston, MA: Butterworth Heinemann. Based on the premise that safety is an important function of industrial supervisors and management, this book gives comprehensive coverage of safety standards. It provides guidance about how to comply with regulations and how to achieve a high level of safety in the workplace.

Robinett, R.D., Dohrmann, C.R., Eisler, G.R., Feddema, J.T., Parker, G.G., Wilson, D.G., and Stokes, D. (2002) *Flexible Robot Dynamics and Controls*, New York: Kluwer Academic/Plenum Publishers. This text represents research and development efforts done by a team of experts at Sandia National Laboratory.

Written for graduate courses, it is also a reference source for engineers in the field. It contains many real-world examples with a strong emphasis on hardware solutions.

Smith, R. and Hawkins, B. (2004) *Lean Maintenance: Reduce Costs, Improve Quality, and Increase Market Share*, Amsterdam; Boston, MA: Elsevier Butterworth Heinemann. This book covers the area of lean maintenance (LM) in detail for those in production management, manufacturing processes, and just-in-time systems. The book presents many examples, methodologies, and checklists, and explanations of the processes involved in the implementation of LM projects.

Journals

Journals are important publications for engineers, researchers, and students in the field of industrial and manufacturing engineering because they present the results of current research, methods, and developments. Articles published in journals are usually indexed in databases such as *Compendex*, *Inspec*, and *Ergonomics Abstracts*. Most articles are now available in electronic format as well as in paper form. The list of titles included in this section is a sample of important journals in this area. This list is limited to scholarly publications, and therefore does not include trade publications.

Advances in Human Factors – Ergonomics (1984–) Amsterdam, the Netherlands: Elsevier BV (0921–2647).

Applied Ergonomics: Human Factors in Technology and Society (1969–) Oxford: Pergamon (0003–6870).

Computers and Industrial Engineering (1977–) Oxford: Pergamon (0360–8352).

Ergonomics: An International Journal of Research and Practice in Human Factors and Ergonomics (1957–) Abingdon, Oxon: Taylor & Francis (0014–0139).

European Journal of Operational Research (1977–) Amsterdam, the Netherlands: Elsevier BV (0377–2217).

Human Factors (1958–) Santa Monica, CA: Human Factors and Ergonomics Society (0018–7208).

Human Factors and Ergonomics in Manufacturing (1991–) Hoboken, NJ: John Wiley & Sons (1090–8471).

IEEE Transactions on Automation Science and Engineering (2004–) New York: Institute of Electrical and Electronics Engineers (1545–5955).

IEEE Transactions on Robotics (1985–) New York: Institute of Electrical and Electronics Engineers (1552–3098).

IEEE Transactions on Systems, Man and Cybernetics, Part A: Systems and Humans (1971–) New York: Institute of Electrical and Electronics Engineers (1083–4427).

IIE Transactions: Industrial Engineering Research and Development (1969–) Philadelphia, PA: Taylor & Francis (0740–817X).

Industrial Engineering and Management (1966–) Mumbai, India: Chary Publications. (0019–8242).

Industrial Engineer: Engineering and Management Solutions at Work (1949–) Norcross, GA: Institute of Industrial Engineers (1542–894X)..

International Journal of Advanced Manufacturing Technology (1985–) Surrey: Springer-Verlag London Ltd (0268–3768).

International Journal of Flexible Manufacturing Systems: Design, Analysis and Operation of Manufacturing and Assembly Systems (1988–) Norwell, MA: Kluwer Academic (0920–6299).

International Journal of Industrial Engineering: Theory, Applications and Practice (1994–) El Paso, TX: International Journal of Industrial Engineering (1072–4761).

International Journal of Industrial Ergonomics (1986–) Amsterdam, the Netherlands: Elsevier BV. (0169–8141).

International Journal for Manufacturing Science and Production (1997–) Tel Aviv, Israel: Freund Publishing House (0793–6648).

International Journal of Manufacturing System Design (1994–) Singapore: World Scientific Publishing (0218–3382).

International Journal of Materials and Product Technology: The Journal of Materials Innovation, Failure Preventive Technology, Product Liability and Technical Insurance (1986–) Bucks: Inderscience Enterprises (0268–1900).

International Journal of New Product Development and Innovation Management (1998–) London: Winthrop Publications (1464–6684).

International Journal of Operations and Production Management (1980–) West Yorkshire: Emerald Group Publishing (0144–3577).

International Journal of Production Research (1961–) Abingdon, Oxon: Taylor & Francis (0020–7543).

Journal of Advanced Manufacturing Systems (2002–) Singapore: World Scientific Publishing (0219–6867).

Journal of Intelligent Manufacturing (1990–) Norwell, MA: Kluwer Academic (0956–5515).

Journal of Manufacturing Systems (1982–) Oxford: Elsevier (0278–6125).

Journal of Manufacturing Technology Management (1982–) West Yorkshire: Emerald Group Publishing (1741–038X).

Journal of Quality in Maintenance Engineering (1995–) West Yorkshire: Emerald Group Publishing (1355–2511).

Journal of Sustainable Product Design: Balancing Economic, Environmental, Ethical and Social Issues in Product Design and Development (1997–) Dordrecht, the Netherlands: Kluwer Academic (1367–6679).

JSME International Journal. Series C, Mechanical Systems, Machine Elements and Manufacturing (1958–) Tokyo, Japan: Japan Society of Mechanical Engineers (1344–7653).

Process Control and Quality: The Science and Technology of Process Quality Measurement Systems – An International Journal (1990–) Leiden, the Netherlands: VSP (0924–3089).

Project Management Journal (1970–) Newtown Square, PA: Project Management Institute (8756–9728).

Psychometrika: A Journal Devoted to the Development of Psychology as a Quantitative Rational Science (1936–) Toronto, Canada: Psychometric Society (0033–3123).

Reliability Engineering and System Safety (1980–) Oxford: Elsevier (0951–8320).

Web resources

The ability of people, organizations, federal and state agencies, and corporate bodies to produce their own documents and make them accessible on their own websites has increased in the past ten years. Many of these websites are providing valuable technical and commercial information at no cost. The use of sophisticated search engines such as *Scirus* makes access to this kind of documentation more effective. Users of the Internet must be aware that in such an open environment it is important to evaluate the quality of the sources found. In this section a selective list of web resources is included; it is not a comprehensive list but rather represents examples of websites with quality information.

Alliance for Innovative Manufacturing (AIM) <http://www.stanford.edu/group/AIM/> (accessed May 17, 2005). This website is an initiative of Stanford University's Graduate School of Business, the School of Engineering, and industry with the purpose of developing new products with practical applications in society.

Best Manufacturing Practices <http://www.bmpcoe.org/> (accessed May 17, 2005). This is a good source for identifying current manufacturing research, and the promotion of exceptional manufacturing practices, methods, innovative technologies, and procedures.

Center for Integrated Manufacturing Studies (CIMS) <http://www.cims.rit.edu/> (accessed May 17, 2005). CIMS uses the expertise of its members from academic, industry, and government resources to improve manufacturing through applied technology and training. This website provides information on CIMS's three centers, and their four major programs:

National Center for Remanufacturing and Resource Recovery (NCR) <http://www.reman.rit.edu/index.asp. (accessed May 17, 2005)

Systems Modernization and Sustainment Center (SMS) <http://www.sms.rit.edu/index.asp> (accessed May 17, 2005)

Sustainable Systems Research Center, SSRC <http://www.rit.edu/~ficwww/ssrc.html> (accessed November 1, 2004)

Imaging Products Laboratory (IPL) <http://www.cims.rit.edu/ipl/index.asp> (accessed May 17, 2005)

Occupational Safety and Ergonomics Excellence Program (OSEE) <http://www.cims.rit.edu/osee/index.asp> (accessed May 17, 2005)

Center for Excellence in Lean Enterprise (CELE) <http://www.lean.rit.edu/index.asp> (accessed May 17, 2005)

Manufacturing Technologies Program (MT) <http://www.cims.rit.edu/mt/index.asp> (accessed May 17, 2005)

Commercial Technologies for Maintenance Activities (CTMA) <http://ctma.ncms.org> (accessed May 17, 2005). This is a collaboration between the National Center for Manufacturing Sciences and the Department of Defense. Organizational members benefit from the use of new manufacturing technologies created under CTMA activities. This website gives details about CTMA projects and services, and a newsletter is also available.

Consortium for Advanced Manufacturing – International (CAM-I) <http://www.cam-i.org/> (accessed May 17, 2005). This organization's main areas of interest or projects are: cost management systems (CMS); next generation manufacturing systems (NGMS); robust quality engineering; standards (STD); and Enterprise Integration Program (EIP). The website offers details about these programs, forums, and other resources.

Engineering Research Center for Reconfigurable Machining Systems <http://www.erc-assoc.org/factsheets/l/html/erc_l.htm> (accessed May 17, 2005). Based at the

University of Michigan, Ann Arbor, this center is dedicated to developing reconfigurable manufacturing systems that will allow rapid and cost-effective applications in industry. This website provides a general overview of the center.

ErgoWeb – Resource Center <http://www.ergoweb.com/resources/> (accessed May 17, 2005). On this website readers can find information about ergonomics such as an ergonomics glossary; a history of ergonomics; concepts; case studies; and reference materials. Included in this latter section are books and journals listings with some abstracts, over 3,000 bibliographic references, and links to various professional organizations in ergonomics.

European Agency for Safety and Health at Work <http://agency.osha.eu.int/> (accessed May 17, 2005). There is a considerable amount of information on this website about the organization in the form of annual reports, newsletters, magazines, reports, fact sheets, forums, conference proceedings, job opportunities, press releases, and events.

Georgia Tech Manufacturing Research Center <http://www.marc.gatech.edu/> (accessed May 17, 2005). This website describes the facilities in manufacturing processes, applications, and technological solutions to manufacturing problems at the MARC center.

Intelligent Manufacturing Systems <http://www.ims.org/> (accessed May 17, 2005). IMS' objective is to develop the next generation of manufacturing and processing technologies. This website provides information about the activities of nearly 300 companies and 200 research institutions participating in this industrial consortium.

ISO 9000 and ISO 14000 <http://www.iso.org/iso/en/iso9000-14000/index.html> (accessed May 17, 2005). Information about the ISO's two most widely known standards is presented on this site. The ISO 9000 series deals primarily with quality management, while the ISO 14000 series is primarily about environmental issues.

Laboratory for Manufacturing and Productivity (LMP) <http://web.mit.edu/org/l/lmp/www/> (accessed May 17, 2005). The LMP is an MIT laboratory engaged in conducting engineering research in manufacturing and in the development of the fundamentals of manufacturing science.

Lean Product Development Initiative (LPDI) <http://lpdi.ncms.org/> (accessed May 17, 2005). This is a program of the National Center for Manufacturing Sciences under its program of Commercial Technologies for Maintenance Activities. Users will learn more about LPDI products, services, and contact sources.

Machine Tool Agile Manufacturing Research Institute (MTAMRI) <http://mtamri.me.uiuc.edu/> (accessed May 17, 2005). The MTAMRI is a virtual research institute dedicated to the development of research projects for the purpose of improving technologies related to the design, manufacturing, and utilization of machine tools. Housed at the University of Illinois Urbana-Champaign, this site lists information about MTAMRI's projects; software test beds; newsletters; workshops; and educational programs.

Manufacturing Engineering Laboratory (MEL) <http://www.mel.nist.gov/> (accessed May 17, 2005). A division of the National Institute of Standards and Technology (NIST), the MEL website gives information about NIST's research and development projects, services and standards, and activities on metrology for industry in the U.S.A.

Manufacturing.net <http://manufacturing.net/> (accessed May 17, 2005). This website provides full-text access to twenty-three trade magazines for people interested in these areas: business and administration, procurement, design, production and operations, plant management and maintenance, and logistics. It offers a wealth of practical information to the manufacturing professional.

Manufacturing Science and Technology at Sandia National Laboratory <http://mfgshop.sandia.gov/1400_ext/1400_ext.htm> (accessed May 17, 2005).

This division of Sandia National Laboratories focuses on the development and application of advanced manufacturing processes in four areas: the manufacturing of engineering hardware; the production of weapon components; the development of robust manufacturing and processes; and the design and fabrication of unique production equipment. The website provides an overview of its operation and contacting resources.

Manufacturing Systems Integration (MISD) – NIST <http://www.mel.nist.gov/msid/> (accessed May 17, 2005). The MSID's purpose is to develop technologies and standards that can be used to implement information-intensive manufacturing systems. Through this link readers can find their products and tools; publications; services; and a description of their major research areas.

NASA-Ames Human Factors Research and Technology Division <http://humanfactors.arc.nasa.gov/> (accessed May 17, 2005). In addition to giving information about its three main areas of research – human information processing; human automation; system safety – this site also has an extensive library of documents on human factors in full text.

NASA-Cognition Lab <http://human-factors.arc.nasa.gov/cognition/coglab.html> (accessed May 17, 2005). This laboratory is a part of the Human-Automation Integration Research Branch at the NASA Ames Research Center. The activities of the lab, research projects, and contacting information are found on this site.

National Center for Manufacturing Sciences <http://www.ncms.org/> (accessed May 17, 2005). This site provides information about the programs and services of this consortium of manufacturers. Among these programs are: government partnerships; the Commercial Technologies for Maintenance Activities (CTMA); environmental health and safety; forums; and the Infragard Manufacturing Industry Association (IMIA).

National Coalition for Advanced Manufacturing (NACFAM) <http://www.nacfam.org/> (accessed May 17, 2005). The NACFAM is involved in several research activities related to public policy recommendations. This site gives information about membership, policy initiatives, research, upcoming events, and publications.

National Institute for Occupational Safety and Health (NIOSH) <http://www.cdc.gov/niosh/homepage.html> (accessed May 17, 2005). The NIOSH's main purpose is to provide programs and policies that prevent work-related illness, injury, disability, and death. This page details NIOSH's goals, contacts, organization, research, accomplishments, programs, services, activities, and publications.

Occupational Safety and Health administration <http://www.osha.org> (accessed May 17, 2005; cited November 1, 2004). Extensive information about OSHA is found here, including full-text documents of industrial safety standards, programs and policies.

Robotics-related Web Servers <http://www-robotics.cs.umass.edu/cgi-bin/robotics/> (accessed May 17, 2005). This is an extensive list of robotics-related websites maintained by the University of Massachusetts Laboratory for Perceptual Robotics.

Royal Society for the Prevention of Accidents (RoSPA) <http://www.rospa.co.uk/occupationalsafety/index.htm> (accessed May 17, 2005). This website provides access to information about this agency on topics such as: conferences and events, products and services, training and consultancy, publications, and a list of related links.

The World Wide Web Virtual Library: Industrial Engineering <http://www.isye.gatech.edu/www-ie/> (accessed May 17, 2005). Housed at Georgia Tech University, this web page provides links to academic programs; publications; courses; conferences; databases; professional societies; software; commercial entities; other links of interest; and related virtual libraries in industrial engineering.

Zentrum Mensch-Maschine-Systeme – Center of Human-Machine-Systems <http://www.zmms.tu-berlin.de/> (accessed May 17, 2005). Housed at the Technische Universität in Berlin, the center supports interdisciplinary research and system development for effective and efficient human-machine-systems. Besides proposals for industrial research partnerships, users can find information about publications, and mailing lists on this site.

Listservs, newsgroups, newsletters, weblogs, and forums are some of the other services available on the Internet. The main purpose of these services is to create a communication platform where professionals working on the same kind of engineering problems form groups to exchange ideas. There are many of these services on the Internet; one way of finding them is by searching in databases such as USENET, Catalist, JISCmail, Liszt, Google Groups, and Yahoo Groups. The most highly recognized of these services are hosted by professional societies or universities. For example, the Consortium for Advanced Manufacturing has a forum on Cost Management Systems (CMS). This organization also has a forum on Next Generation Manufacturing Systems (NGMS). It is advisable to become familiar with the websites of professional organizations, academic departments, federal and state research, and technical agencies of a specific field in order to identify these kinds of Internet services that can be valuable to the advancement of the reader's career. In many instances the services are limited to members of the organization, but others do not require membership in the organization to participate.

Societies, associations, and organizations

Societies have been the most widely used forums of communicating and networking among professionals in industrial and manufacturing engineering. Through publications from various types of journals and proceedings, conferences, expositions, trade shows, seminars, and workshops, societies still remain the pivotal point for exchange and dissemination of research and technical information. All the major societies in industrial and manufacturing engineering maintain websites which have been included in this section. Through these websites, researchers and professionals can obtain information about the society, their focus and activities, and can often access technical information and publications either directly or through purchases. Other information available on most websites also includes calendars of events, conferences, trade shows, job listings, and resources for K-12 students interested in careers in the field of industrial and manufacturing engineering.

American Society for Precision Engineering (ASPE) <http://www.aspe.net/> (accessed May 17, 2005). ASPE promotes the advancement of precision design, manufacturing, and measurement by providing a forum to encourage and enable exchange of ideas and information between industry, academia, and government laboratories, and by providing annual and topical meetings, continuing professional education and training, and publications. Membership is wide and diverse.

American Society for Quality (ASQ) <http://www.asq.org/> (accessed May 17, 2005). The ASQ, founded in 1946, originally as the American Society for Quality Control, is one of the world's leading authorities on quality issues. Membership is very diverse. The ASQ promotes quality worldwide by advancing learning, quality improvement, and knowledge exchange to improve business results; provides networking and resources for improving quality; and works with the media to promote quality-related matters; it also organizes various conferences annually worldwide.

American Society of Mechanical Engineers International <http://www.asme.org/> (accessed May 17, 2005). For more details about this society see Chapter 16 in this book.

American Society of Safety Engineers (ASSE) <http://www.asse.org/> (accessed May 17, 2005). "Founded in 1911, ASSE is the oldest and largest professional safety organization. Its more than 30,000 members manage, supervise and consult on safety, health, and environmental issues in industry, insurance, government and education. ASSE is guided by a 16-member Board of Directors, which consists of 8 regional vice presidents; three council vice presidents; Society president, president-elect, senior vice president, vice president of finance and executive director. ASSE has 12 practice specialties, 150 chapters, 56 sections and 64 student sections" (source: ASSE Web page).

Association for Facilities Engineering (AFE) <http://www.afe.org> (accessed May 17, 2005). "AFE provides education, certification, technical information and other relevant resources for plant and facility engineering, operations and maintenance professionals worldwide" (source: AFE Web page).

Association for Manufacturing Technology <http://www.mfgtech.org/> (accessed May 17, 2005). "AMT actively supports and promotes American manufacturers of machine tools and manufacturing technology. In addition to sponsoring the IMTS (International Manufacturing Technology Show), we provide our members with industry expertise and assistance on critical industry concerns. Our members represent the very best in American manufacturing technology" (source: AMT Web page).

Ergonomics Society – UK <http://www.ergonomics.org.uk/> (accessed May 17, 2005). Founded in 1949, the Ergonomics Society is the only professional society in the UK dedicated to ergonomists and those interested in ergonomics. It provides a forum for networking for professionals in this field, job placements, conferences, journals, workshops, and training.

Human Factors and Ergonomics Society (HFES) <http://www.hfes.org/> (accessed May 17, 2005). Founded in 1957, HFES has grown into an internationally recognized society whose mission is "to promote the discovery and exchange of knowledge concerning the characteristics of human beings that are applicable to the design of systems and devices of all kinds" (source: HFES Web page). The Society has twenty-one technical groups with local and student chapters; it organizes annual meetings and has several periodical publications.

IEEE Components, Packaging and Manufacturing Technology (CPMT)

<http://www.cpmt.org/> (accessed May 17, 2005). "CPMT Society is the leading international forum for scientists and engineers engaged in the research, design and development of revolutionary advances in microsystems packaging and manufacture" (source: CPMT Web page).

IEEE Engineering Management Society (EMS) <http://www.ewh.ieee.org/soc/ems/> (accessed May 17, 2005). "The IEEE Engineering Management Society (EMS) directs its efforts toward advancing the practice of engineering and technology management as a professional discipline, encouraging theory development for managing organizations with a high engineering or technical content, and promoting high professional standards among its members" (source: EMS Web page).

IEEE Robotics and Automation Society (RAS) <http://www.ncsu.edu/IEEE-RAS/> (accessed May 17, 2005). The IEEE-RAS was established in 1989 as one of the thirty-six societies and councils sponsored by the Institute of Electrical and Electronics Engineers. The IEEE-RAS serves the robotics and automation interests in the IEEE.

IEEE Systems, Man, and Cybernetics Society <http://www.ieee.org/ services/join/societies/society36.html> (accessed May 17, 2005). This society is concerned with the integration of communications, control, cybernetics, stochastics, optimization, and system structure toward the formulation of a general theory of systems; development of systems engineering technology; and human factors engineering. The society also organizes conferences and sponsors various publications.

Institute for Operations Research and the Management Sciences (INFORMS) <http://www.informs.org/> (accessed May 17, 2005). "The Institute for Operations Research and the Management Sciences (INFORMS) serves the scientific and professional needs of OR/MS investigators, scientists, students, educators, and managers, as well as the institutions they serve, by such services as publishing a variety of journals that describe the latest OR/MS methods and applications and by organizing professional conferences. The Institute also serves as a focal point for OR/MS professionals, permitting them to communicate with each other and to reach out to other professional societies and to the varied clientele of the profession's research and practice" (source: INFORMS Web page).

Institute of Industrial Engineers (IIE) <http://www.iienet.org/> (accessed May 17, 2005). "IIE is the world's largest professional society dedicated solely to the support of the industrial engineering profession and individuals involved with improving quality and productivity. Founded in 1948, IIE is an international, non-profit association that provides leadership for the application, education, training, research, and development of industrial engineering" (source: IIE Web page). The Institute has chapters worldwide, and membership spans undergraduate and graduate students, engineering practitioners and consultants in all industries, engineering managers, and engineers in education, research, and government. It organizes an annual conference in addition to various workshops, trainings, and seminars. IIE has three societies: Health Systems (SHS), Engineering and Management Systems (SEMS), and Work Science – Ergonomics and Work Measurement (SWS). In addition, IIE has numerous divisions and interest groups.

Institute of Operations Management <http://www.iomnet.org.uk/links.htm> (accessed May 17, 2005). "The Institute of Operations Management is the professional body for persons involved in operations and production management in manufacturing and service industries in the UK" (source: IOM Web page).

Institution of Electrical Engineers, UK – Manufacturing Sector <http://www.iee.org/oncomms/sector/manufacturing/> (accessed May 17, 2005). For more details about this society see Chapter 10 in this book.

Institution of Mechanical Engineers, UK <http://www.imeche.org.uk> (accessed May 17, 2005). For more details about this society see Chapter 10 in this book.

International Ergonomics Association (IEA) <http://www.iea.cc/> (accessed May 17, 2005). "The International Ergonomics Association is the federation of ergonomics and human factors societies from around the world. The mission of IEA is to elaborate and advance ergonomics science and practice and to improve the quality of life by expanding its scope of application and contribution to society" (source: IEA Web page).

Operational Research (OR) Society <http://www.orsoc.org.uk/> (accessed May 17, 2005). "The OR Society, with members in 53 countries, provides training, conferences, publications and information to those working in Operational Research. The Society also provides information about Operational Research to interested members of the general public" (source: OR society Web page).

Production and Operations Management Society (POMS) <http://www.poms.org/> (accessed May 17, 2005). Founded in 1989, POMS is an international society representing and promoting the interests of professionals in the field of production and operations management through integration and dissemination of knowledge, education, conferences and journals.

Society for Industrial and Applied Mathematics (SIAM) <http://www.siam.org/> (accessed May 17, 2005). SIAM membership is broad and includes students and professionals in applied and computational mathematics, numerical analysis, statistics, and engineers working in industry and academia. SIAM strives to advance the application of mathematics and computational science to engineering, industry, science, and society.

Society of Manufacturing Engineers (SME) <http://www.sme.org/> (accessed May 17, 2005). Formed in 1932, SME is a professional organization providing leadership and resources in manufacturing. It is an international society that brings together engineers, companies, educators, students, and others in their need to advance manufacturing in the world. The SME has various technical associations that focus on specific areas such as automated manufacturing; materials; forming and fabrication; machining; manufacturing education; product and process design; rapid technologies; robotics; and manufacturing research.

Technology Transfer Society (TTS) <http://millkern.com/washtts/docs/national.html> (accessed May 17, 2005). Founded in 1975, TTS promotes knowledge and opportunities required to achieve technology transfer. Membership is diverse, and includes professionals and students actively involved in technology transfer. TTS organizes conferences and meetings, in addition to publishing a number of technology transfer journals.

Conferences

Conferences and symposiums are usually organized or sponsored by professional engineering societies. The main purpose of a technical conference is to bring students and experts together to report on the most current results of their projects. Therefore, these kinds of meetings allow for the sharing of ideas and for networking. Presentations made at a conference are usually published as the proceedings of the conference in books, CD-ROMs, or web-based documents. Sometimes the proceedings are published as an issue of a journal. Conference and symposiums may sometimes change their titles or meeting locations and may also have irregularly scheduled meetings. Thousands of conferences are organized every year; this section shows a representative list of recent conferences in the field of industrial and manufacturing engineering. For additional resources, search for other conferences sponsored by the organizations listed below in other years, or use subject keywords and the term "congresses" when searching your local catalog or *WorldCat*.

5th International Conference on Frontiers of Design and Manufacturing (ICFDM 2002), Dalian, China, July 10–12, 2002. Sponsored by: the National Natural Science Foundation of China (NSFC); Shien-Ming Wu Foundation, USA; Chinese Mechanical Engineering Society (CMES); American Society of Mechanical Engineers (ASME); National Science Foundation, U.S.A. (NSF).

9th Design for Manufacturing Conference. Salt Lake City, UT, September 28 to October 2, 2004. Sponsored by: the American Society of Mechanical Engineers.

10th International Conference on Industrial Engineering and Engineering Management (IE&EM 2003), Shanghai, China, August 6–8, 2003. Sponsored by: the Industrial Engineering Institute of China Mechanical Engineering Society (CMES); National Science Foundation of China (NSFC); American Institute of Industrial Engineers (AIIE); Institute of Industrial Engineers of Hong Kong (HKIIE).

11th IFAC Symposium on Information Control Problems in Manufacturing (INCOM 2004), Bahia, Brazil, April 5–7, 2004. Sponsored by: the International Federation of Automatic Control (IFAC).

11th International Conference on Flexible Manufacturing, Dublin, Ireland, July 16–18, 2001.

11th International Product Development Management Conference, Design to Deliver: The Challenge of International Integration, Dublin, Ireland, June 20–22, 2004. Sponsored by: the European Institute for Advanced Studies in Management (EIASM).

14th International Conference on Flexible Automation and Intelligent Manufacturing (FAIM 2004), Toronto, Canada, July 12–14, 2004.

21st European Annual Conference on Human Decision Making and Control, Glasgow, Scotland, June 15–16, 2002. Sponsored by: the Department of Computing Science, University of Glasgow, Scotland.

1997 IEE Colloquium on Next Generation IT in Manufacturing, London, UK, December 11, 1997. Sponsored by: the Institution of Electrical Engineers (IEE).

1998 IEE Colloquium on Open Control in the Process and Manufacturing Industries. London, UK, May 15, 1998. Sponsored by: the Institution of Electrical Engineers (IEE).

2003 ASME International Mechanical Engineering Congress, The Manufacturing Engineering Division, Washington, DC, U.S.A., November 15–21, 2003.

2003 IEEE/RSJ International Conference on Intelligent Robots and Systems (IROS 2003), Las Vegas, NV, U.S.A., October 27–31, 2003. Sponsored by: the IEEE Robotics and Automation Society; the IEEE Industrial Electronics Society; Robotics Society of Japan; Society of Instruments and Control Engineers; New Technology Foundation.

Annual Reliability and Maintainability Symposium – 2004 Proceedings: International Symposium on Product Quality and Integrity, Los Angeles, CA, U.S.A., January 26–29, 2004. Sponsored by: the American Institute of Aeronautics and Astronautics (AIAA); Society of Automotive Engineers (SAE); Institute of Electrical and Electronics Engineers (IEEE); and Institute of Industrial Engineers (IIE).

Bridging the Gap: 35th Annual Conference of the Association of Canadian Ergonomists, Windsor, Ontario, Canada, October 18 – 21, 2004. Sponsored by: the Association of Canadian Ergonomists (ACE).

Change Management and the New Industrial Revolution, IEEE International Engineering Management Conference (IEMC 2001). Albany, NY, U.S.A., October 7–9, 2001. Sponsored by: the IEEE Engineering Management Society.

Eighth IEEE International Symposium on High Assurance Systems Engineering, Tampa, FL, U.S.A., March 25–26, 2004. Sponsored by: the IEEE Computer Society TCDP; National Institute for Systems Test and Productivity; University of Florida.

Energy-efficient manufacturing processes: Technical Sessions presented by the Materials Processing and Manufacturing Division of the Minerals, Metals and Materials Society (TMS). San Diego, CA, USA, March 2–6, 2003.

Environmentally Conscious Manufacturing III, Photonics Technologies for Robotics, Automation, and Manufacturing, Providence, RI, U.S.A., October 27–31, 2003. Sponsored by: SPIE – the International Society for Optical Engineering.

Ergonomics Society's Annual Conference 2004, Swansea, Wales, April 14–16, 2004. Sponsored by: the Ergonomics Society.

Human Factors and Ergonomics Society 48th Annual Meeting, New Orleans, LA, September 20–24, 2004. Sponsored by: the Human Factors and Ergonomics Society Inc.

IEE Colloquium on Responsiveness in Manufacturing, London, UK, February 23, 1998. Sponsored by: the Institution of Electrical Engineers (IEE).

Intelligent Systems in Design and Manufacturing III, Boston, MA, USA, November 5–8, 1998. Sponsored by: SPIE – the International Society for Optical Engineering.

International Conference on Agile Manufacturing, Advances in Agile Manufacturing (ICAM 2003), Beijing, China, December 4–6, 2003. Sponsored by: the International Society of Agile Manufacturing; International Society of Productivity Enhancement.

International Conference on TQM and Human Factors – Towards Successful Integration, Linkoping, Sweden, June 15–17, 1999. Sponsored by: the Centre for Studies of Humans, Technology and Organization, Linkoping, Sweden.

International Symposium on Product Quality and Integrity, Philadelphia, PA, January 22–25, 2001. Sponsored by: the Institute of Electrical and Electronics Engineers Inc.

ISA Integrated Manufacturing Solutions Real-time Manufacturing Strategies, Cleveland, OH, U.S.A., June 25–27, 2002. Sponsored by: the Instrumentation, Systems, and Automation Society (ISA).

ISA Integrated Manufacturing Solutions Supply Chain/Management Strategies, Cleveland, OH, U.S.A., June 25–27, 2002. Sponsored by: the Instrumentation, Systems, and Automation Society (ISA).

Modeling, Simulation, and Control Technologies for Manufacturing, Philadelphia, PA, U.S.A., October 25–26, 1995. Sponsored by: the SPIE – International Society for Optical Engineering.

Modern Trends in Manufacturing, Second International CAMT Conference (Centre for Advanced Manufacturing Technologies), Wroclaw, Poland, February 20–21, 2003. Sponsored by the European Commission Centre of Excellence (COFEXC).

Metrology-based Control for Micro-manufacturing, San Jose, CA, U.S.A., January 25, 2001. Sponsored by: SPIE – the International Society for Optical Engineering.

Network Intelligence: Internet-based Manufacturing. Boston, MA, U.S.A., November 8, 2000. Sponsored by: SPIE – the International Society for Optical Engineering.

Optical Measurement Systems for Industrial Inspections II: Applications in Production Engineering, Munich, Germany, June 20–21, 2001. Sponsored by: SPIE – the International Society for Optical Engineering.

Safety Instrumented Systems for the Process Industry, Houston, TX, U.S.A., March 17–20, 2003. Sponsored by: the ISA Services Inc.

Sensors and Controls for Intelligent Manufacturing II, Boston, MA, U.S.A., October 28, 2001. Sponsored by: SPIE – the International Society for Optical Engineering.

XVI Annual International Occupational Ergonomics and Safety Conference, Toronto, Ontario, Canada, June 10–13, 2002.

Conclusion

Industrial and manufacturing engineering disciplines have expanded dramatically over the past century. With continued global competition for consumer and other products, and the need for high quality at low cost, new technologies continue to be developed. Traditional design and manufacturing techniques have been transformed into computer-assisted technologies, especially with the evolution of microprocessors. With the evolution of the Internet, product design is now an integrated process involving multi-disciplinary approaches, done concurrently with process development. The integration of enterprise data has changed the face of industrial and manufacturing engineering through niche techniques such as rapid prototyping, design automation, supply chain management, automation technologies, and new management techniques. Research in these disciplines continues to focus on newer and leaner manufacturing processes, environmentally friendly designs and processes, zero defects in production, higher productivities, new materials, especially in micro and nanotechnology, and alternative processes that use less or non-conventional energies. As the SME suggests, "innovation, productivity, flexibility, and continuous improvement are key ingredients to success in the constantly evolving world of manufacturing" (SME, 2004). Manufacturing is a key economic activity for every nation as it is one of the leading sources of wealth. It is therefore important that industrial and manufacturing engineers be accorded key literature in their discipline in order to keep the practice competitive and innovative.

References

American National Standards Institute (ANSI) (1989) *Standard Z94.0: Industrial Engineering Terminology*, Norcross, GA: Institute of Industrial Engineers.

Benhabib, B. (2003) *Manufacturing – Design, Production, Automation and Integration*, New York: Marcel Dekker.

Chang, T.C., Wysk, R.A., and Wand, H.P. (1998) *Computer-aided Manufacturing* (2nd edn), Upper Saddle River, NJ: Prentice Hall.

Ergonomics Society of UK (2004) <http://www.ergonomics.org.uk/> (accessed May 14, 2004).

Groover, M.P. (2002) *Fundamentals of Modern Manufacturing – Materials, Processes and Systems*, New York: John Wiley & Sons.

Hradesky, J. (1995) *Total Quality Management Handbook*, New York: McGraw-Hill.

Kalpakjian, S. and Schmid, S.R. (2001) *Manufacturing Engineering and Technology*, Upper Saddle River, NJ: Prentice Hall.

Karwowski, W. and Rodrick, D. (2001) "Physical Tasks: Analysis, Design and Operation," in G. Salvendy (ed.), *Handbook of Industrial Engineering – Technology and Operations Management*, New York: John Wiley & Sons, pp. 1041–1110.

Lincoln Electric Company (1994) *The Procedure Handbook of Arc Welding* (13th edn), Cleveland, OH: Lincoln Electric Company.

Marmaras, N. and Kontoginis, T. (2001) "Cognitive Tasks," in G. Salvendy (ed.) *Handbook of Industrial Engineering – Technology and Operations Management*, New York: John Wiley & Sons, pp. 1013–1040.

Martin-Vega, L.A. (2001) "The Purpose and Evolution of Industrial Engineering," in K.B. Zandin (ed.) *Maynard's Industrial Engineering Handbook*, New York: McGraw-Hill, pp. 1.3–1.19.

NIST/SEMATECH e-Handbook of Statistical Methods <http://www.itl.nist.gov/div898/handbook/> (cited March 11, 2005).

Salvendy, G. (ed.) (2001) *Handbook of Industrial Engineering – Technology and Operations Management*, New York: John Wiley & Sons.

Sink, D.S., Poirier, D.F., and Smith, G.L. (2001) "Full Potential Utilization of Industrial and Systems Engineering in Organizations," in G. Salvendy (ed.) *Handbook of Industrial Engineering – Technology and Operations Management*, New York: John Wiley & Sons, pp. 3–25.

SME (2004) About SME <http://web.archive.org/web/20040414204520/ www.sme.org/cgi-bin/abouthtml.p1?/html/abouthtm&&&SME&> (cited April 11, 2004).

Soska, G.V. (1984) "The Five Disciplines of Manufacturing Engineering Education," *Proceedings of the Second Annual International Robot Conference*. Wheaton, IL: Tower Conference Management, pp 223–226.

Walker, J.M. (1996) *The Handbook of Manufacturing Engineering*, New York: Marcel Dekker.

Zandin, K.B. (ed.) (2001) *Maynard's Industrial Engineering Handbook*, New York: McGraw-Hill.

15 Materials science and engineering

Godlind Johnson

Introduction

Human beings have engaged in some form of "materials engineering" since the beginning of life on Earth; we have even named eras after the materials that were predominantly used in those historic times (e.g., Stone Age, Bronze Age). However, materials science and engineering evolved into a distinct engineering discipline only in the 1960s. The Materials Research Society, founded in 1973, defines the field thus: "Materials Science and Engineering encompasses the study of the structure and properties of any material, as well as using this body of knowledge to create new types of materials, and to tailor the properties of a material for specific uses."

The backgrounds of today's materials science researchers and practitioners range from physics and chemistry to earth sciences, biology, and medicine. Elaborating on the definition above one can outline four distinct but interrelated activities that materials scientists engage in: determining the structure of materials (e.g., by microscopy, powder diffraction), measuring and describing the properties of materials (both mechanical and physical), devising ways of processing materials (i.e., creating new materials, transforming existing materials, and making things out of them), and matching materials with applications or improving performance of materials that are already being used for an application. While any type of material may be the subject of research, the field grew out of metallurgy and ceramics and now includes: metals, polymers, semiconductors, electromaterials, composites, and more recently biomaterials and nanomaterials. Their structural, mechanical, electrical, thermal, magnetic or optical properties, the so-called "engineering properties," are characterized (measured and described). The many methods of characterization that are employed include powder diffraction, electron microscopy, corrosion studies, and more.

The literature of materials science and engineering

Since materials science is such a broad and interdisciplinary field, it is not surprising that the information sources are also many and varied. Handbooks and property guides are especially important. In this guide the most recent sources will be discussed; for slightly older materials the reader should refer to guides such as *Guide to Information Sources in Engineering* by Charles R. Lord (2000). In addition, the emphasis lies on information sources specific to materials science and engineering, excluding those of the related sciences, physics and chemistry, and general engineering.

Searching the library catalog

Considering how this discipline draws on so many different fields and is developing new areas of research and applications rapidly, the best approach to searching the library catalog is to start with keyword searches, combining specific concepts with the terms "materials" or "engineering" (e.g., materials and testing, analysis, microscopy). Once an appropriate title is found, the Subject Headings listed for it may lead to additional possible terms. The following selection of useful Subject Headings also reflects the breadth of subject matter that belongs to the field of materials science:

Acoustical Materials
Biomedical Materials
Building Materials
Characterization of Materials
Chromic Materials
Coatings
Composite Materials
Electric Conductors
Foamed Materials
Friction Materials
Granular Materials
Hard Materials
Inhomogeneous Materials
Laser Materials
Magnetic Materials
Manufacturing Processes
Materials
Materials at High Temperatures
Materials at Low Temperatures
Materials – Thermal Properties
Microstructure
Nonmetallic Materials
Optical Materials
Polymers
Porous Materials
Sintering
Smart Materials
Strategic Materials
Super Lattices Materials

Surfaces (Technology)
Viscoelastic Materials
Welding

The Library of Congress classification system used in most academic libraries places the core of materials science literature in the TA400s, but again, many other T as well as QC, QD call numbers are also relevant:

QC176	Solid state physics
QC350–467	Optics and light
QD71–142	Analytical chemistry
QD241–441	Organic chemistry
QD450–801	Physical and theoretical chemistry
QD901–999	Crystallography
R856–857	Biomedical engineering
TA329–348	Engineering mathematics
TA401–492	Materials engineering and construction. Mechanics of materials
TN600–799	Metallurgy
TP155–156	Chemical engineering
TP785–869	Clay industry, ceramics, glass
TP1080–1185	Polymers and polymer manufacture

Indexes and abstracts

Indexes/abstracts are the tools to discover a body of research or information about a subject/discipline/topic. They may cover a large interdisciplinary field, or a very narrow and specialized field. These indexes give access to the output of research by indexing the content of a carefully selected group of sources. Aspects of materials science are covered in the large physics (*Inspec*), chemistry *(Chemical Abstracts/SciFinder*) and biology (*Biosis*) indexes, in the general engineering indexes (*Applied Science and Technology Index* and *Compendex*), and in a number of very specialized tools that will be included here. Unless noted specifically, the indexes/abstracts listed here exist in print and electronic form, and require subscriptions to be accessed. *The CSA Materials Database with METADEX* includes the most important indexes/abstracts for this discipline.

Aerospace and High Technology Database (1990s–) Bethesda, MD: Cambridge Scientific Abstracts. "Provides citations and abstracts of basic and applied research in aeronautics, astronautics, and space sciences; in addition to journal literature, the database also includes coverage of reports issued by NASA and other U.S. government agencies." Materials, optics, acoustics, and plasmas are among the major areas of coverage. Included in the *CSA Materials Databases with METADEX* (source: CSA website).

Alloys Index (1974–) London: Institute of Materials; Materials Park, OH: ASM International. This is a tool for finding the literature abstracted in *Metals Abstracts* by material name, or alloy classification. It may also be used as a stand-alone index and is included in the *CSA Materials Databases with METADEX*.

Aluminum Industry Abstracts (1992–) Materials Park, OH: Published by Materials Information – a joint service of ASM International and the Institute of Metals. "Provides citations and abstracts to the technical literature on aluminum, including production processes, products, applications, and business developments. Covers scientific and technical journals, government reports, conference proceedings, dissertations, books, and patents." Included in the *CSA Materials Database with METADEX*.

Ceramic Abstracts (1922–) American Ceramic Society; CERAM Research (Firm), Bethesda, MD: Cambridge Scientific Abstracts. "Provides citations and abstracts to the literature on ceramics – which are increasingly becoming the material of choice for high temperature applications, erosive environments, biomedical applications, and other areas where hardness and corrosion resistance are needed. Major areas of coverage include sintering, powder compacts, precursor powders, performance testing, crystal structure, wear, applications, silicon nitride and glass ceramics." Source materials include over 300 journals, conference proceedings, books, patents, standards, and company product literature. The electronic version covers 1990– and includes *World Ceramic Abstracts* (formerly *British Ceramic Abstracts*) and is included in *CSA Materials Database with METADEX*. (source: CSA website)

Chemical Abstracts (1907–) Columbus, OH: American Chemical Society. The electronic version of *Chemical Abstracts* is most commonly searched via *SciFinder Scholar*. Much of the literature on the scientific foundations of materials science may be found here, since chemical literature in the widest sense, with all peripheral disciplines, is indexed in *Chemical Abstracts*. The sources include journals, patents, technical reports, dissertations, conference proceedings, and books in biochemistry; physical, inorganic, and analytical chemistry; applied chemistry and chemical engineering; macromolecular chemistry; and organic chemistry.

Corrosion Abstracts (1990s–) National Association of Corrosion Engineers, Bethesda, MD: Cambridge Scientific Abstracts. In print from 1962– "Provides citations and abstracts in the areas of general corrosion, testing, corrosion characteristics, preventive measures, materials construction and performance and equipment for many industries. Major areas of coverage include alloying, atmospheric corrosion, cathodic protection, corrosion in oil and gas production, corrosion in specific materials, corrosion potential, corrosion prevention, corrosion resistance, cracking, creep, designing for cathodic protection, designing for corrosion control, diffusion, fatigue, and more." Included in *CSA Materials Database with METADEX* (source: CSA website).

CSA Materials Research Database with METADEX (1966–) Bethesda, MD: Cambridge Scientific Abstracts. "Access to these individually described indexes: Aluminum Industry Abstracts, Ceramic Abstracts/World Ceramics Abstracts, Copper Data Center Database, Corrosion Abstracts, Engineered Materials Abstracts, Materials Business File, METADEX, and WELDASEARCH" (source: CSA website).

Compendex (see Chapter 2 in this book).

Engineered Materials Abstracts (1986–) Bethesda, MD: Cambridge Scientific Abstracts. "The growing importance of polymers, ceramics, and composites in a variety of structural and other advanced applications requires the in-depth coverage provided by Engineered Materials Abstracts (EMA). Citations regarding the research, manufacturing practices, properties and applications of these materials have been taken from 1,300 journals, plus dissertations, government reports, conference proceedings, and books indexed by expert editors from Materials Information. Begun in 1986, Engineered Materials Abstracts is an electronic database containing Ceramics, Composites and Polymers subfiles." Included in *CSA Materials Database with METADEX* (source: CSA website).

Inspec (see Chapter 2 in this book).

Liquid Crystal Database (1999–) Edinburgh: Edinburgh Engineering Virtual Library (EEVL) <http://www.eevl.ac.uk/lcd/index.htm> (accessed May 31, 2005). "The Liquid Crystal Database is a bibliographic database which contains references (Title, Author, Journal, Abstract, etc.) to over 700 specialized records. It covers items in niche and mainstream journals and magazines, and also includes details of books, videos and conference proceedings. It is updated regularly and is provided as part of the EEVL service" (source: LCD Home page http://www.eevl.ac.uk/lcd/index.htm).

Materials Business File (1985–) Bethesda, MD: Cambridge Scientific Abstracts. "Materials Business File is the database equivalent of the bulletins *Steels Alert*, *Polymers/Ceramics/Composites Alert* and *Nonferrous Metals Alert*. Together they cover new technologies, new materials and other industry developments reported in the press." Included in *CSA Materials Database with METADEX* (source: CSA website).

Metals Abstracts (1968–) London: Metals Society and Materials Park, OH: American Society for Metals. This is a merger of *Metallurgical Abstracts* (1931–1967) and *Review of Metals Literature* (1945–1967). The electronic equivalent is *METADEX* (1966–) Bethesda, MD: Cambridge Scientific Abstracts. "METADEX is the only comprehensive source for information on metals and alloys: their properties, manufacturing, applications, and development. Information from over 2,000 journals, plus patents, dissertations, government reports, conference proceedings, and books are indexed by expert editors from Cambridge Scientific Abstracts." METADEX includes *Alloys Index*. Included in the *CSA Materials Database with METADEX* (source: CSA website).

Phase Equilibria Diagrams. Cumulative Indexes (1992–) Materials Science and Engineering Laboratory (U.S.), American Ceramic Society, Westerville, OH: American Ceramic Society. "Indexes: *Phase Diagrams for Ceramists (Final compilation)*; *Phase Equilibria Diagrams (Final compilation); Phase Diagrams for Ceramists (Annual); Phase Equilibria Diagrams (Annual)*; and, special unnumbered issues of *Phase Diagrams for Ceramists (Final compilation)* and *Phase Equilibria Diagrams (Final compilation)*."

Platinum Group Metals (PGM) Database <http://www.platinummetalsreview.com/jmpgm/index.jsp> (cited May 31, 2005). A free database of information on more than 400 alloys, including diagrams and graphs with numerical data points, phase diagrams of platinum group metals alloy systems. "Properties of materials are displayed ... giving a comprehensive picture of a material and its capabilities." The database is sponsored by The International Platinum Association, Anglo Platinum and Johnson Mattey.

Polymer Abstracts (1972–) Rapra Technology Limited, Bethesda, MD: Cambridge Scientific Abstracts. Formerly Rapra Abstracts. "Dedicated exclusively to information on rubber, plastics, adhesives and polymeric composites. Includes more than 700,000 records from 1972 to the present, covering commercial, marketing, technical and academic aspects of the rubber and plastics industry." Included in *CSA Materials Database with METADEX* (source: CSA website).

Science Citation Index (see Chapter 2 in this book).

ScienceDirect (1990s–) <www.sciencedirect.com> (cited November 19, 2004), New York, NY: Elsevier Science. This is a searchable full-text database of journals published by Elsevier and other publishers now owned by Elsevier. Many key materials science journals are published by Elsevier.

Weldasearch (1990s–) Welding Institute, Bethesda, MD: Cambridge Scientific Abstracts. "Provides citations and abstracts to journal articles, research reports, books, standards, patents, theses and special publications in the areas of joining metals, plastics and ceramics, metal and ceramic spraying, thermal cutting, brazing, soldering and related topics. Every aspect of welding and allied processes is covered" Included in *CSA Materials Database with METADEX* (source: CSA website).

World Ceramics Abstracts (see Ceramic Abstracts above).

Bibliographies and guides to the literature

While chapters devoted to materials science literature may be found in the more comprehensive guides treated in Chapter 2, very few specific guides or bibliographies have been published recently. The most useful and up-to-date guides may be found on the websites of many of the major engineering libraries, for which some examples are listed below:

MIT Libraries Subject Guide/Materials Science <libraries.mit.edu/guides/subjects/materials/internet.html#props> (accessed May 17, 2005).

Stanford University Engineering Library Subject Guide/Materials Science <http://www-sul.stanford.edu/depts/eng/research_help/guides/materials.html#handbooks> (accessed May 17, 2005).

University of California San Diego Science and Engineering Library Subject Guide/Materials Science <http://libraries.ucsd.edu/sage/subjects//materials_science.html> (accessed May 17, 2005).

University of Wisconsin/Kurt Wendt Engineering Library Subject Guide/Materials Science & Engineering <www.wisc.edu/wendt/help/mseguide.html> (accessed May 17, 2005).

Houldcraft, P.T. (ed.) (1987) *Materials Data Sources*, London: Materials Data Group; Institution of Mechanical Engineers. A bibliographic guide to books containing property data on various classes of materials.

Patten, M.N. (ed.) (1989) *Information Sources in Metallic Materials*, London: Bowker-Saur. However dated, this is an in-depth guide to the literature of each type of metallic material, in eighteen chapters.

Directories

The advent of the WWW has had an impact on the publication of directories. With the increasing ease of finding any kind of information about individuals, companies, or organizations on the Internet, the publication of directories is becoming more and more redundant. Listed here are a few print or online directories that may offer advantages over an Internet search due to their organization and comprehensiveness. Further directory information may be found on the societies' websites, and the Web resources listed in later chapters.

ASTM Directory of Scientific and Technical Consultants and Expert Witnesses (annual) Philadelphia, PA: American Society of Testing and Materials. Online version at <http://www.astm.org/consultants> (accessed May 18, 2005). Includes 250 categories of specialized labs and consulting services.

Industrial Quick Search <http://www.industrialquicksearch.com/> (accessed May 18, 2005). Grand Rapids, MI: IQS, Inc.: "one of the top resources for finding products, services, manufacturers and suppliers. The IQS industrial directory system prides itself in ease of use and the aiding of qualified users in their search for products and services" (source: IQS website).

IndustryLink <http://www.industrylink.com> (accessed May 17, 2005). Missisauga, Canada: Kuhn Marketing Technology. "IndustryLink focuses on providing . . . a wealth of information easily categorized and quickly accessible. . . . Each of the industries covered . . . begins with a section entitled "Resource Sites." These sites are designed to provide an abundance of information related exclusively to their particular industry" (source: Home page).

Industry Search <http://www.industrysearch.com> (accessed May 17, 2005). An online directory of industry portals, trade associations, and over 120,000 manufacturing and supplier companies.

Materials Research Centres: A World Directory of Organizations and Programmes (1995–) (6th edn), London: Cartermill; New York: Stockton Press. About 5,000 industrial, governmental, and academic laboratories in seventy-six countries are included. Listed country with a detailed subject index.

Metallurgy/Materials Education Yearbook <http://www.asminternational.org/Template.cfm?Section=Metallurgy_Materials_Education_Yearbook&Template=/yearbook/yearbook.cfm> (accessed May 17, 2005). Metals Park, Ohio: American Society for Materials. Current information on higher education programs in metallurgy, ceramics, and polymer science. International coverage with emphasis on North America.

Polysort <http://www.polysort.com> (accessed May 17, 2005). "Polysort is an online community for the plastics and rubber industry, providing news, links and networking opportunities. Polysort's Links Directory features more than 5,000 links to plastics and rubber industry suppliers, processors and more" (source: Home page).

Thomas Register of American Manufacturers (see Chapter 14 on Industrial and manufacturing engineering).

Encyclopedias and dictionaries

There is a proliferation of recent, very specialized encyclopedias as well as concise extracts of some of them. They should be useful for students looking for an in-depth overview and for researchers needing information about topics outside their own areas of expertise. Dictionaries were chosen for the quality of their definitions and currency.

Davis, J.R. (ed.) (1992) *ASM Materials Engineering Dictionary*, Metals Park, OH: ASM International. This dictionary covers all material types with detailed definitions, processes, charts, and tables.

Bever, M.B. (ed.-in-chief) (1986) *Encyclopedia of Materials Science and Engineering* (8 vols and 2 supp.), Oxford: Pergamon; Cambridge, MA: MIT Press. This is still the basic, standard reference work of the field, updated periodically by supplementary volumes and by the *Encyclopedia of Advanced Materials.* It also spawned many concise encyclopedias in the early 1990s (e.g., *Concise Encyclopedia of Composites*).

Bloor, D. and Khan, R.W. (1994) *Encyclopedia of Advanced Materials* (4 vols), Oxford; Tarrytown, N.Y: Pergamon Press. Advanced materials are artificially produced to meet the requirements of particular applications. The rapid developments in this field are covered in more depth than was possible in the standard *Encyclopedia of Materials Science and Engineering* (1986) and its supplements. Examples of topics covered are: aerogel catalysts, biomedical polymers, color in ceramics, fast ion conduction in ceramics, hydrogels, metal matrix composites, and oligomers. There is an excellent subject index and a wealth of illustrations.

Buschow, K.H.J. (ed.-in-chief) (2001) *Encyclopedia of Materials: Science and Technology* (10 vols), Amsterdam; New York: Elsevier. Electronic version through ScienceDirect. Over 2,000 articles written by internationally recognized experts are included in this comprehensive encyclopedia. It includes extensive tables, diagrams, and bibliographies.

Cahn, R.W. and Lifshin, E. (eds) (2004) *Concise Encyclopedia of Materials Characterization* (2nd edn), Amsterdam; Oxford: Elsevier. A collection of 116 concise and authoritative articles on most techniques available to characterize materials, ranging from the well established to the very latest, including: atomic force microscopy; confocal optical microscopy; gamma ray diffractometry; thermal wave imaging; x-ray diffraction and time-resolved techniques. It is well illustrated and includes useful references.

Cheremisinoff, N.P. (2001) *Condensed Encyclopedia of Polymer Engineering Terms*, Boston, MA: Butterworth-Heinemann. Provides an overview of commercially available polymers.

Keller, H. and Erb, U. (2004) *Dictionary of Engineered Materials*, Hoboken, NJ: Wiley. Also a Wiley Electronic Book. In over 40,000 entries the authors provide succinct information, including type of material, composition, properties and use of traditional and new materials such as dendrimers, fullerenes and photonics.

Kelly, A. and Zweben, C. (eds-in-chief) (2000) *Comprehensive Composite Materials* (6 vols), Amsterdam; New York: Elsevier. Electronic version through ScienceDirect. Topics covered in the six volumes are fiber reinforcements and general theory of composites, polymer matrix composites, metal matrix composites, carbon/carbon, cement, and ceramic matrix composites, test methods, non-destructive evaluation, and smart materials, and design and applications.

Kelly, A., Cahn, W., and Bever, M.B. (1994) *Concise Encyclopedia of Composites* (rev. edn), Oxford; Tarrytown, NY: Pergamon Press. Culled and updated from the *Encyclopedia of Materials Science and Engineering.*

Mark, H.F. (ed.-in-chief) (2004) *Encyclopedia of Polymer Science and Engineering* (3rd edn) (12 vols), New York; Chichester: Wiley; also a Wiley electronic encyclopedia. This third edition is a completely new version of the *Encyclopedia of Polymer Science and Technology.* The new edition will bring the state of the art up to the twenty-first century, with coverage of nanotechnology, new imaging and analytical techniques, new methods of controlled polymer architecture, biomimetics, and more.

Nalwa, H.S. (2004) *Encyclopedia of Nanoscience and Nanotechnology* (10 vols), Stevenson Ranch, CA: American Scientific Publishers; also available in electronic form. This extensive work brings together the current knowledge about all aspects of nanotechnology. In over 410 chapters it provides an introduction and overview of most recent advances and emerging new aspects of nanotechnology spanning science to engineering to medicine. Extensive bibliographic references and illustrations.

Novikov, V. (2003) *Concise Dictionary of Materials Science: Structure and Characterization of Polycrystalline Materials*, Boca Raton, FL: CRC Press. Entries for about 1,400 terms cover the description, development, and characterization of polycrystalline materials. The author states in the preface that this dictionary is intended to bridge the gap between textbooks and professional literature. The book also contains a three-page list of symbols, two pages of acronyms, and a two-page list of literature for further reading. An English–German/German–English glossary is included as well.

Rosato, D.V. (2000) *Concise Encyclopedia of Plastics*, Boston, MA: Kluwer Academic. The world of plastics is explained in over 25,000 terms. In addition, lists of abbreviations, conversions, industry associations, websites, and more make this dictionary comprehensive and useful.

Salamone, J.C. (ed.-in-chief) (1996) *Polymeric Materials Encyclopedia* (12 vols), Boca Raton, FL: CRC Press. A comprehensive reference work on the synthesis, properties, and applications of polymeric materials.

Salamone, J.C. (ed.-in-chief) (1999) *Concise Polymeric Materials Encyclopedia*, Boca Raton, FL: CRC Press. The most widely applicable articles were culled from the twelve volume *Polymeric Materials Encyclopedia* (see above).

Schwartz, M.M. (ed.) (2002) *Encyclopedia of Materials, Parts, and Finishes* (2nd edn), Boca Raton, FL: CRC Press. Also EngnetBase. This second edition of the encyclopedia covers the new materials that have been invented or modified in recent years (including matrix composites, nano-structures, smart piezoelectric materials, shape memory alloys, and intermetallics), and updates information on basic materials as well. Many tables and figures, but no references.

Schwartz, M.M. (ed.) (2002) *Encyclopedia of Smart Materials* (2 vols), New York: Wiley; also a Wiley electronic book and in Knovel. This indexed two-volume set brings together much of the information known about smart materials; smart materials may be defined as materials that are designed to respond to changes in their environment; applications may be found in almost every area of modern engineering, from biomedical to optical and structural. Each article features materials properties, photographs, charts and tables, and a bibliography for further reading.

Schwarz, J.A., Contescu, C.I., and Putyera, K. (eds) (2004) *Dekker Encyclopedia of Nanoscience and Nanotechnology* (5 vols), New York: Marcel Dekker; also available in electronic form. Another comprehensive encyclopedia recording the state of the art of all aspects of the emerging field of nanoscience.

Schweitzer, P.A. (ed.) (2004) *Encyclopedia of Corrosion Technology* (2nd rev. and enlarged edn), New York: Marcel Dekker. Provides information on all materials affected by corrosion. References for further reading are included with many entries and the encyclopedia is amply illustrated.

Ullmann's Encyclopedia of Industrial Chemistry (2003) (6th edn) (40 vols), New York: Wiley-VCH; also a Wiley Internet encyclopedia. State-of-the-art reference work detailing science and technology in all areas of industrial chemistry. Fully international in scope and coverage. With nearly 10,000 tables, 30,000 figures, and innumerable literature sources and cross-references, *Ullmann's Encyclopedia* offers a wealth of comprehensive and well-structured information on all facets of industrial chemistry.

Handbooks, manuals, and material property databases

Handbooks may be the most valuable resource for the practicing engineer as well as students and researchers, as they present collections of fundamental facts, formulas, tables, and, specific for materials science, the properties of materials. An attempt was made to find at least one recent and comprehensive handbook for each subdiscipline of materials science, either by type of material, or process or application. Special attention was given to the new developments in nanoscience and biomaterials.

ACerS-NIST Phase Equilibria Diagram CD-ROM Database Version 3.0 (2004) Westerville, OH: American Ceramics Society. The Phase Equilibria Diagrams CD-ROM Database Version 3.0 contains more than 19,000 diagrams published in twenty phase volumes produced as part of the ACerS-NIST Phase Equilibria Diagrams Program: volumes I through XIII; Annuals 91, 92 and 93; High Tc Superconductor I & II; Zirconium & Zirconia Systems; and Electronic Ceramics I. This new version offers full commentary text display in addition to diagram display. This makes the CD the same as the printed volumes with many additional features, including high-quality printing and export capability. On-screen plotting can show any line types such as dotted, dash-dot, dash, and solid. Files may be saved in both wmf and bitmap formats.

Amelinckx, S. (ed.) (1996) *Handbook of Microscopy: Applications in Materials Science, Solid-state Physics and Chemistry* (3 vols), New York: Wiley-VCH. A comprehensive treatment of microscopy methods used in the characterization of materials (volumes 1 and 2) and how these techniques can be applied successfully for a given material (volume 3).

ASM International (2002–) *ASM Handbooks Online*, Materials Park, OH: ASM International <http://products.asminternational.org/hbk/index.jsp>. Twenty-one volumes of the ASM handbook series plus the desk editions of *Engineered Materials Handbook* and *Metals Handbook* are presented together in an online format: vol. 1, Properties and Selection: Irons, Steels, and High Performance Alloys; vol. 2, Properties and Selection: Nonferrous Alloys and Special-purpose Materials; vol. 3, Alloy Phase Diagrams; vol. 4, Heat Treating; vol. 5, Surface Engineering; vol. 6, Welding, Brazing, and Soldering; vol. 7, Powder Metal Technologies and Applications; vol. 8, Mechanical Testing and Evaluation; vol. 9, Metallography and Microstructures; vol. 10, Materials Characterization; vol. 11, Failure Analysis and Prevention; vol. 12, Fractography; vol. 13, Corrosion; vol. 14, Forming and Forging; vol. 15, Casting; vol. 16, Machining; vol. 17, Nondestructive Evaluation and Quality Control; vol. 18, Friction, Lubrication, and Wear Technology; vol. 19, Fatigue and Fracture; vol. 20, Materials Selection and Design; vol. 21, Composites; ASM Desk Editions.

ASM Ready Reference Collection (1997–) Materials Park, OH: ASM International. This collection of comprehensive handbooks includes: properties and units for engineering alloys; electrical and magnetic properties of metals; thermal properties of metals.

ASM Specialty Handbooks (1993–) Materials Park, OH: ASM International. So far eleven volumes have been published in this series; they are intended to complement and expand on the corresponding ASM Handbooks. Each volume deals with different types of steel or alloys.

Asmussin, J. and Reinhard, D.K. (2002) *Diamond Films Handbook*, New York: Marcel Dekker. Detailed information on the science, processes, and application of low-pressure diamond deposition, which has been developed over the past two decades to synthesize a form of diamond that is metastable with respect to graphite, as distinguished from the high-pressure synthesis of diamond that is the stable form of carbon.

Bhushan, B. (1998) *Handbook of Micro/Nanotribology* (2nd edn), Boca Raton, FL: CRC Press. The state of the art of this rapidly evolving field is presented in sixteen chapters. Relevant topics range from AFM instrumentation, characterization of solid surfaces, measurement techniques and applications to theoretical modeling of interfaces, and design and construction of magnetic storage devices. Five hundred illustrations and fifty tables are included.

Black, J. and Hastings, G.W. (1998) *Handbook of Biomaterial Properties*, London; New York: Chapman and Hall. This is a compilation of data on natural tissues and fluids and how various implantable materials interact with them. The materials covered range from stainless steels, CoCr-based alloys, titanium and titanium alloys, to thermoplastic polymers and oxide bioceramics. Biocompatibility is discussed with each material and in a chapter on general concepts of biocompatibility.

Brady, G.S. (ed.) (2002) *Materials Handbook: An Encyclopedia for Managers, Technical Professionals, Purchasing and Production Managers, Technicians, and Supervisors* (15th edn), New York: McGraw-Hill. This classic handbook covers essential information for over 15,000 materials. In the latest edition dozens of material families have been updated extensively, including adhesives, activated carbon, fullerenes, heat-transfer fluids, nanophase materials, nickel alloys, olefins, silicon nitride, stainless steels, and thermoplastic elastomers. Property information is embedded in the encyclopedic entry.

Brandrup, J. and Immergut, E.H. (eds) (2004, 1999) *Polymer Handbook* (4th edn) (2 vols), New York: Wiley-Interscience. This is the standard reference for polymer data necessary for theoretical and experimental polymer research. The publisher's online version, *Wiley Database of Polymer Properties*, features added functionality and is updated quarterly.

Bunshah, R.F. (ed.) (2001) *Handbook of Hard Coatings: Deposition Technologies, Properties and Applications*, Park Ridge, NJ: Noyes Publications; Norwich, NY: William Andrews Publications. A comprehensive resource for information on the fabrication, characterization, and applications in the field of hard coatings and wear-resistant surfaces.

Cardarelli, F. (2000) *Materials Handbook: A Concise Desktop Reference*, London; New York: Springer Verlag. This book is divided into fourteen families of materials (e.g., metals and alloys, semiconductors, superconductors, magnetic and electrical materials, ceramics). Physico-chemical properties are supplied for each class of materials in tabular form. Emphasis is given to the most common industrial materials in each class. A bibliography is included.

Chanda, M. and Roy, S.K. (1998) *Plastics Technology Handbook* (3rd edn), Plastic Engineering, vol. 26, New York: Marcel Dekker. Covers processes and property data of industrial polymers. Appendices cover trade names and abbreviations for industrial polymers; formulations; commercial polymer blends and alloys; properties of polymers, rubber compounds, and textile fibers.

Composite Materials Handbook – MIL 17 (2002) (5 vols), Conshohocken, PA: ASTM International; also Knovel. Authoritative source for data on polymer matrix, metal matrix, and ceramic matrix composite materials, updated regularly. Three volumes deal with polymer matrix composites, one with metal matrix composites, and one with ceramic matrix composites.

Corrosion Survey Database. Houston, TX: NACE International. Accessed through Knovel. "This database is a collection of the results from numerous literature references reporting the effects of exposing 87 metal and nonmetal materials to over 1500 different exposure media at various temperatures and concentrations resulting in 28,000 pairs of exposed material and medium. The results are presented in searchable tables, and can be browsed by material or exposure medium" (source: Knovel Home page http://www.knovel.com/knovel2/Toc.jsp?BookID=532).

Engineered Materials Handbook (1987–) (4 vols), Materials Park, OH: ASM International. The concise versions are included in the ASM Handbooks Online. vol. 1, Composites; vol. 2, Engineering plastics; vol. 3, Adhesives and Sealants; vol. 4, Ceramics and Glasses.

ENGnetBASE Engineering Database Online (see Chapter 2 in this book).

Francombe, M.H. (ed.-in-chief) (2000) *Handbook of Thin Film Devices* (5 vols), San Diego, CA: Academic Press. Topics covered in this comprehensive handbook range "from basic device physics and design, through growth and device fabrication, performance characteristics, and applications and integration into subsystems." Each volume is devoted to a class of devices: vol. 1, Hetero-structures for High Performance Devices; vol. 2, Semiconductor Optical and Electro-optical Devices; vol. 3, Superconducting Film Devices; vol. 4, Magnetic Thin Film Devices; vol. 5, Ferroelectric Film Devices (source: back cover).

Frick, J. (ed.) (2000) *Woldman's Engineering Alloys* (9th edn), Materials Park, OH: ASM International. "This standard reference book has again been revised with thousands of new records added or updated to include the latest alloys and the most current manufacturing status. The book also continues to list alloys that are no longer produced but are frequently found in older references and specifications, thus helping you to identify substitutes" (source: ASM Web page).

Gale, W.F. and Totemeier, T.C. (eds) (2004) *Smithells Metals Reference Book* (8th edn), Engineering Materials Selector Series. Oxford; Burlington, MA: Elsevier Butterworth Heinemann. This essential reference work for metal researchers and metallurgists contains essays and data on all aspects of metallurgy. In this new edition chapters were added on: Non-conventional and emerging materials (including) micr/nano-scale materials; Techniques for the modeling and simulation of metallic materials; Supporting technologies for the processing of metals and alloys; and an extensive bibliography of selected sources of further metallurgical information, including books, journals, conference series, professional societies, metallurgical databases, and specialist search tools.

Glocker, D.A. and Shah, S.I. (2002) *Handbook of Thin Film Process Technology*, Bristol, U.K.: Institute of Physics Publishers. A compilation of the current knowledge on thin film processes leading to applications and devices. Includes references with each chapter.

Habashi, F. (ed.) (1997) *Handbook of Extractive Metallurgy* (4 vols), Weinheim; New York: Wiley-VCH. A comprehensive compilation of data on metals, their extraction, their alloys, and their most important inorganic compounds. The four volumes are organized by families of metals according to their industrial importance: vol. 1, *The Metal Industry Ferrous Metals*; vol. 2, *Primary Metals, Secondary*

Metals, Light Metals; vol. 3, *Precious Metals, Refractory Metals, Scattered Metals, Rare Earth Metals*; vol. 4, *Ferroalloy Metals, Alkali Metals, Alkaline Earth Metals*.

Harper, C.A. (ed.-in-chief) (2001) *Handbook of Ceramics, Glasses and Diamonds*, Boston, MA; London: Irwin/McGraw-Hill; electronic version: McGraw-Hill, and netLibrary. Emphasis is on innovative uses of these materials; in addition to a wealth of materials properties, processes, and requirements data, selection and design guidelines are included.

Harper, C.A. (ed.-in-chief) (2001) *Handbook of Materials for Product Design* (3rd. edn), New York: McGraw-Hill; also as an electronic book. This is a compendium of materials data needed for selection or creation of new materials. Information is arranged for easy comparison of options in tables, charts, graphs, and illustrations of data on design, testing, specifications, standards, recyclability, and biodegradability.

Knovel Engineering and Scientific Databases (see Chapter 2 in this book).

Kutz, M. (ed.) (2002) *Handbook of Materials Selection*, New York: Wiley; also as an electronic resource. This is a comprehensive resource for property data as well as selection and evaluation techniques for all groups of materials in use today: metals, ceramics, composites, and plastics. The seven chapters deal with: Quantitative methods; Major materials; Finding and managing materials information and data; Testing and inspection; Failure analysis; Manufacturing; Applications and uses. It is heavily illustrated with photos, graphics, and tables, and an index is included.

Lemaitre, J. (2001) *Handbook of Materials Behavior Models* (3 vols), San Diego, CA: Academic Press. "Gathers together 117 models of behavior of materials. . . . Presents each model's domain of validity, a short background, its formulation, a methodology to identify the materials parameters, advise on how to use it in practical applications as well as extensive references. Covers all solid materials: metals, alloys, ceramics, polymers, composites, concrete, wood, rubber, geomaterials such as rocks, soils, sand, clay, biomaterials, etc. Concerns all engineering phenomena: elasticity, viscoelasticity, yield limit, plasticity, viscoplasticity, damage, fracture, friction, and wear" (source: publisher's website).

Lobo, H. and Bonilla, J.V. (2003) *Handbook of Plastics Analysis (Plastics Engineering)*, New York: Marcel Dekker; also a netLibrary book. Describes the latest analytical developments, technologies, and equipment used to characterize and examine plastic and composite materials.

Mark, J.E. (1999) *Polymer Data Handbook*, New York: Oxford University Press; electronic version available. Concise information on the syntheses, structures, properties, and applications of the approximately 200 most important polymeric materials is presented in a tabular format.

Nalwa, H.S. (ed.) (2003) *Handbook of Organic-inorganic Hybrid Materials and Nanocomposites* (2 vols), Stevenson Ranch, CA: American Scientific Publishers. Brings together up-to-date knowledge on the underlying science and processing and fabrication of hybrid materials (covered in volume 1) and nanocomposites (covered in volume 2). Some of the topics covered include, different synthetic routes allowing multifunctionality with a wide range of composition, sol-gel chemistry, fibers, xerogels, spectroscopic characterization, mechanical, thermal, electronic, optical, catalytic and biological properties, polymer–metal interfaces, and their potential commercial applications. Each chapter ends with extensive references.

Nalwa, H.S. (2002) *Handbook of Nanostructured Materials and Nanotechnology* (5 vols),

San Diego, CA: Academic Press. Provides comprehensive coverage of the dominant technology of this century. The topics of the five volumes are: synthesis and processing; spectroscopy and theory; electrical properties; optical properties; organics, polymers, and biological materials. A one-volume concise edition was published in 2002, focusing on synthesis and fabrication as well as the electrical and optical properties of nanoscale materials.

Nalwa, H.S. (ed.) (2002) *Handbook of Thin Film Materials* (5 vols), San Diego, CA: Academic Press. This is a comprehensive reference work on thin film science and technology, covering semiconductors, superconductors, ferroelectrics, nanostructured materials, and magnetic materials. sixty-five articles contain state-of-the-art knowledge and data on a vast number of thin film materials, their deposition, processing, and fabrication techniques, spectroscopic characterization, optical characterization probes, physical properties, and structure–property relationships. Ample illustrations and extensive references are included: vol. 1, Deposition and Processing of Thin Films; vol. 2, Characterization and Spectroscopy of Thin Films; vol. 3, Ferroelectric and Dielectric Thin Films; vol. 4, Semiconductor and Superconductor Thin Films; vol. 5, Nanomaterials and Magnetic Thin Films – Index in each volume. *Handbook of Thin Film Devices* is a companion volume.

Plastics Design Library Handbook Series (1990–), Norwich, NY: William Andrews, Inc. in partnership with Knovel Corporation. So far twenty-six volumes have been published in this comprehensive handbook series providing data on all kinds of plastics, their evaluation, characterization, processing, and manufacturing. All volumes exist in hard copy as well as in electronic form as part of Knovel (see entry). A list of the currently existing titles follows: *Chemical Resistance of Plastics and Elastomers* (3rd edn, 2002); *Coloring Technology for Plastics* (1999); *Conductive Polymers and Plastics in Industrial Applications* (1999); *Dynamic Mechanical Analysis for Plastics Engineers* (1998); *Effect of Creep and Other Time Related Factors on Plastics and Elastomers* (1991); *Effect of Sterilization Methods on Plastics and Elastomers* (1994); *Effect of Temperature and Other Factors on Plastics* (1990); *Effect of UV Light and Weather on Plastics and Elastomers* (1994); *Fatigue and Tribological Properties of Plastics and Elastomers* (1995); *Film Properties of Plastics and Elastomers – A Guide to Non-wovens in Packaging Applications* (2nd edn, 2004); *Fluoroplastics*; vol. 1, Non-melt Processible Fluoroplastics, Fluoroplastics; (2000); vol. 2, Melt Processible Fluoropolymers (2003); *Handbook of Molded Part Shrinkage and Warpage* (2003); *Handbook of Plastics Joining; Imaging and Image Analysis Applications for Plastics* (1997); *Medical Plastics – Degradation Resistance and Failure Analysis* (1998); *Metallocene Technology in Commercial Applications* (1999); *Metallocene-catalyzed Polymers – Materials, Properties, Processing and Markets* (1998); *Permeability Properties of Plastics and Elastomers – A Guide to Packaging and Barrier Materials* (2nd edn, 2003); *Permeability and Other Film Properties of Plastics and Elastomers* (1995); *Plastics Failure – Analysis and Prevention* (2001); *Polypropylene – The Definitive User's Guide and Databook* (1998); *Rotational Molding Technology* (2002); *Rubber Formulary* (1999); *Specialized Molding Techniques* (2001); *Weathering of Plastics – Testing to Mirror Real Life Performance* 1999).

Poole, C.P. (1999) *Handbook of Superconductivity*, San Diego: Academic Press. This reference book draws together much of the information that has been published in the journal literature, including: thermal, electrical, magnetic, and mechanical properties, phase diagrams, spectroscopic crystallographic structures for many superconductor types, and coherence length and depth tabulations. Traditional superconductors are treated as well as high temperature superconductors.

Prelas, M.A. (ed.) (1998) *Handbook of Industrial Diamonds and Diamond Films*, New

York: Marcel Dekker. Both mined and synthetic diamonds are discussed as engineering materials for applications in modern technology. Integrates science, technology, and applications of diamond.

Pruett, K.M. (2000) *Chemical Resistance Guide for Plastics: A Guide to Chemical Resistance of Engineering Thermoplastics, Fluoroplastics, Fibers, and Thermoset Resins*, La Jolla, CA: Compass.

Ramachandran, V.S. and Beaudoin, J.J. (2001) *Handbook of Analytical Techniques in Concrete Science and Technology. Building Materials Science Series*, Norwich, NY: Noyes Publications. A handbook of analytical techniques for new types of concrete and related materials. The techniques range from chemical and thermal analysis, to IR and nuclear magnetic resonance spectroscopy, to scanning electron microscopy, x-ray diffraction, computer modeling and more.

Revie, R.W. and Uhlig, H.H. (eds) (2000) *Uhlig's Corrosion Handbook* (2nd edn), New York: Wiley; also a Wiley electronic resource and included in Knovel. After fifty years, this classic corrosion reference book has been completely updated with the latest information, including the development of many non-metallic materials, their corrosion behavior, and engineering approaches to their corrosion control.

Roberge, P.R. (2000) *Handbook of Corrosion Engineering*, New York: McGraw-Hill; included in Knovel. Similar content as *Uhlig's*, but deals with metals only. Appendices include corrosion economics, chemical compositions of engineering alloys, and thermodynamics data and e-pH diagrams.

Rosato, D.V. (ed.) (2001) *Plastics Engineering, Manufacturing and Data Handbook* (2 vols), Boston, MA: Kluwer Academic, for the Plastics Institute of America. This comprehensive reference work is presented in two volumes: vol. 1, Fundamentals and Processes; vol. 2, Design, Testing, Marketing, and Regulations. The interrelationships of materials-to-processes are explained for technical and non-technical readers alike. Information is included on static properties (tensile, flexural), dynamic properties (creep, fatigue, impact), and physical and chemical properties. The text is illustrated by 1,060 figures and 415 tables.

Saunders, N. (1998) *CALPHAD (Calculation of Phase Diagrams): A Comprehensive Guide*, Oxford; New York: Pergamon. This is an in-depth treatment of the method of computer coupling of phase diagrams and thermochemistry, which makes it possible to calculate the phase behavior of multi-component materials. See also the journal *CALPHAD* that publishes new developments quarterly.

Seshan, K. (2002) *Handbook of Thin-film Deposition Processes and Techniques: Principles, Methods, Equipment and Applications* (2nd edn), Norwich, NY: Noyes Publications; electronic version is part of Knovel. Brings together information on physical vapor deposition techniques.

Shackelford, J.F. and Alexander, W. (eds) (2001) *CRC Materials Science and Engineering Handbook* (3rd edn), Boca Raton, FL: CRC Press; electronic version is part of ENGnetBASE. Possibly the most comprehensive materials science handbook containing property data for the major groups of materials (metals, alloys, polymers, ceramics and glass, composites), as well as a section of comparative tables for selecting materials for specific properties. Data have been verified by professional societies such as ASM International and the American Ceramic Society.

Shah, V. (1998) *Handbook of Plastics Testing Technology* (2nd edn), New York: Wiley Interscience. Six chapters are devoted to mechanical, thermal, electrical, weathering, optical, and chemical properties and respective testing. The remainder of the book deals with other plastic testing issues and techniques. Includes tables, illustrations, and extensive references.

Smithells, C.J., Brandes, E.A., and Brook, G.B. (eds) (1998) *Smithells Light Metals Handbook*, Oxford; Boston, MA: Butterworth-Heinemann. Drawing on the data within *Smithells Metals Reference Book*, the editors have created a new book dedicated to aluminum, magnesium, and titanium, the most commonly used light metals. An extensive section on binary phase diagrams is included, as well as standards and international materials specifications.

ThermoDex: an index of selected thermodynamic data handbooks <http:thermodex.lib.utexas.edu> (accessed May 17, 2005). Austin, TX: Mallet Chemistry Library, University of Texas at Austin. This database contains records for selected printed and web-based compilations of thermochemical and thermophysical data. Searching for properties linked to types of compounds will return lists of sources that may contain the data.

Utracki, L.A. (ed.) (2002) *Polymer Blends Handbook* (2 vols), Dordrecht; Boston, MA: Kluwer Academic. Covers the fundamental principles and technology of thermoplastic blends. Appendices include lists of miscible commercial polymer blends.

Van Recum, A. (1998) *Handbook of Biomaterials Evaluation: Scientific, Technical, and Clinical Testing of Implant Materials* (2nd edn), Philadelphia, PA: Taylor & Francis. A valuable reference work for researchers and practitioners of working on testing and evaluating biocompatible materials.

Villars, P. (1997) *Pearson's Handbook: Crystallographic Data for Intermetallic Phases* (3rd edn) (2 vols), Materials Park, OH: ASM International. This is the standard reference work for crystal data, representing over 27,000 compounds.

Wessel, J.K. (2004) *Handbook of Advanced Materials*, Hoboken, NJ: Wiley; also a Wiley electronic book. This book brings together the most recent information about advanced materials and their properties. Materials range from polymer composites, continuous fiber ceramic composites to intermetallics and light metal alloys. Chapters on standards and codes, non-destructive evaluation, and rapid prototyping are included.

Whitehouse, D.J. (2003) *Handbook of Surface and Nanometrology* (rev. edn), Bristol: Institute of Physics. Explains the physics of surface metrology and nanometrology in nine chapters: General philosophy; Surface characterization; Processing; Instrumentation; Traceability; Surface metrology in manufacture; Surface geometry; Nanometrology; Conclusion and overview. Chapters are illustrated with practical examples and include references.

Zaitsev, A.M. (2001) *Optical Properties of Diamond: A Data Handbook*, Berlin; New York: Springer. A comprehensive compilation of data on the optical properties of diamonds.

Monographs and textbooks

From the countless books that have been and are being published relating to materials science, a few classic textbooks were chosen to be listed here, as well as the most important monographic and conference proceedings series. The Subject Headings or call numbers outlined in the introduction should be used to discover materials in library catalogs.

Annual Review of Materials Research (2001–) Palo Alto, CA: Annual Reviews. Formerly *Annual Review of Materials Science* (1971–2000).

Ashby, M. (2005) *Materials Selection in Mechanical Design* (3rd edn), Boston, MA: Butterworth-Heinemann.

Askeland, D.R. and Phule, P.P. (2003) *The Science and Engineering of Materials* (4th edn), Pacific Grove, CA: Brooks/Cole Thomson-Learning.

Cahn, R.W., Haasen, P., and Kramer, E.J. (eds-in-chief) (1991–) *Materials Science and Technology: A Comprehensive Treatment* (18 vols), New York: VCH Publishers. An in-depth topic-oriented publication devoted to this enormous interdisciplinary field.

Callister, W.D. (2003) *Materials Science and Engineering: An Introduction* (6th edn), New York: John Wiley.

Treatise on Materials Science and Technology (1972–) (vol. 1–31, then unnumbered), San Diego, CA: Academic Press. Each volume is an in-depth treatment of a topic, such as structural ceramics, metal matrix composites, auger electron spectroscopy.

Materials Research Society Symposia Proceedings (1980s–) Pittsburgh, PA: Materials Research Society. Proceedings of the Society's meetings are published in book form annually. Also available as an online license.

Journals

The journal literature is the most important vehicle for scholarly communication for the discipline of materials science, especially for the new subdisciplines of biomaterials and nanotechnology. For the selection of the journals listed below ISI's *Journal Citation Reports* were consulted. The list is limited to English-language titles and excludes trade journals.

Acta Materialia (1995–) (1359–6454) Elsevier Science.

Advanced Materials (1989–) (0935–9648) Wiley-VCH.

American Ceramic Society Bulletin (1946–) (0002–7812) American Ceramic Society.

Annual Review of Materials Research (1971–) (1531–7331) Annual Reviews.

Applied Surface Science (1985–) (0169–4332) North-Holland.

Biomaterials (1980–) (0142–9612) IPC Science and Technology Press.

Biomaterials Science and Engineering (1994–) (1076–6286) Mary Ann Liebert.

Chemistry of Materials (1989–) (0897–4756) American Chemical Society.

Corrosion (1945–) (0010–9312) National Association of Corrosion Engineers.

Critical Reviews in Solid State and Materials Sciences (1980–) (1040–8436) CRC Press.

Ferroelectrics (1970–) (0015–0193) Gordon and Breach.

IEEE Transactions on Components and Packaging Technology (1978–) (1521–3331) Institute of Electrical and Electronics Engineers.

IEEE Transactions on Nanotechnology (2002–) (1536–125X) Institute of Electrical and Electronics Engineers.

Inorganic Materials (1965–) (0020–1685) Consultants' Bureau.

International Journal of Nanoscience (2002–) (0219–581X) World Scientific.

International Polymer Science and Technology (1974–) (0307–174X) Rubber and Plastics Research Association of Great Britain.

JOM: The Journal of the Minerals, Metals and Materials Society (1949–) (1047–4838) The Society (TMS).

Journal of Applied Polymer Science (1959–) (0021–8995) John Wiley.

Journal of Biomaterials Applications (1986–) (0885–3282) Technomic Publications.

Journal of Biomaterials Science – Polymer Edition (1989–) (0920–5063) VSP.

Journal of Biomedical Materials Research, A, B (1967–) Part A (1549–3296) Part B (1552–4973) John Wiley (Official journal of the Biomaterials Society).

Journal of Composite Materials (1967–) (0021–9983) Technomics.

Journal of Electronic Materials (1972–) (0361–5235) Metallurgical Society of AIME.

Journal of Engineering Materials and Technology (1973–) (0094–4289) (ASME) *Journal of Magnetism and Magnetic Materials* (1975–) (0304–8853) North Holland.

Journal of Materials Chemistry (1991–) (0959–9428) Royal Society of Chemistry.

Journal of Materials Engineering and Performance (1992–) (1059–9495) ASM International (Merger of: *Journal of Materials Shaping Technology* and *Journal of Heat Treating).*

Journal of Materials Research (1986–) (0884–2914) Materials Research Society.

Journal of Materials Science (1966–) (0022–2461) Springer Verlag.

Journal of Materials Science. Materials in Electronics (1990–) (0957–4522) Chapman and Hall.

Journal of Micromechanics and Microengineering (1990–) (0960–1317) Chapman and Hall.

Journal of Nanoscience and Nanotechnology (2001–) (1533–4880) American Scientific Publishers.

Journal of Non-crystalline Solids (1968–) (0022–3093) North Holland.

Journal of Phase Equilibria and Diffusion (2004–) (1547–7037) ASM International; Continues *Journal of Phase Equilibria.*

Journal of Polymer Science, Part A, B, C (1986–) Part A (0887–624X) Part B (0887–6266) John Wiley.

Journal of the American Ceramic Society (1918–) (0002–7820) American Ceramic Society.

Journal of the Electrochemical Society (1948–) (0013–4651) Electrochemical Society.

Journal of the European Ceramic Society (1989–) (0955–2219) Elsevier Science.

Journal of the Mechanics and Physics of Solids (1952–) (0022–5096) Pergamon Press.

Journal of Thermal Spray Technology (1992–) (1059–9630) ASM International.

Journal of Vacuum Science and Technology (1964–) Part A (0734–2101) Part B (1071–1023) American Institute of Physics.

Key Engineering Materials (1986–) (1013–9826) Trans Tech Publications.

Macromolecular Bioscience (2001–) (1616–5187) VCH.

Materials Chemistry and Physics (1983–) (0254–0584) Elsevier.

Materials Performance (1974–) (0094–1492) National Society for Corrosion Engineers.

Materials Science and Engineering, A, B (1988–) Part A (0921–5093) Part B (0921–5107) Elsevier Sequoia.

Materials Science and Engineering, C (1998–) (0928–4931) Elsevier.

Materials Science Forum (1984–) (0255–5476) Trans Tech Publications.

Materials World (1993–) (0967–8638) The Institute of Materials.

Metallurgical and Materials Transactions A (1994–) (1073–5623) AMS International; formerly *Metallurgical Transactions A.*

MRS Bulletin (1976–) (0883–7694) Materials Research Society.

Nano Letters (2001–) (1530–6984) American Chemical Society.

Nanotechnology (1990–) (0957–4484) Institute of Physics.

Plastics Engineering (1973–) (0091–9578) Society of Plastics Engineers.

Polymer (1960–) (0032–3861) Butterworth Scientific.

Polymers for Advanced Technologies (1990–) (1042–7147) VCH.

Progress in Materials Science (1961–) (0079–6425) Pergamon Press.

Scripta Materialia (1996–) (1359–6462) Elsevier Science; continues *Scripta.*

Semiconductor Science and Technology (1986–) (0268–1242) Institute of Physics.

Smart Materials and Structures (1992–) (0964–1726) Institute of Physics.
Surface Coatings and Technology (1986–) (0257–8972) Elsevier Sequoia.
Thin Solid Films (1967–) (0040–6090) Elsevier Sequoia.

Patents and standards

As in other engineering disciplines, standards are essential for designing new materials and applications, and the body of intellectual property literature is increasing rapidly, and contains a wealth of technical information. Most standard developing organizations host their own websites and sell their standards online; therefore only the organization most relevant to materials science, ASTM, is listed here. In addition to the patent databases listed below, several of the databases listed under Indexes and abstracts include relevant patents; *Chemical Abstracts/SciFinder* is especially useful for finding materials-related patents (see Chapter 2 in this book for more information on standards).

ASTM Standards <http://www.astm.org/> (accessed May 17, 2005). West Conshohocken, PA: American Society of Testing and Materials. This is the primary source of standards relating to materials science. Publisher of *Annual Book of ASTM Standards.*

Websites

Many of the sources listed so far are proprietary in some way and have to be paid for by a library or by an individual user. In this section websites were selected that are freely accessible and either constitute gateways to a host of information on the Internet or are themselves information-rich sites. Most society websites are treasure troves of information – they are listed in the following section. University departments of materials science and engineering and government agencies and offices often host information-rich websites as well.

Biomaterials Network <http://www.biomat.net> (accessed May 17, 2005). Maintained by Biomaterials Laboratory, Instituto de Engenharia Biomédica (INEB), University of Porto, Portugal. The major goals of Biomat.net consist in providing an organized and meaningful biomaterials communication resource for scientists, researchers, members of the business community, government, academia, and the general public, and acting as a resource center to disclose resources, organizations, research activity, educational initiatives, scientific events, journals, books, articles, funding opportunities, industrial developments, market analyses, jobs and every other initiative related to biomaterials science and associated fields, such as tissue engineering.

Center for Computational Materials Science <http://cst-www.nrl.navy.mil/> (accessed May 17, 2005). Created and maintained by the National Research Laboratory (NRL). Included are the Crystal Lattice Structures and the Electronic Structure Calculations databases. Current research area summaries and other scientific resources may also be found on this website.

Composites Corner <http://www.advmat.com> (accessed May 17, 2005). This site is maintained by the Turner Moss Company and provides hundreds of links to advanced composite material sites, including links to academics, societies, government sites, and suppliers.

Composite Materials <http://composite.about.com/mbody.htm> (accessed May 17, 2005). This comprehensive website is a complete source to find information and resources for composite materials and the composites industry.

Crystallography Online <http://www.iucr.ac.uk/cww-top/crystal.index.html> (accessed May 17, 2005). Provides links to current information concerning crystallography and other topics of interest to crystallographers. This site is part of the International Union of Crystallography pan-crystallography information service.

Department of Energy (DOE) <http://www.doe.gov> (accessed May 17, 2005). Provides energy-related information from the Department of Energy. Covers full-text DOE documents and regulations.

EEVL – The Internet Guide to Engineering, Mathematics and Computing <http://www.eevl.ac.uk/engineering/index.htm> (accessed May 17, 2005). EEVL is a service, created and run by a team of information specialists from a number of universities and institutions in the U.K. Its mission is to provide access to quality networked information in engineering, mathematics, and computing. The sections related to materials science list over 900 relevant websites.

MatPro: Materials and Processes Database <http://amptiac.alionscience. com/MATPRO/> (accessed May 17, 2005). This website is designed by AMPTIAC (Advanced Materials and Processes Technology Information Analysis Center) "to help locate websites, companies, colleges and universities, reference books and other information sources relating to advanced materials" (source: MatPro Web page).

MatWeb: The Online Materials Resource <http://www.matweb.com> (accessed May 17, 2005). This searchable database covers properties of over 20,000 materials. Categories include thermoplastics, thermosets, ferrous metals, nonferrous metals, and ceramics. Searchable by materials name or designation, to match property requirements, or to find metals based on their constituent elements.

MEMS and Nanotechnology Clearinghouse <http://www.memsnet.org> (accessed May 17, 2005). This information resource, hosted by the MEMS and Nanotechnology Exchange, includes an extensive materials index with property information culled from the *CRC Materials Science and Engineering Handbook.*

NIST Scientific and Technical Databases Online <http://www.nist.gov/srd/online.htm> (accessed May 17, 2005).This is the home of the National Institute of Standards and Technology's over eighty databases covering a broad range of substances and properties from many different disciplines. Many consist of data relevant to materials scientists (e.g., the *NIST Ceramic WebBook* and *Phase Diagrams and Computational Thermodynamics*). The databases may be searched individually or through the user-friendly NIST Data Gateway <http://srdata.nist.gov/gateway> (cited November 22, 2004).

Plastics.com <http://www.plastics.com> (accessed May 17, 2005). "Plastics.com, Inc., serves the needs of the professionals representing all segments of the global plastics industry by providing content and services designed to engage, inform and educate its members." (source: Plastics.com Web page).

PSIgate – Materials Gateway <http://www.psigate.ac.uk/newsite/materials-gateway> (accessed May 17, 2005). This Materials Science Gateway is part of the

Physical Science Information Gateway (PSIgate), the physical sciences hub of the UK-based Resource Discovery Network (RDN). PSIgate provides free access to high-quality Internet resources for students, researchers, and practitioners in the physical sciences. Each resource in the main PSIgate Catalog has been selected by information professionals and subject specialists. A full description of each resource is provided, together with a range of other information and direct access to the resource itself.

Associations, organizations, and societies

Professional societies provide an extremely important forum for the exchange of ideas and the dissemination of research results, and thereby are a major factor in the advancement of a discipline. Each of the societies listed produces a variety of publications (journals, books, technical reports) and holds regular meetings, conferences or workshops; and all have extensive websites providing access to a wealth of information, including job openings and access to experts.

Abrasive Engineering Society (AES) <http://www.abrasiveengineering.com> (accessed May 17, 2005). Founded in 1957, "AES is dedicated to promoting technical information about abrasives minerals and their uses. That includes abrasives grains and products such as grinding wheels, coated abrasives (sandpaper) and countless other items made from synthetic minerals as well as the myriad of tools and products that serve as accessories" (source: AES Web page).

American Ceramic Society (ACeRS) <http://www.acers.org/> (accessed May 17, 2005). "The American Ceramic Society (ACerS) is a 100-year-old non-profit organization that serves the informational, educational, and professional needs of the international ceramics community. The Society's more than 7,500 members comprise a wide variety of individuals and interest groups that include engineers, scientists, researchers, manufacturers, plant personnel, educators, students, marketing and sales professionals, and others in related materials disciplines" (source: ACS Web page).

American Concrete Institute (ACI) <http://www.aci-int.org> (accessed May 17, 2005). "ACI is a technical and educational consensus organization whose purpose is to further engineering and technical education, scientific investigation and research, and development of standards for the design and construction of concrete structures. The Institute gathers, correlates, and disseminates information for the improvement of the design, construction, manufacture, use and maintenance of concrete products and structures." A searchable database of research reports posted by members is at <http://www.concrete.org/PUBS/LIBRARY/LI-IDX.HTM> (cited November 11, 2004).

American Electroplaters and Surface Finishers Society, Inc. (AESF) <http://www.aesf.org> (accessed May 17, 2005). "The American Electroplaters and Surface Finishers Society, Inc. (AESF), is an international, individual-membership, non-profit professional society. Founded in 1909, the AESF has 78 Branches and more than 7,000 members, worldwide. The Society is regarded and respected as the foremost finishing authority in the world. The primary mission of the AESF is to advance the science of surface finishing, to benefit industry and society through education, information and social involvement" (source: AESF Web page).

American Foundry Society (AFS) <http://www.afsinc.org> (accessed May 17, 2005). "The American Foundry Society was founded in 1896, and its mission today is to provide and promote knowledge and services that strengthen the metalcasting industry for the ultimate benefit of its customers and society. . . . With the direction of its volunteer committee structure, the professional staff of AFS provides support in the areas of technology, management and education to further activities that will enhance the economic progress of the industry" (source: AFS Web page). The Cast Metal Institute is AFS's educational arm.

American Institute for Mining, Metallurgical and Petroleum Engineers (AIME) <http://www.aimeny.org/> (accessed May 17, 2005). AIME was organized in 1871 to advance and disseminate, through the programs of the Member Societies, knowledge of engineering and the arts and sciences involved in the production and use of minerals, metals, energy sources and materials for the benefit of humankind, and to represent AIME and the Member Societies within the larger engineering community. Its member societies are: the Association for Iron and Steel Technology (AIST); Society for Mining, Metallurgy, and Exploration (SME); Society of Petroleum Engineers (SPE); The Minerals Metals and Materials Society (TMS); Women's Auxiliary to AIME (WAIME).

American Iron and Steel Institute (AISI) <http://www.steel.org> (accessed May 17, 2005). AISI began in 1855 as the American Iron Association and has served as the voice of the North American steel industry in the public policy arena and advances the case for steel in the marketplace as the preferred material of choice. AISI also plays a lead role in the development and application of new steels and steel-making technology. AISI comprises thirty-one member companies, including integrated and electric furnace steel-makers, and 118 associate and affiliate members who are suppliers to or customers of the steel industry. AISI's member companies represent approximately 75 percent of both U.S. and North American steel capacity.

American Society for Testing of Materials (ASTM) <http://www.astm.org> (accessed May 17, 2005). "Founded in 1898 as the American Society for Testing of Materials, ASTM International is one of the largest voluntary standards development organizations in the world – a trusted source for technical standards for materials, products, systems, and services." (source: ASTM Web page).

American Vacuum Society (AVS) <http://www.avs.org> (accessed May 17, 2005). "Founded in 1953, AVS is organized into 10 technical divisions and 3 technical groups that encompass a range of established as well as emerging science and technology areas. There are also 18 local-area chapters that promote communication and networking for professionals and students within a geographical region. AVS is comprised of approximately 6,000 members worldwide and welcomes all member and non-member scientists, technologists, students, and educators to participate in its national and regional events" (source: AVS Web page).

American Welding Society (AWS) <http://www.aws.org> (accessed May 17, 2005). The goal of AWS is "to advance the science, technology and application of welding and related joining disciplines. From factory floor to high-rise construction, from military weaponry to home products, AWS continues to lead the way in supporting welding education and technology development to ensure a strong, competitive and comfortable way of life" (source: AWS Web page).

ASM International <http://www.asm-intl.org> (accessed May 17, 2005). "ASM International is the society for materials engineers and scientists, a worldwide network dedicated to advancing industry, technology and applications of metals and materials." Since its foundation in 1913, the Society has expanded to include 40,000 members in over 100 countries. ASM's reference materials cover every aspect of materials science and are indispensable for materials professionals. ASM has five member societies: Heat Treating Society, Thermal Spray Society, International Metallographic Society, Electronic Device Failure Analysis Society, and Society of Carbide Tool Engineers.

Association for Iron and Steel Technology (AIST) <http://www.aistech.org> (accessed May 17, 2005). AIST was established (from AISE and ISS) in January 2004, with the goal "of advancing the technical development, production, processing and application of iron and steel. The best practices of both predecessor organizations were incorporated into AIST, and we now have a strong, international, member-based technical organization that can sustain itself in an environment of continual change... AIST is committed to presenting superior technical meetings, conferences, exhibits and publications to better serve those involved in the iron and steel community, including steel manufacturers, suppliers, consumers and academics" (source: AIST Web page).

Cryogenic Society of America (CSA) <http://www.cryogenicsociety.org/> (accessed May 17, 2005). "CSA was formed in 1964 and derived most of its initial members from the aerospace industry. However, the membership is now diversified to include engineers, physicists, other scientists, sales representatives, technicians, all levels of management, systems designers and operators and a host of other occupations . . . with a wide range of academic degrees and from more than 12 countries. In 1971, the Helium Society was incorporated into CSA" (source: CSA Web page). The objectives of CSA include: to disseminate information concerning low temperature processes and techniques, and to promote research and development of low temperature processes by meetings, professional contacts, papers, reports, and publications.

Electrochemical Society (ECS) <http://www.electrochem.org> (accessed May 17, 2005). "Since its inception as the American Electrochemical Society in 1902, the ECS has grown and continued to respond to the changing technical needs and interests of our membership. The ECS has a long tradition of fundamental objectives based on the advancement of the theory and practice of electrochemical and solid state science by the dissemination of information through individual membership, corporate membership, student services," and their publications (source: ECS Web page).

Electronic Device Failure Analysis Society (EDFAS) <http://www.edfas.org> (accessed May 17, 2005). EDFAS is "unique, because it is the only international professional organization dedicated to the specialty of Electronic Device Failure Analysis; broad reaching because our specialty draws its practitioners – both engineers and technicians – from many fields: Electronics Engineering, Materials Science, Chemistry, Physics, Computer Science, Metallurgy, and others" (source: EDFAS Web page). An affiliate of AMS.

Institute of Materials, UK (IOM3) <http://www.iom3.org> (accessed May 17, 2005). "The Institute of Materials, Minerals and Mining (IOM^3) was officially recognized by the UK's Privy Council on 26 June 2002. Created from the merger of The Institute of Materials (IOM) and The Institution of Mining and Metallurgy (IMM), the Institute intends to develop to be the leading international professional body for the advancement of materials, minerals and mining to governments, industry, academia, the public and the professions" (source: IOM3 Web page).

International Metallographic Society (IMS) <http://www.internationalmetallographicsociety.org> (accessed May 17, 2005). IMS was founded in 1967. Its mission is "to benefit the art and science of metallography and materials characterization. IMS members are involved in material characterization, performance, behavior and fabrication. Of special interest to members are techniques and the equipment needed for microstructural examination, analysis and evaluation. Some techniques of interest include light, electron and acoustical microscopy; quantitative/computer-aided microstructural analysis; metallography, ceramography, resinography; and allied sciences for physical and chemical analysis" (source: IMS Web page).

International Organization of Materials, Metals and Minerals Societies (IOMMMS) <http://www.iommms.org> (accessed May 17, 2005). "Our mission is to promote and facilitate communication and cooperation among the worldwide materials, metals and minerals societies and institutes to the benefit of their members and the enhancement of the profession and industries associated with them" (source: IOMMMS Web page).

International Society of Coating Science and Technology (ISCST) <http://www.iscst.org> (accessed May 17, 2005). "ISCST is a technical society, which was formed in December of 1996 by an international group of academic and industrial engineers who are active in the many aspects of coating technology. This technology is growing in importance and it was believed that the global coating technical community would benefit by having its own society." The specific purposes of ISCST include: "To provide forums for presenting advances in the mechanisms and technologies used in the preparation, application, solidification, and microstructuring of coated films; and to encourage technology transfer, professional development, and networking of scientists who are engaged in the coating process sciences" (source: ISCST Web page).

The International Thermal Spray Association (ITSA) <http://www.spray-itsa.com> (accessed May 17, 2005). "The history of the ITSA is closely interwoven with the history of thermal spray development in this hemisphere. Founded in 1948, and once known as Metallizing Service Contractors (MSC), the ITSA Group has been closely tied to almost all major advances in technology, equipment and materials, industry events, education, standards and market development in North and South America" (source: ITSA Web page).

International Union of Crystallography (IUCR) <http://www.iucr.ac.uk> (accessed May 17, 2005). "Since its admission to the International Council of Scientific Unions, the International Union of Crystallography has been instrumental in encouraging activities involving the crystallographic community worldwide. The formation of the Union itself was, in turn, inspired by pre-existing international links between crystallographers, which had already resulted in a number of important collaborative projects. The Union built on these traditions, especially with respect to its unique feature, that is, its very active publication programme" (source: IUCR Web page). It was founded in 1946.

International Union of Materials Research Societies (IUMRS) <http://www.iumrs.org> (accessed May 17, 2005). "The International Union of Materials Research Societies was established in 1991 as an international association of technical groups or societies which have an interest in promoting interdisciplinary materials research. Current membership includes countries from around the world" (source: IUMRS Web page).

Materials Research Society (MRS) <http://www.mrs.org> (accessed May 17, 2005).

"The MRS is dedicated to goal-oriented basic and applied research on materials of technological importance. They emphasize an interdisciplinary approach to materials science and engineering, inspiring cooperation among scientists from industry, academia and government" (source: MRS Web page).

Metal Powder Industries Federation (MPIF) <http://www.mpif.org> (accessed May 17, 2005). "MPIF is a 'not-for-profit' trade association formed by the P/M industry to promote the advancement of the metal powder producing and consuming industries" (source: MPIF Web page). MPIF is a federation of six member associations.

National Association of Corrosion Engineers (NACE) <http://www.nace.org> (accessed May 17, 2005). "The National Association of Corrosion Engineers was established in 1943 by eleven corrosion engineers in the pipeline industry. These founding members were involved in a regional cathodic protection group formed in the 1930s, when the study of cathodic protection was introduced. With more than 60 years of experience in developing corrosion prevention and control standards, NACE International has become the largest organization in the world committed to the study of corrosion" (source: NACE Web page).

Society for the Advancement of Material and Process Engineering (SAMPE) <http://www.sampe.org> (accessed May 17, 2005). "Material and Process Engineering is the technology by which materials are developed or selected and manufacturing processes chosen which will convert those materials into products which meet the design, performance, producibility, quality, and cost effectivity criteria required. SAMPE provides information on new materials and processing technology either via technical forums, journal publications, or books" (source: SAMPE Web page).

Society for Mining, Metallurgy, and Exploration (SME) <http://www.smenet.org> (accessed May 17, 2005). "A member society of the American Institute of Mining, Metallurgical and Petroleum Engineers, SME's roots date back to 1871 when a handful of mining engineers founded AIME. Since its inception, SME has continued to evolve over the years to stay abreast of industry changes and to reflect the ever-broadening interests of its members. SME is organized into five distinct divisions – coal & energy, environmental, industrial minerals, mineral and metallurgical processing, and mining and exploration" (source: SME Web page).

Society of Glass Technology (SGT) <http://www.sgt.org> (accessed May 17, 2005). SGT was founded in 1916. Its objects are: "To encourage and advance the study of the history, art, science, design, manufacture, after treatment, distribution and end use of glass of any and every kind. These aims are furthered by meetings, publications, the maintenance of a library and the promotion of association with other interested persons and organizations." (source: SGT Web page).

Society of Plastics Engineers (SPE) <http://www.4spe.org> (accessed May 17, 2005). "The Society of Plastics Engineers, Inc. was originally incorporated by the State of Michigan on January 6, 1942 as the Society of Plastics Sales Engineers. . . . SPE has become the recognized medium of communication amongst scientists and engineers engaged in the development, conversion and applications of plastics" (source: SPE Web page).

Steel Founders' Society of America (SFSA) <http://www.sfsa.org> (accessed May 17, 2005). SFSA is a trade association representing steel foundries, founded in 1902.

Surface Analysis Society of Japan (SASJ) <http://www.sasj.gr.jp> (accessed May 17, 2005). "SASJ promotes national and international collaborative researches, holds symposia and publishes related documents in order to improve the authenticity and the standardization of surface analysis techniques. Besides, SASJ promotes construction of Database for surface analysis" (source: SAS Web page). It was founded in 1995.

The Minerals, Metals and Materials Society (TMS) <http://www.tms.org> accessed May 17, 2005). TMS "is a rare professional organization that encompasses the entire range of materials and engineering, from minerals processing and primary metals production to basic research and the advanced applications of materials. Included among its nearly 10,000 professional and student members are metallurgical and materials engineers, scientists, researchers, educators, and administrators from more than 70 countries on six continents. . . . society provides forums for the exchange of information; promotes technology transfer; promotes the education and development of current and future professionals; represents the profession in the accreditation of educational programs and in the registration of professional engineers (a U.S.-grounded activity); encourages professionalism, ethical behavior, and concern for the environment; and stimulates a worldwide sense of unity in the profession" (source: TMS Web page).

The Thermal Spray Society: ASM International (TSS) <http://www.asm-intl.org/tss/> (accessed May 17, 2005). "TSS members are dedicated to learning more about thermal spray performance, processes, properties and applications; exchange information with other members via forums, programs and services; and the aerospace-proven advantages of thermal spray to other industries" (source: TSS Web page).

Reference

Lord, C.R. (2000) *Guide to Information Sources in Engineering*, Englewood, CO: Libraries Unlimited.

16 Mechanical engineering

Mel DeSart and Aleteia Greenwood

Introduction

If something moves because work (defined as force times distance) is done, chances are very good that mechanical engineering concepts are involved. Mechanical engineering is the branch of engineering that encompasses the generation and application of heat and mechanical power, and the design, production, and use of machines and tools. One of the oldest of the engineering disciplines, mechanical engineering concepts have been in use since prehistoric times. From the discovery and use of the most basic tools (lever, wheel and axle, inclined plane (or wedge)) to the most complex machines of the modern world, mechanical engineering concepts impact upon virtually everything we do.

From those tools and the concepts associated with them came some of the first simple machines (e.g., the pulley, bellows, screw, and the first use of gears). These were followed by the water wheel, windmill, mechanical clock, lathe, the first pumps, and the printing press with moveable type, among hundreds of others.

The Newcomen steam-engine, invented in 1712, is regarded as one of the greatest inventions in the history of mechanical engineering, and led directly to the Industrial Revolution; but the steam-engine that Newcomen invented Watt in turn transformed into the tool that led to much of the industrial development that followed. While railways or tramways existed prior to the steam-engine, the development of the high-pressure steam-engine led directly to the rapid expansion of rail transportation. Other important advances during the Industrial Revolution included the refining of the centrifugal pump, the application of power to spinning and weaving textiles, and, in turn, the development of the factory system.

The late 1800s brought far more varied developments with the invention of the bicycle, the steam turbine, the internal combustion engine, the automobile, the first power-station to produce electricity, and the techniques for making steel. The ability to make steel, in turn, increased the capabilities of thousands of other machines. During this period the foundations of fluid mechanics also appeared, including the discovery of the two different modes of motion of moving fluids – laminar and turbulent. Investigations of drag or resistance to motion in both water and air fed directly into later aerodynamic and hydrodynamic design considerations.

The early twentieth century was the period when other engineering disciplines began to spring from the work of mechanical engineers. The aerodynamic studies of the late 1800s led to the development of gliders and hence to the most important invention of the first half of the twentieth century, the airplane. The helicopter and autogyro were also developed during this period. The improvements in the automobile, and in particular to its assembly, led directly into the blossoming of manufacturing engineering as a discipline. While the power take-off was invented in the late 1800s, its expanded use led to the popularity of the tractor in agriculture and the development of many other more specialized powered agricultural implements, from whence came agricultural engineering as a discipline.

A number of other important developments of the first half of the twentieth century focused on heavy machinery, as earth-moving equipment such as the bulldozer, excavator, road grader, and so on came into being, as well as a number of the tools of war, including the creation of specialized ships (including landing craft) and various forms of armored vehicles, the most mass produced being the tank.

The 1940s to date saw the invention of the jet engine, robots and robotics, the development of nuclear power and construction of power reactors, the merging of electrical and mechanical concepts in the creation of control systems, and the development of the rocket engine, which allowed human beings to venture beyond the boundaries of their home planet for the first time.

For a more detailed discussion of the history of mechanical engineering, the authors offer the following suggested texts.

American Society of Mechanical Engineers. History and Heritage Committee (1997) *Landmarks in Mechanical Engineering*, West Lafayette, IN: Purdue University Press. While the Semler book (below) focuses as much or more on the people behind the achievements, this work from ASME highlights the landmark developments or inventions of mechanical engineering, organized by broad topical area.

Burstall, A.F. (1965) *A History of Mechanical Engineering*, Cambridge, MA: MIT Press. The most comprehensive of the three works listed, this volume comprises nine chapters, from "The Prehistoric Period, before 3000 B.C." up through "Dawn of the Nuclear Age and Space Travel, 1940–1960."

Lawton, B. (2004) *Various and Ingenious Machines*, London: Brill. A two-volume set on the early history of mechanical engineering from prehistory to the beginnings of industrialization: vol. 1, Power Generation and Transport; vol. 2, Manufacturing and Weapons Technology.

Semler, E.G. (1963–1966) *Engineering Heritage: Highlights from the History of Mechanical Engineering*, London: Heinemann, on behalf of the Institution of Mechanical Engineers. Two volumes comprising short pieces on particular people or inventions that shaped the history of mechanical engineering.

Searching the catalog

One particularly wonderful aspect of online catalogs is the ability to search by keyword. Usually searching by keyword in a library catalog will search the title, author, publisher, subject heading, series, notes, and table of contents fields. Library of Congress Subject Headings may also be searched. The following Library of Congress Subject Headings and corresponding classification areas are useful for mechanical engineering.

Bearings (Machinery) TJ 1061
Biomechanics TJ 211
Deformations (Mechanics) TA 417–TA 418
Dynamics TJ 170, TL 243
Electronic control TJ 163
Engineering design T 353, TA 174
Engineering graphics T 351, T 353
Fasteners TJ 1320
Fluid dynamics TA 357, TJ 260
Fluid mechanics TA 357
Fracture mechanics TA 409
Friction TJ 1075
Gearing TJ 184
Heat engineering TJ 260–TJ 265
Heat transmission TJ 260, TJ 265
Internal combustion engines TJ 782, TJ 785
Lubrication and lubricants TJ 1075, TJ 1077
Machine design TJ 230
Machinery, dynamics of TJ 170
Machinery, kinematics of TJ 175
Mass transfer TA 357, TA 418, TJ 260
Materials fatigue TA 418.38, TA 409
Mechanical engineering TJ 144–TJ 159
Mechanical movements TJ 181, TJ 1075
Mechanical wear TJ 170, TJ 1075
Mechanics TA 350, TA 352
Mechanics, Applied TA 350, TA 405
Mechatronics TJ 163.12
Power (Mechanics) TJ 163.6–TJ 163.9
Prime movers TJ 250
Railroad engineering TF 200
Rheology TA 418.14
Servomechanisms TJ 214
Statics, Applied TA 351
Strains and stresses TA 351, TA 405, TA 407
Strength of materials TA 405
Thermodynamics TJ 265, TJ 756

Tribology TJ 1075–TJ 1081
Vibration TA 355–TA 356, TJ 177, TL 246
Viscosity TA 357.5 V6, TJ 1077

Article indexes

Compendex (1884–) Hoboken, NJ: Elsevier Engineering Information, Inc. (see Chapter 2 in this book).

Engineering Index (1884–) Hoboken, NJ: Elsevier Engineering Information, Inc. (see Chapter 2 in this book).

Indexes to . . . Publications: Transactions of the ASME. Society Records (1965?–1981) Easton, PA: American Society of Mechanical Engineers. This annual publication contains indexes to all ASME papers and publications.

ISMEC Bulletin (1973–1987) Bethesda, MD: Cambridge Scientific Abstracts. This print bibliography/index of mechanical engineering literature was succeeded by *ISMEC Mechanical Engineering Abstracts*.

ISMEC Mechanical Engineering Abstracts (1988–1992) Bethesda, MD: Cambridge Scientific Abstracts. This bimonthly print index, with annual cumulated indexes, was succeeded by *Mechanical Engineering Abstracts*.

Mechanical Engineering Abstracts (1981–2002) Bethesda, MD: Cambridge Scientific Abstracts, and Hoboken, NJ: Engineering Information, Inc. This is a "closed" database, with content from 1981 through 2002, when *Mechanical Engineering Abstracts* was succeeded by *Mechanical and Transportation Engineering Abstracts*. The database contains 227,382 records as of its closure at the end of 2002. Also published in paper form from 1993 to 2002 under the same title. The print edition was published bimonthly with an annual index. Succeeded *ISMEC Mechanical Engineering Abstracts*.

Mechanical and Transportation Engineering Abstracts (1966–) Bethesda, MD: CSA. This database contains over 400,000 records drawn from the indexing of over 2,600 journals and other publications. As with many CSA databases, indexed publications are designated either core, priority, or selective, with the percentage of content indexed decreasing from core down through selective. Coverage is primarily from 1966 to date, although some older records exist in the database. Some 30,000 to 35,000 records are added to the database annually in bimonthly updates.

NTIS (1964–) Springfield, VA: National Technical Information Service. The NTIS database contains records for over 2,000,000 government technical reports received by NTIS since 1964. Most other databases that cover mechanical engineering content do not cover government technical reports, making *NTIS* a complementary resource to the other subject-specific indexes and abstracts listed here (see Chapter 2 in this book for more information).

SAE Publications and Standards Database (1906–) Warrendale, PA: Society of Automotive Engineers. Covers the current and past publications of SAE, including its technical papers (from 1906 forward), magazine articles, books, standards, specifications, and research reports. Supports all aspects of mobility, including mechanics of engines, lubrication, engine design and technology, vehicle body engineering, and fluid mechanics in engines.

Seventy-seven Year Index, Technical Papers, 1880–1956, New York: American Society of Mechanical Engineers (ASME). A three-part index of the papers published in

ASME publications. Separate indexes are included for the *Mechanical Engineering, Journal of Applied Mechanics*, and to the *ASME Transactions.* A sixty-year index, covering 1880 to 1939, was also published.

Technology Research Database, Bethesda, MD: CSA (see Chapter 2 in this book).

Databases and datasets

ENGnetBASE: Engineering Electronic Library, Boca Raton, FL: CRC Press. Electronic collection of handbooks and other titles published by CRC Press. Contains dozens of titles in mechanical engineering and related areas (see Chapter 2 in this book).

Knovel, Norwich, NY: Knovel Corporation. Electronic collection of books and databases enhanced with productivity tools. Contains dozens of titles in mechanical engineering and related areas (see Chapter 2 in this book).

SAE Digital Library, Warrendale, PA: Society of Automotive Engineers. Database offers customizable content by the purchaser. Options include the SAE technical papers from 1998 to date, Aerospace Standards (AS), Aerospace Materials Specifications (AMS), and Ground Vehicle Standards (J-reports).

Bibliographies and guides to the literature

For the most part, the most recent guides in print, devoted specifically to mechanical engineering literature resources, appeared in the late 1980s. Chapters specific to mechanical engineering may be found in: *Scientific and Technical Information Sources* by Ching-chih Chen, *Reference Sources in Science, Engineering, Medicine, and Agriculture* by H. Robert Malinowsky, *Encyclopedia of Physical Sciences and Engineering Information Sources* by Steven Wasserman, Martin A. Smith, and Susan Mottu. There are a multitude of bibliographies and guides to the literature accessible through the Internet. Usually these sites are specific to the institutions for which they are made, and point to resources available at or through those institutions. Internet sites that cover a wider range of resources have been included in the section on Search engines and Internet sites.

Veilleux, D. (2003) *The Automotive Bibliography: 13,000 Works in English, Czech/Slovak, Danish, Dutch, Finnish, French, German, Italian, Norwegian, Polish, Portuguese, Slovenian, Spanish and Swedish*, Jefferson, NC: McFarland & Company. Comprises books on motor vehicles and motorization published to 2000. A wide range of items are catalogued, including monographs, theses, biographies, encyclopedias, other reference books such as handbooks, company and government publications, and buyers', collectors', spotters' and identification guides.

Directories

GradSchools.com: The most comprehensive online source of graduate school information <http://www.gradschools.com/listings/menus/mech_eng_menu.html> (accessed May 18, 2005). This URL provides access specifically to mechanical engineering graduate school programs that offer Master's and Ph.D. degrees. Provides access to universities by State. Also includes links to mechanical engineering programs available worldwide. Provides links directly to the mechanical engineering department of universities (see Chapter 11 on Engineering education in this book).

Peterson's Graduate Programs in Engineering and Applied Sciences (1997–) Princeton, NJ: Peterson's. Arranged alphabetically by discipline. Look under both Mechanical engineering and Mechanics. Entries include entrance and degree requirements, expenses and financial support, programs of study, and faculty research specialties. Coverage includes North America. Also available online at <http://www.petersons.com/GradChannel/> (accessed May 18, 2005) (see Chapter 11 on Engineering education in this book).

Thomas Register of American Manufacturers and Thomas Register Catalog File (1905–), New York: Thomas Publishing Company (see Chapter 14 on industrial and manufacturing engineering in this book).

Who's Who in Science and Engineering (1992–) Wilmette, IL: Marquis Who's Who. Look in the Professional area index for "Engineering: mechanical" to find mechanical engineers in North America and internationally. Arranged alphabetically by state, then alphabetically by country (see Chapter 2 in this book).

Worldwide Automotive Supplier Directory 2005/2006 (10th edn), Warrendale, PA: Society of Automotive Engineers. Includes over 10,000 companies, and more than 250 products. Locate suppliers geographically or by major product categories such as body (exterior and interior); body (seating and passenger); chassis; engineering design; manufacturing equipment, power train, and testing/instrumentation. Also available online through SAE.

Dictionaries and encyclopedias

Dictionaries – English

Goodsell, D. (1995) *Dictionary of Automotive Engineering* (2nd edn), Warrendale, PA: Society of Automotive Engineers. A concise work, useful for automotive engineers in the industry as well as engineering students. Includes definitions for over 3,000 terms as well as detailed drawings for 100 of the entries.

Nayler, G.H.F. (1996) *Dictionary of Mechanical Engineering* (4th edn), Warrendale, PA: Society of Automotive Engineers. Includes definitions to over 4,500 terms. Entries cover traditional as well as newer areas in mechanical engineering such as micromachining and nanotechology. Some entries accompanied by illustrations.

Perrot, P. (1998) *A to Z of Thermodynamics*, New York: Oxford University Press. A clear, comprehensive guide with entries that describe the fundamental concepts, principles, and applications of thermodynamics. Some entries include short etymological considerations, historical notes, and discussion of classical paradoxes. Accessible to undergraduate students.

South, D.W. and Ewert, R.H. (1995) *Encyclopedic Dictionary of Gears and Gearin*, New York: McGraw-Hill. Contains alphabetically arranged, clear, concise definitions for common terms, trade names, and abbreviations. Covers the fundamentals of gearing as well as advanced concepts. Some terms are accompanied by illustrations.

Dictionaries – multilingual

Federation of European Heating and Airconditioning Associations; Representatives of European Heating and Airconditioning Associations (1994) *The International Dictionary of Heating, Ventilating, and Air Conditioning* (2nd edn), London; New York: Spon Press. Comprises two major parts. The main part is made up of approximately 4,000 alphabetically arranged terms in English, with translations. The second part contains alphabetical indexes for the other eleven languages included in the dictionary: French, German, Italian, Danish, Finnish, Dutch, Spanish, Swedish, Hungarian, Polish, and Russian. Each alphabetical index is linked to serial numbers that refer the user to the translation in the main part.

Schellings, A. (1998) *Elsevier's Dictionary of Automotive Engineering in English, German, French, Dutch, and Polish*, Amsterdam: Elsevier. A unique resource that comprises technical terms relating to vehicles, materials, assemblies, and components parts. All terms are listed in English. Under each English word is the corresponding other language terms, German, French, Dutch, and Polish, labeled as to which language.

Encyclopedias

Cheremisinoff, N.P. (ed.) (1986–1990 + supplements) *Encyclopedia of Fluid Mechanics*, Houston, TX: Gulf Publishing.

vol. 1. *Flow Phenomena and Measurement*
vol. 2. *Dynamics of Single-fluid Flows and Mixing*
vol. 3. *Gas-liquid Flows*
vol. 4. *Solids and Gas-solids Flows*
vol. 5. *Slurry Flow Technology*
vol. 6. *Complex Flow Phenomena and Modeling*
vol. 7. *Rheology and Non-Newtonian Flows*
vol. 8. *Aerodynamics and Compressible Flows*
vol. 9. *Polymer Flow Engineering*
vol. 10. *Surface and Groundwater Flow Phenomena*
Supplement 1 *Applied Mathematics in Fluid Dynamics*
Supplement 2 *Advances in Multiphase Flow*
Supplement 3 *Advances in Flow Dynamics*

Comprehensive coverage of all aspects of fluid mechanics and dynamics. Clearly written with extensive use of diagrams and reference lists at the end of each entry.

Hewitt, G.F., Shires, G.L., and Polezhaev, Y.V. (eds) (1997) *International Encyclopedia of Heat and Mass Transfer*, Boca Raton, FL: CRC Press. Entries are written by approximately 300 contributors from around the world, including previously unreachable authors from Russia and Eastern European countries. Contributors are from the academic and industrial sectors. The encyclopedia is intended to support an undergraduate as well as an experienced practitioner. Entries are highly specialized where necessary, and accessible by the general reader where possible. Terms are cross-referenced as well as linked by the terms "following from" and "leading to."

Handbooks and manuals

General

Avallone, E.A. and Baumeister, T. III (1996) *Mark's Standard Handbook for Mechanical Engineers* (10th edn), New York: McGraw-Hill. The gold standard for mechanical engineering handbooks. This is the title one starts with to build a mechanical engineering reference collection. *Mark's* covers both theory and applications, with a slightly wider focus on materials than many other general ME handbooks. Extensive index at the end of the volume.

Beitz, W. and Kuttner, K-H. (eds) (1994) *Handbook of Mechanical Engineering/Dubbel* (1st English edn), London: Springer-Verlag. English translation of a classic German mechanical engineering handbook. Presents comprehensive coverage of all mechanical engineering areas, but in a very dense format (two columns per page, very small print).

Carvill, J. (1993) *Mechanical Engineers Data Handbook*, Boston, MA: Butterworth-Heinemann. According to the author's preface, this work is designed to be a "compact but comprehensive source of information." The title is large format, contains more illustrations in support of concepts presented, and still concludes nearly 350 pages of data. However, text is kept to a minimum and the focus is on equations and data.

Hicks, T.G. (ed.) (2006) *Handbook of Mechanical Engineering Calculations* (2nd edn), New York: McGraw-Hill. While Hicks' pocket guide that follows focuses on listing formulas for specific settings and applications, this volume is a much more extensive, step-by-step guide to what formulas should be used and calculations in dozens of specific applications. Contains hundreds of pages of supporting text.

Hicks, T.G. (2003) *Mechanical Engineering Formulas Pocket Guide*, New York: McGraw-Hill. As the title suggests, the focus of this small-format guide is on formulas used by mechanical engineers in various applications. Explanatory text is kept to a minimum.

Kreith, F. (ed.) (2004) *CRC Handbook of Mechanical Engineering* (2nd edn), Boca Raton, FL: CRC Press. The mechanical engineering entry to the successful CRC Press handbooks series. At roughly 2,500 pages, perhaps the largest and most extensive of all mechanical engineering handbooks. Covers areas such as transportation, communication, and information systems, and patent law not covered by most other ME handbooks.

Kutz, M. (ed.) (2005) *Mechanical Engineers' Handbook* (3rd edn), New York: Wiley. Wiley's entry into mechanical engineering handbooks. Similar in size and scope to the CRC handbook. A perhaps distinguishing feature of this work is that the chapter on electronic information resources is co-written by an engineering librarian.

Lindeburg, M.R. (2001) *Mechanical Engineering Reference Manual: For the PE Exam* (11th edn), Belmont, CA: Professional Publications. Although containing much of the same basic mechanical engineering information as other handbooks in this section, the focus of this work is on those concepts needed for successful completion of the U.S. Professional Engineering (PE) exam in mechanical engineering.

Marghitu, D.B. (ed.) (2001) *Mechanical Engineers' Handbook*, San Diego: Academic Press. Almost a combination of handbook and textbook, the presentation is perhaps more at a novice level than that of some of the other works. Also distinguishing this volume is that many chapters offer the reader problems and worked solutions.

Matthews, C. (2005) *ASME Engineer's Data Book* (2nd edn), New York: ASME Press.

Matthews, C. (2000) *Engineer's Data Book* (2nd edn), London and Bury St. Edmunds: Professional Engineering Publishing. While both titles are pocket-sized, written by the same author, and focus on mechanical engineering equations and data, each title is surrounded by background material and supporting documentation pertinent to its country of origin.

Pope, J.E. (ed.) (1997) *Rules of Thumb for Mechanical Engineers: A Manual of Quick, Accurate Solutions to Everyday Mechanical Engineering Problems*, Houston, TX: Gulf Publishing. Large format and soft cover, this volume is very applications oriented. The focus, across sixteen chapters, is on short cuts, calculations, and practical methods for use in different areas of mechanical engineering.

Smith, E.H. (ed.) (1998) *Mechanical Engineer's Reference Book* (12th edn), Oxford; Boston, MA: Butterworth-Heinemann. Hicks' work (above) is the pocket guide supported by IMechE, while this volume edited by Smith is the full handbook entry. Extensive, with a definite U.K. focus in much of the supporting material.

Automotive

Bauer, H. (ed.) (2004) *Automotive Electrics/Automotive Electronics* (4th edn), Stuttgart: Robert Bosch. Contains a compilation of content on automotive electrical and electronics systems culled from the German-published "Bosch Technical Instruction" series.

Bauer, H. (ed.) (2005) *Automotive Handbook* (6th edn), Stuttgart: Robert Bosch. A pocket-sized reference work on all aspects of automotive technology, including relevant physics and math concepts. Focused primarily on passenger cars and commercial vehicles.

Caines, A.J. and Haycock, R.F. (2004) *Automotive Lubricants Reference Book* (2nd edn), Warrendale, PA: Society of Automotive Engineers. As the title suggests, this work covers content on lubrication that in any way impacts upon the automotive industry. That includes typical automotive oils, greases, and lubricating fluids, but also gas turbine oils, railroad oils, and industrial lubricants used in automotive plants. Contains fifteen appendices on various topics, including an extensive glossary.

Cebon, D. (1999) *Handbook of Vehicle–Road Interaction*, Lisse, Exton, PA: Swets & Zeitlinger Publishers. Where mechanical and civil engineering meet in the automotive world. This work covers the interaction between vehicles and the surfaces they move upon. Of particular importance to the automotive engineer is how aspects of vehicle dynamics affect stability, performance, and damage to the vehicle or road surface.

Challen, B. and Baranescu, R. (1999) *Diesel Engine Reference Book* (2nd edn), Woburn, MA: Butterworth-Heinemann. Covers all aspects of the design and performance of diesel engines of all sizes. Includes an extensive section on environmental aspects.

Fenton, J. (1998) *Handbook of Automotive Body and Systems Design*, London:

Professional Engineering. Covers numerous aspects of vehicle design, first, in general areas such as aerodynamics, body trim and fittings, and ergonomics; second, it addresses numerous aspects of electrical and related systems design. Covers road cars, specialist (racing) cars, trucks, and more specialized vehicles such as tankers, refrigerated vehicles, light vans, ambulances, and others.

Fenton, J. (1998) *Handbook of Automotive Body Construction and Design Analysis*, London: Professional Engineering. Contains detailed descriptions of body fabrication and assembly techniques, panel cutting and forming, finishing, various materials used in automotive construction, and significant content on passenger safety concerns.

Fenton, J. (1998) *Handbook of Automotive Powertrain and Chassis Design*, London: Professional Engineering. The focus here is twofold: what powers a vehicle and the framework on which that power plant resides. Various types of power plants and designs are discussed, including gasoline, diesel, natural gas, electric, and hybrids. Then content switches to discussions of suspensions, steering and handling, wheels and braking, drive axles and drive trains, and transmissions.

Fitch, J.W. (1994) *Motor Truck Engineering Handbook* (4th edn), Warrendale, PA: Society of Automotive Engineers. This work focuses specifically on motor trucks – those trucks used in all areas of the trucking industry.

Jurgen, R. (ed.) (1999) *Automotive Electronics Handbook* (2nd edn), New York: McGraw-Hill. This work covers every aspect of electrical systems as they apply to automobiles, from lights and displays to the sensors and actuators that control and monitor the functioning of most systems within the modern automobile. Most chapters have their own glossary.

SAE Handbook (2005) New York: Society of Automotive Engineers. The definitive reference work for anyone working in the automotive industry in the United States. Contains most of SAE's automotive-related standards in three volumes. Contains numeric and subject indices.

Heat transfer, HVACR, fluid mechanics/dynamics/flow

ASHRAE Handbook. Fundamentals (2005) Atlanta, GA: American Society of Heating, Ventilating and Air-Conditioning Engineers.

ASHRAE Handbook. Heating, Ventilating and Air-conditioning Applications (2003) Atlanta, GA: American Society of Heating, Ventilating and Air-Conditioning engineers.

ASHRAE Handbook. Heating, Ventilating and Air-conditioning Systems and Equipment (2004) Atlanta, GA: American Society of Heating, Ventilating and Air-conditioning Engineers.

ASHRAE Handbook. Refrigeration (2002) Atlanta, GA: American Society of Heating, Ventilating and Air-conditioning Engineers.

This set of four handbooks is a comprehensive work on the various facets of heating, ventilating, air-conditioning, and refrigerating. Each volume comes in both SI and inch-pound editions, and one volume of the four-volume set is updated annually.

Cheremisinoff, N.P. (ed.) (1986–) *Handbook of Heat and Mass Transfer*, Houston, TX: Gulf Publishing. A four-volume set – volume 1 covers heat transfer operations, volume 2 mass transfer and reactor design, volume 3 addresses catalysis, kinetics, and reactor engineering, and volume 4 covers advances in reactor design and combustion science.

Hewitt, G.F. (ed.) (2002) *HEDH: Heat Exchanger Design Handbook* (rev. edn), New York: Begell House. This four-volume set is divided into five broad topical areas: theory; fluid mechanics and heat transfer; thermal and hydraulic design of heat exchangers; mechanical design of heat exchangers; and physical properties. This last section is particularly data-focused.

Johnson, R.W. (ed.) (1998) *Handbook of Fluid Dynamics*, Boca Raton, FL: CRC Press. Covers all aspects of fluid dynamics, but with a particular emphasis on theoretical, computational, and experimental aspects. Chapters related to applications are a secondary focus of this work, not a primary one.

Kreith, F. (ed.) (2000) *CRC Handbook of Thermal Engineering*, Boca Raton, FL: CRC Press. This handbook focuses on specific current topics in thermal engineering rather than on fundamentals and theory.

Kuppan, T. (2000) *Heat Exchanger Design Handbook*, New York: Marcel Dekker. Discusses design, construction, and appropriate modes of operation for various types of heat exchangers. Offers examples of applications in a number of different industries, and addresses appropriate standards and codes.

Miller, R.W. (ed.) (1996) *Flow Measurement Engineering Handbook* (3rd edn), New York: McGraw-Hill. As the name suggests, the focus here is on flow measurement. A series of twelve appendices makes up over a quarter of the volume.

Rohsenow, W.M., Hartnett, J.P., and Cho, Y.I. (eds) (1998) *Handbook of Heat Transfer Fundamentals*, New York: McGraw-Hill. Focuses heavily on the physics of heat transfer rather than on applications.

Saleh, J.M. (ed.) (2002) *Fluid Flow Handbook*, New York: McGraw-Hill. Focusing on fluid dynamics and fluid flow, this work offers practical examples to illustrate various theoretical principles. Useful in a number of engineering disciplines.

Schetz, J.A. and Fuhs, A.E. (eds) (1996) *Handbook of Fluid Dynamics and Fluid Machinery*, New York: John Wiley. Volume 1 of this three-volume set focuses on the fundamentals of fluid dynamics; volume 2 on experimental and computational fluid dynamics; volume 3 on applications. The emphasis on machinery and the size and coverage of the work set it off from others in the field.

Stoeker, W.F. (1998) *Industrial Refrigeration Handbook*, New York: McGraw-Hill. This volume covers the different types of refrigeration systems, particularly multi-state systems. Equipment such as reciprocating and screw compressors, evaporators, chillers, and condensers are described in detail, as are appropriate applications. Piping, vessels, valves, and controls appropriate to industrial refrigeration are described, as are the types and properties of various refrigerants.

Thumann, A. and Mehta, D.P. (2001) *Handbook of Energy Engineering* (5th edn), Lilburn, GA: Fairmont Press. Although focused primarily on power generation, chapters dealing with HVAC systems and equipment will be of interest to some in mechanical engineering.

Turner, W.C. (ed.) (2004) *Energy Management Handbook* (4th edn), Lilburn, GA: Fairmont Press. Chapters on the various systems used to generate energy, on HVAC, and on energy storage systems, and in particular the focus on economics, highlight the mechanical engineering aspects of this work.

Hydraulics and pneumatics

Barber, A. (1997) *Pneumatic Handbook* (8th edn), Oxford: Elsevier Science. Comprehensive work on all aspects of compressed air. Unlike most handbooks in this chapter, the *Pneumatic Handbook* contains dozens of advertisements from companies working in the compressed air industry.

Hunt, T. and Vaughn, N. (1996) *Hydraulic Handbook* (9th edn), Oxford: Elsevier Advanced Technology. Focuses primarily on the mechanical engineering aspects of hydraulics rather than on those of concern to the civil engineer. As with a number of other Elsevier-published handbooks, this works also contains advertisements from companies in the hydraulics industry.

Parr, E.A. (1998) *Hydraulics and Pneumatics: a Technician's and Engineer's Guide* (2nd edn), Oxford; Boston, MA: Butterworth-Heinemann. A relatively brief overview of hydraulics and pneumatics. Subjects covered are hydraulic pumps and pressure regulation, air pressure, treatment and regulation, control valves of various types, actuators, hydraulic and pneumatics accessories, process control pneumatics, and safety, fault-finding, and maintenance.

Rollins, J.P. (ed.) (1989) *Compressed Air and Gas Handbook* (5th edn), Englewood Cliffs, NJ: Prentice-Hall. This work on applications and uses of compressed air and other gases is culled from the accumulated experiences of member companies of the Compressed Air and Gas Institute.

Totten, G.E. (ed.) (2000) *Handbook of Hydraulic Fluid Technology*, New York: Marcel Dekker. Covers the narrowly focused area of hydraulic fluids and associated technology in great detail. A brief discussion on hardware leads off the volume, followed by chapters on fluid properties, compatibility issues, and testing and analysis. The last half of the volume contains chapters on various hydraulic fluids. Eleven appendices are included.

Yeaple, F.D. (1996) *Fluid Power Design Handbook* (3rd edn), New York: Marcel Dekker. This volume is laid out very simply and cleanly. Major sections address hydraulics, including fluids, equipment, and applications; pneumatics, focusing on equipment and applications; accessories, such as filters, tubing, piping hoses, fittings, and seals; control and performance issues; and system design.

Machinery and machining

Boyes, W.E. and Bakerjian, R. (eds) (1989) *Handbook of Jig and Fixture Design* (2nd edn), Dearborn, MI: Society of Manufacturing Engineers. A handbook that deals not with the materials being machined or the process of machining, but rather with the devices that position, support, and hold an item during machining or other processes (fixtures) and the devices that provide additional guidance to a cutting or machining tool (jigs).

Drozda, T.J. and Wick, C. (eds) (1983–) *Tool and Manufacturing Engineers Handbook*, Dearborn, MI: Society of Manufacturing Engineers. This nine-volume set, released over a period of sixteen years, covers: machining; forming; materials, finishing and coating; quality control and assembly; manufacturing management; design for manufacturability; continuous improvement; plastic part manufacturing; and materials and parts handling in manufacturing.

Lingaiah, K. (2001) *Machine Design Databook* (2nd edn), New York: McGraw-Hill. One of a number of related handbooks from McGraw-Hill. This work, as its name suggests, focuses on formulas, calculations, and other data needed for various machine design applications. Background or supporting text is kept to a bare minimum.

Machinability Data Center (1980) *Machining Data Handbook* (3rd edn), Cincinnati, OH: The Machinability Data Center. A classic work in the field, this two-volume set is a compendium of data for the machining and finishing of materials into various types of machine parts, both simple and complex.

Machinery's Handbook (2004) (27th edn), New York: Industrial Press, Inc.

Machinery's Handbook Guide (2004) (27th edn), New York: Industrial Press, Inc. The *Handbook* covers the properties, testing and manufacturing of tools and basic types of machine parts. The *Guide* explains in detail how to use the tables and formulas in the Handbook.

Metal Cutting Tool Handbook (1989) New York: Industrial Press, for the Metal Cutting Tool Institute. Addresses all forms of mechanical metal cutting, but does not address any other forms (e.g., laser, water jet).

Parmley, R.O. (ed.) (2005) *Machine Devices and Components Illustrated Sourcebook*, New York: McGraw-Hill. Different from most of the handbooks listed in this chapter, this mammoth work offers detailed illustrations of various mechanical parts, their assembly, and their uses in both conventional and innovative applications.

Rothbart, H.A. (ed.) (2006) *Mechanical Design Handbook* (2nd edn), New York: McGraw-Hill. This work focuses on the fundamentals and design considerations that need to be taken into effect by anyone interested in designing and constructing machines of various types.

Sclater, N. and Chironis, N.P. (2001) *Mechanisms and Mechanical Devices Sourcebook* (3rd edn), New York: McGraw-Hill. Similar in format to the Parmley title above, while that title focuses on mechanical components, this work focuses on more complex creations – the combination of components which result in a mechanism or other more complex mechanical devices.

Shigley, J.E. and Mischke, C.R. (eds) (2004) *Standard Handbook of Machine Design* (3rd edn), New York: McGraw-Hill. While the McGraw-Hill work by Lingaiah (above) is a databook, with the emphasis on data, this work is a handbook, with much more explanatory text and background material. Subject area coverage, on the other hand, is quite similar.

Suchy, I. (2005) *Handbook of Die Design* (2nd edn), New York: McGraw-Hill. Covers various types of dies and their functions and construction. Describes various types of metalworking machinery and operations, including blanking, piercing, bending, forming, and drawing. Significant attention is given to materials, their properties, and surface finishing.

Walsh, R.A. (1999) *McGraw-Hill Machining and Metalworking Handbook* (2nd edn), New York: McGraw-Hill. This handbook covers all typical areas of interest to machinists or metalworkers, but also standards and standardizing organizations, and safety. Contains a brief list of societies and other organizations with specific connections to various machining and metalworking areas.

Miscellaneous

Ahmad, A. (ed.) (1997) *Handbook of Optomechanical Engineering*, Boca Raton, FL: CRC Press. Optomechanical engineering addresses where and how mechanical systems interact and impact with modern optics. Topics include optical mounts, adjustment systems, thermal and thermoelastic issues, materials and their properties, and manufacturing.

Barber, A. (ed.) (1996) *Handbook of Noise and Vibration Control* (6th edn, revised), Oxford: Elsevier Advanced Technology. Includes sections on defining types and causes of vibration and noise, ways of measuring them, and suggestions for remediation. Unlike many Elsevier handbooks, does not contain advertisements.

Bishop, R.H. (ed.) (2002) *The Mechatronics Handbook*, Boca Raton, FL: CRC Press. Mechatronics is the integration of mechanical, electrical, and computer systems. This volume details the design and materials of mechatronic systems and their various applications in microelectromechanical systems, sensors and actuators, controls, and so on.

Bleier, F.P. (1998) *Fan Handbook: Selection, Application and Design*, New York: McGraw-Hill. The basics of stationary and moving air are defined, followed by detailed descriptions and application information about various types of fans, ventilators, and blowers. Appropriate standards are covered, as is information on installation, safety, maintenance, trouble-shooting, and problem-solving.

Boyce, M.P. (2002) *Gas Turbine Engineering Handbook* (2nd edn), Boston, MA: Gulf Professional Publications. Begins with an overview of gas turbine design, including discussions of the major components, such as compressors, turbines, and combustors, along with auxiliary components and accessories, such as bearings, seals, and gears. Installation, operation, and maintenance are also addressed.

Brown, M.W. (1995) *Seals and Sealing Handbook* (4th edn), Oxford: Elsevier Advanced Technology. Addresses all aspects of seals and sealing technology. As with other Elsevier handbooks, contains advertisements from companies in the industry.

Cho, H. (ed.) (2003) *Opto-mechatronic Systems Handbook: Techniques and Applications*, Boca Raton, FL: CRC Press. Mechatronics is the area where mechanical, electrical, and computer technologies blend and merge. Opto-mechatronics adds optics to mechantronics. This volume addresses the burgeoning new field of opto-mechantronics and applications of those technologies.

Dudley, D.W. (1994) *Handbook of Practical Gear Design* (rev. edn), Lancaster, PA: Technomic Publishing. This work not only deals with the design and manufacturing of various types of gears; it also branches into the design of tools to make gears and has an extensive section on gear failures.

Gad-el-Hak, M. (ed.) (2002) *The MEMS Handbook*, Boca Raton, FL: CRC Press. Covers one of the newest areas of research for mechanical and other engineers: microelectromechanical systems. Included are discussions of scaling, mechanical properties, fabrication, and lubrication of MEMS, as well as many current and projected applications.

Giampaolo, T. (2003) *The Gas Turbine Handbook* (2nd edn), Lilburn, GA: Fairmont Press. This work examines the various types of gas turbines and their applications, operation, and troubleshooting. Thermodynamics and heat transfer, mechanical drive application, and acoustics and noise control and prevention are addressed.

Harris, C.M. and Piersol, A.G. (eds) (2001) *Harris' Shock and Vibration Handbook* (5th edn), New York: McGraw-Hill. For any machine with relatively high-speed moving parts, vibration is a serious consideration. This work addresses all aspects of vibration and its control, some of which are directly pertinent to the mechanical engineer.

Hoffman, D.M., Singh, B. and Thomas, J.H. III (eds) (1998) *Handbook of Vacuum Science and Technology*, San Diego, CA: Academic Press. This work addresses the materials, design, and construction of pumps and related systems for the creation of vacuum. It also covers measurement techniques and applications.

Kobayashi, A.S. (ed.) (1993) *Handbook on Experimental Mechanics*, New York, VCH. This work focuses almost entirely on the testing and properties of the materials used by mechanical engineers and others.

Parmley, R.O. (ed.) (1997) *Standard Handbook of Fastening and Joining* (3rd edn), New York: McGraw-Hill. This work addresses the multitude of processes by which materials or components can be connected. Included are: fasteners, such as threads; pins; rings; pipe connections; expansion joints; various types of welding; concrete, lumber, and steel connections; locks; electrical connections; adhesives; rivets; rope and cable connections; shafts and couples; seals; and even rope splicing and tying.

Piotrowski, J. (1995) *Shaft Alignment Handbook* (2nd edn), New York: Marcel Dekker. This title deals with the narrow field of shaft alignment, a field crucial to the proper functioning of pumps, gears, turbines, fans, engines, compressors, and so on. Particular emphasis is placed on measuring proper function and on techniques to correct misalignments.

Sawyer, J.W. (ed.) (1985) *Sawyer's Gas Turbine Engineering Handbook* (3rd edn), Norwalk, CN: Turbomachinery International Publications. This three-volume work addresses theory and design, selection and various applications, and accessories and support for those in the gas turbine industry.

Skousen, P.L. (2004) *Valve Handbook* (2nd edn), New York: McGraw-Hill. While Smith and Zappe (below) focus solely on valves, this volume covers both valves and actuators. Valve selection criteria are detailed, as are the various types of valves and actuators. Sizing, purchasing issues, and common problems are also discussed.

Smith, P. and Zappe, R.W. (2004) *Valve Selection Handbook* (5th edn), Houston, TX: Gulf Publishing. This work is very applications-oriented. Initial pages are devoted to fundamentals, but the majority of the volume is devoted to detailed descriptions of various valve types, how they are to be installed, and appropriate uses for each. Appendices on ASME safety codes, properties of fluids, and various standards pertaining to valves are included.

Stokes, A. (1992) *Gear Handbook: Design and Calculations*, Oxford: Butterworth-Heinemann. The emphasis in this work is on the "calculations" portion of the title. This relatively brief volume contains all the formulae necessary to design, manufacture, and inspect a wide variety of gear types. Produced in association with the Society of Automotive Engineers.

Taylor, J.I. (2000) *The Gear Analysis Handbook*. Tampa, FL: Vibration Consultants, Inc. Narrow in focus, this work describes methods of gear analysis and appropriate tools, data collection, gear physics and specifications, and various gear problems and their causes.

Taylor, J.I. (2003) *The Vibration Analysis Handbook* (2nd edn), Tampa, FL: Vibration Consultants, Inc. While the other two books with Taylor as author or co-author focus on vibration in specific locations (gears and bearings), this work is more general in nature. It discusses machinery vibration in general, time and frequency analysis techniques, how to evaluate the condition of machinery, plus chapters on gears, bearings, and press roll and nip problems.

Taylor, J.I. and Kirkland, D.W. (2004) *The Bearing Analysis Handbook*, Tampa, FL: Vibration Consultants, Inc. Bearing defects, problem analysis and data collection, and case histories documenting various types of defects are highlighted in this volume.

Townsend, D.P. (ed.) (1992) *Dudley's Gear Handbook* (2nd edn), New York:

McGraw-Hill. Broader in scope than the title by Stokes (above), the emphasis here is on the manufacture, functions, and problems associated with different gear types.

Walsh, R.A. (2000) *Electromechanical Design Handbook* (3rd edn), New York: McGraw-Hill. This volume addresses designing when the designer must incorporate some combination of mechanical, electrical, and electronic components into the finished product. Emphasis is geared heavily toward the practical and less to theory.

Ocean/marine/ship

El-Hawary, F. (ed.) (2001) *Ocean Engineering Handbook*, Boca Raton, FL: CRC Press. Geared primarily toward electrical engineers, this book also covers topics such as vehicle control, and positioning and thrust control systems, both relevant to mechanical engineering.

Hunt, E.C. (ed.) (1999–) *Modern Marine Engineers Manual* (3rd edn), Centreville, MD: Cornell Maritime Press. This two-volume set covers those topics of concern to the marine engineer, with many of those topics tied directly to mechanical engineering. Included are: engineering materials; steam power plants; bearings and lubrication; steam generation; steam and gas turbines; transmission systems; heat exchangers; piping systems; engines; HVAC and refrigeration systems; and vibration analysis.

Taggart, R. (ed.) (1980) *Ship Design and Construction*, New York: Society of Naval Architects and Marine Engineers. Although much of this work focuses on the design of ships for particular applications, mechanical engineering topics such as propulsion and navigation systems, acoustics and vibration, HVAC and refrigeration are also addressed.

Pressure vessels and piping

Bednar, H.H. (1991) *Pressure Vessel Design Handbook* (2nd edn), Malabar, FL: Krieger Publishing. This book describes design considerations for many types and shapes of pressure vessels. Particular consideration is given to various types of stresses and loading, including localized stresses from supports, attachments, and so on. Contains a brief glossary.

Bernstein, M.D. and Yoder, L.W. (1998) *Power Boilers: A Guide to Section 1 of the ASME Boiler and Pressure Vessel Code*, New York: American Society of Mechanical Engineers. Focuses entirely on design considerations related to power boilers and their attachments. Materials, piping, fabrication and welding, testing, inspection, certification, quality control, safety, and potential causes of damage or failure are all covered.

Dickenson, T.C. (ed.) (1999) *Valves, Piping and Pipelines Handbook*, Oxford: Elsevier Advanced Technology. As the title suggests, this book focuses entirely on valves and pipes and does not address pressure vessels. Covers over twenty types of valves, various metal and plastic pipes, and contains a large section on determining pipe flow performance.

Helguero, M.V. (1986) *Piping Stress Handbook* (2nd edn), Houston, TX: Gulf Publishing. As the title suggests, this book addresses specifically the various types and sources of stresses that can affect pipes and piping systems, including internal pressure, corrosion, weight, pressure from external supports, and so on. Focus is on data rather than text.

Kohan, A.L. (1998) *Boiler Operator's Guide* (3rd edn), New York: McGraw-Hill. This book covers different types of boilers and boiler systems, appropriate construction materials and their properties, fabrication and repair methods, and connections, appurtenances, and auxiliaries. Operational issues, such as combustion, fuels, burners, and controls, are addressed, as are inspections, problems, repairs, and maintenance. Each chapter ends with a series of questions and answers.

Matthews, C. (2001) *Engineer's Guide to Pressure Equipment*, London: Professional Engineering. Another pocket guide from Matthews (see "General" for two more). This work focuses on pressure equipment of all kinds, not just vessels or containers. Thus, pipes, flanges, valves, nozzles, and other fittings are considered. Contains a list of pertinent websites.

McAllister, E.W. (ed.) (2005) *Pipe Line Rules of Thumb Handbook: Quick and Accurate Solutions to Your Everyday Pipe Line Problems* (5th edn), Houston, TX: Gulf Publishing. Very "how to" in focus, this handbook is designed primarily for those working with, rather than designing, piping systems. Format poses questions or concerns and then supplies answers.

Megyesy, E.F. (2001) *Pressure Vessel Handbook* (12th edn), Tulsa, OK: Pressure Vessel Publishing. Although called a handbook, this work is probably better described as a databook. Text is at a minimum throughout the work, with the focus on data elements for design and construction, geometry and layout, measures and weights.

Moser, A.P. (2001) *Buried Pipe Design* (2nd edn), New York: McGraw-Hill. The book focuses entirely on considerations related to buried pipe systems. These include loading and other external forces, pressure considerations, designs of various types of piping systems, choice of materials, and installation.

Moss, D.R. (2004) *Pressure Vessel Design Manual: Illustrated Procedures for Solving Major Pressure Vessel Design Problems* (3rd edn), Amsterdam; Boston, MA: Elsevier, Gulf Professional Publishers. Designed as more of a "how to" work than Bednar's book, this work is heavily illustrated and is procedurally oriented. Contains over a dozen appendices on various topics.

Nayyar, M.L. (ed.) (2000) *Piping Handbook* (7th edn), New York: McGraw-Hill. At over 2,400 pages, this works covers components, materials, codes, and standards, manufacturing/fabrication, installation, dozens of design considerations, and piping systems for various applications, primarily related to metal piping. Contains a small section on non-metallic piping, and ten appendices, most of which are data tables.

Nayyar, M.L. (2002) *Piping Databook*, New York: McGraw-Hill. While Nayyar's handbook (above) is heavily text-oriented, this work is virtually all data related to piping and the piping industry. Included are units and conversions, materials, and various organization specifications for types of pipe, fittings, flanges, and joints. Much more content is devoted to non-metallic piping than what appears in the handbook.

Pipe Characteristics Handbook (1995) Tulsa, OK: Pennwell Books. Although billed as a handbook, virtually the entire work is two sets of data tables: 200 pages on maximum allowable working pressure as a percentage of specified minimum yield, and less than twenty pages on properties of pipe.

Sixsmith, T. and Hanselka, R. (1997) *Handbook of Thermoplastic Piping System Design*, New York: Marcel Dekker. This work covers in detail the advances in plastic pipe and piping systems and applications where modern plastics may be used in place of more traditional metal pipe. Includes over seventy pages of basic reference data, and nearly 200 pages of chemical resistance data.

Pumps and pumping

Dickenson, T.C. (ed.) (1995) *Pumping Manual*, Oxford, UK: Elsevier Advanced Technology. This manual covers pump performance and characteristics, twenty different types of pumps, appropriate materials, construction and operation, pump ancillaries (e.g., engines, motors, seals, bearing), and a variety of pumping applications. Contains advertisements from companies in the pumping industry.

Karassik, I.J. (ed.) (2001) *Pump Handbook* (3rd edn), New York: McGraw-Hill. Covers fewer types of pumps than the Dickenson title (above), but in greater detail. Contains additional material on pump drivers, controls and valves, pumping systems, intakes and suction piping, and a variety of pumping applications.

Rayner, R. (ed.) (1995) *Pump Users Handbook*, Oxford: Elsevier Advanced Technology. As the title suggests, this work is very applications-oriented. Criteria for pump selection are covered, as well as for installation, testing, and start-up. Different types of pumps and drives are documented, followed by various pumping applications.

Rishel, J.B. (2006) *HVAC Pump Handbook* (2nd edn), New York: McGraw-Hill. As the title suggests, this book covers all aspects of HVAC pumps and pumping. Includes design basics, selection criteria, installation and operation, performance issues and measures, and the use of pumps in open and closed HVAC cooling systems and in HVAC hot water systems.

Tribology/lubrication

Bailey, C.A. (ed.) (1996) *Lubrication Engineers Manual* (2nd edn), Warrendale, PA: Association for Iron and Steel Technology. A general manual for anyone involved with lubrication or maintenance. Includes sections on fundamentals, test methods, statistics, performance, and others.

Bhushan, B. and Gupta, B.K. (1991) *Handbook of Tribology: Materials, Coatings and Surface Treatments*, New York: McGraw-Hill. Although this work touches on the basics of friction, wear, and lubrication, the vast majority of the text deals with different types of coatings and their application/deposition, and with treating surfaces by various methods.

Booser, E.R. (ed.) (1983–1994) *CRC Handbook of Lubrication: (Theory and Practice of Tribology)*, Boca Raton, FL: CRC Press. This three-volume set is somewhat odd in design in that volumes 1 and 2 were issued in 1983 and 1984, respectively; then volume 3 was issued in 1994 and covers new developments and updates to the field in the decade since publication of the original two volumes. Volume 1 focuses on application and maintenance, volume 2 on theory and design, and volume 3 on monitoring, materials, synthetic lubricants, and applications.

Booser, E.R. (ed.) (1997) *Tribology Data Handbook*, Boca Raton, FL: CRC Press. Designed as a follow-up or companion volume to the book listed immediately above, the focus of this work is on data. Among the topics covered are properties of various lubricants, lubrication specifications for different types of equipment, material properties, industrial application processes, and friction, wear, and surface characterization.

Neale, M.J. (ed.) (1995) *Tribology Handbook* (2nd edn), Oxford: Butterworth-Heinemann. This work is very applications-directed and touches only marginally on theory. Performance characteristics are primary in presentation. Various types of bearings, drives, seals, lubricants, and lubrication systems are addressed, as well as failures, environmental effects, and maintenance and repair.

Monographs and textbooks

There are hundreds of textbooks covering mechanical engineering and its subsets. Those included here are just some of the textbooks used currently in classrooms. Monographs were chosen as representative of classical works in the field of mechanical engineering. Both textbooks and monographs were selected based on book reviews in *Books in Print* and in consultation with mechanical engineering professors.

Beer, F.P., Johnston, E.R. Jr, and DeWolf, J.T. (2006) *Mechanics of Materials* (4th edn), New York: McGraw-Hill. This book covers mechanics of materials in general, as well as giving a brief review of statics, an expanded discussion of the behavior of materials on stress-strain tests that covers materials such as concrete and composites, and increased coverage of the behavior of fiber-reinforced composites. Presents concepts clearly and in detail, and includes illustrations and diagrams to explain those concepts.

Beer, F.P., Johnston, E.R. Jr, Eisenberg, E.R., Clausen, W.E., and Staab, G.H. (2006) *Vector Mechanics for Engineers: Statics and Dynamics* (8th edn), New York: McGraw-Hill. This book is designed for beginning courses in statics and dynamics. Each chapter contains sample problems with solutions, end-of-chapter problems, and computer problems, as well as a review and summary.

Den Hartog, J.P. (1985) *Mechanical Vibrations* (4th edn), New York: Dover Publications. Originally published: 1956, New York: McGraw-Hill. This book covers the fundamentals of mechanical vibrations. It may be used by practicing engineers as well as for classroom instruction. An elementary knowledge of dynamics and calculus is necessary, but differential equations are explained in detail. Examples have been drawn from real-life experiences of the author and his friends. Each chapter includes problems that illustrate typical practical situations with answers included at the back.

Den Hartog, J.P. (1987) *Advanced Strength of Material*, New York: Dover Publications. Reprint. Originally published 1952. New York: McGraw-Hall. Intended for the intermediate student studying strength of materials, this title covers torsion, rotating disks, membrane stresses in shells, bending of flat plates, and two-dimensional theory of elasticity and buckling. Includes problem sets and answers.

Dorf, R.C. and Bishop, R.H. (2005) *Modern Control Systems* (10th edn), Upper Saddle River, NJ: Prentice Hall. This work introduces control systems, including a brief history of control theory and practice. Topics covered include mathematical models of systems, state variable models, feedback control system characteristics, performance of feedback control systems, stability of linear feedback systems, design of feedback, state variable, and robust control systems. Each chapter lists problems at the end.

Fermi, E. (1956) *Thermodynamics*, New York: Dover Press. Based on lectures given by Fermi in 1936, this book is about pure thermodynamics. Fermi describes the state of a system and its transformations, the first and second laws of thermodynamics, thermodynamic potentials, and thermodynamics of dilute solutions. This is an elementary treatment of thermodynamics, but the reader should be familiar with the fundamentals of thermometry and calorimetry.

Finnemore, E.J. and Franzini, J.B. (2001) *Fluid Mechanics with Engineering Application* (10th edn), New York: McGraw-Hill. This book is intended to be used as a textbook for a beginning course in fluid mechanics for engineering students. Readers should be familiar with differential and integral calculus as well as thermodynamics. Topics include properties of fluids, fluid statics, kinematics of fluid flow, basic hydrodynamics, and fluid measurements. The text includes numerous examples of how the basic principles of fluid mechanics can be applied to particular engineering problems.

Gere, J.M. (2004) *Mechanics of Materials (with CD-ROM)* (6th edn), London: Brooks/Cole. This title describes the fundamentals of mechanics of materials. Principal topics are analysis and design of structural members subjected to tension, compression, torsion, and bending, as well as stress, strain, elastic behavior, inelastic behavior, and strain energy. Transformations of stress and strain, combined loadings, stress concentrations, deflections of beams, and stability of columns are also covered. Includes many problem sets, with answers at the back.

Hibbeler, R.C. (2004) *Mechanics of Materials* (6th edn), Upper Saddle River, NJ: Prentice Hall. This book presents the theory and applications of mechanics of materials in a fluid manner, with cohesive organization, illustrations, exercises, examples, and free body diagrams. Clear and comprehensive, this text examines the physical behavior of materials under load.

Incropera, F.P. and DeWitt, D.P. (2006) *Fundamentals of Heat and Mass Transfer* (5th edn), New York: John Wiley. This book provides the basics of heat and mass transfer in a clear presentation style, with a problem-solving methodology that is easy to understand. Contains hundreds of problems and examples illustrating real-life engineering processes.

Lewis, E.V. (ed.) (1988) *Principles of Naval Architecture* (2nd edn), Jersey City, NJ: Society of Naval Architects and Marine Engineers. This comprehensive three-volume set, SNAME's leading reference work, covers load modeling, flows around hulls, propeller design, vibration, stability and control, and motion in waves. Vol. 1, Stability and Strength; vol. 2, Resistance, Propulsion; vol. 3, Sea-keeping and Controllability.

Meirovitch, L. (2001) *Fundamentals of Vibrations*, Boston, MA: McGraw-Hill. This book covers all basic concepts at a level appropriate for undergraduate engineering students and is excellent for introductory vibrations courses. Meirovitch, one of the leading authorities in the field of vibration, emphasizes analytical developments and computational solutions. Includes physical explanations, problems at the end of each chapter, worked-out examples, and illustrations.

Mott, R.L. (2004) *Machine Elements in Mechanical Design* (4th edn), Upper Saddle River, NJ: Prentice Hall. This book provides a current, thorough, practical approach to designing machine elements in the context of complete mechanical design. It includes some of the primary machine elements such as belt drives, chain drives, gears, shafts, keys, couplings, seals, and rolling contact bearings. It also covers plain surface bearings, linear motion elements, fasteners, springs, machine frames, bolted connections, welded joints, electric motors, controls, clutches, and brakes.

Mott, R.L. (2002) *Applied Strength of Materials* (4th edn), Upper Saddle River, NJ: Prentice Hall. This work is useful for undergraduate, introductory-level courses. Coverage is comprehensive in the key areas of strength of materials. Emphasizes applications and problem-solving, as well as design of structural members, mechanical devices, and systems. Includes coverage of current tools, trends, and analysis techniques; also example problems.

Munson, B.R., Young, D.F., and Okiishi, T.H. (2002) *Fundamentals of Fluid Mechanics* (4th edn), New York: John Wiley. Intended as an undergraduate textbook, this edition includes a chapter on turbomachines as well as over one hundred examples from everyday life of fluid-flow phenomena. The accompanying CD-ROM contains the full text of the book as well as video segments and extra problems.

Nise, N.S. (2004) *Control Systems Engineering* (4th edn), Hoboken, NJ: Wiley. This text focuses on the practical application of control systems engineering. It explains how to analyze and design real-world feedback control systems, and shows how to create control systems that support today's advanced technology. Examples help give a practical view of each stage in the design process. A methodology with clearly defined steps is presented for each type of design problem.

O'Hanlon, J.F. (2003) *A User's Guide to Vacuum Technology* (3rd edn), New York: John Wiley. A detailed, practical guide to current vacuum technology that emphasizes the fundamentals while covering operation procedures, understanding, and selection of equipment used in semiconductor, optics, packaging, and related coating technologies.

Shigley, J.E., Mischke, C.R., and Budynas, R. (2004) *Mechanical Engineering Design* (7th edn), Boston, MA: McGraw-Hill. A classic text that covers the significant machine components encountered in a machine design course. Emphasis is on developing good design and problem-solving skills. Basic concepts are presented in a clear and concise manner. Includes examples from industry, case studies and problem sets, as well as an Online Learning Center where students have access to support material such as MATLAB for Machine Design, as well as various tutorials.

Timoshenko, S.P. (1983) *Strength of Materials, Part 1 and Part 2* (3rd edn), Malabar, FL: R.E. Krieger Publishing. Part 1, Elementary Theory and Problems, covers the fundamentals of strength of materials such as stress, deformation, and strain. Part 2, Advanced Theory and Problems, includes later developments that are of practical importance in the fields of strength of materials, and theory of elasticity. Both books include problems and solutions. Many of the problems illustrate how to apply the theory to solve practical design problems.

Timoshenko, S.P. and Young, D.H. (1968) *Elements of Strength of Materials* (5th edn), Princeton, NJ: Van Nostrand. Timoshenko, a pioneer in the field of applied mechanics, influenced the teaching of mechanics throughout the world through his research and textbooks. This book covers the fundamentals of the elements of strength of materials, such as stress, deformation, and strain.

Ullman, D.G. (2003) *The Mechanical Design Process* (3rd edn), Boston, MA: McGraw-Hill. This book describes the steps in the design process, from understanding the process, to concept evaluation, product generation, launching and supporting the product, to properties of materials most commonly used in mechanical design, and human factors in design. Clearly laid out with numerous illustrations. Includes exercises at the end of each chapter.

Van Ness, H.C. (1983) *Understanding Thermodynamics*, New York: Dover Press. This book covers the basic concepts of thermodynamics. It is not intended to cover thermodynamics in the same way as a textbook, but is meant more as a supplement to assist students through the difficult early stages of a beginning course in thermodynamics. Topics covered include the first law of thermodynamics, the concept of reversibility, heat engines, power plants; the second law of thermodynamics; and statistical mechanics. It is written in an engaging, informal style.

White, F.M. (2003) *Fluid Mechanics* (5th edn), Boston, MA: McGraw-Hill. This book is excellent for graduate-level studies. Covers advanced thermodynamics of fluid properties, viscous flow and heat transfer, stability of boundary layers, incompressible and compressible turbulent flow, and compressible turbulent boundary layers. Includes a review of each topic with examples.

Major publishers include John Wiley & Sons, McGraw-Hill Higher Education, Prentice Hall, Oxford University Press, Cambridge University Press.

Journals

Mechanical engineering journals cover a broad range of topics from heating and refrigeration, to automobiles, to thermodynamics. To determine the most significant journals in the mechanical engineering field the following resources were consulted: Journal Citation Reports (using the categories Mechanics; and Engineering, Mechanical); *Guide to Information Sources in Engineering* by Charles Lord; *Scientific and Technical Information Sources* by Ching-chih Chen; *Encyclopedia of Physical Sciences and Engineering Information Sources* by Steven Wasserman, Martin Smith, and Susan Mottu; and *Core List of Books and Journals in Science and Technology* by Russell Powell and James Powell Jr

Annual Review of Fluid Mechanics (1969–) Palo Alto, CA: Annual Reviews (0066–4189)

Applied Thermal Engineering (1980–) Amsterdam: Elsevier (1359–4311).

ASHRAE Journal (1959–) New York: American Society of Heating, Refrigeration and Air-Conditioning Engineers (0001–2491).

Automotive Engineering International (1998–) Warrendale, PA: Society of Automotive Engineers (1543–849X).

Combustion and Flame (1963–) Amsterdam: Elsevier (0010–2180).

Computers and Fluids (1973–) Amsterdam: Elsevier (0045–7930).

Experiments in Fluids (1983–) New York: Springer-Verlag (0723–4864).

Heat Transfer Engineering (1979–) Philadelphia, PA: Taylor & Francis (0145–7632).

IEEE/ASME Transactions on Mechatronics (1996–) Piscataway, NJ: Institute of Electrical and Electronics Engineers (1083–4435).

International Journal of Heat and Fluid Flow (1979–) New York: Elsevier (0142–727X).

International Journal of Heat and Mass Transfer (1960–) Amsterdam: Elsevier (0017–9310).

International Journal of Impact Engineering (1983–) Amsterdam: Elsevier (0734–743X).

International Journal of Mechanical Sciences (1960–) Amsterdam: Elsevier (0020–7403).

International Journal of Multiphase Flow (1974–) Amsterdam: Elsevier (0301–9322).
Journal of Applied Mechanics. Transactions of the ASME (1933–) New York: ASME International (0021–8936).
Journal of Biomechanical Engineering. Transactions of the ASME (1977–) New York: ASME International (0148–0731).
Journal of Dynamic Systems Measurement and Control. Transactions of the ASME (1971–) New York: ASME International (0022–0434).
Journal of Energy Resources Technology. Transactions of the ASME (1979–) New York: ASME International (0195–0738).
Journal of Engineering for Gas Turbines and Power. Transactions of the ASME (1984–) New York: ASME International (0742–4795).
Journal of Engineering Materials and Technology. Transactions of the ASME (1973–) New York: ASME International (0094–4289).
Journal of Engineering Mechanics (1983–) New York: American Society of Civil Engineers (0733–9399).
Journal of Fluid Mechanics (1956–) London: Taylor & Francis (0022–1120).
Journal of Fluids Engineering. Transactions of the ASME (1973–) New York: ASME International (0098–2202).
Journal of Heat Transfer. Transactions of the ASME (1959–) New York: ASME International (0022–1481).
Journal of Manufacturing Science and Engineering. Transactions of the ASME (1996–) New York: ASME International (1087–1357).
Journal of Microelectromechanical Systems (1992–) Piscataway, NJ: Institute of Electrical and Electronics Engineers (1057–7157).
Journal of Micromechanics and Microengineering (1991–) Bristol: United Kingdom. Institute of Physics (0960–1317).
Journal of Non-Newtonian Fluid Mechanics (1976–) Amsterdam: Elsevier (0377–0257).
Journal of Offshore Mechanics and Arctic Engineering. Transactions of the ASME (1987–) New York: ASME International (0892–7219).
Journal of Pressure Vessel Technology. Transactions of the ASME (1974–) New York: ASME International (0094–9930).
Journal of Solar Energy Engineering. Transactions of the ASME (1980–) New York: ASME International (0199–6231).
Journal of Sound and Vibration (1964–) Amsterdam: Elsevier (0022–460X).
Journal of Tribology. Transactions of the ASME (1984–) New York: ASME International (0742–4787).
Journal of Turbomachinery. Transactions of the ASME (1986–) New York: ASME International (0889–504X).
Journal of Vibration and Acoustics. Transactions of the ASME (1990–) New York: ASME International (1048–9002).
Mechanical Engineering (1906–) New York: American Society of Mechanical Engineers (0025–6501).
Mechanical Systems and Signal Processing (1987–) Amsterdam: Elsevier (0888–3270).
Mechanics of Materials (1982–) Amsterdam: Elsevier (0167–6636).
Numerical Heat Transfer. Part A, Applications (1989–) Philadelphia, PA: Taylor & Francis (1040–7782).
Numerical Heat Transfer. Part B, Fundamentals (1989–) Philadelphia, PA: Taylor & Francis (1040–7790).

Proceedings of the Institution of Mechanical Engineers. Part A, Journal of Power and Energy (1990–) Birmingham, AL: Professional Engineering Publishing (0957–6509).

Proceedings of the Institution of Mechanical Engineers. Part B, Journal of Engineering Manufacture (1989–) Birmingham, AL: Professional Engineering Publishing (0954–4054).

Proceedings of the Institution of Mechanical Engineers. Part C, Journal of Mechanical Engineering Science (1983–) Birmingham, AL: Professional Engineering Publishing (0954–4062).

Proceedings of the Institution of Mechanical Engineers. Part D, Journal of Automobile Engineering (1989–) Birmingham, AL: Professional Engineering Publishing (0954–4070).

Proceedings of the Institution of Mechanical Engineers. Part E, Journal of Process Mechanical Engineering (1989–) Birmingham, AL: Professional Engineering Publishing (0954–4089).

Proceedings of the Institution of Mechanical Engineers. Part F, Journal of Rail and Rapid Transit (1989–) Birmingham, AL: Professional Engineering Publishing (0954–4097).

Proceedings of the Institution of Mechanical Engineers. Part G, Journal of Aerospace Engineering (1989–) Birmingham, AL: Professional Engineering Publishing (0954–4100).

Proceedings of the Institution of Mechanical Engineers. Part H, Journal of Engineering in Medicine (1989–) Birmingham, AL: Professional Engineering Publishing (0954–4119).

Proceedings of the Institution of Mechanical Engineers. Part I, Journal of Systems and Control Engineering (1991–) Birmingham, AL: Professional Engineering Publishing (0959–6518).

Proceedings of the Institution of Mechanical Engineers. Part J, Journal of Engineering Tribology (1994–) Birmingham, AL: Professional Engineering Publishing (1350–6501).

Proceedings of the Institution of Mechanical Engineers. Part K, Journal of Multi-body Dynamics (1999–) Birmingham, AL: Professional Engineering Publishing (1464–4193).

Proceedings of the Institution of Mechanical Engineers. Part L, Journal of Materials Design and Applications (1999–) Birmingham, AL: Professional Engineering Publishing (1464–4207).

Proceedings of the Institution of Mechanical Engineers. Part M, Journal of Engineering for the Maritime Environment (2002–) Birmingham, AL: Professional Engineering Publishing (1475–0902).

SAE Transactions (1893–) Warrendale, PA: Society of Automotive Engineers (0096–736X).

Tribology International (1968–) Amsterdam: Elsevier (0301–679X).

Tribology Letters (1995–) Norwell, MA: Kluwer Academic Publishers (1023–8883).

Wear (1957–) Amsterdam: Elsevier (0167–6636).

Patents and standards

There are no databases or other resources geared specifically toward patents in mechanical engineering (see Chapter 2 in this book for patent information).

However, there are a number of organizations that produce standards in ME and related areas. General standards resources which offer mechanical engineering standards are listed in Chapter 2 in this book. Standardizing organizations producing codes and standards related specifically to mechanical engineering are listed below.

ASME Standards, New York: American Society of Mechanical Engineers. With over 600 codes and standards being currently produced and distributed, ASME is the pre-eminent standardizing organization in mechanical engineering in the world, and has been producing codes and standards for nearly a century. Of particular note is the *ASME Boiler and Pressure Vessel Code*. Produced since 1914, the Code is now adopted by forty-nine of the fifty states in the U.S. and by all the provinces in Canada.

ASTM Standards, West Conshohocken, PA: American Society for Testing and Materials. With over 12,000 standards in fifteen sections published in over seventy volumes, the *Annual Book of ASTM Standards* is an invaluable resource for mechanical engineers. Available in print, CD-ROM, or online.

SAE Standards, Warrendale, PA: Society of Automotive Engineers. SAE has produced over 8,300 standards, primarily in two broad areas: aerospace and ground vehicle. Aerospace content is in turn divided primarily between Aerospace Materials Specifications and Aerospace Standards. Ground vehicle standards are primarily known as J-Reports, with most published annually in the *SAE Handbook*. Both aerospace and ground vehicle standards are available in print, on CD-ROM, or online (see Chapter 3 and 20 in this book).

Search engines and important websites

Topical websites can readily be divided into two groups: meta-sites, which collect links to more specific topic-specific sites, and those topic-specific sites. Representative meta-sites are listed immediately below, followed by a select list of topic-specific sites.

Meta-sites (following sites accessed May 17, 2005)

Australasian Virtual Engineering Library: Mechanical Engineering <http://avel.library.uq.edu.au/browse/browse_tree_204.html> (accessed May 18, 2005)

BUBL Links (all accessed May 18, 2005)

Fluid Mechanics <http://www.bubl.ac.uk/link/f/fluidmechanics.htm>

Mechanical Engineering <http://bubl.ac.uk/link/m/mechanicalengineering.htm>

Pneumatics <http://www.bubl.ac.uk/link/p/pneumatics.htm>

Edinburgh Engineering Virtual Library (EEVL): Engineering <http://www.eevl.ac.uk/engineering/index.htm> (accessed May 18, 2005), click on "Mechanical Engineering and Related Industries."

Engineering Links: Mechanical <http://www.englinks.com/mecheng.html> (accessed May 18, 2005).

Galaxy Directory: Mechanical Engineering <http://www.galaxy.com/galaxy/

Engineering-and-Technology/Mechanical-Engineering.html> (accessed May 18, 2005).

Lycos Directory: Science: Technology: Mechanical Engineering <http://dir.lycos.com/Science/Technology/Mechanical_Engineering/> (accessed May 18, 2005).

ViFaTec (Virtuelle Fachbibliothek Technik) <http://vifatec.tib.uni-hannover.de/fit/index.php3?L=e> (accessed May 18, 2005), click "browse" next to the mechanical engineering category).

World Wide Web Virtual Library: Mechanical Engineering <http://dart.stanford.edu/vlme/> (accessed May 18, 2005).

Yahoo! Directory: Engineering: Mechanical Engineering <http://dir.yahoo.com/Science/engineering/mechanical_engineering/> (accessed May 18, 2005).

Individual subject sites

Animated Engines <http://www.keveney.com/Engines.html> (accessed May 18, 2005). As the title suggests, this site provides access to animations of how nearly twenty different types of engines function.

Automotive Intelligence <http://www.autointell.com/> (accessed May 18, 2005). One of a number of sites collecting news and information on the automotive industry.

E-fluids.com <http://www.efluids.com> (accessed May 18, 2005). A meta-site in its own right, e-fluids.com brings together a variety of resources related to flow engineering, fluid mechanics research, education, and related topics.

iCrank.com <http://www.icrank.com> (accessed May 18, 2005). The self-professed "perfect starting page for a mechanical engineer," iCrank.com is a site designed by mechanical engineers to provide links to information and tools that MEs will likely need to use on an ongoing basis. iCrank.com, like a number of other sites listed here, offers users the option of submitting URLs for sites to be added to the portal.

MEMS and Nanotechnology Clearinghouse <http://www.memsnet.org/> (accessed May 18, 2005). A clearinghouse site for information on micro-electro-mechanical (microelectromechanical) systems (MEMS) and nanotechnology. Offers background information via a "Beginner's Guide" and a glossary, as well as content for the engineering student or engineer, including industry news, a conferences and workshops calendar, list of job openings, and more.

Shock and Vibration Information Analysis Center <http://iac.dtic.mil/iac_dir/SAVIAC.html> (accessed May 18, 2005). This site, supported by the U.S. Department of Defense and managed by the U.S. Army Engineer Research and Development Center, is a "focal point for research and analysis in the field of shock and vibration technology."

Thermal Connection <http://www.tak2000.com> (accessed May 18, 2005). Thermal Connection collects tools, data, and links relevant to anyone working in heat transfer and related fields.

Vibrationdata.com <http://www.vibrationdata.com> (accessed May 18, 2005). A subscription-based site offering tutorials, software, and other resources related to acoustics, shock and vibration, signal processing, and other areas.

WWW Tribology @ Sheffield <http://www.shef.ac.uk/mecheng/tribology/> (accessed May 18, 2005). Based at the University of Sheffield, this site offers both information on the research and teaching of the tribology team at Sheffield as well as tools and information resources in tribology and lubrication.

Associations, organizations, and societies

Acoustical Society of America (ASA) <http://asa.aip.org> (accessed May 18, 2005). "The premier international scientific society in acoustics, dedicated to increasing and diffusing the knowledge of acoustics and its practical applications" (source: ASA website).

American Bearing Manufacturers Association (ABMA) <http://www.abma-dc.org> (accessed May 18, 2005). "The American Bearing Manufacturers Association (ABMA) is a non-profit association consisting of American manufacturers of anti-friction bearings, spherical plain bearings or major components thereof. The purpose of ABMA is to define national and international standards for bearing products and maintain bearing industry statistics" (source: ABMA website).

American Gear Manufacturers Association (AGMA) <http://www.agma.org> (accessed May 18, 2005). AGMA's membership is largely companies/corporations involved in some aspect of the gear, coupling, or mechanical power transmission industries.

American Society for Nondestructive Testing (ASNT) <http://www.asnt.org> (accessed May 18, 2005). "ASNT exists to create a safer world by promoting the profession and technologies of nondestructive testing . . . through publication, certification, research and conferencing" (source: ASNT website).

American Society of Heating, Refrigerating, and Air-conditioning Engineers <http://www.ashrae.org> (accessed May 18, 2005). "ASHRAE will advance the arts and sciences of heating, ventilation, air conditioning, refrigeration and related human factors to serve the evolving needs of the public and ASHRAE members" (source: ASHRAE website).

American Society of Mechanical Engineers (ASME) <http://www.asme.org> (accessed May 18, 2005). "Founded in 1880 as the American Society of Mechanical Engineers, today's ASME is a 120,000-member professional organization focused on technical, educational and research issues of the engineering and technology community. ASME conducts one of the world's largest technical publishing operations, holds numerous technical conferences worldwide, and offers hundreds of professional development courses each year. ASME sets internationally recognized industrial and manufacturing codes and standards that enhance public safety" (source: ASME website).

American Welding Society (AWS) <http://www.aws.org> (accessed May 18, 2005). "The American Welding Society (AWS) was founded in 1919 as a multifaceted, non-profit organization with a goal to advance the science, technology and application of welding and related joining disciplines" (source: AWS website).

Association of Energy Engineers (AEE) <http://www.aeecenter.org> (accessed May 18, 2005). "AEE is your source for information on the dynamic field of energy efficiency, utility deregulation, facility management, plant engineering, and environmental compliance" (source: AEE website).

Combustion Institute <http://www.combustioninstitute.org> (accessed May 18, 2005). "The Combustion Institute is an educational non-profit, international, scientific society whose purpose is to promote and disseminate research in combustion science" (source: Institute website).

Compressed Gas Association (CGA) <http://www.cganet.com> (accessed May 18, 2005). "Since 1913, the Compressed Gas Association has been dedicated to the development and promotion of safety standards and safe practices in the industrial gas industry" (source: CGA website).

Institution of Mechanical Engineers (IMechE) <http://www.imeche.org.uk> (accessed May 18, 2005). "Established in 1847, the Institution of Mechanical Engineers (IMechE) is the leading body for professional mechanical engineers. With a world-wide membership now in excess of 75,000 engineers, the IMechE is the United Kingdom's qualifying body for Chartered and Incorporated mechanical engineers" (source: IMechE website).

International Union of Theoretical and Applied Mechanics (IUTAM) <http://www.iutam.net> (accessed May 18, 2005). "The International Union of Theoretical and Applied Mechanics (IUTAM) was formed in 1946 with the object of creating a link between persons and national or international organizations engaged in scientific work (theoretical or applied) in solid and fluid mechanics or in related sciences" (source: IUTAM website).

NACE International <http://www.nace.org> (accessed May 18, 2005). "NACE International was originally known as 'The National Association of Corrosion Engineers' when it was established in 1943. With more than 60 years of experience in developing corrosion prevention and control standards, NACE International has become the largest organization in the world committed to the study of corrosion" (source: NACE website).

National Fluid Power Association (NFPA) <http://www.nfpa.com> (accessed May 18, 2005). The NFPA's mission is to "advance hydraulic and pneumatic motion control technology" and "To act for and represent, on a worldwide basis, companies active in the U.S. fluid power industry" (source: NFPA website).

SAE International – the Society of Automotive Engineers <http://www.sae.org> (accessed May 18, 2005). "The Society of Automotive Engineers has more than 84,000 members – engineers, business executives, educators, and students from more than 97 countries – who share information and exchange ideas for advancing the engineering of mobility systems. SAE is your one-stop resource for standards development, events, and technical information and expertise used in designing, building, maintaining, and operating self-propelled vehicles for use on land or sea, in air or space" (source: SAE website).

Society for Experimental Mechanics (SEM) <http://www.sem.org> (accessed May 18, 2005). "The Society for Experimental Mechanics is composed of international members from academia, government, and industry who are committed to interdisciplinary application, research and development, education, and active promotion of experimental methods to: (a) increase the knowledge of physical phenomena; (b) further the understanding of the behavior of materials, structures and systems; and (c) provide the necessary physical basis and verification for analytical and computational approaches to the development of engineering solutions" (source: SEM website).

Society of Naval Architects and Marine Engineers (SNAME) <http://www.sname.org> (accessed May 18, 2005). "The Society of Naval Architects and Marine Engineers is an internationally recognized non-profit, technical, professional society of individual members serving the maritime and offshore industries and their suppliers. SNAME is dedicated to advancing the art, science and practice of naval architecture, shipbuilding and marine engineering" (source: SNAME website).

Society of Tribologists and Lubrication Engineers (STLE) <http://www.stle.org> (accessed May 18, 2005). STLE's mission is "to advance the science of tribology and the practice of lubrication engineering in order to foster innovation, improve the performance of equipment and products, conserve resources and protect the environment" (source: STLE website).

Vibration Institute <http://www.vibinst.org> (accessed May 18, 2005). "A nationally recognized not-for-profit organization dedicated to the exchange of practical information about vibration and condition monitoring" (source: Institute website).

Acknowledgments

The authors wish to thank Bonnie Osif for the opportunity to contribute to this work.

Aleteia wishes to thank Mel DeSart because it was a pleasure collaborating; Jay Bhatt for his very informative *Mechanical Engineering Resource Guide*; and Randy Reichardt for discussion, assistance, and advice.

Mel wishes to thank Aleteia for her energy, good humor, and willingness to co-author; and reference assistants Genevieve Williams, Phoebe Ayers, Sarah Lester, and Karen Andring for date and edition checking.

References

Chen, C. (1987).\ *Scientific and Technical Information Sources* (2nd edn), Cambridge, MA: MIT Press.

Journal Citation Reports [electronic resource] Philadelphia, PA: Institute for Scientific Information.

Lord, C. (2000) *Guide to Information Sources in Engineering*, Englewood, CO: Libraries Unlimited.

Malinowsky, H.R. (1994) *Reference Sources in Science, Engineering, Medicine, and Agriculture*, Phoenix, AZ: Oryx Press.

Powell, R. and Powell, J. Jr (1994) *Core List of Books and Journals in Science and Technology*, Phoenix, AZ: Oryx Press.

Wasserman, S., Smith, M.A., and Mottu, S. (1989) *Encyclopedia of Physical Sciences and Engineering Information Sources*, Detroit, MI: Gale Research.

17 Mining engineering

Jerry Kowalyk

Introduction and scope of the chapter

Mining involves the recovery of solid, liquid, or gas mineral substances from the Earth for direct use or conversion into a usable product. Recovery of energy minerals in the form of natural gas and petroleum has developed into a separate industry with its own technology. That aspect of mining engineering is discussed in Chapter 19.

This guide points to information resources available in either print or online for mining engineering. Where available, URLs have been provided to readers for further exploration of the pertinent Internet resource. Further mineral processing through refinement and fabrication techniques is the property of metallurgical, material, and chemical engineering, and is not within the scope of this chapter.

History of mining engineering

Mining ranks as one of the basic industries of early civilization and, together with early hunting, gathering, and agriculture is concurrent with the very development of civilization itself. From the gathering of surface deposited raw materials during the Paleolithic era, 300,000 years ago in Africa, to the first underground mines of the Stone Age 40,000 years ago, and on to contemporary underground excavations, mining has supplied the basic resources used by human civilization.

In ancient history, mining work was performed predominantly by slaves (prisoners-of-war, criminals, or political prisoners). As surficial deposits became exhausted, the late Roman Empire and early Middle Ages saw a more complicated extension of mines underground. Slaves were replaced by skilled artisans and guild organizations. Ownership and the financial support of mines through investment funding encouraged the formation of organized company enterprises, with guilds taking on the form of modern trade labor organizations.

The discovery and refinement of metallurgical processes in the eighteenth century made possible the transformation of labor to more efficient factory systems. With the Industrial Revolution of the late eighteenth century, the quest for mineral resources became internationalized as nations projected economic, political, and military power abroad to satisfy an increasing mechanized demand for raw minerals and sources of energy. As globalization increased, it may be argued that the history of many major wars of the twentieth century and the early decades of this century may be attributable to national governments identifying their own national and economic security with resource security.

Mining is not pursued as an isolated activity. Minerals discovered through prospecting and exploration must first submit to geological investigation and economic evaluation before mining engineers can plan, develop, and proceed with resource recovery and extraction itself. Following exhaustive mineral exploitation, environmental engineers conclude the mining process with closure and reclamation of the mine site with a view to sustainable future uses for the land, such as wildlife refuges, real estate development, or waste disposal sites.

Within the realm of mining activity proper, the mining engineer is concerned with the methods, tools, and processes for recovering mineral resources from the relative near surface of the Earth or extracting those resources from underground excavations. The extraction process must be executed in a logical set of unit operations consisting of organized drilling, blasting, loading, and hauling in a continuous, productive cycle. Throughout the production cycle myriad detailed supplementary (auxiliary) operations are carried out to ensure success. Power, pumping, lighting, ventilation, air-conditioning, water, communication, lighting, and waste disposal need to be supplied and maintained to sustain the entire operation up to the day of its closure.

This chapter has been compiled to help bring an awareness of the published and unpublished sources of information in mining engineering. An exhaustive list of resources is not provided; only the more important titles have been included. The business, economic, and financial components of the petroleum industry are not part of the scope of this guide, though some titles listed herein may point out links to those aspects.

History of mining monographs

Standard monographic works in English, which outline the history of the human mining experience, are:

Craddock, P. and Lang, J. (eds) (2002) *Mining and Metal Production Through the Ages*, London: British Museum.

Davies, O. (1935) *Roman Mines in Europe*, Oxford: Clarendon Press.

Dibner, B. (1958) *Agricola on Metals*, Norwalk, CT: Burndy Library.

Gregory, C.E. (1980) *A Concise History of Mining*, Oxford: Pergamon Press.

Lynch, M. (2002) *Mining in World History*, London: Reaktion.

Rickard, T.A. (1932) *A History of American Mining*, New York: McGraw-Hill.

Rickard, T.A. (1932) *Man and Metals; A History of Mining in Relation to the Development of Civilization*, New York: McGraw-Hill.

Shepherd, R. (1993) *Ancient Mining*, London: Elsevier Applied Science.

Spence, C.C. (1993) *Mining Engineers and the American West; The Lace-boot Brigade, 1849–1933*, Moscow: University of Idaho.

Temple, J. (1972) *Mining: An International History*, New York: Praeger.

History of mining websites

Bir Umm Fawakhir: Insights into Ancient Egyptian Mining <http://www.tms.org/pubs/journals/JOM/9703/Meyer-9703.html> (accessed May 18, 2005). An archaeological survey of the site at Bir Umm Fawakhir in Egypt.

Bronze Age <http://www.encyclopedia.com/html/b/bronzea1g.asp> (accessed May 18, 2005). Encyclopedia.com, a free web-based encyclopedia, provides users with more than 57,000 frequently updated articles from the *Columbia Encyclopedia* (6th edn). The article on the Bronze Age includes imbedded hyperlinks to related topics as well as to other "prehistory" discussions in the Encyclopedia about the Copper Age, Coal Mining, Iron Age, and more.

Copper: The Red Metal <http://www.unr.edu/sb204/geology/copper2.html> (accessed May 18, 2005). A thorough, scholarly discussion on the early history of copper mining, smelting, metallurgy, and mining technique. Geographic themes abound with discussions about copper in Sub-Saharan Africa; the Bronze Age in Asia Minor, Europe, Asia; the Bronze Age to the Fall of Rome, copper in the Middle Ages and Renaissance, in the Western Hemisphere, and copper mining in the U.S. The article is complete with references.

History of Mining: The Miner's Contribution to Society <http://www.dmtcalaska.org/course_dev/intromining/01history/notes01.html> (accessed May 18, 2005).

History of Parys and Mona Copper Mines <http://www.rhosybolbach.freeserve.co.uk/history.htm> (accessed May 18, 2005). A useful starting point for visitors to this website is site index to "subjects within the Parys Underground website." An A–Z listing of hyperlinks connect to topics such as mining techniques, the Middle Ages, the Dark Ages, Bronze Age copper recovery, Bronze Age man, old maps, slavery and copper, and Roman copper recovery.

*Mining History Network** <http://www.exeter.ac.uk/~RBurt/MinHistNet/> (accessed May 18, 2005). The Mining History Network is hosted at Exeter University in the U.K. The website includes an A–Z listing of mining historians, bibliographies of British and North American mining history, and links to related websites. Details of a number of mining history organizations are also available including Exeter Mining History Research Group, Australian Mining History Association, Northern Mine Research Society, Japan Mine Research Society, Early Mining Research Group, and the Peak District Mines Historical Society. *From: *The Internet Guide to Engineering, Mathematics and Computing* (URL: http://www.eevl.ac.uk/show_full.htm?rec=libmom.857962570).

A Pictorial Walk Through the 20th Century <http://www.msha.gov/century/century.htm> (accessed May 18, 2005).

Information sources and guides to the literature of mining engineering and technology

Anthony, L.J. (ed.) (1988) *Information Sources in Energy Technology*, London; Boston: Butterworths.

Chen, C. (1987) *Scientific and Technical Information Sources* (2nd edn), Cambridge, MA: MIT Press.

Hurt, C.D. (1998) *Information Sources in Science and Technology* (3rd edn), Englewood, CO: Libraries Unlimited.

Lord, C.R. (2000) *Guide to Information Sources in Engineering*, Englewood, CO: Libraries Unlimited.

Macleod, R.A. and Corlett, J. (2005) *Information Sources in Engineering* (4th edn), New York: Bowker-Saur.

Malinowsky, H.R. (1980) *Science and Engineering Literature: a Guide to Reference Sources* (3rd edn), Littleton, CO: Libraries Unlimited.

Malinowsky, H.R. (1994) *Reference Sources in Science, Engineering, Medicine, and Agriculture*, Phoenix: Oryx.

Mildren, K.W. and Hicks, P.J. (eds) (1996) *Information Sources in Engineering* (3rd edn), London: Bowker-Saur.

Mount, E. (1976) *Guide to Basic Information Sources in Engineering*, New York: Halsted Press.

Mount, J. *Books about Mining* <http://home.att.net/~newbooks/miningbooks.html> (accessed May 18, 2005). An alphabetical list of books about mining, prospecting, mines, miners, mining history, economic minerals, mining engineering, and some geological, geotechnical, and petroleum engineering that are currently in print: compiled by a retired paleontologist/geologist and geosciences librarian, Jack Mount. Each title links to additional information and availability for purchase at Amazon.com. There is a separate section devoted to books on mining history.

New York Public Library, Engineering Societies Library (1975) *Bibliographic Guide to Technology*, Boston, MA: G.K. Hall.

Wasserman, S.R., Smith, M.A., and Mottu, S. (eds) (1989) *Encyclopedia of Physical Sciences and Engineering Information Sources: A Bibliographic Guide to Approximately 16,000 Citations for Publications, Organizations, and Other Sources of Information on 425 Subjects Relating to the Physical Sciences and Engineering*, Detroit, MI: Gale.

Wood, D.N., Hardy, J.E., and Harvey, A.P. (eds) (1989) *Information Sources in the Earth Sciences* (2nd edn), London; New York: Bowker-Saur.

Searching the library catalog

When looking for books on mining engineering, search by LC subject or keyword in your local Library Catalog. If you know of a particular title or author you would like to find, search by that specific field in the catalog. Examples of relevant Subject Headings for mining engineering include the following.

Mining engineering; blasting; tunneling

To browse for books on mining engineering in a library that shelves according to the Library of Congress Classification scheme, the following ranges of call numbers will lead the reader to the correct area of the Library.

Metallurgy; mining engineering:	TN 1–997
Mineral deposits; prospecting:	TN 263–271

Practical mining operations; safety:	TN 275–325
Mining machinery:	TN 331–347
Mining of particular metals:	TN 400–580
Mineral and metal industries:	HD 9506–9624

Abstracts and indexes

Effective use of bibliographic databases requires an understanding of search strategy development and techniques. Online help and user guides normally accompany databases.

Key databases for Mining Engineering include:

AESIS, Australia's Geoscience, Minerals, and Petroleum Database, developed by The Australian Mineral Foundation, is available on Thomson Dialog (Bluesheet File 105). AESIS indexes all available Australian information related to mines and mining. Published and unpublished resources cover blasting, drilling, extractive metallurgy, mineral exploration, mineral industry, mineral processing, mineral resources, mineralogy, geology, and petroleum subject areas. The database specializes in material written by Australians, or about Australia or Papua New Guinea, and contains some 10,000 references in the English language to worldwide technical literature.

Applied Science and Technology Abstracts (1983–) (Thomson Dialog (Bluesheet File 99). Subject coverage includes geology, metallurgy, mineralogy, and mining engineering.

Arctic and Antarctic Region (mid-1800s–) <http://www.nisc.com/> (accessed May 18, 2005). The world's largest collection of international polar databases. Includes references to scientific periodicals, monographs, proceedings of conferences and symposia, government reports, theses, dissertations, pipeline documents, consultants' reports, and monographs. Subject areas covered are diverse, but those generally relevant to mining engineering are cold regions engineering, exploration, geosciences (climate, geography, geology), health and natural sciences.

Arctic Science and Technology Information System (ASTIS) <http://www.aina.ucalgary.ca/astis/> (all accessed May 18, 2005). ASTIS contains over 54,000 records describing publications and research projects about northern Canada. ASTIS is maintained by the Arctic Institute of North America <http://www.ucalgary.ca/aina/> at the University of Calgary, and is part of the Canadian Polar Information Network <http://www.polarcom.gc.ca/english/>.

Cambridge Scientific Abstracts Databases <http://www.csa.com/> (accessed May 18, 2005). Mining engineering subject coverage emphasis is on the environmental management of mining activity with databases such as *Pollution Abstracts* (1981–); *Water Resources Abstracts* (1967–) and *TOXLINE* (1999–). Articles on toxicology including chemicals, pharmaceuticals, pesticides, environmental pollutants, mutagens, and teratogens.

Canada Energy Minerals and Metals Information Centre (EMMIC) <http://www.nrcan.gc.ca/es/msd/emmic/web/index.html> (all accessed May 18, 2005). EMMIC is a section within the Energy Sector's Management Services Division of Natural Resources Canada. EMMIC provides library and information services to the Energy Sector, Minerals and Metals Sector, Corporate Services Sector and Executive offices as well as information to other libraries via interlibrary loan and to outside institutions and the public.

About Us (http://www.nrcan.gc.ca/es/msd/emmic/web/about_e.html) (accessed March 11, 2005). For subject guides and library/information resources at EMMIC on mining and mineral processing <http://www.nrcan.gc.ca/es/msd/emmic/web/Subject_Guides/subjectminingminerals_e.html>.

As an additional bonus, EMMIC users may access the ETDE's World Energy Base database free of charge. ETDE's Energy Database contains the world's largest collection of energy literature of more than four million records with coverage from 1995 to the present. The database contains bibliographic references and abstracts to journal articles, reports, conference papers, books and theses.

Canadian Research Index (Microlog) (1982–) <http://www.micromedia.ca/Products_Services/Microlog.htm>. Indexes all depository Canadian government publications at the municipal, provincial, and federal levels. Includes scientific and technical reports, statistical reports, policy papers, and annual reports. Microlog includes published documents of the Canada Centre for Mineral and Energy Technology (CANMET) at <http://www.nrcan.gc.ca/es/etb/cetc/cetc01/htmldocs/home_e.html.>. Database subject coverage includes agriculture; energy; environment; fisheries; forestry; life sciences; natural resources, earth sciences, physical sciences, technology, and telecommunications.

Environment Abstracts on LexisNexis Environmental (1975–) <http://www.lexisnexis.com/academic/1univ/envir/default.asp) Content from mining-related journals includes titles such as: *African Mining Monitor, Coal Week, Coal Week International, Engineering and Mining Journal, Metals Week, Mining Annual Review, Mining Journal, Mining Magazine, Waste News, Waste Treatment Technology News*. The principal focus of this database is American law, Environmental codes and regulations, Case law, Regulatory agency decisions, and Case law and agency actions.

GEOBASE – Bibliographic database for the earth, geographical, and ecological sciences. An earth science database of bibliographic information and abstracts covering earth sciences, ecology, geology, geography, geomechanics, human geography, mines and mining oceanography, and related disciplines. Material includes refereed scientific papers; trade journals and magazine articles; product reviews, directories, books, conference proceedings, and reports. GEOBASE is currently available worldwide via Dialog <http://www.dialogweb.com>; EINS <http://www.eins.org>; OCLC <http://www.oclc.org>, and ScienceDirect <http://www.sciencedirect.com/> (all accessed May 18, 2005).

GeoBase (free, online geospatial database) <http://www.geobase.ca/geobase/en/> (accessed May 18, 2005). *GeoBase* is a Canadian federal, provincial, and territorial government initiative overseen by the Canadian Council on Geomatics (CCOG) to ensure the provision and access to quality geospatial data with free and unrestricted use. *GeoBase* comes with full metadata and provides references, and contexts to a wide variety of thematic data for government, business, and individual applications. The data address concerns such as sustainable resource development (forestry, mines, energy, water); public safety, protection, and sanitation (emergency response, disaster management, tracking SARS), and environmental protection (greenhouse gas effects, global warming, natural risks, flooding).

GeoRef (1785–) Compiled by The American Geological Institute <http://www.agiweb.org/georef/index.html> (accessed May 18, 2005). *GeoRef* offers international bibliographic coverage of all aspects of earth sciences, geology, and geophysics from 1785 to the present including archaeology, economics, education, engineering and environmental geology, expeditions, fossils, geologic hazards (avalanches, hurricanes, sunspots, volcanoes), instruments, land use, landfills, magnetic fields, and maps, medical and military geology, and soils. The database contains over 2.5 million references to books, theses, journals, reports, maps, conferences, and government documents in all areas of the earth sciences. *GeoRef* includes pure and applied research in applied geology, exploration geophysics, mineral exploration, petroleum and environmental organic chemistry, earth science, and economic and engineering geology.

Geoscan <http://www.nrcan.gc.ca/ess/esic/geoscan_e.html> (accessed May 18, 2005). A free database of 40,000 Geological Survey of Canada publications issued since 1986. Some are free; others may be ordered for a fee as photocopies.

IMMAGE (commercial database of IMM Information Services) The IMMAGE database is compiled by the Institution of Mining and Metallurgy in the UK and consists of 75,000 references (expanding by 4,000 a year), to published papers, books, reports, and so on, covering technical research and development and operating practices of the international minerals industry. Subject coverage includes tunneling and underground excavation, including rock mechanics, instrumentation and automation, economic geology, mining technology and extraction processing in non-ferrous metals and industrial mineral fields. Web access is available through the European Information Network Services <www.eins.org>. IMMAGE is also available on CD-ROM, or in print as IMM ABSTRACTS.

METADEX (Metals Abstracts) – CSA Internet Database Service. *METADEX* contains over 950,000 references from 2,000 journals, patents, dissertations, reports, books, and conference proceedings (1966–) covering all aspects of metallurgical science, including corrosion, metallography, mechanical and physical properties, testing and quality control, physical properties, extraction and refining, machining, welding, metal matrix composites, powder technology, surface finishing, heat treatment, testing and quality control, alloy production and development, thermal treatment, and application and end uses.

MINABS Online (The Mineralogical Society) <http://www.minabs.com/>. *Mineralogical Abstracts* was an essential research tool for seventy years with abstracts covering mineralogy, petrology, geochemistry, crystallography, gemmology, meteoritics, and allied subjects. Since January 2004 this printed journal has been replaced with an electronic journal known as *MINABS Online*. Like its print predecessor, MINABS Online is a specialized abstracting publication covering international research in mineralogy, crystallography, geochemistry, petrology, environmental mineralogy, and related topics in earth science.

ScienceDirect (1995+) <http://www.sciencedirect.com/>. This general science database contains articles from journals published by Elsevier. Examples of Elsevier mining journals indexed are: *International Journal of Coal Geology; International Journal of Rock Mechanics and Mining Sciences; International Journal of Rock Mechanics and Mining Science and Geomechanics Abstracts; Mining Science and Technology; Minerals Engineering.*

Journals

Journal literature (magazines, periodicals, serials, and trade literature publications) all help researchers to keep abreast of trends, current issues, and techniques in the industry. Some prominent mining engineering journal titles, ranked high by their impact factor over the past three years in ISI's *Journal Citation Reports*, are listed below. Scope notes for these subject categories have been quoted also, in full, from the JCR website <http://www.isinet.com/products/evaltools/jcr/>.

The "Description" fields of information were quoted from *Ulrich's Periodical Directory* (URL: http://www.ulrichsweb.com/ulrichsweb/). In *Ulrich's*, subject categories assigned to these periodical titles were a variant combination of "Engineering, geological," "Metallurgy and metallurgical engineering," or "Mining and mineral processing."

Engineering, geological

"Engineering, Geological includes multidisciplinary resources that encompass the knowledge and experience drawn from both the geosciences and various engineering disciplines (primarily civil engineering). Resources in this category cover geotechnical engineering, geotechnics, geotechnology, soil dynamics, earthquake engineering, geotextiles and geomembranes, engineering geology, and rock mechanics."

Metallurgy and metallurgical engineering

"Metallurgy & Metallurgical Engineering includes resources that cover the numerous chemical and physical processes used to isolate a metallic element from its naturally occurring state, refine it, and convert it into a useful alloy or product. Topics in this category include corrosion prevention and control, hydrometallurgy, pyrometallurgy, electrometallurgy, phase equilibria, iron-making, steel-making, oxidation, plating and finishing, powder metallurgy, and welding."

Mining and mineral processing

"Mining & Mineral Processing includes resources on locating and evaluating mineral deposits; designing and constructing mines; developing mining equipment; supervising mining operations and safety; and extracting, cleaning, sizing, and dressing mined material. Relevant topics in this category include exploration and mining geology, rock mechanics, geophysics, and mining science and technology."

Samples of top-ranked journal titles

Canadian Mining Journal. Southam Business Communication Inc. (0008–4492) <http://www.canadianminingjournal.com>. Leading mining and exploration journal in Canada. Covers mineral exploration trends, metal prices and new geological models, underground mine developments and operating performances, and new technology and services.

CIM BULLETIN. Canadian Institute of Mining, Metallurgy and Petroleum (0317–0926) <http://www.cim.org/mainEn.cfm>. Technical data, papers, and information on mining practice, research mineral engineering subjects to promote the technological interests of people involved in the development of the industry in Canada with special feature articles on new mining developments and technologies. Contents also include monthly features on career opportunities, industry equipment, and society and division newsletters.

Engineering Geology: an International Journal. Elsevier BV (0013–7952) Contains original studies, case histories, and comprehensive reviews in the field of engineering geology. Special features include advertising, bibliographies, charts, illustrations, and book reviews.

Engineering and Mining Journal. New York: McGraw-Hill (0095–8948) <http://www.mining-media.com/emj/index.html>. Covers exploration, deposit discoveries and development, milling, smelting, refining, and other extractive processing of metals and nonmetallics, including coal; provides current price quotes for over forty minerals on a single day each month in "Prices, American Metal Market"; supplemented by "Mining Equipment Digest," a catalog of mining machinery, equipment, and supplies for the mineral industry.

International Journal of Mineral Processing. Elsevier Science BV (0301–7516) <http://www.elsevier.com/locate/minpro>. Covers all aspects of the processing of solid-mineral materials such as metallic and non-metallic ores, coals and other solid sources of secondary materials, and so on. Special features in the journal include bibliographies, charts, illustrations, and book reviews.

International Journal of Rock Mechanics and Mining Sciences. Pergamon Elsevier Science (1365–1609) <http://www.elsevier.com/locate/ijrmms>. Original research, new developments, and case studies in rock mechanics and rock engineering, for mining and civil applications. Special features include advertising, abstracts, charts, illustrations, and book reviews.

International Journal of Surface Mining, Reclamation and Environment. A.A. Balkema (1389–5265) <http://www.balkema.nl>. Examines all aspects of surface mining technology and waste disposal systems relating to coals, oil sands, industrial minerals, and metalliferous deposits. Includes computer applications and automation processes.

Mining Journal. Mining Journal Ltd. (0026–5225) <http://www.mining-journal.com/>. Weekly international coverage of political, financial, and technical news affecting mining industry activity. Special features include advertising, illustrations, and book reviews.

SME Resource Guide. Society for Mining, Metallurgy and Exploration (1087–0113) <http://www.smenet.org/digital_library/library_mining.cfm>. Directed at engineering professionals in the mining and mineral processing industries. Special features include advertising, illustrations, trade literature, and book reviews.

Transactions of the Institution of Mining and Metallurgy Section A: Mining Technology. Maney Publishing (0371–7844).

Transactions of the Institution of Mining and Metallurgy Section B: Applied Earth Science. Maney Publishing (0371–7453).

World Mining Equipment. Bristol Lane Voss (0746–729X) <http://www.wme.com/>. Covers mines and mining equipment, and decision-makers with purchasing power in such companies. Special features include advertising and book reviews.

Journal titles related to mining environmental management

IEA coal research newsletter. International Energy Agency (not supplied).

International Journal for Numerical and Analytical Methods in Geomechanics (0363–9061).

International Journal of Earth Sciences. Springer-Verlag (1437–3254).

International Journal of Geomechanics. CRC Press (1532–3641).

Ironmaking and Steelmaking. Maney Publishing (0301–9233).

Journal of Geotechnical and Geoenvironmental Engineering. ASCE (1090–0241).

Mine Water and the Environment. Journal of International Mine Water Association; formerly *International Journal of Mine Water.* Springer Scientific Germany (1025–9112).

Minerals and Metallurgical Processing. Society for Mining, Metallurgy and Exploration (0747–9182).

Northern Miner online <http://www.northernminer.com/>. Covers the worldwide mining activities of mining companies based in North America or listed on North American stock exchanges.

Handbooks and manuals

Handbooks and manuals are a valuable source for quickly needed information like facts, figures, technical specifications, data, properties, design and testing criteria, explosive formulas and calculations, recommended procedures along the production cycle, and log interpretation criteria.

ASM Handbook. ASM International <http://products.asminternational.org/hbk/index.jsp>. Twenty-one volumes providing major reference information to the industry about mining extraction, metallurgical processing, and fabrication of metals; testing, inspection, and failure analysis; microstructural analysis and materials characterization; corrosion and wear phenomena in machinery and equipment, some of it used in the mining industry.

Bickel, J.O. (ed.) (1996) *Tunnel Engineering Handbook* (2nd edn), New York: Chapman & Hall. A comprehensive "state-of-the-art" overview written for practicing engineers on the classification types, planning, design, layout, construction methods, and rehabilitation of tunnels above and below ground.

Caterpillar Performance Handbook (1988–) Peoria, IL: Caterpillar Tractor Company. Handbook and manual of earth-moving machinery, industrial power trucks, and farm tractors.

Fuerstenau, M.C. and Han, K.N. (eds) (2003) *Principles of Mineral Processing*, Littleton, CO: Society for Mining, Metallurgy, and Exploration. Examines all aspects of minerals processing, from handling raw materials to separation strategies, to remediation of waste products. The book relates recent developments in engineering, chemistry, computer science, and environmental science to explain how these disciplines contribute to the production of minerals and metals efficiently and economically from ores.

Gertsch, R.E. (ed.) (1998) *Techniques in Underground Mining: Selections from Underground Mining Methods Handbook*, Littleton, CO: Society for Mining, Metallurgy, and Exploration. Sections in the book review support methods used in mine planning and mining including pillars, room-and-pillar methods in open-stope mining, sublevel caving, and block and panel caving.

Hartman, H.L. (1992) *SME Mining Engineering Handbook*, Littleton, CO: Society for Mining, Metallurgy, and Exploration. Also available on CD-ROM. Published by SME in 1998. This is the one, indispensable, essential reference work anyone in the mining industry must have, even if it is the only book they have in their collection. A distillation of every aspect of mining engineering. Provides professional practitioners with an authoritative reference and design source. The book covers all branches of mining – metal, coal, and non-metal, and all locales of mining – surface, underground, and hybrid with a concentration on mining in the United States. Numerous references are also devoted to international practices.

International Society of Explosives Engineers (2000) *ISEE Blasters Handbook* (17th edn) (formerly the ETI and, prior to that, DuPont Blasters Handbook), Cleveland: ISEE. A manual describing explosives and standard, practical methods of use.

Karassik, I.J. (ed.) (2001) *Pump Handbook* (3rd edn), New York: McGraw-Hill. Provides practical data on the design, application, specification, purchase, operation, trouble-shooting, and maintenance of pump technology used in every type of engineering application throughout industry, including mining. This is an essential reference and guide to the design, application, specification, purchase, operation, and maintenance of pumps from advanced seals to basic design.

Marcus, J.J. (1997) *Mining Environmental Handbook: Effects of Mining on the Environment and American Environmental Control*, Edge River, NJ: World Scientific. This handbook examines the effects of mining on the environment and environmental laws that deal with mining in the United States both currently and in future perspective.

Minerals Handbook (annual) London: Van Nostrand Reinhold. Incorporates information for each of over fifty metals and minerals; an essential reference for minerals industry analysts and economists and for any mining or exploration company. Discusses world reserves and reserve base, production capacity, adequacy of reserves, consumption, end-use patterns, value of contained metal, substitutes, technical possibilities, and marketing arrangements.

Sirois, L.L. and MacDonald, R.J.C. (eds) (1983–) *Minerals, Metals and Mining Technologies*, Ottawa: CANMET. Describes metallurgical research and research in the mineral resources of Canada.

Stack, B. (1982) *Handbook of Mining and Tunnelling Machinery*, New York: John Wiley. The author describes the history, evolutionary development, and interrelationship of tunneling machines employed in Australia, the U.K., U.S.A., and Germany from earliest times to the present. Coverage is given for the entire range of associated drilling machines, including "horizontal and vertical, raise drills and shaft-boring machines, shields, mechanized shields, slurry machines, rock tunnelling machines . . . reaming machines, incline-boring machines and the various boom-type machines such as road-headers and mounted impact breakers, etc." (source: Preface, p. xxvii).

Taggart, A.F. (1954) *Handbook of Mineral Dressing: Ores and Industrial Minerals*, New York: John Wiley. A reference work and guide to industrial minerals and cement. Covers crushing, grinding, screen sizing, classification, washing and scrubbing, gravity concentration, flotation and electrical methods of concentration, dewatering, filtration, drying, storage and mill transport, sampling and testing, design and construction of ore treatment plants; with seventy pages of mathematical and other tables.

Dictionaries

Dictionaries and encyclopedias provide definitions and clarify concepts. The terminology peculiar to mining is set out in a number of works. Samples of standard references to this unique vocabulary may be found in the following.

Blackburn, W.H. (ed.) (1997) *Encyclopedia of Mineral Names*, Ottawa, Ontario, Canada: Mineralogical Association of Canada/Association Mineralogique du Canada.

Clark, A. (1993) *Hey's Mineral Index: Mineral Species, Varieties and Synonyms* (3rd edn), London; New York: Chapman & Hall. An alphabetical listing of all known minerals, their chemical composition, and type of locality. Also includes an additional finding index arranged by mineral chemical composition.

Dictionary of Mining, Mineral, and Related Terms (1997) (2nd edn) Compiled by the American Geological Institute, Alexandria, VA: American Geological Institute. This dictionary focuses solely on mining and geological terminology related to mining, including industrial minerals, pollution, and environmental mining terminology. The CD-ROM version of this dictionary was compiled and edited by the staff of the U.S. Bureau of Mines. Washington, DC: U.S. Bureau of Mines.

Dorian, A.F. (1993) *Elsevier's Dictionary of Mining and Mineralogy in English, French, German and Italian.* French & European Publications, Inc., Amsterdam; New York: Elsevier.

Glossary of Mining Terms <http://www.dep.state.pa.us/dep/deputate/minres/dms/website/training/glossary.html> (accessed May 18, 2005). Compiled by the Pennsylvania Department of Environmental Protection, Bureau of Deep Mine Safety, the glossary contains thousands of terms related mainly to coal-mining.

Jackson, J.A. (1997) *Glossary of Geology* (4th edn), Alexandria, VA: American Geologic Institute.

Mining Glossary (Internet resource) <http://www.mininglife.com/Glossary/a.asp> (accessed May 18, 2005). Rock blasting terms and symbols: a dictionary of symbols and terminology in rock blasting and related areas such as drilling, mining, and rock mechanics. Editor-in-chief Agne Rustan, Rotterdam; Brookfield, VT: Balkema, 1998.

Encyclopedias

There is no entire encyclopedia devoted specifically to mining engineering. As a process technology, mining engineering applies the lessons learned from many disciplines such as the geosciences, geochemistry, materials engineering, environmental sciences, and applied automotive mechanics.

Carr, D. and Herz, N. (1989) *Concise Encyclopedia of Mineral Resources*, Oxford; New York: Pergamon Press; Cambridge, MA: MIT Press. "Based on material from the *Encyclopedia of Materials Science and Engineering*, first published in 1986, with revisions and updated material"(title-page verso). This work is both a dictionary of terminology and an encyclopedic discussion of concepts in mining and mineral resources, non-metallic minerals, and advances in materials science and engineering.

Dasch, E.J. (ed.) (1996) *Macmillan Encyclopedia of Earth Sciences* (2 vols), New York: Macmillan Reference.

Frye, K. (ed.) (1981) *The Encyclopedia of Mineralogy*, Stroudsburg, PA: Hutchinson Ross. Actually a basic mineralogical dictionary and Volume 4B from the series title: Encyclopedia of Earth Sciences.

Marshall, C.P. and Fairbridge, R.W. (eds) (1999) *Encyclopedia of Geochemistry*, Dordrecht; Boston, MA: Kluwer Academic.

Middleton, G.V. (ed.) (2003) *Encyclopedia of Sediments and Sedimentary Rocks*, Dordrecht; Boston, MA: Kluwer Academic.

Roberts, W.L., Campbell, T.J., and Rapp, G.R. (eds) (1990) *Encyclopedia of Minerals* (2nd edn), New York: Nostrand Reinhold.

Directories

Directories provide access to information about the petroleum industry and the location of people, companies, products, services, manufacturers, and suppliers of components to the industry. In recent years, online access has proven to be of greater use in providing up-to-date information on a timely basis than have traditional hard copy publications.

American Mines Handbook. Toronto, Ontario: Business Information Group (0840–8610) Published 1989–2003 (merged/incorporated into: *Canadian and American Mines Handbook*). Lists information on public mining companies registered in the United States. Includes mines and projects summary, five-year stock range table, and statistical tables.

CAMESE Compendium of Canadian Mining Suppliers. Don Mills, Ontario: Canadian Mining Journal in cooperation with the Canadian Association of Mining Equipment and Services for Export (1997–) (1485–8401) <http://www.camese.org> (accessed May 18, 2005). Canadian Association of Mining Equipment and Services for Export – "CAMESE is a national, sectoral export trade association. It exists to support Canadian mining suppliers in global marketing, and to assist foreign buyers in finding Canadian sources for mining equipment and services."

Canadian and American Mines Handbook 2004–2005 (2004–) Toronto, Ontario: Business Information Group. Formed by the union of *Canadian Mines Handbook* and *American Mines Handbook*, this new annual handbook and directory lists over 1,200 Canadian and 300 U.S. and foreign mining companies with updates on mines, mining area maps, and historic stock prices.

C.I.M. Directory (1990–) Montréal: Canadian Institute of Mining, Metallurgy, and Petroleum. The directory lists the names, addresses, and companies of all members of the Canadian Institute of Mining, Metallurgy, and Petroleum with access to over 1,500 active metallurgists and materials engineers throughout Canada and the world.

Financial Times Business Global Mining Directory (2000–) London: FT Business. Provides details of the world's non-ferrous metals industry with in-depth operational analyses, ownership and subsidiary information, plus three-year financial results to over 300 exploration and mining companies. Includes lists of over 700 service and supply companies, as well as over 4,000 key industry contact personnel.

Industry Canada. *Canadian Mining, Metal and Mineral Industries* <http://www.nrcan.gc.ca/mms/lien/gov_e.htm> (accessed May 18, 2005). Directory of Canadian government links to federal, provincial, and territorial government agencies responsible for the mining industry within their particular jurisdiction.

International Mining Directory (2004) London; Sterling, VA: Kogan Page. A compendium of detailed information about 2,500 mining and mine equipment companies, consultants and service companies worldwide, as well as the largest, most comprehensive equipment buyers' guide in the mining industry. At the end of the directory is a brief, five-page glossary containing definitions of mining terminologies and technology.

Mining-technology.com – the website for the mining industry <http://www.mining-technology.com/)> (accessed May 18, 2005). The SPG Media Limited (UK) website provides international news coverage by geographical region of current mining industry projects extracting energy minerals, ferrous, base and precious metals, industrial minerals, and gemstones. Includes a comprehensive "Products and services" catalog and directory to mining industry equipment worldwide; an A–Z company index listing mining contractors and suppliers alphabetically, as well as industry news releases about the latest mining technology products, and the job market.

Walker, S. (2000) *Major Coalfields of the World*, London: IEA Coal Research. Provides a clear and comprehensive analysis of the world's coalfields, considering the geology, structure, stratigraphy, quality, industry structure and performance, transport infrastructure, production costs, and market potential of each major coalfield.

Government ministries

Begin searches for government information resources on mining and minerals by visiting the Home page website of the jurisdiction in question. There, search for the relevant departmental agencies with stated responsibility for energy and natural resources. Government sites often include valuable statistical and commodity pricing data.

Australia

Geoscience Australia <http://www.agso.gov.au/> (accessed May 18, 2005). Originally the Bureau of Mineral Resources, Geology and Geophysics, this site has been created by a Commonwealth government department with a broad range of responsibilities in the area of geosciences. Another excellent, all-inclusive, up-to-date website linking visitors to governmental institutions and private industries participating in Australian mining may be found at the website of the next entry.

Australia Mining and Exploration (AME) <http://www.reflections.com.au/MiningandExploration/index.html> (accessed May 18, 2005). "Portal about Australasia's mining & exploration industry . . . covers all aspects of the mining industry from exploration through to mining, processing and transport including information on mining and exploration companies announcements and reports, commodities, calendar of events and conferences, and details of government and mining related organizations, publications, drilling companies, geological and mining consultants, mining plant and equipment, assayers, financiers, share brokers, accountants, etc. Also includes sections on the Stock market with daily market reports, stock prices and charts on indices, commodities and companies" (source: AME website).

Canada

In Canada, information at the federal government level is compiled under one agency, Natural Resources Canada <http://www.nrcan.gc.ca> (all accessed May 18, 2005) and is accessible through their basic web root directory. The principal goal of providing electronic resources through NRCAN is to increase Canadian competitiveness for attracting industrial investments in mineral and energy exploration and development sustained through timely knowledge transfers. Government information about all aspects of Canadian mining activity (acts and regulations, statistics, geodetic surveys, scientific and technological innovation, cartographic, geological, environmental, business investment opportunities, investment tax credit information for exploration, mining legislation and regulations, standards, and online maps, educational resources) may be obtained.

Mapping Federal-Provincial-Territorial Mining Knowledge <http://mmsd1.mms.nrcan.gc.ca/maps/intro_e.asp>. Provides information on major exploration projects and producing mines in Canada. The site maps exploration projects, producing mines and resource-dependent communities in Canada. All plug-ins required for viewing infrastructures of maps for producing mines, exploration projects and deposits on this site are included for downloading. Maps products are searchable by commodity. Also available are country profile reports of gross domestic product (GDP), inflation rates, economic profiles, and resource commodities produced in a country of interest.

The Map Image Rendering Database for Geoscience (MIRAGE) project <http://rgsc.nrcan.gc.ca/mirage/index_e.php>. Begun in July 1998, the aim of the project is to make ESS geoscience maps accessible through the Internet. Thousands of paper maps were scanned and image-linked to their metadata records. Users may search this metadata to discover geoscience maps that they can view and download. The scanned images are stored as both MrSID compressed images and as Adobe Acrobat files.

United States

United States Geological Survey (USGS) <http://www.usgs.gov/> (all accessed May 18, 2005). Self-described as the "Federal source for science about the Earth, its natural and living resources, natural hazards, and the environment." From the Home page a link to USGS Minerals Information at <http://minerals.usgs.gov/minerals/> provides subscription information and full-text links to the following major publications of the USGS:

Mineral Commodity Summaries <http://minerals.usgs.gov/minerals/pubs/mcs/>. Published on an annual basis, this report is the earliest government publication to furnish estimates covering non-fuel mineral industry data. Data sheets contain information on the domestic industry structure, government programs, tariffs, and five-year salient statistics for over ninety individual metallic and non-metallic minerals and materials.

Mineral Industry Surveys <http://minerals.usgs.gov/minerals/pubs/commodity/ mis.html>. Mineral Industry Surveys (MIS) are periodic online statistical and economic publications designed to provide timely statistical data on production, distribution, stocks, and consumption of significant mineral commodities. These publications are issued monthly, quarterly, or annually.

Minerals Yearbook (1994–) <http://minerals.usgs.gov/minerals/pubs/myb.html>. The *Minerals Yearbook* is an annual publication that reviews the mineral and material industries of the United States and foreign countries; contains statistical data on minerals, including economic and technical trends; chapters list approximately ninety commodities and over 175 countries.

Bureau of Mines Minerals Yearbook (1932–1993) <http://minerals.usgs.gov/-minerals/pubs/usbmmyb.html>. Links to the image scan, full-text contents of the *Minerals Yearbooks* from 1932–1993 courtesy of the University of Wisconsin Ecology and Natural Resources Collection <http://webcat.library.wisc.edu:3200/EcoNatRes/>.

Mine and Mineral Processing Plant Locations <http://minerals.usgs.gov/minerals/pubs/mapdata/>. Provides the locations of 1,879 coal-mines and facilities, eight uranium mines, and 1,965 mines and processing plants for seventy-four types of non-fuel minerals and materials are shown on large lithologic maps. The localities account for most of the fuel and non-fuel minerals and materials produced in the United States in 1997 other than crushed stone, sand and gravel, and common clay.

Metal Industry Indicators <http://minerals.usgs.gov/minerals/pubs/mii/>. *Metal Industry Indicators* (*MII*) is a monthly newsletter that analyzes and forecasts the economic health of five metal industries: primary metals, steel, copper, primary aluminum, and aluminum mill products.

Nonmetallic Mineral Products Industry Indexes <http://minerals.usgs.gov/minerals/pubs/imii/>. The USGS has prepared leading and coincident indexes for the *Nonmetallic Mineral Products Industry* (NAICS 327). The former name for this industry was the Stone, Clay, Glass, and Concrete Products Industry (SIC 32) under the Standard Industrial Classification (SIC) system. The SIC has been replaced by the North American Industry Classification System (NAICS). These indexes are similar to the ones in *Metal Industry Indicators*. The latest report for these indexes (November 2004) is available in pdf format. Historical data for these new indexes are available back to 1948. See also the section on "Statistics" for a discussion on the "Statistical Compendium" (US Bureau of Mines) and the USGS "Commodity Statistics and Information."

Mineral Resource Data System (MRDS) <http://tin.er.usgs.gov/mrds/>. Formally cited as: Mason, G.T. Jr, and R.E. Arndt (1996) *Mineral Resources Data System* (*MRDS*): U.S. Geological Survey Digital Data Series DDS-20, Reston, VA: U.S. Geological Survey.

As of 2004, MRDS, contained 111,955 variable-length records of metallic and non-metallic mineral resources of the world. A record contains descriptive information about mineral deposits and mineral commodities. The types of information in the database include deposit name, location, commodity, deposit description, geologic characteristics, production, reserves, potential resources, and references. A description of the primary "metallic and nonmetallic mineral resources throughout the world" [and the United States] including information by "deposit name, location, commodity, deposit description, geologic characteristics, production, reserves, potential resources, and references" (source: MRDS website).

MRDS also includes a featured link to "Mineral Resources On-Line Spatial Data" <http://mrdata.usgs.gov/>. The site functions as an enhanced data download facility to offer users the choice to select data by geographic area rather than download an entire dataset. At present this enhanced data selection is available for the geochemical data, MRDS, MAS/MILS, active mines, and radiometric ages. All of the data are available free of charge.

National Environmental Publications Internet Site (NEPIS) <http://www.epa.gov/nepis/> (accessed May 18, 2005). Keyword searchable index and full-text site for over 10,000 U.S. EPA books, reports, and documents.

National Library for the Environment: Congressional Research Service Reports <http://www.ncseonline.org/NLE/> (accessed May 18, 2005). Index and full-text site for over 1,300 reports on environmental or resource issues prepared for the U.S. Congress. A subject search on the term "Mining" leads searchers to a full-text collection of informational research reports about issues relating to legislative acts and natural conservation regulations for members of the House and Senate.

Within both the National Institute for Occupational Safety and Health (NIOSH) at <http://www.cdc.gov/niosh/homepage.html>, and the Centers for Disease Control and Prevention (CDC) at <http://www.cdc.gov/> (both accessed May 18, 2005) are hyperlinks to a number of free databases and search engines leading to full bibliographic citations for thousands of publications about mine safety, mine construction and stability, hazardous materials handling, equipment and instrumentation, mining by-products, and training programs for safety and health.

Common Information Service System (CISS) <http://www.cdc.gov/> (accessed May 18, 2005). CISS is an information system provided as a public service by NIOSH Mining Safety and Health Research (formerly the U.S. Bureau of Mines). Thousands of publications are stored in this searchable database in bibliographic form and include abstracts.

NIOSH (National Institute for Occupational Safety and Health – Mining Site Index) <http://www.cdc.gov/niosh/mining/default.htm> (all accessed May 18, 2005). A bibliographic database of occupational safety and health publications, documents, grant reports, and other communication products related to mining and supported in whole or in part by the National Institute for Occupational Safety and Health (NIOSH). Also includes statistics on mining accidents, fact sheets on minerals and coal, and materials on injury prevention, ergonomics, hearing loss, fire prevention, ventilation, explosives, dust, aging miners, and many other topics.

NIOSHTIC-2 <http://www.cdc.gov/niosh/nioshtic-2/>. NIOSHTIC-2 is a bibliographic database of occupational safety and health publications, documents, grant reports, and other communication products supported by the National Institute for Occupational Safety and Health.

Electronic Library of Construction Occupational Safety and Health (eLCOSH) <http://www.cdc.gov/elcosh/index.html> (accessed May 18, 2005). *eLCOSH* offers links to articles and information related to hazards, job sites, trades, and training.

See also the website for Mine Safety and Health Administration (U.S. Department of Labor) <http://www.msha.gov/> and MSHA's Data Retrieval System (DRS) <http://www.msha.gov/drs/drshome.htm> (accessed May 18, 2005). MSHA permits those interested to visit the site to retrieve mine overviews, accident histories, violation histories, inspection histories, inspector and operator dust samplings, and employment/production data.

Associations, organizations, and societies

Generally the websites listed offer a single convenient location wherein may be found descriptions for the mission statements, scope and objectives of the associations, their organizational divisions including statements of benefits and reasons for joining, membership forms and membership directories, by-laws, newsletters, lists of monographic and serial publications produced by the association in support of scientific research, technology or legal aspects of the mining industry and its impact on economies and the environment. As a bonus, all websites will generally also include valuable resource links to member companies, related organizations and associations, government agencies, schools and universities, commodities information, production statistics, and links useful to the mining industry or in support of that industry. Society websites may provide readers with updated discussions of current issues and events which affect the mining industry, or issues where an association is engaging representation and support on behalf of its constituents. The following lists of major associations, institutes, and societies are selective samples at best.

Australia

The Australasian Institute of Mining and Metallurgy <http://www. ausimm. com.au/> (accessed May 18, 2005). "The Australasian Institute of Mining and Metallurgy represents the interests of member professionals in all areas of mining, exploration and minerals processing industries located in a network of Branches throughout the Asia Pacific region. Within its member base of private enterprise, government, research, education and the support sector, including consultancies, manufacturers and suppliers, the AusIMM promotes continuing programs of research and educational development" (source: About The AusIMM <http://www.ausimm.com/about/about.asp> accessed March 11, 2005).

Australian Mineral Industries Research Association (AMIRA) <http://www.amira. com.au/> (accessed May 18, 2005). AMIRA International is an industry association and not-for-profit, private sector company which manages collaborative research for more than eighty member companies in the minerals industry in Australasia, Asia, Europe, Africa, and North and South America. The association provides a forum for the minerals industry to meet, network, and cooperate in areas of common interest. AMIRA develops and manages jointly funded research projects on a fee-for-service basis on behalf of its members.

CSIRO <http://www.csiro.au/> (accessed May 18, 2005). CSIRO is Australia's Commonwealth Scientific and Industrial Research Organization. The CSIRO Division of Exploration and Mining supplies R&D to the Australian exploration and mining industry; undertakes major research initiatives in designing mines and mining equipment; technologically assisted prospecting techniques, and resource development in socially, economically, and environmentally sustainable and acceptable ways.

Canada

The following list is selective. A more comprehensive list of the larger Canadian mining associations, institutes, mining companies, and metal producers may be found on the Web at <http://www.nrcan.gc.ca/mms/lien/mac_e.htm> (all accessed May 18, 2005).

Canadian Institute of Mining, Metallurgy and Petroleum (CIM/ICM) <(http://www.cim.org/mainEn.cfm>. The Canadian Institute of Mining, Metallurgy and Petroleum is the result of individuals in the mining industry seeking a vehicle for lobbying for safety laws and workers' protection, as well as a method of ensuring the communication of ideas. CIM has three main objectives: the facilitation of exchange of knowledge and technology, fraternity, and the recognition of excellence through conferences, publications, and awards. The 2001 to 2004 Strategic Plan stated as one of its goals the provision of technical leadership in the development and use of industry standards. The CIM Standing Committee on Reserve Definitions, the Guidelines for the Estimation of Mineral Resources and Mineral Reserves Committee, the Mineral Property Valuation Committee, and the CIM/CSA Working Committee are continuing to participate in the development of global standards for resource and reserve definition.

Coal Association of Canada <http://www.coal.ca/>. "Headquartered in Calgary, Alberta, the Coal Association of Canada represents companies engaged in the exploration, development, use and transportation of coal. Its members include major coal producers and coal-using utilities, the railroads and ports that ship coal, and industry suppliers of goods and services. The Coal Association of Canada provides a forum to discuss and coordinate the views of its members on matters of common interest. It provides a respected voice which enhances the viability of the industry by advocating the clean use of coal through technology development and communication with members, government and public-sector stakeholders" (source: About us <http://www.coal.ca/about.htm> accessed March 8, 2005).

Mineralogical Association of Canada (MAC) <http://www.mineralogicalassociation.ca/> (accessed May 18, 2005). "The Mineralogical Association of Canada (MAC) was formed in 1955 as a non-profit scientific organization to promote and advance the knowledge of mineralogy and the allied disciplines of crystallography, petrology, geochemistry and mineral deposits" (source: About us <http://www.mineralogicalassociation.ca/index.php?p=2>).

The Mining Association of Canada <http://www.mining.ca/> (accessed May 18, 2005). The Mining Association of Canada (MAC) is the national organization of the Canadian mining industry. It comprises companies engaged in mineral exploration, mining, smelting, refining, and semi-fabrication. Member companies account for the majority of Canada's output of metals and major industrial materials. The primary role of MAC is the presentation of industry information and views to the federal government and parliamentary committees. MAC provides information about the mining industry to the media, and to schools, libraries, and other publics. Most companies with producing mines and plants are members.

International

International Council on Mining and Metals (ICCM) (based in London, UK) <http://www.icmm.com/> (accessed May 18, 2005). ICMM is made up of representatives from private industry, national, and international mining associations. "ICMM members offer strategic industry leadership towards achieving continuous improvements in sustainable development performance in the mining, minerals and metals industry. ICMM provides a common platform for the industry to share challenges and responsibilities as well as to engage with key constituencies on issues of common concern at the international level, based on science and principles of sustainable development." (source: "ICMM's Mission" <http://www.icmm.com/about.php>).

Institute of Mining and Metallurgy (IMM) <http://www.iom3.org/index.htm> (accessed March 18, 2005). "The Institute of Materials, Minerals and Mining (IOM3) was officially recognized by the UK's Privy Council on 26 June 2002. Created from the merger of The Institute of Materials (IOM) and The Institution of Mining and Metallurgy (IMM), the Institute intends to develop to be the leading international professional body for the advancement of materials, minerals and mining to governments, industry, academia, the public and the professions" (source: IMM website).

United Kingdom

Institute of Materials, Minerals and Mining (London) <http://www.iom3.org/index.htm> (accessed March 18, 2005). The Institute is an international, professional body for people working in the materials, minerals, and mining communities. The Institute's divisional structure is based on that of the former Institute of Materials, enlarged to include four new divisions relating to activities in the minerals and mining sectors – Mining Technology, Applied Earth Sciences, Mineral Processing, and Petroleum and Drilling Engineering. The Mining Technology Division of the Institute exists to support all professional, technical, and educational aspects of the mining industry worldwide by providing free access to the Institute's Information Service, which includes information on materials, minerals, and mining through institute publications, conferences, and library services, a telephone help line and on-site specialist support.

Mining History Network <http://www.ex.ac.uk/~RBurt/MinHistNet/> (accessed March 18, 2005). The Mining History Network is both an information resource for mining historians and a focus to improve opportunities for continuing communication between researchers. The MHN supervisor is Dr Roger Burt (R.Burt@ex.ac.uk), Department of Economic and Social History, University of Exeter.

United States

In addition to the selective listing below, more exhaustive lists of U.S. regional and state mining associations may be found at <http://www.miningusa. com/associat/associ.asp>.

American Institute of Mining, Metallurgical, and Petroleum Engineers (AIME) <http://www.aimeny.org/> (accessed May 18, 2005). Founded in 1871, the AIME, with headquarters in Littleton, Colorado, is, according to its mission statement, "a Nonprofit Corporation organized and operated to advance and disseminate, through the programs of Member Societies, knowledge of engineering and the arts and sciences involved in the production and use of minerals, metals, energy sources and materials." AIME activities involve sponsorship of regularly scheduled member meetings/events, conferring of awards/scholarships, providing library research services, maintaining mining-related industry links, and supporting charitable giving. AIME is composed of five separately incorporated units, AIME Institute Headquarters, and four autonomous Member Societies – Society for Mining, Metallurgy, and Exploration (SME), The Minerals, Metals and Materials Society (TMS), Association For Iron and Steel Technology (AIST), and Society of Petroleum Engineers (SPE). There is also a Women's Auxiliary to AIME (WAAIME).

American Society of Appraisers (ASA) <http://www.appraisers.org/> (accessed May 18, 2005). ASA is a multi-disciplinary organization of accredited appraisal professionals recognized by the U.S. Congress as a source of appraisers and appraisal standards. The website provides a directory to valuation expertise for mine and quarry machinery and specialty equipment, oil and gas machinery and equipment, business valuation, and real and personal property.

American Society of Mining and Reclamation (ASMR) <http://ces.ca.uky.edu/asmr> (accessed May 18, 2005). The primary goal of ASMR is to expand opportunities for technology transfer to meet the needs of reclamationists worldwide. To this end, the Society seeks to keep its membership informed about the latest developments in reclamation technology both nationally and internationally. Another goal of the Society is to provide written and verbal technology transfer between members of similar organizations and interests in Canada, China, Australia, Great Britain, and the United States through the formation of the International Affiliation of Land Reclamationists (IALR). In future, the Society plans to use its website to keep members informed of its activities, and as a supplement to its regularly published newsletter.

Mining and Metallurgical Society of America (MMSA) <http://www.mmsa.net/> (accessed May 18, 2005). "The Mining and Metallurgical Society of America (MMSA) is a professional organization dedicated to increasing public awareness and understanding about mining and why mined materials are essential to modern society and human well being. MMSA delivers this message to two different public sectors: (1) public policy makers including elected officials and regulatory agencies; and (2) to all levels of the educational system." In support of the latter activity, MMSA provides teachers at all levels, from elementary through university classes with teaching materials and funded programs to improve public education about mining and minerals (source: About MMSA <http://www.mmsa.net/about.htm> accessed March 18, 2005).

Society for Mining Metallurgy, and Exploration (SME) <http://www.smenet.org/> (accessed May 18, 2005). The SME is an international society of professionals in the minerals industry. A member society of the American Institute of Mining, Metallurgical and Petroleum Engineers (AIME), the SME is organized into five distinct divisions – coal and energy, environmental, industrial minerals, mineral and metallurgical processing, and mining and exploration. The range of programs and services offered to SME members facilitates professional development and information exchange through Society publications, professional registration, peer review of technical papers, college accreditation programs, meetings and exhibits, public education, and short courses.

Society of Economic Geologists, Inc (SEG) <http://www.segweb.org/> (accessed March 18, 2005). The SEG is an international organization with interests in the field of economic geology. Membership includes representatives from industry, academia, and government institutions. The objectives of the Society are "to advance the science of geology through the scientific investigation of mineral deposits and mineral resources and the application thereof to exploration, mineral resource appraisal, mining and mineral extraction. . . . To disseminate information arising from investigations of mineral deposits and mineral resources through SEG publications, meetings, symposia, conferences, field trips, short courses, workshops and lecture series" (source: About SEG <http://www.segweb.org/History_Mission.htm>).

United States National Mining Association (NMA) <http://www.nma.org/> (accessed May 18, 2005). The NMA is a 501(c) 6, non-profit trade association and represents the political presence of the American mining industry in Washington, DC before Congress, the Administration, federal agencies, the judiciary, and the media. Membership includes coal, metal, and industrial mineral producers, mineral processors, equipment manufacturers, state associations, bulk transporters, engineering firms, consultants, financial institutions, and other companies that supply goods and services to the mining industry.

Metal prices, indexes

Average World Steel Transaction Prices ($US/ton) <http://www.meps.co.uk/world-price.htm> (accessed March 18, 2005). The latest monthly steel price updates are available to MEPS subscribers through an annual subscription.

Metal Bulletin Prices and Data (London) <http://www.metalbulletin.com/welcome_2003.asp>) (accessed March 18, 2005). With data available through annual subscription, the MetalBulletin.com website covers international prices for non-ferrous metals, iron and steel and scrap and secondary with the latest news on mergers, acquisitions, tenders, financials, and import/export news. The prices archive contains over 800 prices series going back to 1992, allowing a researcher to track price trends and create graphs online.

Metal Price Charts on InfoMine – Precious Metals, Base Metals <http://www.infomine.com/investment/metalschart.asp> (accessed March 18, 2005). The InfoMine website provides, aggregates, integrates, and syndicates access to worldwide mining and mineral exploration information for investors in the mining and exploration industry. Interactive commodity and currency charts provide prices of metals in the currencies of countries that mine the commodities over a selectable 15 day to 3 year range. The revenue model for the InfoMine website is based on the dual system of subscription, and advertising. While there is considerable free content on InfoMine, to get the most out of the site, users may need to subscribe. From About <http://www.infomine.com/about/>

Mineral Industry Indicators <http://minerals.usgs.gov/minerals/pubs/mii/> (accessed March 18, 2005). Monthly newsletter that analyzes and forecasts the economic health of five U.S. metal industries: primary metals, steel, copper, primary aluminum, and aluminum mill products.

Industry profiles and yearbooks

Canadian Minerals Yearbook: Review and Outlook <http://www.nrcan.gc.ca/mms/cmy/pref_e.htm> (1994–) (accessed May 18, 2005). Provides full-text access to Canadian mineral statistics from 1886 to 2003. Each year, the Minerals and Metals Sector (MMS) of Natural Resources Canada undertakes a comprehensive review of developments and outlook for the mineral industry.

Canadian Mines Handbook (1931–2000) Don Mills: Southam Mining Group (0068–9289). An annual directory with company profiles of 2,000 Canadian mining companies, industry, and government contacts and statistical information about metal/mineral production and prices. Provides details on mines, projects, smelters, and refineries. Includes mining area maps and stock range tables.

Metals and Minerals Annual Review (1990–1998) London: Mining Journal Ltd. Reviews the state of metals markets and trends in the industry. Provides detailed coverage about precious metals and minerals, steel industry metals, and industrial and energy minerals. Continues as *Mining Annual Review*.

Mining Annual Review (2001–) London: Mining Journal Ltd. <http://www.mining-journal.com/> (accessed May 18, 2005). An annual report on global mining industry activity through over 130 country reports, seventy commodity reports plus technical articles written by experts on developments in technology in mineral exploration, surface mining, underground mining, and in mineral and coal processing.

Statistics

Australian Bureau of Statistics (ABS) <http://www.abs.gov.au/> (accessed May 18, 2005). In the "Search" window, type terms such as "mining" or mining statistics" to obtain a list of search results linking to statistical resources on mineral and petroleum exploration in Australia. The ABS site provides key performance indicators, publications, and articles about the mining industry with data sourced from the ABS and statistical publications of mining-related international organizations.

Canadian Minerals Yearbook (1994–) <http://www.nrcan.gc.ca/mms/cmy/pref_e.htm> (accessed May 18, 2005). Provides full-text access to Canadian mineral statistics from 1886 to 2003. Each year, the Minerals and Metals Sector (MMS) of Natural Resources Canada undertakes a comprehensive review of developments in the mineral industry.

Coal and Coke Statistics/Statistics Canada <http://dsp-psd.pwgsc.gc.ca/Collection-R/Statcan/45-002-XIB/45-002-XIB-e.html> (accessed May 18, 2005). "Production, imports, exports, stocks and disposition of coal by province together with supply and disposition of coke in Canada are provided in this publication. Documents are available full text in Adobe Acrobat PDF format" (source: Statcan website).

Commodity Statistics and Information <http://minerals.usgs.gov/minerals/pubs/commodity/> (accessed May 18, 2005). Statistics and information on the worldwide supply of, demand for, and flow of minerals and materials essential to the U.S. economy, the national security, and protection of the environment. Includes an alphabetical index to publications, contacts and links to information about dozens of specific commodities.

Crowson, P. (2001) *Minerals Handbook: Statistics and Analyses of the World's Minerals Industry* (biennial) New York: Van Nostrand Reinhold. Intended as a comprehensive introductory guide for the non-specialist, this book provides basic data on fifty-two minerals and metals to permit informed debate on mining and mineral policies. Figures provided list domestic production, trade and consumption, geographical sources of net imports, shares of world production and consumption, historic growth of consumption, end-use patterns, and estimates on world reserves and reserve bases.

Energy, Mines and Resources Canada (1977–) *Statistical Review of Coal in Canada*, Ottawa: Energy, Mines and Resources Canada. Annually reviews production, consumption, and export statistics of Canadian metallurgical grade bituminous coal and sub-bituminous coal used for domestic electricity generation. Includes statistics on coal imported into Canada from U.S. and South American export markets.

Gold Institute <http://www.goldinstitute.org/> (accessed May 18, 2005). The Institute website offers links to anything and everything related to the gold industry in particular or mining in general. The information is divided into a number of categories covering American, Canadian, and Australian associations and organizations, international exchanges, gold price data and statistics on world gold mine production with supplementary links to mining technology-related websites.

Mine Safety and Health Administration Statistics <http://www.msha.gov/stats/statinfo.htm> (accessed May 9, 2006). U.S. Department of Labor site that includes statistics on coal, metal, and non-metal mining industries. Citations, violations, standards, fatalities, employment, injury, and illness statistics are listed. A mine data retrieval database allows searching by mine name, operator, location, or ID number.

Mineral Industry Surveys and Minerals Yearbook (US) – (Vol. I, Metals and Minerals) <http://minerals.usgs.gov/minerals/pubs/myb.html> (accessed May 18, 2005). U.S. Geological Survey Statistical data on worldwide production, distribution, stocks, and consumption of mineral commodities.

Minerals and Mining Statistics On-Line/Natural Resources Canada <http://mmsd1.mms.nrcan.gc.ca/mmsd/intro_e.asp> (accessed May 18, 2005). "The Minerals and Mining Statistics Division (MMSD) is part of the Minerals and Metals Sector (MMS) of Natural Resources Canada. MMSD is responsible for the collection, analysis and dissemination of comprehensive statistics on the Canadian minerals industry describing ore reserves, exploration, and mineral production and use, and for the conduct, in collaboration with Statistics Canada and the provincial and territorial governments, of various surveys of mine complex development and production" (source: "About Us").

Mineral Resources Program (U.S. Geological Survey) <http://minerals.usgs.gov/> (accessed May 18, 2005). The Mineral Resources Program funds science to provide and communicate current information on the occurrence, quality, quantity, and availability of mineral resources. Links are provided to announcements of USGS Mineral Research Grants, database and documentation for the National Geochemical Survey, historical statistics for mineral commodities across the United States, current commodity statistics, and quantitative global mineral resource assessments with links to mineral resource spatial data online. Some of the documents on this website are in pdf format and require the free Adobe Acrobat Reader.

Office of Surface Mining from the U.S. Department of the Interior <http://www.osmre.gov/statisti.htm> (accessed May 9, 2006). Statistics on active and abandoned surface mines. Use the site index to locate some unexpected information, including links to pages on wildlife in mines, court decisions, and lands unsuitable for mining. Site is updated daily.

Primary Iron and Steel/Statistics Canada <http://dsp-psd.pwgsc.gc.ca/Collection-R/Statcan/41-001-XIB/41-001-XIB-e.html> (accessed May 18, 2005). "This on-line publication provides current data on the Canadian iron and steel industry. It includes monthly production and shipment statistics in metric tons for pig iron, steel primary forms and steel castings at the Canada level. It also includes monthly shipment data in metric tons on rolled steel products broken down by product and consuming industry, again at the Canada level. The December issue includes a list of reporting firms" (source: Statcan website).

Statistical Compendium (U.S. Bureau of Mines) <http://minerals.usgs.gov/minerals/pubs/stat/> (accessed May 18, 2005). This Statistical Compendium provides a consistent set of official, long-term (twenty years or more) data series through 1990 for selected commodities to facilitate analysis of long-term trends in these mineral sectors. The website provides an index list of mineral commodity data published in the Statistical Compendium.

Educational subject guides to mining and mining engineering on selected websites

BUBL LINK – Catalogue of Internet Resources – Mining <http://www.bubl.ac.uk/link/m/mining.htm> (accessed May 18, 2005). BUBL is a free Internet-based information service for the U.K. higher education community, but is widely used worldwide by non-librarians as well. BUBL is located and run from the Centre for Digital Library Research of the University of Strathclyde, Glasgow, Scotland. The BUBL subject guide to "Mining" includes hundreds of links to Internet resources and services; descriptions are searchable and may be browsed by alphabetical order or Dewey Decimal Classification.

Country Mine <http://www.infomine.com/countries/> (accessed May 18, 2005). Complete mining information on the world mining industry by country. Includes mines and mining properties in the country, a snapshot of the country's mining activities, and links to geological and other maps.

Earth Sciences Information Resources – Internet Links, Branner Earth Sciences Library and Map Collections at Stanford University <http://library.stanford.edu/depts/branner/research_help/index.html> (accessed May 18, 2005). An excellent meta-collection of Internet resources for the earth sciences, including links especially of interest to mining engineers, such as professional associations and societies, government agencies of the United States (USGS), the European continent, Asia and the Pacific, Canada, Mexico and the Americas, links to global databases (including the poles), software tools and utilities, and academic institutions with programs in mining, petroleum engineering, and the geological sciences.

Galaxy – Engineering and Technology – Mining <http://www.galaxy.com/galaxy/Engineering-and-Technology/Mining/> (accessed May 18, 2005). Galaxy is a searchable Internet directory of contextually relevant information. Information is researched, compiled, classified, and organized by human Internet librarians who are Masters of Library Sciences graduates with experience in specific topic areas resulting in subject depth and concentrated, relevant coverage. Mining-related links are given to lists of mining companies, industry employment opportunities, geological surveys, government agencies, industry information, investment vehicles, trade magazines, organizations, and prospecting sites.

Materials Science Gateway. PSIgate (Physical Sciences Information Gateway) <http://www.psigate.ac.uk/newsite/materials-gateway.html> (accessed May 18, 2005). PSIgate provides free access to high-quality Internet resources for students, researchers, and practitioners in the physical sciences. Each resource has been selected by information professionals and subject specialists for relevance and quality. A description of each resource is provided with direct access to the resource itself. In the "Find" window type a word or short phrase such as "mining" to retrieve hundreds of web links to mineral mining sites and research.

Mindat.org <http://www.mindat.org/> (accessed May 18, 2005). The site states that it is "The largest mineral database and mineralogical reference website on the internet." The site contains worldwide data on mineral properties, localities, and other mineralogical information worldwide. Visitors are invited to search minerals by properties, by chemistry, to browse alphabetical and chemical indexes of minerals, or to link to mineral dealers.

Mines, Mining, and Mineral Resources – Library of Congress Science Reference Services. Science Tracer Bullets Online. Tracer Bullet 94-2 <http://www.loc.gov/rr/scitech/tracer-bullets/minestb.html> (accessed May 18, 2005). This guide reviews the literature on mines, mining, and mineral resources in the collections of the Library of Congress with the exception of materials on coal, uranium, petroleum, metallurgy, alloys, and gemstones.

Mining and Mineral Processing – EEVL (Internet Guide to Engineering, Mathematics, and Computing) <http://www.eevl.ac.uk/engineering/> (accessed May 18, 2005). EEVL is the Internet Guide to Engineering, Mathematics, and Computing. The site provides access to qualitative, evaluated international, networked engineering, mathematics and computing resources. The target audience is students, staff, and researchers in higher education, working, studying, or looking for information in engineering, mathematics, and computing. From the Engineering Home page, select the link for "Mining and mineral processing."

Mining and Minerals Pathfinder – Library/Information Resources at Natural Resources Canada on Mining and Minerals Processing. <http://www.nrcan.gc.ca/es/msd/emmic/web/Subject_Guides/subjectminingminerals_e.html> (accessed May 18, 2005). Links to Canadian publications, mineral and metal sector policies, dictionaries and other reference works, lists of mining associations, institutes, mining companies and metal producers, electronic journals, statistics and other data, conferences and meetings.

Mining Education. Mining Internet services <http://www.miningusa.com/> (accessed May 18, 2005). Advertising itself as providing information to the mining industry as a courtesy, the website includes links to articles on what coal-miners do, the history of coal, a glossary of coal-mining terminology, directories and addresses of journals, publishers, associations and companies in the transportation, and coal and metal-mining industries.

Conclusion

Mining engineering is a subfield requiring very specialized information from a variety of sources specific to the field, as well as knowledge of aspects of chemistry, chemical, civil, mechanical, and petroleum engineering.

18 Nuclear engineering

Mary Frances Lembo

Introduction, history, and scope of the discipline

On December 2, 1942, the first successful self-sustaining atomic chain reaction was achieved at the University of Chicago, thus ushering in the Nuclear Age and the discipline of nuclear engineering. Rather than being a theoretical endeavor of physicists, "nuclear fission propelled the subject into the military arena, leading to the enormous Manhattan Project in the US that produced the Hiroshima and Nagasaki bombs" (McKay, 1984: preface). The demands of an undertaking on the scale of the Manhattan Project presented new and unique challenges to the engineering industry. Developing processes to create and refine plutonium and uranium, constructing buildings and reactors to facilitate these processes and designing nuclear weapons were all immediate challenges made even more imperative by the military threat of Germany and Japan during World War II.

After the nuclear bombs dropped on Hiroshima and Nagasaki effectively ended World War II, scientists were in a position to devote their energies to non-military uses for nuclear energy. Atomic technology, originally developed for its destructive uses, was now being tapped for peaceful applications such as producing electricity and medical isotopes. However, nuclear production also resulted in the challenge of cleaning up wartime facilities and safe storage of waste by-products, as well as maintaining active nuclear power plants.

The field of nuclear engineering covers the entire range of nuclear energy production including: design, construction, operation and maintenance of nuclear power and naval propulsion reactors, reactor safety, development of nuclear weapons, disposal of radioactive wastes and the production of radioisotopes (*Encyclopaedia Britannica*, 1993: 423–424).

As in other engineering disciplines, the journal literature plays a major role for nuclear engineers, as do handbooks and standards. The technical report literature also plays an important role. Significant publishing agencies key to nuclear engineering include the American Nuclear Society in the United States and the International Atomic Energy Agency worldwide.

Searching the library catalog

The following is a list of useful Library of Congress Subject Headings in nuclear engineering and related fields. Using the phrase "Nuclear engineering" in combination with other keyword terms should provide a targeted topic search.

Nuclear engineering:

Nuclear energy
Nuclear facilities
Nuclear fission
Nuclear fuel
Nuclear fuel elements
Nuclear fuels
Nuclear industry
Nuclear power plants
Nuclear reactors
Radioactive decontamination
Radioactive waste disposal
Radioactive wastes
Spent nuclear fuels

The Library of Congress (LC) call numbers classify the nuclear literature into two main areas: the TK9000+ range, which emphasizes the applied aspect of nuclear engineering, and the QC770+ range, which demonstrates the nuclear physics aspect of the discipline.

TK9001–9401	Nuclear engineering. Atomic power
QC770–798	Nuclear and particle physics. Atomic energy. Radioactivity

In addition, health effects of nuclear radiation is covered in RA1231.

Abstracts and indexes

The discipline of nuclear engineering is covered in the general indexing and abstracting services such as *Compendex* and *Inspec.* The following are services that are more targeted toward nuclear engineering. These indexes and abstracts will assist users in locating literature in nuclear engineering covered in journal publications, conference proceedings, and technical reports.

Energy Citations Database (1948–) Oak Ridge, TN: Office of Scientific and Technical Information, US Department of Energy. *Energy Citations* contains bibliographic records for energy and energy-related research from the U.S. Department of Energy and its predecessor agencies. This database is publicly available at <http://www.osti.gov/energycitations/> (accessed May 18, 2005). It includes the information found in *Information Bridge* as well as *Nuclear Science Abstracts* and continues *Energy Research Abstracts.*

Energy Research Abstracts (1977–1995) Oak Ridge, TN: Technical Information Center, US Department of Energy. Continues *ERDA Energy Research Abstracts* and *Nuclear Science Abstracts.* Energy Research and Development Administration (ERDA) was the predecessor agency to the U.S. Department of Energy. In addition, the scope was expanded to include broader coverage of non-nuclear energy information.

Energy Science and Technology (1974–) Oak Ridge, TN: Office of Scientific and Technical Information, US Department of Energy. Available online from Dialog, Gov.Research_Center, and STN, but is restricted to the International Energy Agency (IEA) Energy Technology Data Exchange (EDTE) member countries. The database corresponds, in part, to *Energy Research Abstracts* and *INIS Atomindex.* Continues *Nuclear Science Abstracts* but with broader coverage to include non-nuclear energy information.

Information Bridge (1995–) Oak Ridge, TN: Office of Scientific and Technical Information, US Department of Energy. This source contains the full text of Department of Energy (DOE) research and development reports generally from 1995 onward, but reports are being added retrospectively as they become available in electronic format. The database allows for full-text searching and is publicly available at <http://www.osti.gov/bridge/> (accessed May 18, 2005). The information found in *Information Bridge* is also incorporated into the more retrospective *Energy Citations Database.*

INIS (International Nuclear Information System) Database (1970–) Vienna: International Atomic Energy Agency (IAEA). Supersedes *INIS Atomindex.* The *INIS* database contains worldwide literature on the peaceful uses of nuclear science and technology. Beginning in 1992, the database also includes the economic and environmental aspects of all non-nuclear energy sources. Access to the database is limited to INIS member states, and specific agencies and organizations within the member states. Access is via the Internet at <http://www.iaea.org/ programmes/inis/database/inis_database.htm> (accessed May 18, 2005). Information from *INIS* is also included in *Energy Science and Technology Database* but is restricted to the International Energy Agency (IEA) Energy Technology Data Exchange (EDTE) member countries.

Nuclear Science Abstracts (1948–1976) Oak Ridge, TN: Oak Ridge Directed Operations, Technical Information Division. *Nuclear Science Abstracts* covers the nuclear science and technology literature, and includes scientific and technical reports of the predecessor agencies of the U.S. Department of Energy (the Atomic Energy Commission and the Energy Research and Development Administration). Available online from Dialog and Gov.Research_Center. The information from this database is also included in *Energy Citations Database.*

Waste Management Research Abstracts (WMRA) (1960s?–) Vienna: International Atomic Energy Agency (IAEA). WMRA is a collection of research summaries dealing with the topic of radioactive waste, including environmental restoration and decommissioning of nuclear facilities. The collection of these abstracts began in the late 1960s. Originally published annually, there are currently twenty-three printed volumes in this collection. In 1997, the IAEA began an online system, the International Research Abstract Information System (IRAIS), so that researchers can submit abstracts via the Internet. Their website states that although "the information contained in IRAIS/WMRA covers a wide range of programs in many countries, the system should not be interpreted as providing a complete survey of planned or on-going research in IAEA Member States"

<http://www.iaea.org/cgi-bin/irais.showwmt.pl?wmwmra.wmt> (accessed May 18, 2005).

Bibliographies and guides to the literature

Although no bibliography on nuclear engineering has been published recently, the following provide solid background and historical references for the field:

Anthony, L.J. (1966) *Sources of Information on Atomic Energy*, International Series of monographs in library and information science, Vol. 2, Oxford; New York: Pergamon Press. While dated, this book provides an introduction to atomic energy, information sources from a variety of countries, and a discussion of published sources in the areas of atomic energy, high energy physics, nuclear power and engineering, ionizing radiation and radioisotopes, and controlled nuclear fusion and plasma physics. A subject and author index is included along with an index to organizations and to periodicals.

Information International Associates, Inc., Oak Ridge National Laboratory (1980–1997) *Nuclear Facility Decommissioning and Site Remedial Actions: A Selected Bibliography* (18 vols), Oak Ridge, TN: Martin Marietta Energy Systems. This multi-volume bibliography contains citations and abstracts of documents relevant to environmental restoration, nuclear facility decontamination and decommissioning, uranium mill tailings management, and site remedial actions. The most current volume (18) is available in pdf format at: <http://www.osti.gov/-dublincore/ecd/servlets/purl/569081-h1cLbB/webviewable/> (accessed May 18, 2005).

International Atomic Energy Agency, Commission of the European Communities, OECD Nuclear Energy Agency (1990) *Water Cooled Reactor Technology: Safety Research Abstracts*, No. 1. IAEA/WCRT/SRA/1, Vienna: International Atomic Energy Agency (IAEA). This collection of abstracts focuses on the subject areas related to nuclear safety such as pressurized light water-cooled and moderated reactors; boiling light water-cooled and moderated reactors; light water-cooled and graphite moderated reactors; pressurized heavy water-cooled and moderated reactors; and gas-cooled graphite moderated reactors.

Mullay, M. and Schlicke, P. (1999) *Walford's Guide to Reference Material*, Vol. 1, *Science and Technology* (8th edn), London: Facet Publishing. This guide, organized by the Universal Dewey Decimal Classification system, contains references to a full variety of resources, including journal articles, online database services, journal articles, encyclopedias, dictionaries, handbooks, and so on. Within the section on engineering are several pages devoted to nuclear and atomic engineering.

United Nations. Department of Political and Security Council Affairs. Atomic Energy Commission Group (1949–1953) *An International Bibliography on Atomic Energy*, Lake Success, NY: Atomic Energy Commission Group, Department of Security Council. This bibliography comes in two volumes and is international in scope. Volume 1 deals with political, economic, and social aspects and includes two supplements. It tends to be more selective rather than comprehensive. Volume 2 deals with scientific aspects and includes one supplement. This volume focuses on the peaceful purposes of atomic energy.

Directories

Although changes in personnel, location, or other data cannot be reflected quickly in print format, the following resources can act as a starting point for locating directory-type information. Societies and professional organizations often have membership directories that may be available via their websites.

American Nuclear Society (2004) *World Directory of Nuclear Utility Management* (16th edn), La Grange Park, Ill: American Nuclear Society. This directory contains listings for important personnel at nuclear plant sites, such as plant managers, maintenance superintendents, radioactive waste managers, and public relations contacts. The plant listings are arranged by country and include status information (operating, under construction, or decommissioned). International organizations and contact information are also included.

Business Press International (2004) *World Nuclear Industry Handbook*, Sutton, Surrey: Business Press International. Issued as a supplement to the November issue of *Nuclear Engineering International*, this publication provides up-to-date contact information as well as products and services information.

International Atomic Energy Agency (1959–1976) *Directory of Nuclear Reactors*, Vienna: International Atomic Energy Agency (IAEA). This ten-volume set discusses electric power and research reactor projects. Information includes name and location of reactor, data on reactor physics, core, fuel element, reflector and shielding, safety and containment, and a brief bibliography for each reactor.

International Atomic Energy Agency (1998) *Directory of Nuclear Research Reactors*, Vienna: International Atomic Energy Agency (IAEA). This reference contains technical information on research reactors known to the IAEA at the end of October 1998. IAEA collected the information compiled in this directory by sending questionnaires to the reactor operators. Data that IAEA could not verify is listed in Part III of the directory.

International Atomic Energy Agency (1996) *The Nuclear Fuel Cycle Information System: A Directory of Nuclear Fuel Cycle Facilities*, Vienna: International Atomic Energy Agency (IAEA). This publication provides data on civilian nuclear fuel cycle facilities. The information for this directory was obtained by questionnaires to IAEA member states and from additional published sources. Information includes name and location of facility, start-up and shutdown dates, and so on. An introduction to the nuclear fuel cycle industry and processes, and a glossary of terms and references are also included.

International Atomic Energy Agency (2004) *Nuclear Power Reactors in the World*, Reference Data Series no. 2, Vienna: International Atomic Energy Agency (IAEA). This small pamphlet-style book is a collection of tables providing specific reactor data, such as Nuclear power reactors in operation and under construction, December 31, 2001, Scheduled construction starts during 2002, Reactor units and net electrical power, 1970 to 2002, and more. Information is gathered by questionnaires to member states.

International Atomic Energy Agency (2002) *Nuclear Research Reactors in the World*, Reference Data Series no. 3, Vienna: International Atomic Energy Agency (IAEA). This small pamphlet-style book is a collection of tables providing specific reactor data on research reactors that are operational, under construction, shut down, or decommissioned. The data on research reactors are now available from the IAEA Research Reactor Database at <http://www.iaea.org/worldatom/rrdb/> (accessed May 18, 2005).

Longman Publishing Co. (1996) *World Energy and Nuclear Directory: Organizations and Research Activities in Atomic and Non-atomic Energy* (5th edn), Harlow, Essex: Longman; Detroit, MI: distributed in the USA and Canada by Gale Research. This directory provides contact information as well as budget, staff size, and summaries of research interests for research and development centers around the world in the areas of energy and nuclear technology. The entries are arranged alphabetically by the language of that country.

Encyclopedias and dictionaries

Encyclopedias and dictionaries can give useful background information for researchers who are beginning in nuclear engineering or who need to get up to speed on a topic outside their area of expertise. Multi-language dictionaries provide assistance to researchers working with documents in foreign languages or with teams from other countries.

Alter, H. (1986) *Glossary of Terms in Nuclear Science and Technology*, La Grange Park, IL: American Nuclear Society. Prepared by ANS-9, the American Nuclear Society Standards Subcommittee on Nuclear Terminology and Units. This concise glossary includes brief definitions of nuclear terms and includes an appendix on a classification system for reactors.

Atkins, S.E. (2000) *Historical Encyclopedia of Atomic Energy*, Westport, CT: Greenwood Press. This encyclopedia provides an alphabetical listing and brief explanations of key events, people, organizations, and sites significant in the history of nuclear energy. A chronology of atomic energy, an index, and a selected biography are also included.

Clason, W.E. (1958) *Elsevier's Dictionary of Nuclear Science and Technology, in Six Languages: English/American, French, Spanish, Italian, Dutch, and German*, Amsterdam; New York; Elsevier Publishing. This polyglot dictionary lists words in English, defines them, and then gives the translation for the five languages listed in the title. It also has a section for each language referring back to the entry number on the chart. A brief bibliography is included at the end.

International Atomic Energy Agency (2002) *IAEA Safeguards Glossary*, International Nuclear Verification Series No. 3, Vienna: International Atomic Energy Agency (IAEA). This glossary is divided into thirteen parts, each of which covers a different area related to IAEA safeguards. Definitions and explanations are given for each term or concept. There is also a translation section where the terms are translated into the six official IAEA languages (Arabic, Chinese, English, French, Russian, and Spanish) as well as German and Japanese.

International Atomic Energy Agency (2003) *Radioactive Waste Management Glossary*, Vienna: International Atomic Energy Agency (IAEA). This concise publication provides definitions of terms used in radioactive waste management.

Office of Nonproliferation and National Security, US Department of Energy, Office of Emergency Management (1996) *Nuclear Terms Handbook*, Washington, DC: Department of Energy. This concise handbook contains not only a glossary of nuclear terms, but also includes frequently used acronyms and abbreviations, a materials description and uses list, and a world list of nuclear power plants with related maps. A bibliography is included at the end of the book.

Handbooks and manuals

Handbooks and manuals are essential to the study of nuclear engineering because they provide key facts, equations and formulas, and fundamental information in a concise and easy-to-use format for the engineer. The following are a list of reference handbooks important to the study of nuclear engineering.

Argonne National Laboratory (1963) *Reactor Physics Constants*, ANL-5800 (2nd edn), Washington, DC: U.S. Atomic Energy Commission. This handbook contains data on fission properties, selected cross-sections, constants for thermal homogeneous reactors, and shielding constants. A subject index is available at the end of the book.

Baum, E.M., Knox, H.D., and Miller, T.R. (2002) *Chart of the Nuclides* (16th edn), New York: Knolls Atomic Power Laboratory, Lockheed Martin. Available as either a wall chart or a textbook version, this publication shows the key nuclear properties of the known stable and radioactive forms of the elements. In chart format, the nuclides are arranged with the atomic number along the vertical axis and the neutron number along the horizontal axis. Descriptive information includes a history of the development of the periodic table, descriptions of the type of data on the chart, and unit conversion factors and fundamental physics constants.

Etherington, H. (1958) *Nuclear Engineering Handbook*, New York: McGraw-Hill. This classic publication presents data used in all aspects of nuclear engineering. Divided into fourteen sections, the book contains information regarding mathematical data and tables, nuclear data, experimental techniques, radiological protection, fluid and heat flow, reactor materials, and isotopes.

Frisch, O.R. (1958) *The Nuclear Handbook*, Princeton, NJ: Van Nostrand. This book is intended as a desk reference source for those interested in the science and technology of nuclear physics. Material is presented in a concise format. Topics covered include radiation effects and protection, elements and isotopes, natural radioactivity, nuclear materials, particle accelerators, and nuclear reactors.

Nero, A.V. (1979) *A Guidebook to Nuclear Reactors*, Berkeley, CA: University of California Press. The first part of this text provides a general introduction to nuclear power plants, including basic reactor design features, environmental interactions, and nuclear power plant emissions. Part 2 discusses commercial nuclear power plants. Uranium resources and other nuclear materials are discussed in Part 3. Part 4 looks into advanced reactor systems. Appendices include abbreviations and units, reactions, and the nuclear fuel cycle. A glossary and an index are included.

Poenaru, D.N. and Greiner, W. (1996) *Handbook of Nuclear Properties*, Oxford Studies in Nuclear Physics 17, Oxford; New York: Oxford University Press. This handbook begins with information on atomic masses and shell model interpretations of nuclear masses. Additional information includes nuclear deformations and nuclear stability. Tables on fundamental constants, energy conversion factors, particle properties, alpha-particle emitters, and a table of nuclides are also included.

U.S. Atomic Energy Commission (1960–1964) *Reactor Handbook* (2nd edn), New York: Interscience Publishers. This four-volume set provides data on the materials, fuel processing, and physics of nuclear engineering. Volume 1 discusses fuel materials, cladding and structural materials, control, moderator, coolant and shielding materials. Volume 2 provides background on fuel processing including aqueous and non-aqueous separations, reconversions, and radioactive waste disposal. Volume 3 presents the physics of nuclear reactors, including the theory of neutron transport and reactor dynamics. In addition, this volume discusses shielding information, including neutron attenuation, heat generation in shields, and sources of neutrons and gamma rays. Volume 4 presents the engineering or nuclear reactors with discussions on fluid flow and heat transfer, reactor operations and safety, and types of reactors.

Wick, O.J. (1980) *Plutonium Handbook: A Guide to the Technology*, LaGrange Park, IL: American Nuclear Society. The *Plutonium Handbook* takes a multi-disciplinary approach to examining the element plutonium. It contains basic information on the physics, metallurgy, engineering, chemistry, and health and safety of plutonium. An author and subject index is included at the end of the book.

Monographs and textbooks

The following are a list of monographs and textbooks that provide background for a basic understanding in the areas of nuclear engineering, nuclear waste issues, reactor physics theory, and power plant design.

Bell, G.I. and Glasstone, S. (1970) *Nuclear Reactor Theory*, New York: Van Nostrand Reinhold. This book serves as an introduction to nuclear reactor theory for physicists, mathematicians, and engineers, and explains the physical and mathematical concepts used in nuclear reactors. The chapters include references and exercises. An appendix at the end includes selected mathematical functions.

Duderstadt, J.J. and Hamilton, L.J. (1976) *Nuclear Reactor Analysis*, New York: John Wiley & Sons. Designed for the nuclear engineering student, this book provides the basic scientific principles of nuclear fission chain reactions and applications in nuclear reactor design. References and problems are included at the end of each chapter along with appendices, which include selected nuclear data, selected mathematical formulas, and nuclear power reactor data.

Glasstone, S. and Sesonske, A. (1994) *Nuclear Reactor Engineering* (4th edn), New York: Chapman & Hall. This two-volume publication covers the principles of nuclear engineering and applications for the design and operation of nuclear power plants. The first volume discusses the fundamentals of nuclear engineering, including nuclear fission, radioactivity, neutron transport behavior, and the basics of nuclear design. Volume 2 focuses on the applications and other advanced topics including energy transport and fuel management as well as environment, health, and safety concerns. Innovations in plant design and challenges to the nuclear power industry are also discussed. References and exercises are included at the end of most chapters.

Lamarsh, J.R. (1972) *Introduction to Nuclear Reactor Theory*, Reading, MA: Addison-Wesley Publishing. This publication is based on a one-year course in nuclear reactor theory at Cornell and New York universities. The book gives readers a fundamental understanding of the principles regarding the operation of a nuclear reactor. References and exercises are included at end of each chapter. Appendices cover conversion factors, selected constants, isotopes of importance in nuclear engineering, and properties of selected molecules.

Lamarsh, J.R. and Baratta, A.J. (2001) *Introduction to Nuclear Engineering* (3rd edn), Upper Saddle River, NJ: Prentice Hall. This book, now in its third edition, was originally based on the class notes and lectures of the late Dr. John R. Lamarsh. It contains an overview of the field of nuclear engineering and discusses the basics of atomic and nuclear physics, nuclear reactor theory and reactor design, and both U.S. and non-U.S. nuclear reactor design. Information on nuclear reactor safety is also included. References and practice exercises are included at the end of most chapters.

Murray, R.L. (2000) *Nuclear Energy: An Introduction to the Concepts, Systems and Applications of Nuclear Processes* (5th edn), New York: Elsevier Science and Technology Books. This publication provides an overview of the field of nuclear energy and its uses. It focuses on the processes, uses, and future outlook of nuclear energy, and nuclear non-proliferation. New trends in the industry are discussed including probabilistic safety analysis (PSA), electric power industry deregulation, performance characteristics of nuclear power plants, storage and disposal of radioactive wastes, and developments in decontamination and decommissioning. The basic concepts of energy, nuclear reactions, and effects of nuclear energy on humans are discussed. Exercises are included at the end of each chapter and a list of selected references is given at the end of the book. The author studied under J. Robert Oppenheimer at UC Berkeley and made contributions to the uranium separation process for the Manhattan Project at Berkeley and at Oak Ridge.

Murray, R.L. and Manke, K.L. (eds) (2003) *Understanding Radioactive Waste* (5th edn), Richland, Washington, DC: Battelle Press. Written for the general public, this publication provides the reader with the background information necessary to understand the complexities of radioactive waste and its relationship to the political climate.

Olander, D.R. (1976) *Fundamental Aspects of Nuclear Reactor Fuel Elements*, TID-27611-P1, Oakridge, TN: Technical Information Center. Energy Research and Development Administration. This book was based on lectures in graduate courses in the Department of Nuclear Engineering, University of California, Berkeley. The first part of the book reviews selected aspects of statistical thermodynamics, crystallography, chemical thermodynamics, and physical metallurgy as it applies to nuclear reactor fuel elements. The second part shows how these principles apply to issues of nuclear fuel elements. Each chapter includes a section on nomenclature, references, and problems.

Rust, J.H. (1979) *Nuclear Power Plant Engineering*, Buchanan, GA: Haralson Publishing. This text focuses on the technological and engineering aspects of nuclear power plant design and nuclear energy. This book is based on the author's lectures on nuclear power at the University of Virginia and the Georgia Institute of Technology. Chapters include references and exercises. Appendices at the end cover conversion factors, properties of reactor coolants, and materials.

Waltar, A.E. and Reynolds, A.B. (1981) *Fast Breeder Reactors*, New York: Pergamon Press. This book discusses the basic principles and methods as well as design features of fast breeder reactors. Each chapter ends with references and problems to be solved. Appendices include fast reactor data, comparison of homogeneous and heterogeneous core designs, and a list of symbols.

Journals and series

The journal literature plays a significant role in the communication of research and development in nuclear engineering. The following is a list of key journals in the field of nuclear energy and engineering. The list has been compiled by consulting with the Institute for Scientific Information's (ISI)

Journal Citation Reports and by investigating the collection of the Pacific Northwest National Laboratory's Hanford Technical Library. Many of the journals listed below are available in electronic format.

Advances in Nuclear Science and Technology (1962 –) Plenum Press. Monographic series (0065–2989).

Annals of Nuclear Energy (1975–) Pergamon Press. Continues *Annals of Nuclear Science and Engineering* (0306–4549).

Annals of the ICRP (1977–) International Commission on Radiological Protection. Pergamon Press. Continues *Advances in Radiological Protection* (0146–6453).

Atomic Data and Nuclear Data Tables (1973–) New York: Academic Press. Merges *Atomic Data* and *Nuclear Data Tables* (0092–640X).

Atomic Energy/Atomnaia Energiia (1993–) Consultants Bureau. Translated from the Russian. Continues *Soviet Atomic Energy* (1063–4258).

Bulletin of the Atomic Scientists (1946–) Chicago, IL: Educational Foundation for Nuclear Science. Continues *Bulletin of the Atomic Scientists of Chicago* (0096–3402).

Fusion Engineering and Design (1987–) North-Holland Publishing. Continues *Nuclear Engineering and Design/fusion* (0920–3796).

Fusion Science and Technology: An International Journal of the American Nuclear Society (2001–) American Nuclear Society. Continues *Fusion Technology* (1536–1055).

IEEE Transactions on Nuclear Science (1963–) Institute of Electrical and Electronics Engineers. Continues *IRE Transactions on Nuclear Science* (0018–9499).

Journal of Nuclear Materials Management (JNMM) (1986–) Institute of Nuclear Materials Management. Continues *Nuclear Materials Management* (0893–6188).

Journal of Nuclear Materials. Journal des Matériaux Nucléaires (1959–) North Holland Publishing. Articles in English, French, or German (0022–3115).

Journal of Nuclear Science and Technology (1964–) Nihon Genshiryoku Gakkai. Atomic Energy Society of Japan (0022–3131).

Kerntechnik (1987–) C. Hanser. Continues *Atomkernenergie/Kerntechnik* (0932–3902).

Nuclear Energy (1978–) London: British Nuclear Energy Society. Continues *Journal of the British Nuclear Energy Society* (0140–4067).

Nuclear Engineering and Design; An International Journal Devoted to the Thermal, Mechanical and Structural Problems of Nuclear Energy (1965–) North-Holland Publishing. Continues *Nuclear Structural Engineering* (0029–5493).

Nuclear Engineering International (1968–) London: Heywood-Temple Industrial Publications. Continues *Nuclear Engineering* (London) (0029–5507).

Nuclear Fusion. Fusion nucléaire (1960–) Vienna: International Atomic Energy Agency (0029–5515).

Nuclear News (1959–) La Grange Park, IL: American Nuclear Society (0029–5574).

Nuclear Plant Journal (1987–) Glen Ellyn, IL: EQES. Continues *Nuclear Plant Safety* (0892–2055).

Nuclear Science and Engineering: the Journal of the American Nuclear Society (1956–) New York: Academic Press (0029–5639).

Nuclear Technology (1971–) La Grange Park, IL: American Nuclear Society. Continues *Nuclear Applications and Technology* (0029–5450).

Progress in Nuclear Energy (1977–)New York: Elsevier (0149–1970).

Radiation Measurements (1977–) Oxford: Pergamon Press. Continues *Nuclear Tracks and Radiation Measurements* (1350–4487).

Radiation Protection Dosimetry (1981–) Ashford, Kent: Nuclear Technology Publishing (0144–8420).

Radiochimica Acta (1962–) New York: Academic Press (0033–8230).
Transactions of the American Nuclear Society (1958–) New York: Academic Press (0003–018X).

Patents and standards

Patents and, in particular, standards play a major role in the nuclear engineering literature. Look for significant patents from the following companies: Babcock & Wilcox, British Nuclear Fuels, Combustion Engineering, DuPont, Framatome, General Electric Company, Japan Atomic Energy Research Institute, Siemens, and Westinghouse Electric Corporation.

The following associations include standards pertinent to nuclear engineering:

American Nuclear Society (ANS) Standards <http://www.ans.org/standards/> (accessed May 18, 2005). ANS has over ninety standards specific to the nuclear industry.

American Society of Civil Engineers issues the following standards for nuclear facility construction:

Seismic Analysis of Safety-related Nuclear Structures and Commentary, ASCE 4-98. (2000) Reston, VA: American Society of Civil Engineers. This publication provides the requirements for the analysis of new nuclear structure designs or the evaluation of existing structures to determine their reliability in the event of an earthquake.

A Summary Description of Design Criteria, Codes, Standards, and Regulatory Provisions Typically Used for the Civil and Structural Design of Nuclear Fuel Cycle Facilities (1988) Reston, VA: American Society of Civil Engineers. Although this title is not specifically a standard, it is useful in understanding the government regulations and industry codes and standards to those who are constructing a new nuclear facility. It brings together information on the requirements from a variety of different resources.

American Society for Testing and Materials (ASTM) *Section 12 – Nuclear, Solar, and Geothermal Energy.* This section of the ASTM standards contains over 300 active standards in the area of nuclear energy.

American Society of Mechanical Engineering (2004) *ASME Boiler and Pressure Vessel Code. The ASME Boiler and Pressure Vessel Code* provides requirements for the design and fabrication of boilers and pressure vessels and nuclear power plant components during construction. *The Code* is regularly updated by addenda.

IEEE Nuclear Standards <http://www.ieee.org> (accessed May 18, 2005). The IEEE has over 130 standards related to nuclear and plasma science, accelerator technology, fusion, nuclear instruments, plasma science and applications, radiation effects, and reactor instruments and controls.

International Atomic Energy Agency (IAEA) <http://www.iaea.org/Publications/Standards/index.html> (accessed May 18, 2005). The IAEA publishes standards in the areas of nuclear safety, radiation protection, radioactive waste management, transport of radioactive materials, safety of nuclear fuel cycle facilities, and quality assurance.

U.S. Nuclear Regulatory Commission. *Regulatory Guide*, Washington, DC: U.S. Nuclear Regulatory Commission. Office of Standards Development. The Nuclear Regulatory Commission (NRC) *Guides* provide the requirements that bind those people or organizations who receive a license from the NRC to operate nuclear facilities. The *Guides* are issued in ten broad divisions with individual titles for each division (Power reactors, Research and test reactors, Fuels and materials facilities, Environmental and siting, Materials and plant protection, Products, Transportation, Occupational health, Antitrust and financial review, and General). An electronic version is available at <http://www.nrc.gov/reading-rm/doc-collections/reg-guides/> (accessed May 18, 2005).

Search engines and important websites

Although most scholarly publications are not freely available via the Internet, the field of nuclear engineering has a slight advantage. Due to the efforts of the Department of Energy's (DOE) Office of Science and Technology Information (OSTI), much of the government-sponsored publicly available DOE research is obtainable on the Internet. See the Abstracts and indexes sections for links to *Information Bridge* and *Energy Citations Database*. In addition, society websites may contain pertinent information and the major societies in nuclear engineering host websites which may be found in the Societies section.

International Atomic Energy Agency (IAEA) <http://www.iaea.org/> (accessed May 18, 2005). The IAEA was founded in 1957 to promote the peaceful use of atomic energy. It is an organization within the United Nations and is very active in safeguards, verification, and monitoring of nuclear materials. In working with other organizations, IAEA supports research and development in nuclear technology that will aid food, environmental, and health issues facing developing countries. IAEA technical documents may be found at http://www-pub.iaea.org/MTCD/publications/tecdocs.asp (accessed May 18, 2005).

Lawrence Berkeley National Laboratory (LBNL), Isotopes Project <http://ie.lbl.gov/>. The purpose of the Isotopes Project is to gather, evaluate, and disseminate data regarding neutron capture and radioactive decay as well as developing new techniques to distribute the data.

National Nuclear Data Center (NNDC) <http://www.nndc.bnl.gov/index.jsp> (accessed May 18, 2005). The NNDC is a worldwide resource for data on nuclear physics, nuclear research, and applied nuclear technologies. In addition, the NNDC focuses on data compilation and evaluation for nuclear information.

Nuclear Age Timeline <http://web.em.doe.gov/timeline/> (accessed May 18, 2005). The Timeline, created by the Department of Energy's Office of Environmental Management, gives an historical outlook to the development of the nuclear age from the development of the x-ray to the clean-up of the nuclear weapons complexes (1895–1993). Links to references and selected resources are provided as well.

Nuclear Regulatory Commission (NRC) <http://www.nrc.gov> (accessed May 18, 2005). The NRC is an independent agency tasked with regulating the civilian use of nuclear materials. The site includes an Electronic Reading Room at <http://www.nrc.gov/reading-rm.html> which provides access to NRC documents and reports. The NRC also contains a docket collection, which provides specific licensing information for power and research reactors in the United States. For assistance in accessing this information, contact the NRC Public Document Room.

Associations, organizations, and societies

Societies and associations play an important role in the discipline of nuclear engineering. The following societies create standards and technical documents, as well as publishing conference proceedings and journals to disseminate nuclear engineering literature. Society websites are another useful source of information where you can find out more about the society, purchase materials, or view their publications. Upcoming conference dates are often listed on these pages, as are professional development opportunities and employment announcements.

American Nuclear Society (ANS) <http://www.ans.org/> (accessed May 18, 2005). The ANS was founded in 1954 and facilitates the exchange of information with members in the development and safe application of nuclear science and technology.

British Nuclear Energy Society (BNES) <http://www.bnes.com/> (accessed May 18, 2005). The BNES acts as a forum to provide information to members on nuclear energy issues, debate nuclear issues, and offer education to the public on the use of nuclear energy. Founded in 1962, the Society also promotes nuclear energy training in the United Kingdom.

Canadian Nuclear Association – Association Nucléaire Canadienne (CNA) <http://www.cna.ca/> (accessed May 18, 2005). The CNA was established in 1960 to promote the peaceful use of nuclear technology within Canada.

European Nuclear Society (ENS) <http://www.euronuclear.org/> (accessed May 18, 2005). Established in 1975, the ENS is an association of twenty-six nuclear societies from twenty-five countries. Its goals are to promote the peaceful use of nuclear energy through the promotion of science and nuclear engineering.

IEEE Nuclear and Plasma Sciences Society (NPSS): http://www.ewh.ieee.org/soc/nps/ (accessed May 18, 2005). When the Society was first established in 1949, it was known as the Professional Group on Nuclear Science within the Institute of Radio Engineers (IRE). When the Institute of Electrical and Electronic Engineers (IEEE) was formed in 1963 with the merger of the American Institute of Electrical Engineers (AIEE) and the IRE, the professional group became the Nuclear Science Group. In 1972, Plasma Science was added and the group was promoted to a society. Their goal is to encourage the advancement of nuclear and plasma sciences.

Institute of Nuclear Materials Management (INMM) <http://www.inmm.org/> (accessed May 18, 2005). INMM provides guidance and professional development opportunities in the management of nuclear materials.

The Institution of Nuclear Engineers <http://www.inuce.org.uk/> (accessed May 18, 2005). The Institution originated in 1959 and promotes the advancement of nuclear engineering and related fields.

International Commission on Radiation Units and Measurements (ICRU) <http://www.icru.org/> (accessed May 18, 2005). Established in 1925, this organization has developed recommendations on "(1) quantities and units of radiation and radioactivity; (2) procedures suitable for the measurement and application of these quantities in diagnostic radiology, radiation therapy, radiation biology, and industrial operations; and (3) physical data needed in the application of these procedures, the use of which tends to assure uniformity in reporting" (source: ICREU Web page).

National Council on Radiation Protection and Measurements (NCRP) <http://www.ncrponline.org/> (accessed May 18, 2005). The NCRP provides information, guidance, and recommendations on radiation protection and measurements. The Council also facilitates cooperation among organizations concerned with the scientific and technical aspects of radiation protection and measurements.

Nuclear Energy Agency/Agence pour l'énergie nucléaire (NEA) <http://www.nea.fr/> (accessed May 18, 2005). As an agency within the Organization for Economic Cooperation and Development (OECD), the NEA assists member countries in maintaining and developing technologies for safe, environmentally sound, and peaceful use of nuclear energy.

World Nuclear Association (WNA) <http://www.world-nuclear.org/> (accessed May 18, 2005). The aims of the WNA are to promote the peaceful use of nuclear energy. In addition, the WNA focuses on all facets of the nuclear fuel cycle from mining to fuel fabrication to the safe disposal of spent nuclear fuel.

Conclusion

The discipline of nuclear engineering has traditionally had close ties with physics and chemistry. As the understanding of the physics of nuclear reactors advanced, and the knowledge of nuclear physics moved beyond the theoretical, the discipline of nuclear energy began to grow on its own. Although no new nuclear reactors have been built in the United States since 1979, this may change, since there is a renewed interest in nuclear energy as a viable alternative to fossil fuels for energy generation. In the meantime, existing reactors continue to function and need to be kept in working order. Legacy issues of cleaning up wartime facilities and waste by-products are still challenges for today's nuclear engineer. Finding new technologies to deal with these past issues is as crucial as continuing to search for safer and cleaner methods of producing nuclear energy. To move forward, nuclear engineers must have access to the literature of the discipline – including both current and historical publications.

Acknowledgments

Many thanks to Patricia Cleavenger, Nancy Doran, David J. Senor, and John M. Tindall for their support and assistance in writing this chapter.

References

Encyclopaedia Britannica (1993) "Nuclear Engineering," In *The New Encyclopaedia Britannica: Macropaedia – Knowledge in Depth*, Chicago, IL: Encyclopaedia Britannica, Inc.

Lee, T.S. (2000) *Selective Guide to Literature on Nuclear Engineering*, Engineering

Literature Guides, No. 27, Washington DC: Engineering Libraries Division, American Society for Engineering Education <http://eld.lib.ucdavis.edu/fulltext/NuclearEng.pdf> (accessed October 10, 2004).

McKay, A.(1984) *The Making of the Atomic Age*, New York: Oxford University Press.

World IQ.com. *Definition of Nuclear Reactor* <http://www.wordiq.com/definition/Nuclear_reactor> (accessed October 10, 2004).

19 Petroleum engineering and refining

Randy Reichardt

Introduction and scope of the chapter

The petroleum industry includes both upstream and downstream activities. Petroleum engineers work upstream, being responsible for the exploration and development of oil and gas fields and wells. Petroleum engineers' work focuses on three areas: drilling, reservoir and production engineering, in the search for unprocessed crude oil, also known as petroleum.

Petroleum refining and processing are considered downstream activities, and generally fall within the subject scope of the chemical engineer. Crude oil, transported from the ground to the refinery, is used to produce many products, including petroleum gas, gasoline, kerosene, heavy oil (fuel oil), coke, asphalt, tar, and other by-products.

This chapter will highlight key resources that provide coverage in either upstream, downstream, or both disciplines. An exhaustive list of resources for each category is not provided; only the most important titles have been included. The business, economic, and financial components of the petroleum industry are not part of the scope of this guide, although selected titles listed in this chapter do cover some of these aspects.

History of petroleum engineering

As noted in the American Petroleum Institute's *All About Petroleum*, the use of petroleum dates back thousands of years. Asphalt, bitumen, and pitch were used to inlay mosaics in walls, line water canals, and grease chariots. The Chinese first discovered underground deposits of oil, and built bamboo pipelines to transport oil and natural gas. By AD 1500, the Chinese were drilling wells greater than 2,000 feet deep (American Petroleum Institute, 2004).

In one sense, petroleum engineering is an offshoot of mining engineering. Mining of the Earth's resources dates back millennia, and has always involved the extraction and preparation of coal, minerals, and metals. Petroleum engineering involves extraction as well, but of oil and gas only, and is a relatively new profession.

The petroleum industry began to emerge in the United States in the mid-1800s. The first American oil company, Pennsylvania Rock Oil Company, was formed in 1854 (American Petroleum Institute, 2004). The first well in the United States drilled specifically to recover oil became known as the Drake Well, and was created on August 28, 1859, at Oil Creek, near Titusville PA (American Petroleum Institute, 1961: 170–175). The Drake Well heralded the beginning of the petroleum engineering industry in the U.S.A.

However, the first commercial oil well in North America was dug and began operations one year earlier, in Canada in 1858. Many consider James Miller Williams to be the father of the North American petroleum industry, having dug a well in the Enniskillen Township in Ontario in August 1858, specifically searching for oil (Morritt, 1993: 25–34; Cronin, 1955: 15–20). Williams set up two pumps to bring the oil to the surface, and soon the well was producing between five and one hundred barrels a day. The Oil Museum of Canada now resides on the site of North America's first commercial oil well, in Oil Springs, Ontario (Oil Museum of Canada, 2004).

The development of the industry soon followed in Russia in the 1870s, and by 1900 had expanded to the Middle East and other parts of Asia (Giebelhaus, 1996). In the twentieth century, the petroleum engineering and refining industries expanded to most parts of the planet. In North America, the industry grew in Canada and Mexico, as new oil fields and wells were discovered. The British Petroleum Company was formed in 1901. Elsewhere in the world, discoveries of oil deposits led to the introduction of petroleum engineering operations in Africa, Asia, The Middle East, and Latin America. In 1960, countries from these regions joined to form the Organization of Petroleum Exporting Countries (OPEC). OPEC member nations in 2004 were Algeria, Indonesia, Iran, Iraq, Kuwait, Libya, Nigeria, Qatar, Saudi Arabia, the United Arab Emirates, and Venezuela. In 2003, OPEC countries accounted for 78.3 percent of world proven crude oil reserves by region (OPEC, 2003).

A very detailed history of petroleum engineering in the United States was published in 1961 by the American Petroleum Institute (American Petroleum Institute, 1961). A newer edition has not been issued. Published in 1975, *Trek of the Oilfinders: A History of Exploration for Petroleum*, written by Edgar Wesley Owen (Owen and Doth, 1975), is primarily of interest to petroleum geologists. However, the first chapter of the book, "The Earliest Oil Industry," offers a brief but concise history of oil and petroleum, from the ancient record through to the Drake Well. *Petrochemicals: The Rise of an Industry* was published in 1988. The author, Peter H. Spitz, notes that the petrochemical industry is an American phenomenon, created by oil and chemical companies in the 1930s and early 1940s, but drew much of its early technology from the Germans, who had built a substantial chemical industry since the early 1800s (Spitz, 1988).

In addition to the publications referenced above, the following titles and website cover various aspects and time periods of the history of the petroleum and petrochemical industries.

Bamberg, J.H. (1994) *The History of the British Petroleum Company, Vol. 2: The Anglo–Iranian Years, 1928–1954*, Cambridge: Cambridge University Press.

Chapman, K. (1991) *The International Petrochemical Industry: Evolution and Location*, Oxford: Blackwell.

Chastko, P. (2004) *Developing Alberta's Oil Sands: From Karl Clark to Kyoto*, Calgary: University of Calgary Press.

Ferrier, R.W. (1982) *The History of the British Petroleum Company, Vol. 1: The Developing Years, 1901–1932*, Cambridge: Cambridge University Press.

Jones, G. (1981) *The State and Emergence of the British Oil Industry*, London: Macmillan.

Morritt, H. (1993) *Rivers of Oil: The Founding of North America's Petroleum Industry*, Kingston, ON: Quarry Press.

Pees, S.T. *Oil History* <http://www.oilhistory.com/> (accessed March 3, 2005).

Shah, S. (2004) *Crude: The Story of Oil*, New York: Seven Stories Press.

Stackenwalt, F.M. (1998) "Dmitrii Ivanovich Mendeleev and the Emergence of the Modern Russian Petroleum Industry, 1863–1877," in *Ambix: The Journal of the Society for the Study of Alchemy and Early Chemistry*, Vol. 45, Part 2: 67–84.

Williamson, H.F. and Daum, A.R. (1959) *The American Petroleum Industry: the Age of Illumination 1859–1899*, Evanston IL: Northwestern University Press.

Williamson, H.F. (ed.) (1963) *The American Petroleum Industry: the Age of Energy 1899–1959*, Evanston, IL: Northwestern University Press.

Yergin, D. (1991) *The Prize: The Epic Quest For Oil, Money and Power*, New York: Simon & Schuster.

Searching the catalog

Most online library catalogs have a keyword search function, allowing the user to type in any word or phrase when searching for material in the library system. Keyword searching generally covers titles, subject headings, notes, tables of contents, series, publisher, and author fields. When searching a library catalog, the following Library of Congress Subject Headings are among the most useful and relevant to both petroleum engineering and petroleum refining and processing:

Bitumen	Natural Gas Pipelines	Oil Wells
Coal-Tar Products	Oil Field Flooding	Petroleum
Enhanced Oil Recovery	Oil Fields	Petroleum Chemicals
Gas Reservoirs	Oil Reservoir Engineering	Petroleum Coke
Heavy Oil	Oil Sands	Petroleum Engineering
Hydrocarbons	Oil Well Drilling	Petroleum Geology
Natural Gas	Oil Well Pumps	
Petroleum Industry and Trade	Petroleum Refineries	Secondary Recovery of Oil
Petroleum Pipelines	Petroleum Reserves	Thermal Oil Recovery
Petroleum Products	Petroleum – Refining	Petroleum, Synthetic

The Library of Congress classification system includes a number of locations within its schedules for petroleum engineering and petroleum refining. The main classification ranges include the following:

HD 9560–9579	Petroleum Industry and Trade
HD 9581	Pipelines. Petroleum Pipelines
HD 9581–9582	Natural Gas
TN 850	Bitumen
TN 860–879.5	Petroleum. Petroleum Engineering
TN 871	Enhanced Oil Recovery
TN 879.5–879.6	Petroleum Pipelines
TP 690–692.5	Petroleum Refining. Petroleum Products

Abstracts and indexes

The literature of petroleum engineering and refining can appear in many different formats: scholarly and trade journal articles, conference papers, patents, websites, standards, handbooks and manuals, monographs, government reports, theses and dissertations, gray literature, encyclopedias and dictionaries, articles in newspapers and magazines, and so forth. In 1987, Pearson and Ellwood listed fifteen indexing and abstracting services (in print), and sixty-seven databases, divided into bibliographic, textual, and databases containing only numbers and/or data (Pearson *et al.*, 1987). No attempt has been made to provide similar coverage in this section. Only core resources are listed, together with a number of related but important titles. The most up-to-date and comprehensive list of petroleum industry databases may be found in each issue of the *Gale Directory of Databases*. In the 2005, Part 1, edition of the Directory, eighty-four databases are listed in the category, "Petroleum and petroleum technology" (Meuckeneheim, 2005).

The following abstracting and indexing services provide good coverage of publications in petroleum engineering and/or refining and processing. Databases offering related coverage are also listed.

EnCompassLIT and *EnCompassPAT* <http://www.eiencompass.com/c/s/C> (accessed May 18, 2005) Hoboken: Elsevier Engineering Information, Inc. (1964–) Previously known as ApiLiT and ApiPAT, these databases are now available via Elsevier Engineering Information as part of the *EnCompassWEB* platform. Subject coverage is the downstream petroleum industry. *EnCompassLIT* is a technical literature database that began in 1964, containing over 730,000 records as of November 2004. Entries are selected from over 300 scientific and trade journals and magazines, conference papers, and technical reports. An additional 1,200 titles from secondary sources provide further sources for the db. *EnCompassPAT* also began in 1964, and covers the patent literature of downstream petroleum activity. Patents are selected from forty patenting authorities across the globe. *EnCompassLIT* and *EnCompassPAT* are also available via Dialog (as *Ei EnCompassLit* and *Ei EnCompassPat*), and Questel•Orbit (as *ALIT* and *WPAM*).

Petroleum Abstracts <http://www.pa.utulsa.edu/> (accessed May 18, 2005) Tulsa: Petroleum Abstracts (1961–) Provides coverage of oil and gas exploration, production, transportation storage, environment, safety and health, relevant to the upstream sector of the petroleum industry. Entries are created from over 300 journals, 200 conference proceedings, patents, government documents, reports, dissertations, maps, and other sources. PA is available in a number of formats, including via db vendors such as Dialog, Questel/Orbit and STN, where it is known as TULSA. It is available via the Internet as the *Petroleum Abstracts Discovery* database. The *Petroleum Abstracts Bulletin* is published fifty times a year, and is sent by e-mail, with 400 to 600 abstracts per issue. Two *Dialog OnDisc Petroleum Abstracts* CD-ROM versions are also available, covering the most recent twenty or ten years of PA. Also available via Dialog and Questel•Orbit.

Petroleum Abstracts Discovery is searchable by full text, author, source, title, corporate source, and abstract number. No thesaurus is provided. Searches can be refined. Selection of search results for e-mailing, printing, or downloading is not available.

SPE eLibrary <http://www.spe.org/spe/jsp/basic/0,2396,1104_1561_0,00.html> (accessed May 18, 2005) Richardson, TX: Society of Petroleum Engineers (1958–) Contains full text of over 35,000 papers (as of July 2004) presented at SPE conferences and peer-reviewed SPE journals. Conference and meeting papers are added to the db beginning the first day of that conference or meeting. SPE journal papers are added six months after publication. Previously, technical papers were available on microfiche from 1955, and on CD-ROM, but these formats have been discontinued. Access requires an SPE-issued log-in and password. Searchable by year, SPE paper number, title, source (name of SPE meeting or periodical), author, organization, or keyword. Papers can be selected and added to a cart. E-mail is sent with the URL for each selected paper, allowing the searcher to download the full text within a 144-hour time period.

Other relevant databases include:

Chemical Abstracts (CA). Columbus: Chemical Abstracts Service. *Chemical Abstracts* is also known as *CA Search: Chemical Abstracts* (Dialog), *CAPlus* (*SciFinder Scholar*). While mentioned in Chapter 7 of this book (Chemical engineering), it is important to note that CA provides good coverage related to petroleum technology, including fossil fuels and derivatives, natural gas exploration, reservoirs, petroleum refining and processing, and hydrocarbons.

Compendex. New York: Elsevier Engineering Information Inc. Covered in more detail in Chapter 2 of this book, *Compendex* nonetheless provides solid coverage of the literature of the upstream and downstream petroleum industries.

Fuel and Energy Abstracts. Guilford, Surrey: Elsevier, Vol. 36 (1995–) Available via ScienceDirect, also published bimonthly in hard copy. Provides summaries of world literature on scientific, technical, environmental, and commercial aspects of fuel and energy. Covers over 800 international publications, monographs, conference proceedings, reports, surveys, and statistical analyses. Number of abstracts per issue can vary from 500 to 800. The online indexing is inconsistent. Individual abstracts are indexed in most issues. However, in a smaller number of issues, the articles are buried under headings such as "Liquid fuels," Heat pumps," and "Fuel science and technology," and cannot be retrieved with a keyword search.

GeoRef. American Geological Institute (1785–) Over 2.2 million records on North America from 1785, and other areas of the world since 1933. While primarily a geosciences database, good coverage is provided in areas including: petroleum engineering, exploration, geology, maps, products, reserves and resources, reservoir properties, oil and gas fields, and pipelines.

International Petroleum Abstracts. London, Energy Institute (1995–) Coverage of upstream and downstream oil industry topics. Articles abstracted are of a technical nature. Available only to members of the Institute of Petroleum (which merged with the Institute of Energy in 2003 to form the Energy Institute).

Also of note is the *Heavy Oil, Enhanced Recovery, and Oil Sands (HERO)* database, from the Alberta Oil Sands Information Services (AOSIS). AOSIS was established in 1975, its purpose being "to acquire, organize and supply public information on Canadian and international developments in heavy oil, enhanced oil recovery and oil sands research." HERO was developed over the years, covering topics including research, geology, exploration, recovery, upgrading, transportation, economic assessment, and others. The database contains over 32,000 records, from various sources including technical journals, patents, reports, books, and theses. A "final copy" of *HERO* is available on CD-ROM, and AOSIS will also run customized literature searches of the database for clients. For more information, contact AOSIS at: Alberta Oil Sands Information Services (AOSIS), Suite 2540, 801–6th Avenue S.W., Calgary, Alberta T2P 3W2, Canada. Tel: (403) 297–5221; Fax: (403) 297–3638.

Databases of importance to the field of petroleum engineering, discussed elsewhere in this book, include *NTIS*, *Dissertation Abstracts*, and *Web of Science*.

Bibliographies and guides to the literature

The most recent guides devoted specifically to petroleum literature resources appeared in the late 1980s. None that could be identified has been published since that time. The 193-page guide by Pearson and Ellwood was thorough and detailed in its coverage, with over 420 entries in numerous categories, together with extensive lists of publishers and associations. Charles R. Lord's *Guide to Information Sources in Engineering* was published in 2000, but did not place significant emphasis on information resources in petroleum engineering. In 2005, the fourth, revised edition of *Information Sources in Engineering* was published with a twelve-page chapter devoted to petroleum and offshore engineering (MacLeod and Corlett, 2005). The titles listed below, despite being outdated, remain important for historical reasons, and for identifying earlier editions of important titles, as well as titles that are now out of print, or have ceased publication.

Anthony, L.J. (ed.) (1988) *Information Sources in Energy Technology*, London: Butterworths. Coverage of information sources in a number of energy resources, this title is divided into three distinct components: energy in general, fuel technology, and specific energy resources. Petroleum and oil are covered in the chapter on liquid fuels, and natural gas in the chapter on gaseous fuels. Rather than present key resources in alphabetical, numbered, or bulleted lists, titles in each section are identified and discussed in paragraphs, rendering the book somewhat less than user-friendly.

Myers, A. (ed.) (1993) *Petroleum and Marine Technology Information Guide: A Bibliographic Sourcebook and Directory of Services*, London: E & FN Spon. This title is the fourth edition of a work that began as a bibliography of offshore technology in 1981. The third edition was renamed and expanded coverage to include marine technology, and online databases in the oil and gas industry. The fourth edition was renamed and revised again, adding sources on petroleum exploration, reservoirs, and production to its bibliography. Primary focus of the work is on the offshore and marine industry. Included are a bibliography by subject, organizations providing information services, online databases and CD-ROMs, and an industry directory relevant to petroleum and marine technology.

Pearson, B.C. and Ellwood, K.B. (1987) *Guide to Petroleum Reference Literature*, Littleton, CO: Libraries Unlimited Inc. An exhaustive survey of the literature of petroleum at the time of its publication. Coverage included guides to the literature, bibliographies, indexing, and abstracting services, dictionaries, encyclopedias and yearbooks, handbooks, manuals, and basic texts, directories, statistical sources, databases (including bibliographic, data and full text), periodicals, professional and trade associations, and publishers.

Stark, M. (ed.) (1988) *Bibliography of Petroleum Information Resources*, Washington, DC: Petroleum and Energy Resources Division, Special Libraries Association. This bibliography was created by at least eight members of the Petroleum and Energy Resources Division of SLA. Coverage included handbooks, databases, indexes and abstracts, directories, journals, maps and atlases, standards, legislation and regulations, journals, and theses and dissertations.

Dictionaries

Association of Desk and Derrick Clubs (comp.) (2001) *D&D Standard Oil and Gas Abbreviator* (5th edn), Tulsa, OK: PennWell. An "indispensable tool" in the oil and gas industry, useful in deciphering over 11,500 abbreviations and acronyms. The dictionary is divided into the following sections: abbreviations with definitions, definitions with abbreviations, abbreviations for logging tools and services, federal environmental acronyms, pipe coating terminology and definitions, mnemonics, abbreviations for companies, associations and organizations, and miscellaneous information and symbols. The fifth edition includes a searchable CD-ROM with the full text of the book, with an additional chapter on universal conversion factors, and a Michigan stratigraphic chart.

Dictionary for the Petroleum Industry (2001) (3rd edn, rev.), Austin, TX: Petroleum Extension Service, Continuing and Extended Education, University of Texas at Austin. Includes over 8,500 definitions of terms used in all aspects of the industry including petroleum geology, exploration, drilling, production, pipelines, refining and processing, and management. Includes contact information for industry associations and government agencies.

Hyne, N.J. (1991) *Dictionary of Petroleum Exploration, Drilling and Production*, Tulsa, OK: PennWell. Comprehensive dictionary of terminology used in upstream petroleum. Features include graphs, charts, diagrams, and photos. Includes an appendix with addition information such as: drilling and completion records, detailed diagrams of a rotary drilling rig, cable tool drilling rig, and a crank counterbalanced beam pumping unit, geological time scale, giant oil and gas fields, geological features, drill stem test symbols, flow sheet symbols, and U.S. land subdivisions.

Langenkamp, R.D. (1994) *Handbook of Oil Industry Terms and Phrases* (5th edn), Tulsa, OK: PennWell. Over 4,200 entries, with 1,000 new entries since the fourth edition. It appears that Langenkamp has coordinated the publication of this title with his related work, *The Illustrated Petroleum Reference Dictionary*: entries in both current, 1994 editions are identical, with the latter publication featuring over 200 illustrations.

Langenkamp, R.D. (1994) *Illustrated Petroleum Reference Dictionary* (4th edn), Tulsa, OK: PennWell. Over 4,200 entries, including 1,280 new entries since the third edition, along with over 200 illustrations. Covers both technical and slang terminology. Included, as "adjuncts" to the dictionary, are two other complete works: *The D&D Standard Oil Abbreviator*, and *Universal Conversion Factors*, compiled and edited by Stephen Gerolde. The contents of the dictionary appear to be identical to the fifth edition of Langenkamp's *Handbook of Oil Industry Terms and Phrases*, the difference being the inclusion of illustrations.

Miles, J.A. (1989) *Illustrated Glossary of Petroleum Geochemistry*, Oxford: Clarendon Press.

Dictionaries – multilingual

Kedrinksy, V.V. (2001) *English–Russian Dictionary of Petroleum Chemistry and Processing* Кедринский В. В. Англо-русский словарь по химии и переработке , нефти: Словарь содержит около 60000 терминов Moscow: Russo.

Moureau, M. and Brace, G. (1993) *Comprehensive Dictionary of Petroleum Science and Technology: English–French, French–English* (Dictionnaire Des Sciences et Techniques du Pétrole: Anglais–Français, Français–Anglais). Paris: Éditions Technip.

Proubasta, M-D. (2006) *Glossary of the Petroleum Industry: English–Spanish and Spanish–English* (Glosario de la Industria Petrolera) (4th edn), Tulsa, OK: PennWell. Includes more than 20,000 technical terms, covering the oil and gas industry as well as related fields. The fourth, revised edition of this work is scheduled for publication in early 2005.

Directories

Directories provide immediate access to information about the petroleum industry, including people, companies, products, services, manufacturers and suppliers of components, and cover both the upstream and downstream components of the petroleum industry. The advent of online access has helped alleviate the problem of information in print going out of date quickly, especially in directories that are published annually. Directories are also included in many of the sites listed in the section on Search engines and portals.

Alberta Oil and Gas Directory (annual) Edmonton, AB: Armadale Publications. Includes section called CANADA-Z Oil Gas Mining. Available online at <http://www.global-serve.net/> (accessed May 18, 2005). This publication doubles as a directory for the Alberta and Canada-wide petroleum industry. Most of the Canadian petroleum and natural gas resources are found in Alberta, which is why the guide focuses on Alberta, and then the rest of Canada.

Annuaire européen de pétrole – European Oil and Gas Yearbook. Hamburg: Urban-Verlag. Began publishing in 1973, ceased with the 2002 edition.

Arab Oil and Gas Directory. Arab Petroleum Research Center. Available in print and electronic versions <http://www.arab-oil-gas.com/indexns.htm> (accessed May

18, 2005). Covers all aspects of the oil, gas, and petrochemical industries in Algeria, Bahrain, Egypt, Iran, Iraq, Jordan, Kuwait, Lebanon, Libya, Morocco, Oman, Qatar, Saudi Arabia, Sudan, Syria, Tunisia, United Arab Emirates, Yemen, as well as OAPEC and OPEC. Provides surveys, information, data, maps, statistics, addresses, information on projects, and activities of foreign oil companies operating in the Middle East and North Africa and financial and business analyses.

Canadian Oil Register (annual) Calgary, AB: Nickle's Energy Group. Available online at <http://www.canadianoilregister.com> (accessed May 18, 2005), requires subscription. Also known as *Nickle's Canadian Oil Register*. Provides listings of more than 3,200 companies related to the Canadian energy industry, including oil and gas producers, explorer and developers, service and supply companies, consultants, engineers, pipeline contractors, designers, construction and fabricators, geophysical data brokers and contractors, drilling contractors, pipeline companies, transportation, and oil field construction companies. Includes an Internet index of websites.

Canadian Oilfield Service and Supply Directory (annual) Edmonton, AB: JuneWarren Publishing <http://www.cossd.com/> (accessed May 18, 2005). Called "The Buyer's Guide for the Canadian Oil and Gas Industry," this annual publication is divided into four sections: White – all companies alphabetically and by location; Blue – oil company producers, developers and explorers; Green – over 160 environmental categories; and Yellow – over 1,820 specialized oil field categories. Maps are also included.

Composite Catalog of Oilfield Equipment and Supplies. Houston, TX: Gulf Publishing. Available in print, on CD and DVD, and searchable online at <http://www.worldoil.com/CCatalog/CCatalog_Main.asp.> (accessed May 18, 2005). Access by products/services, detailed product information from leading manufacturers and suppliers, and a directory with complete contact information for each supplier listed in the catalog.

Global Oil and Gas Directory 2002. (2001) London: FT Business. Includes international oil and gas production and exploration companies, international service and supply companies, buyers' guide by service provided, who's who, financial service companies, and international oil and gas associations. FT Business merged with Platt's (McGraw-Hill) in 2001, and a newer edition has not been published.

Gulf Coast Oil Directory (2004) Houston, TX: Atlantic Communications <http://www.oilonline.com/atcom/>. A directory of the downstream petroleum industry in the U.S. Gulf Coast region. For details, see the *Houston/Texas Oil Directory* below.

Hart's Gulf States Petroleum Directory. Houston, TX: Hart Energy Publishing, LP. A directory of offshore operations in the Gulf of Mexico, and onshore activity stretching from South Texas to Florida. Industry sectors covered include exploration, production, land, pipelining, refining, gas processing, drilling, well servicing, information technology, and offshore equipment and supplies. Buyers' guide is included.

Hart's Rocky Mountain Petroleum Directory. Houston, TX: Hart Energy Publishing, LP. A directory of offshore operations in the Rocky Mountain states from New Mexico to Canada, and in Kansas, Nebraska, and Oklahoma. Coverage identical to *Hart's Gulf States Petroleum Directory*, except, of course, for offshore.

Houston/Texas Oil Directory (2004) Houston, TX: Atlantic Communications <http://www.oilonline.com/atcom/> (accessed May 18, 2005). A directory of the petroleum industry for Texas and Oklahoma. Subject coverage is identical to the *Gulf Coast Oil Directory*, and includes: blowout and firefighting, drilling contractors, geologists, geophysical and seismographic services, petroleum and consulting engineers, pipeline contractors and operators, well testing, completion and wireline services, and a buyer's guide to oil field products.

Industry Canada. *Oil and Gas Industry Company Directories* <http://strategis.ic.gc.ca/>. Links to Canadian pipeline, and onshore and offshore oil and gas company directories.

Offshore Oil and Gas Directory. Kent: CMP Information Services <http://www.cmpdata.co.uk/oilgas/> (accessed May 18, 2005). Lists over 7,000 global companies in 12,000 locations, with 3,200+ classifications. Divided into sections covering exploration and production, drilling, offshore engineering, fabrication and project management, and manufacturers, suppliers, and service contracting.

Oil and Gas Directory. Houston, TX: The Geophysical Directory <http://informationservices.com/oil_and_gas_directory/index.htm> (accessed May 18, 2005). In about 2000, PennWell ceased publishing its numerous annual individual oil and gas directories. These included: Canadian, global, Gulf Coast, natural gas petroleum, U.S.A., worldwide offshore petroleum, worldwide pipeline and contractors, and worldwide refining and gas processing. *The Oil and Gas Directory* may be the one publication able to fill the void left by the collective absence of these individual directories. The directory provides regional (i.e., U.S.A.) and worldwide coverage of exploration, well drilling and production. Geographic areas beyond the U.S.A. include Canada, Latin America, Europe, Africa, the Middle East, and Asia-Pacific. The thirty-fourth edition (2004) included listings for 15,576 companies: 10,728 U.S., 1,400 Canadian, 3,448 outside of North America. Thirty supply and service sections are listed, and there is a separate section for oil and gas companies. Regional indexes are also included.

PC Directory (2001–) Tulsa, OK: Midwest Publishing Company <http://www.midwestpub.com/index.html> (accessed May 18, 2005). The Midwest Publishing Company of Tulsa, Oklahoma, has produced directories for the energy industries since 1943. In 2001, it ceased publishing the directories in print, and began offering a software package called PC Directory. Described as "a custom, self-contained software that resides directly on your desktop or laptop computer," the software is delivered on CD-ROM, loaded on to a PC and connects to the Internet. Updates to the product listings are provided when required. When not connected to the Internet, the PC Directory continues to function as an online reference tool. Directories available include pipeline transmission and gathering, refining and gas processing, petrochemical, drilling and well servicing, offshore and international exploration and production, and the North American exploration and production industry.

Petroleum Equipment Directory (annual) Tulsa: Petroleum Equipment Institute <http://www.pei.org/search> (accessed May 18, 2005). Provides listings of Canadian and U.S. Institute members, with contact and product information, covering the petroleum marketing and liquid handling equipment industry. Available in print and CD-ROM, and searchable online at the above URL.

Petroleum Supply Americas (biannual) Lakewood, NJ: OPIS/STALSBY <http://www.opisnet.com/directories/psa.asp> (accessed May 18, 2005). Provides coverage of the downstream petroleum industry for North, Central, and South America. Divided into sections by company, personnel, products handled, and country-state-city. Types of industries listed include banking/financial, consultants, crude oil, ethanol production facilities, feedstocks/intermediates, natural gas liquids, oil refiners, petrochemicals, and transportation/storage.

Petroleum Supply Europe (biannual) Lakewood, NJ: OPIS/STALSBY <http://www.opisnet.com/directories/pse.asp> (accessed May 18, 2005). Provides coverage of the downstream petroleum industry for Europe, Asia, Africa, Australia, and the Middle East. Layout similar to previous title.

Petroleum Terminal Encyclopedia (annual) Lakewood, NJ: OPIS Directories <http://www.opisnet.com/directories/pte.asp> (accessed May 18, 2005). U.S. and international coverage, with indexes by state/country, company, and type of terminal. Information for each terminal may include all of the following: address, physical location, corporate headquarters, terminal type, methods used for supply and outloading, outloading features, pipelines used by terminal, high and low water depths and berth length, total storage capacity, and individual storage capacity by product.

PetroProcess HSE Directory. Houston, TX: Atlantic Communications <http://www.oilonline.com/atcom/> (accessed May 18, 2005). A guide to the downstream petroleum industry for the United States of America. The focus of the directory is processing plants and refineries, and includes companies that supply products or services to these plants. Health, safety, and environment products and services are included.

Pipeline Intelligence Directory. Sugarland, TX: Pipeline Intelligence Company <http://www.pipelineintelligence.com/> (accessed May 18, 2005). Provides coverage of the pipeline industry by companies, contractors and subcontractors, and suppliers. Each section of the print directory is also available on diskette.

Subsea Oil and Gas Directory. Subsea.Org <http://www.subsea.org/> (accessed May 18, 2005). Online directory and industry catalog for the subsea oil and gas industry. Also provides industry news, links and conference information, as well as an overview of the subsea industry.

Who's Who in Natural Gas and Power. Lakewood, NJ: OPIS/STALSBY. Similar in layout to other OPIS directories, this publication covers both the natural gas and electric power and utility companies. Detailed listings cover all segments of the natural gas industry, including consulting, equipment, processor, producer, training and education, and transportation.

Encyclopedias

There is no "traditional" encyclopedia devoted specifically to petroleum engineering and refining. As Pearson and Ellwood note, the *International Petroleum Encyclopedia* functions as a yearbook, with a focus on statistics, data, and maps. Good coverage of processes related to downstream petroleum activities may be found in encyclopedias such as *Ullmann's* and *Kirk-Othmer* (covered in Chapter 7 (Chemical engineering) in this book). Given the extensive coverage of the upstream and downstream petroleum industry provided throughout many handbooks and manuals as well as the aforementioned encyclopedias, perhaps the industry does not see the need for a full-blown encyclopedia covering the subject area.

Busby, R.L. (ed.) (2004) *International Petroleum Encyclopedia* Tulsa, OK: PennWell. Published annually by PennWell. Provides world coverage by geographic regions of the year's activities in petroleum and natural gas. Figures for each country include refining capacity, oil production, oil reserves, and gas reserves. Key statistics are provided, and statistical tables cover oil production and consumption, oil refining, natural gas, and petroleum prices. Historical figures are included. Maps are also provided, showing locations of oil fields, oil sands, gas fields, crude, natural gas and products pipelines, refineries, and tanker terminals. This is an important resource for historical and current data, and for the latest information on developments in the petroleum industry.

Desbrandes, R. (1985) *Encyclopedia of Well Logging*, Houston, TX: Gulf Publishing. Translation of the French edition. Functioning as a manual and an encyclopedia, this title covers well-logging measurements from drilling to production. Well-logging techniques covered include electric and dielectric, nuclear, acoustic and wireline. Formation evaluation is discussed, and well-logging in specific environments such as relief wells, shallow resources, geothermal, and oil muds are covered. Although twenty years old as of writing, still a valuable resource on well-logging, and the only one of its kind at this time.

Government resources

When searching for government resources on petroleum and energy, the recommendation is to begin with the Home page or website of the jurisdiction in question, and search for the relevant department, such as one dealing with energy, natural resources, petroleum, natural gas, and the like. For example, if you were looking for federal government information in North America on petroleum, oil and/or natural gas, your search could begin with the following federal departments:

Geological Survey Canada: Natural Resources Canada <http://www.nrcan.gc.ca/inter/index.html> (accessed May 18, 2005). Includes links to the Petroleum Resources Branch, made up of the Oil Division, Natural Gas Division, and Frontier Lands Management Division.

United States: Department of Energy <http://www.energy.gov> (accessed May 18, 2005). Includes links to The Office of Fossil Energy, Strategic Center for Natural Gas and Oil, Office of Natural Gas, Office of Petroleum, and the National Energy Technology Laboratory. Statistical information is covered elsewhere in this chapter.

Individual states and provinces will have equivalent departments or ministries. In Texas, the Bureau of Economic Geology at the University of Texas, Austin, began operations in 1909, having succeeded the Texas Geological Survey and the Texas Mineralogical Survey. In addition to functioning as a research unit at the University of Texas, Austin, it is also the State Geological Survey and the Regional Lead Organization for the Petroleum Technology Transfer Council <http://www.beg.utexas.edu/> (accessed May 18, 2005).

Alberta is the major oil-producing province in Canada, and information relating to its petroleum-based activities is available from its Department of Energy, as well as from the Alberta Energy and Utilities Board, an independent, quasi-judicial agency of the government of Alberta. Among its responsibilities is the regulation of the development of energy resources in Alberta, including oil, natural gas, oil sands, coal, and electrical energy, and the pipelines and transmission lines to move the resources to market. Additional information available at their websites <http://www.energy.gov.ab.ca and http://www.eub.gov.ab.ca/> (accessed May 18, 2005).

Government sites often include valuable statistical and pricing data. See the Statistics and data section (below) for more information.

Handbooks and manuals

Handbooks and manuals are valuable to the petroleum and refining engineer, supplying figures, data, properties, design and testing criteria, formulas and calculations, refining processes, well-log interpretation criteria, and other much-needed quick reference information.

Ahmed, T.H. (2001) *Reservoir Engineering Handbook* (2nd edn), Boston, MA: Gulf Professional Publishers. This book "explains the fundamentals of reservoir engineering and their practical applications in conducting a comprehensive field study." Divided into fifteen chapters, coverage includes reservoir fluid behavior and properties, fundamentals of reservoir fluid flow, oil and gas well performance, and oil recovery mechanisms, and methods for the prediction of oil reservoir performance. Also available online via Knovel.

American Society of Mechanical Engineers. Shale Shaker Committee (2005) *Drilling Fluids Processing Handbook*, Amsterdam: Elsevier. An updated version of an earlier work, *Shale Shakers and Drilling Fluids Systems*, published in 1999. The revised work expands on the earlier title to include many other aspects of drilling control, and was written by twenty-one experts in drilling and drilling fluids. It is divided into twenty detailed chapters, and includes an extensive glossary and index. Coverage includes drilling fluids, tank arrangement, shale shakers, mud cleaners, and other pieces of equipment used in the field. Each chapter is divided into subsections, which are in turn subdivided, providing for fast access to any part of the book.

Bradley, H.B. (ed.) (1992) *Petroleum Engineering Handbook*, Richardson, TX: Society of Petroleum Engineers. Third printing of one of the major handbooks of the field. It is divided into fifty-nine chapters, covering every aspect of petroleum operations from production engineering to reservoir management.

Chaudhry, A.U. (2003) *Gas Well Testing Handbook*, Amsterdam: Gulf Professional Publishing; Chaudhry, A.U. (2004) *Oil Well Testing Handbook*, Amsterdam: Gulf Professional Publishing. The preface of each of these two titles states that the purpose is "to provide a practical reference source for knowledge regarding state-of-the-art gas well testing technology" (former title) and "state-of-the-art oil well testing technology" (latter title). Both titles have similar formats, and virtually identical forewords and prefaces, with only a few words changed between them. The focus of *Gas Well Testing Handbook* is the "theory and practice of gas well testing, pressure transient analysis techniques, and analytical methods required to interpret well behavior in a given reservoir and evaluate well behavior in a given reservoir and evaluate reservoir quality, simulation efforts, and forecast producing capacity." The target audience includes graduate students, reservoir and simulation engineers, technologists, geophysicists and geologists, and managers. Its main emphasis is on practical field application, and includes over 100 examples from the field, used to illustrate basic analysis methods. *Oil Well Testing Handbook* is divided into sixteen chapters, beginning with the introduction, which covers the role of oil well tests and information in the petroleum industry, history, oil well test data acquisition, analysis and management, and other basic concepts. Many different methods and techniques for conducting well testing and analysis are covered in detail, each of which has been "field tested."

Each chapter in both books includes references and suggestions for additional reading, and both titles include extensive bibliographies. The two bibliographies feature most of the same references. and are not formatted consistently (i.e., some data are missing from some references). One concern worth noting: the bibliographies include only six references from the 1990s, the most recent from 1994. One reference cites a standard work from 1972, the ninth edition of the *Engineering Data Book*, which has since been revised three times, the twelfth edition having been published in 2004 in two versions: FPS and SI. However, a search on *Compendex*, run in May 2005 and covering the period from 1992 to 2005, returned 1,181 citations with the *Compendex* controlled vocabulary term (CV) "oil well testing," and seventy-seven citations with a combination of the CV terms, "natural gas wells" and "well testing." Perhaps the field has not changed substantially enough to warrant more recent references, which calls into question the books' "state-of-the-art" designation. Ultimately it will be up to the practicing professional engineer to confirm or deny this assertion.

Cholet, H. (ed.) (2000) *Well Production Practical Handbook*, Paris: Editions Technip. This is intended as a reference guide for oil field operators and petroleum engineers, but also provides solutions to practical problems. Includes guidelines, recommendations, formulas, and charts. Topics covered include: well technology, well productivity evaluation and control, stimulation, horizontal and multilateral wells, and production improvement. Information in the book is based on a wide variety of sources, including papers published by SPE.

Devereux, S. (1998) *Practical Well Planning and Drilling Manual*, Tulsa, OK: PennWell. This manual is divided into three sections: well design, well programming, and practical well site operations and reporting. The author notes three differences between this title and others of a similar nature. First, office aspects (design and programming) have been separated from the rig aspects (operations and control). Second, reference information, such as design data, formulations, or strength tables, is not included. Finally, emphasis is placed on the practical rather than the theoretical.

Fink, J.K. (2003) *Oil Field Chemicals*, Amsterdam: Gulf Professional Publishing. A comprehensive and detailed summary of chemicals used in the oil field. The compilation was prepared by critically examining over 20,000 references from the literature, primarily from petroleum abstracts and patent databases. Only materials that are publicly accessible have been included. The author notes that the information presented is not complete, and that developments from the past ten years have been screened for inclusion. The book is divided into twenty-two chapters, organized according to applications of parallel job processes, beginning with drilling, through to demulsifiers. Two indices, one chemical and one general, are provided.

Gabolde, G. and Nguyen, J.-P. (1999) *Drilling Data Handbook* (7th edn), Paris: Editions Technip. With its sturdy, tough soft binding and small size, this is designed for use in the field as well as the classroom. Provides data for drilling and production, and is divided into thirteen tabbed sections. Data are primarily in tabular form, with some graphs and diagrams. Topics covered include drill string standards, casing, tuning line pipe standards, drilling bits and downhole motors, drilling mud, cementing, direction drilling, and more.

Grace, R.D. (2003) *Blowout and Well Control Handbook*, Amsterdam: Gulf Professional Publishing. This handbook covers blowout containment and well control procedures, for the drilling or petroleum engineer. Coverage includes equipment in well control operations, classic pressure control procedures while drilling or tripping, special conditions, problems, and procedures in well control, fluid dynamics and special services in well control, relief well design and operation, and contingency planning. A case study of the 1985 blowout of the E.N. Ross No. 2 well, in Rankin Country, Mississippi, is presented. The book closes with a brief history and overview of the Al-Awda Project: The Oil Fires of Kuwait.

Lapeyrouse, N.J. (2002) *Formulas and Calculations for Drilling, Production, and Workover* (2nd edn), Amsterdam: Gulf Professional Publishing. The primary purpose of this book is to serve as a reference source to those who do not routinely use formulas and calculations. It is divided into five sections: basic formulas, basic calculations, drilling fluids, pressure control, and engineering calculations. Conveniently coil-bound to open flat on a desk or in the oil field.

Lyons, W.C., Guo, B., and Seidel, F.A. (2001) *Air and Gas Drilling Manual* (2nd edn), New York: McGraw-Hill. The second edition of this work has been written as an engineering practice book for both engineers and earth scientists working in air and gas drilling. Divided into three sections, the authors cover basic technology, air and gas drilling fundamentals, and deep well operations. Five appendices are included.

Lyons, W.C. and Plisga, G.J. (eds) (2005) *Standard Handbook of Petroleum and Natural Gas Engineering* (2nd edn), Burlington, MA; Oxford: Gulf Professional Publishing. A revised edition of the first edition, a two-volume set published in 1994. The 1994 edition began as a project to revise and rewrite the fifth edition of the *Practical Petroleum Engineer's Handbook*, published in 1970. When the authors (twenty-seven in total) realized that revisions would be inadequate, they chose to write a new handbook in the style of handbooks of other major engineering fields. As such, the initial chapters cover basic mathematics, general engineering and science, and auxiliary equipment. Specific petroleum engineering chapters follow, covering drilling and well completions, reservoir engineering, production engineering, and petroleum economics. The new, second edition features contributions from seventy-five authors, most of whom are practicing engineers in industry. Published information from the American Petroleum Institute and the Society of Petroleum Engineers was used in the preparation of this handbook.

Meyers, R.A. (ed.) (2004) *Handbook of Petroleum Refining Processes* (3rd edn), New York: McGraw-Hill. Detailed coverage of sixty-one petroleum refining processes, divided into fifteen technologies, such as catalytic cracking, hydrotreating, and visbreaking and coking. A valuable reference tool for engineering students working on term projects involving process design.

Meyers, R.A. (ed.) (2005) *Handbook of Petrochemicals Production Processes*, New York: McGraw-Hill. This book is mentioned in this chapter because it is a companion volume to *Handbook of Petroleum Refining Processes*, and shares a similar format. Provides detailed descriptions of fifty-three petrochemical production processes, divided into eighteen technologies. The petrochemicals described within represent the most economically important petrochemicals, including ethylbenzene, ethylene, phenols and acetone, propylene and light olefins, polyethylene, polypropylene, and polystyrene.

Mian, M.H. (1991) *Petroleum Engineering Handbook for the Practicing Engineer* (2 vols), Tulsa, OK: PennWell. Volume 1 covers engineering economics, basic rock and fluid properties, well-log interpretation, reservoir engineering and evaluation, and secondary oil recovery. Volume 2 covers principles of transient test analysis, transient testing of oil and gas wells, drilling technology, and production technology.

Muhlbauer, W.K. (2004) *Pipeline Risk Management Manual* (3rd edn), Oxford: Gulf Professional Publishing. The new, expanded third edition of this title now includes offshore pipelines and distribution systems and cross-country liquid and gas transmission lines. Divided into three sections, covering risk evaluation at a glance, customizing the basic risk assessment model, and risk management. Over fifty examples are included to help illustrate the concepts and models.

The Petroleum Handbook (1983) (6th edn), Amsterdam: Elsevier. Given that the fifth edition was published in 1966, it may not be surprising that the sixth edition, published in 1983, remains the most current available. The book is overdue for a revised edition, as certain sections of the 1983 edition, such as the overview of the world petroleum industry, are outdated. The handbook is of a technical nature, but written to be of use both to experts in the field and those outside of the petroleum industry who seek general information about the discipline. Coverage includes exploration and production, petroleum chemistry, manufacture and marketing of oil products, natural gas and gas liquids, oil supply and trading, petrochemicals, unconventional raw materials and synfuels, R&D, and environmental conservation.

Maps and atlases

Maps and atlases are important in petroleum engineering as reference tools, and can cover different topics such as well locations, pipelines, oil fields, gas fields, refineries, gas plants, oil sands deposits, sour gas, and tanker terminals. In addition to the selected sources listed below, petroleum-related maps and atlases are often available from the appropriate state or provincial agency or department responsible for energy or economic geology.

American Association of Petroleum Geologists (AAPG) <http://bookstore. aapg. org/source/orders/index.cfm> (accessed May 18, 2005). The AAPG publishes a number of maps and atlases, including some in digital/GIS format. One example is the Digital, GIS Atlas of O&G Fields, a four-module product covering the U.S. Gulf Coast (100 fields), east South America and western Africa (53 fields), and Southeast Asia and Oceania (32 fields).

GEOSCAN Database. Geological Survey of Canada <http://ess.nrcan.gc.ca/esic/ geoscan_e.php (accessed May 18, 2005). A bibliographic database of over 40,000 records covering GSC publications. Maps are included, and searches may be restricted to maps about petroleum-related features.

National Geologic Map Database. United States Geological Survey <http://ngmdb.usgs.gov/> (accessed May 18, 2005). Included in the database is a geoscience map catalog of over 67,000 maps, data, and related products from over 300 publishers, covering both U.S. state and territory areas. The advanced search function allows for limiting a search by resources, including petroleum, and coal.

Oilfield Publications Limited (OPL) <http://www.oilpubs.com/> (accessed May 18, 2005). OPL publishes atlases, maps, GIS mapping, and wall charts of interest to the international offshore oil and gas industry. OPL also publishes vessel registers, books, online data, databases, and CD-ROMs.

Oilweek <http://www.junewarren.com/PUBLICATIONS/OILWEEK/oilweek.html> (accessed May 18, 2005). Canadian oil and gas industry trade journal publishes maps and charts related to the industry, including oil sands and heavy oil, sour gas, gas plants.

PennWell MAPSearch <http://www.mapsearch.com/home.cfm> (accessed May 18, 2005). Provides maps, databases and GIS products on oil, gas, petroleum and electric power, covering the United States and Canada. MAPSearch products are published in hard copy or CD-ROM, or licensed as GIS or CAD format. The most detailed references available are five major US/Canada atlases on crude oil, LPG/NPL (liquefied petroleum gas/natural gas liquid), natural gas, refined products, and petrochemicals/olefins. Each atlas contains information on the pipelines and facilities for the commodity covered. State maps, and systems maps are also published. PennWell publishes the *Crude Oil Atlas*, *Natural Gas Atlas*, and *Refined Product Atlas*, each available on CD-ROM.

The Petroleum Economist <http://www.petroleum-economist.com/> (accessed May 18, 2005). Publishes large, detailed maps that are included with a subscription to the journal, and are suitable for wall mounting. Coverage is either world or by selected geographic region. In September 2003, the journal published the *World Energy Atlas 2004*, which includes maps of all major oil and gas fields, and pipelines, gas processing and storage facilities, deepwater fields, major oil refineries by capacity, liquid natural gas facilities, and tanker terminals, together with enhanced topographic information throughout (*World Energy Atlas 2004*. London: Petroleum Economist).

Monographs and textbooks

Where to begin, if one wanted to build a solid collection of monographs on All Things Petroleum? Rather than list a few dozen key titles to build or upgrade a petroleum engineering and refining collection, the significant publishers whose output is dedicated primarily to the petroleum industry are presented and discussed.

Editions Technip <http://www.editionstechnip.com/> (accessed May 18, 2005). Based in Paris, France, Editions Technip was founded in 1956 by Institut Français du Pétrole (IFP), and publishes on all aspects of the oil and gas industry, as well as related disciplines such as hydrocarbon chemistry. In the early years, Editions Technip authors came only from the IFP, but are now chosen in addition from among industry specialists, and French universities and research centers. Over 940 titles were available as of 2004, including original works in English, English translations of its top French titles, and international symposia. Editions

Technip publishes the journal *Oil and Gas Science and Technology, Revue de l'Institut Français du Pétrole.*

Gulf Publishing Company <http://www.gulfpub.com/> (accessed May 18, 2005). Gulf Publishing of Houston, Texas, publishes various series of technical books and manuals in a wide range of energy topics. Petroleum-related titles cover drilling, exploration, gas processing, offshore, petrochemicals, pipelines and flow, process engineering, petroleum engineering, production, and refining. Gulf also publishes two major trade journals: *World Oil* and *Hydrocarbon Processing.*

OCGI, Inc <http://www.ogci.com/> (accessed May 18, 2005). OGCI, Oil & Gas Consultants, Inc., is a training and consulting company, which publishes a number of important titles used in its courses, covering many basic aspects of petroleum engineering including: reservoir engineering, petroleum geology, decision analysis, production operations, risk and decision analysis, and structural styles in petroleum exploration.

PennWell <http://www.pennwell.com/> (accessed May 18, 2005). Based in Tulsa, Oklahoma, PennWell publishes a wide range of titles in business, well-logging, geology and geophysics, drilling, refining and processing, reservoir engineering, offshore, production, and pipelines and storage, as well as a series of subject-related maps, and atlases of the continental U.S. and Canada. Two important reference titles published annually by PennWell, and mentioned elsewhere in this chapter, are the *Oil and Gas Journal DataBook*, and the *International Petroleum Encyclopedia.* PennWell's set of eighteen "nontechnical" titles is a must for any library collection focusing on the petroleum industry. Examples in the series include: *Well Logging in Nontechnical Language*, *Petrochemicals in NonTechnical Language*, *Deepwater Petroleum Exploration and Production: A Nontechnical Guide*, and *Petroleum Production in Nontechnical Language.*

PETEX: The Petroleum Extension Service, Continuing & Extended Education, University of Texas at Austin <http://www.utexas.edu/cee/petex/> (accessed May 18, 2005). PETEX develops, produces, and distributes publications, computer-based training, and audio-visual materials designed for field-level personnel in the petroleum industry. Coverage includes drilling, personnel and rig management, offshore technology, well servicing, pipelines, and production.

Society of Petroleum Engineers <http://www.spe.org/> (accessed May 18, 2005). The Society of Petroleum Engineers publishes titles in four series: monograph, textbook, reprint, and conference proceedings. Topics include completions, drilling, economics, enhanced oil recovery, health, safety and environment, formation evaluation, management, production and facilities, reservoir, as well as a set of reference titles.

Other publishers issuing selected titles in petroleum engineering, refining, processing, and related topics include CRC Press, Marcel Dekker, Butterworth-Heinemann, American Association of Petroleum Geologists, American Institute of Chemical Engineers, Oilfield Publications Limited, Professional Engineering Publishing, Elsevier Science, McGraw-Hill, Oxford University Press, Blackwell, A.A. Balkema, Academic Press, and the International Association of Drilling Contractors.

Monographic series consist of individual titles sharing a broader subject theme or topic, and number in the hundreds of volumes. The following monographic series include titles that would be of interest to the petroleum engineering and refining professional. It should be noted that titles within chemical engineering monographic series about petroleum refining and processing will also be of interest.

AAPG Memoir. Tulsa, OK: American Association of Petroleum Geologists (0271–8510).
AAPG Studies in Petroleum Geology. Tulsa, OK: American Association of Petroleum Geologists (0271–8510).
Developments in Petroleum Science. Amsterdam: Elsevier (0376–7361).

Journals and magazines

A scholarly journal will contain articles of interest written by and for experts and researchers in a specific field or discipline. An article in a scholarly journal reports in-depth research, and includes a detailed literature review of the topic in question to that point in time. Each article is peer-reviewed, meaning it has been examined by one or more subject experts in the field before it is accepted for publication. The peer-review process is anonymous: the editor of the journal, before accepting the article, submits it to one or more scholars, who vet the manuscript, checking it for accuracy, validity, and so on. The scholars examining the paper are not provided with the author(s)' identity, nor is the author(s) made aware of the names of the reviewers.

The following peer-reviewed titles are recommended for collections in petroleum engineering, refining, and processing. The list includes journals that appear in the 2003 ISI Journal Citation Reports, under the heading, "Engineering, Petroleum," as well as in *Ulrich's Periodicals Directory*, under the heading "Petroleum and Gas." Titles selected from *Ulrich's* are those for which indexing and abstracting is provided by at least one A&I service. The journals in the list publish some or all articles in English.

Scholarly journals

AAPG Bulletin (1917–) Tulsa, OK: American Association of Petroleum Geologists (0149–1423).
Advances in Petroleum Geochemistry (1984–) London: Academic Press (0739–8352).
Applied Energy (1975–) Oxford UK: Pergamon (0306–2619).
Bulletin of Canadian Petroleum Geology (1953–) Calgary, AB: Canadian Society of Petroleum Geologists (0007–4802).
Chemical and Petroleum Engineering (1965–) New York: Kluwer Academic/Plenum (0009–2355).
Chemistry and Technology of Fuels and Oils (1965–) New York: Kluwer Academic/Plenum (0009–3092).
CIM Bulletin (1898–) Montreal, PQ: Canadian Institute of Mining, Metallurgy, and Petroleum (0317–0926).
Energy Sources (1973–) Taylor & Francis (0090–8312).
Fuel (1922–) New York: Elsevier (0016–2361).
Fuel Processing Technology (1978–) New York: Elsevier (0378–3820).

Institute of Energy. Journal (1927–) London: Institute of Energy (0144–2600).
International Gas Engineering and Management (1961–) Leicestershire: Institution of Gas Engineers (1465–7058).
International Journal of Offshore and Polar Engineering (1991–) Golden, CO: International Society of Offshore and Polar Engineers (1053–5381).
Journal of Petroleum Geology (1978–) Beaconsfield, Bucks: Scientific Press (0141–6421).
Journal of Petroleum Science and Engineering (1987–) New York: Elsevier (0920–4105).
Journal of Synthetic Lubrication: Research, Development and Application of Synthetic Lubricants and Functional Fluids (1984–) UK: Leaf Coppin (0265–6582).
Journal of the Japan Petroleum Institute (Sekiyu Gakkaishi) (1958–) Japan Petroleum Institute (1346–8804).
Journal of Tribology (1880–) New York: American Society of Mechanical Engineers (0742–4787).
Marine and Petroleum Geology (1984–) New York: Elsevier (0264–8172).
Natural Resources Research (1992–) New York: Kluwer (1520–7439).
Oilfield Review (1950–) Chester UK: Schlumberger (0742–4787).
Petroleum Chemistry (Neftekhimiya) (1962–) Moscow, Russia: IAPC "Nauka/Interperiodica" (0965–5441).
Petrophysics (1962–) Houston, TX: Society of Petrophysicists and Well Log Analysts (1529–9074) (*Note:* Not listed as peer-reviewed in *Ulrich's*, but papers accepted for publication are refereed by three reviewers.).
SPE Drilling and Completion (1961–) Richardson, TX: Society of Petroleum Engineers (1064–6671).
SPE Reservoir Evaluation and Engineering (1998–) Richardson, TX: Society of Petroleum Engineers (1094–6470).
Spill Science and Technology Bulletin (1994–) Oxford: Pergamon Press (1353–2561).

Trade journals

Petroleum, oil, and gas trade journals and magazines provide subscribers and society members with the most current information, news, and developments in the field. Trade journals can provide weekly and monthly data, prices, and statistics, information about forthcoming conferences, the latest society or association news, and reviews of the newest products and services available to the industry. A search on *Ulrich's Periodicals Directory*, for titles under the heading, "Petroleum and Gas," and restricted to "Serial type: Trade," returned 400 titles. Many of the trade publications listed are state- or country-specific. What follows is a very selective list of trade journals for the petroleum industries.

AAPG Explorer (1979–) Tulsa, OK: American Association of Petroleum Geologists (0195–2986).
American Gas (1918–) Washington, DC: American Gas Association (1043–0652).
Drilling Contractor (1944–) Houston, TX: International Association of Drilling Contractors (0046–0702).
Energy World (1973–) London: Energy Institute (0307–7942).
GTI Journal (1976–) Des Plaines, IL: Gas Technology Institute (1539–6495).
Hart's E&P (1973–) Houston, TX: Hart Energy Publishing (1527–4063).

Hydrocarbon Processing (1922–) Houston, TX: Gulf Publishing (0887–0284).
Journal of Canadian Petroleum Technology (1962–) Montreal, PQ: Canadian Institute of Mining, Metallurgy and Petroleum (0021–9487).
JPT: Journal of Petroleum Technology (1949–) Richardson, TX: Society of Petroleum Engineers (0149–2136).
New Technology Magazine (1937–) Calgary, AB: Nickle's Energy Group (1480–2147).
Offshore (1954–) Tulsa, OK: PennWell (0030–0608).
Oil and Gas Inquirer (1989–) Edmonton, AB: JuneWarren Publishing (1204–4741).
Oil and Gas Journal (1902–) Tulsa, OK: PennWell (0030–1388).
Oil and Gas Science and Technology: Revue de l'Institut Francais du Petrole (1946–) Paris: Editions Technip (1294–4475).
Oilweek Magazine (1948–) Edmonton, AB: JuneWarren Publishing (1200–9059).
OPEC Bulletin (1967–) Vienna: Organization of the Petroleum Exporting Countries (0474–6279).
Pipeline and Gas Journal (1970–) Houston, TX: Oildom Publishing (0032–0188).
PipeLine and Gas Technology (2002–) Houston, TX: Hart Energy Publishing (1540–3688).
SPE Journal (1996–) Richardson, TX: Society of Petroleum Engineers (1086–055X)
SPE Production and Facilities (1961–) Richardson, TX: Society of Petroleum Engineers (1064–668X).
World Oil (1916–) Houston, TX: Gulf Publishing (0043–8790).
World Refining (1991–) Houston, TX: Hart Energy Group (1524–9840).

Patents

Patents are critical to all engineering disciplines. For many reasons, patent searching is important to the engineer. For example, 80 percent of patent data is not published elsewhere, and will contain copious amounts of detail, research data and results, and technical drawings of inventions. In the publication cycle, patents will precede conference papers, which will in turn precede publication in scholarly journal. Searching the patent literature can save the engineer time and money. When planning to develop new technology, a patent search may reveal that the technology already exists, thus avoiding duplication of effort. Examining the patent literature can lead to new ideas and challenges in research and development, allow for the tracking of the work of competitors, and even predict forthcoming hot areas of R&D.

Search engines and portals

This is a very selective set of websites which provide links to resources on petroleum engineering and refining, such as buyer's guides, product showcases, auctions, job openings, discussion forums, company listings, industry news, and more. It is by no means exhaustive.

Most, if not all, major petroleum industry players have websites, as do petroleum engineering departments at academic and technical institutions and colleges, and publishers of print and electronic resources, including monographs, encyclopedias, scholarly and trade journals, newsletters, and statistics. A number of the sites below include membership fee or subscription-only access to either part of, or the entire site. Finally, while browsing throughout each portal or directory listed below, users will, at some point, discover that some links have changed or disappeared – such is the life of a website.

Canadian Wellsite <http://www.canadian-wellsite.com/> (accessed May 18, 2005). Designed specifically for the Canadian oilpatch. Some sections require subscription access. Free access is available to the oil field directory, classifieds, events, software, and discussion group.

EEVL: The Internet Guide to Engineering, Mathematics and Computing <http://www.eevl.ac.uk/engineering/index.htm> (accessed May 18, 2005) (see general engineering searches within the Engineering section). Can be restricted to petroleum and offshore engineering, where the searcher will find links to over 250 industry- and association-related sites.

FCC Network <http://www.thefccnetwork.com/> (accessed May 18, 2005). An online network covering fluid catalytic cracking and related petroleum refining processes. Registration is free, and includes access to a monthly newsletter, technical papers and catalyst reports, tips and techniques for trouble-shooting, tools and strategies for FCC performance optimization, and submission of questions to an advisory panel.

Hydrocarbon Online <http://www.hydrocarbononline.com/> (accessed May 18, 2005). Provides links to services, suppliers, markets for buying and selling, career development, news and community for the hydrocarbon processing industry.

Oil and Gas International <http://www.oilandgasinternational.com/> (accessed May 18, 2005). Fee-based site covering worldwide exploration and production news, information, and analysis.

Oil and Gas on the Internet. Cypress, TX: Competitive Analysis Technologies <http://catsites.com/> (accessed May 18, 2005). A subscription-based product, available in print, CD-ROM, and online editions. Updated daily online. December 2004 editions featured 5,600 sites (upstream) and 3,900 sites (downstream).

Oil and Gas Online <http://www.oilandgasonline.com> (accessed May 18, 2005). From VertMarkets, the company which produces Hydrocarbon Online, with the same features as that site. Coverage of the international upstream petroleum industry.

Oil.com <http://www.oil.com/> (accessed May 18, 2005). One of many sites produced by the WorldNews Network. Provides links to sites covering oil prices, news and industry sites, refineries, and related energy sites.

Oilfield Directory <http://www.oilfielddirectory.com/> (accessed May 18, 2005). Provides extensive coverage of the petroleum industry, including links to news feeds and oil prices, a global product and services directory, an equipment section, job bank, and discussion forum. Companies interested in more detailed listings of their services can pay an annual membership fee for additional access.

OilOnline <http://www.oilonline.com/> (accessed May 18, 2005). One of the oldest oil and gas portals on the Web, having been online since December 1, 1995. Links are divided into industry news, key indicators, careers, industry information, equipment and services, and an online store.

PETROassist.com <http://www.petroassist.com/index.asp> (accessed May 18,

2005). A "technical and commercial source for the petroleum industry," containing an oil and gas company directory, a directory of service and supply companies, training and technical resources, and industry news. The directories do not have a search capability. Focus of the site is on upstream activity.

Refining Online <http://www.refiningonline.com/> (accessed May 18, 2005). A web portal for the petroleum refining industry. Registration is free. Features include an interactive Q&A, a refining industry search engine, free calculation software, a technical knowledge base, events calendar, and membership directory.

RIGZONE <http://www.rigzone.com/> (accessed May 18, 2005). The "gateway to the oil and gas industry," RIGZONE focuses on the upstream oil and gas industry. The information available on the site is categorized into the following "zones": news and analysis, insight and expertise, oil and gas directory, forthcoming events, meetings, and conferences, a data center on offshore rig activity, career center, and equipment market. Members can subscribe to four newsletters. For a subscription fee, members can subscribe to the RIGZONE News Professional Edition. The industry directory is perhaps the best available online at the time of publication, providing detailed subject access by company, region, product, service, e-commerce, government and education, associations and societies, and news and information. RIGZONE is a well-designed, easy-to-use site.

WorldOil.com <http://www.worldoil.com/> (accessed May 18, 2005). Home page for the journal *World Oil*, but also includes the Composite Catalog of Oilfield Equipment and Services and an industry directory. The Info Center includes a marine rig directory, statistics and tables, forecasts and research, and an energy events calendar.

Associations, societies, and organizations

Below is a selective list of major associations, institutes, and societies relevant to petroleum engineering and refining. There are many more than can be listed here. For example, RIGZONE lists 108 societies and associations, and OilOnline lists over 125. All of the following websites were accessed on May 18, 2005.

American Association of Drilling Engineers <http://www.aade.org/>

American Association of Petroleum Geologists (AAPG), <http://www.ifp.fr/IFP/en/aa.htm>

American Gas Association <http://www.aga.org/>

American Institute of Chemical Engineers (AIChE) <http://www.aiche.org/>

American Petroleum Institute (API) <http://www.api.org>

American Society of Gas Engineers <http://www.asge-national.org/>

Canadian Association of Petroleum Producers (CAPP) <http://www.capp.ca/>

Canadian Institute of Mining, Metallurgy and Petroleum (CIM) <http://www.cim.org/>

Canadian Petroleum Products Institute.<http://www.cppi.ca/links.htm>

Energy Institute (UK) <http://www.energyinst.org.uk/>

Gas Technology Institute <http://www.gastechnology.org/>

Independent Petroleum Association of America <http://www.ipaa.org/>

Institut Français du Pétrole (IFP) <http://www.ifp.fr/IFP/en/aa.htm>

Instituto Mexicano del Petróleo (IMP) <http://www.imp.mx/>

International Energy Agency <http://www.iea.org/>
International Association of Drilling Contractor <http://www.iadc.org/>
International Association of Oil and Gas Producers <http://www.ogp.org.uk/index.html>
International Society of Offshore and Polar Engineers <http://www.isope.org/>
NACE International – The Corrosion Society <http://www.nace.org/nace/>
National Petrochemical and Refiners Association <http://www.npradc.org/>
National Petroleum Council <http://www.npc.org/>
North America Energy Standards Board <http://www.gisb.org/>
Offshore Engineering Society (UK) <http://www.oes.org.uk/>
Society of Petroleum Engineers (SPE) <http://www.spe.org/>
Society of Petrophysicists and Well Log Analysts <http://www.spwla.org/>

Standards

Standards are an essential part of an engineer's work. A standard establishes parameters for design, capacity, or property characteristics, which permit interchangeability of parts and materials. Engineering companies cannot compete locally or in the global marketplace without integrating standardization into every phase of their operations. thinkstandards.net is an example of a good website with basic information about standards, including definitions, a brief history, the benefits and ROI of standards, and a list of links to standards resources.

Standards of importance to the petroleum industry include those produced by the following:

American Petroleum Institute <http://www.api.org> (accessed May 18, 2005). The primary standards development and issuing body for petroleum engineering and refining. The API publishes hundreds of standards, recommended practices, specifications, codes and technical publications, reports and studies, for both the upstream and downstream areas. Mandatory for any petroleum engineering collection.

American Society for Testing and Materials. *ASTM Annual Book of Standards* <http://astm.org/> (accessed May 18, 2005). ASTM publishes the *Annual Book of Standards*, a seventy-volume set containing over 12,000 standards. Section 05 of the set is *Petroleum Products and Lubricants*, over 630 standards, including methods to measure properties of natural and liquefied petroleum, crude petroleum, and pure light hydrocarbons.

International Standards Organization (ISO) <http://www.iso.org/> (accessed May 18, 2005). ISO publishes approximately 120 standards dealing with various aspects of the petroleum and natural gas industry, including offshore structures, steel pipe for pipelines, pipeline transportation systems, subsea production systems, drilling and production equipment, drilling fluids, casing, and tubing.

Standards applicable to aspects of petroleum engineering, refining, processing, and testing are also issued by the following organizations and associations:

AFNOR: Association Française de Normalisation
AIChE: American Institute of Chemical Engineers
ASME: American Society of Petroleum Engineers

AWS: American Welding Society
BSI: British Standards Institute
GPA: Gas Processors Association
ISA: The Instrumentation, Systems and Automation Society
NFPA: National Fire Protection Association
TEMA: Tubular Exchange Manufacturers Association

Special mention is made of three important compilations of methods, procedures, and correlations involving petroleum and gas.

Engineering Data Book (2004) (12th edn), Tulsa, OK: Gas Processors Suppliers Association, Available in SI (International System of Units), and FPS (foot-pound-second) versions. Published in two loose-leaf binders to allow for updates. Useful for the field or plant engineers who are determining operating and design parameters. Also of use to design engineers as a general reference tool for accepted engineering practice in estimating, preliminary design, feasibility studies, and on-site operating decisions. Students studying engineering design in areas such as design of processing plants or refineries will find useful and practical information, data, and procedures within. Divided into twenty-six sections, covering equipment, storage, and processes.

Standard Methods for Analysis and Testing of Petroleum and Related Products and British Standards 2000 Parts (annual) London: Energy Institute. Published annually in two volumes. A compilation of test methods for the analysis and testing of petroleum and petroleum products, using traditional and modern instrumentation techniques. Joint methods with BSI, EN, ISO, and ASTM are included. The methods are used for quality control, and are important for national (U.K.) and international trading of petroleum and petroleum products.

Technical Data Book – Petroleum Refining (1997) Washington, DC: API Publishing Services. A loose-leaf, three-volume set providing physical and thermodynamic data and correlations needed by the petroleum refining industry for design of equipment and process evaluation. Divided into fifteen chapters, each chapter is devoted to a single property or group of related properties, such as hydrocarbon characterization, critical properties of pure hydrocarbons, defined mixtures, natural gases, and petroleum fractions, viscosity, combustion, and eleven others. Unfortunately (or not), the sixth edition was the last available in print. The databook has been replaced by the American Petroleum Institute Technical Database, an interactive software application that includes over 130 API standard methods in pdf format.

Statistics and data

Statistics and data are crucial to the petroleum industries. From hourly commodity prices to annual estimates of oil reserves, the need for this kind of information is ongoing and relentless. The American Petroleum Institute provides petroleum industry statistics via Access API, an online subscription service available to API members and non-members. Included in the service are weekly bulletins of refinery inputs, production, imports and inventories, monthly statistical reports, inventories of natural gas liquids and liquefied refinery gases, and imports and exports of crude oil and petroleum products.

The *Basic Petroleum Data Book*, mentioned below, is available online via Access API <http://www.api.org/axs-api/> (accessed May 18, 2005). Pricing for the individual reports is high, and designed for the single, corporate subscriber.

Alberta Energy and Utilities Board <http://www.eub.gov.ab.ca/BBS/products/publications/statseries/default.htm> (accessed May 18, 2005). Publishes a number of statistical series on Alberta's energy industries. Statistical coverage includes gas plants, mineable oil sands, active oil sands schemes, crude bitumen production, drilling activity, and reserves and supply demand outlook.

American Petroleum Institute (biannual) *Basic Petroleum Data Book*, Washington, DC: American Petroleum Institute. Covers primarily U.S. statistics, with some world data. Sections include energy, crude oil reserves, exploration and drilling, production, financial, prices, demand, refining, imports, exports, offshore, transportation, natural gas, OPEC, environmental, and miscellaneous. Historical data are included. The API also publishes the weekly *Statistical Bulletin* and the Monthly *Statistical Report*.

Beck, R.J. (2004) *Worldwide Petroleum Industry Outlook* (20th edn), Tulsa, OK: PennWell. Provides statistics and forecasts in a number of areas, including worldwide supply and demand, U.S. supply and demand, capital expenditures, exploration, drilling and production, refining and petrochemicals, transportation, natural gas, other energy sources, and OPEC. Beck offers commentary on current and historical issues for short- and long-term outlooks, and observations of the current political and economic conditions, and their impact over the next ten years.

BP Statistical Review of World Energy (annual) London: British Petroleum Co. Published annually in print and available online. Coverage includes oil and natural gas, and includes statistics for reserves, production, consumption, prices, stocks, refining and trade movements. The online version offers downloads in different formats such as PDF, Excel and Powerpoint, and includes an "energy charting tool," which can create charts and graphs, which in turn can be exported for further analysis.

Busby, R.L. (ed.) *International Petroleum Encyclopedia* (see Encyclopedias above).

Canadian Association of Petroleum Producers (annual) *Statistical Handbook for Canada's Upstream Petroleum Industry*. First published in 1955, and data updated annually. Provides historical summaries of the progress of the petroleum industry in Canada. Coverage includes land, exploration, and drilling, reserves, production, expenditures/revenue, prices, demand/consumption, refining, imports/exports. Sections covering transportation, energy, and world data have been discontinued. A CD-ROM version is included with the hard copy.

Oil and Energy Trends: Annual Statistical Review (1979–) Oxford: Blackwell. Gathers together the relevant data on the world oil industry, including oil and gas reserves, active and drilled wells, refinery capacity and production, oil and oil products demand and prices.

Oil and Gas Journal Data Book (annual) Tulsa, OK: PennWell. Content is reproduced from selected extracts of the previous year's volumes. Production and drilling reports, surveys, forecasts, rig counts, and construction updates are featured, and commentary and analysis included. A comprehensive index of the year's issues provides access by subject and author, and titles of each issue are listed.

Oil and Gas Journal Online Research Center <http://orc.pennnet.com/home.cfm> (accessed May 18, 2005). A fee-based service that provides reports, statistics and research information on energy, oil, and gas. Oil and gas data include prices, imports, exports, stocks (inventories), demand and consumption, and storage. The OCJ Energy Database has over 100,000 data series. The Oil and Gas Journal Online Research Center also publishes a series of electronic energy industry directories, covering various aspects of the oil and natural gas, and electric power industries. These include pipeline, refining and gas processing, petrochemical, liquid terminals, and gas utility. The directories are not available online, but are designed to be downloaded to one computer at a time. When connected to the Internet, users can update listings when made available.

OPEC. *Annual Statistical Bulletin* <http://www.opec.org/library> (accessed May 18, 2005). Contains tables, charts, and graphs covering the world's reserves of oil and gas, crude oil and product output, exports, refining, tankers, and other data. OPEC publications may be downloaded free of charge from the OPEC website. Print copies are available for a subscription charge.

Twentieth Century Petroleum Statistics (annual) Dallas: DeGolyer & MacNaughton. Covers primarily U.S. statistics, with some world data. Despite the title, it does include twenty-first-century data. U.S. statistics include crude reserves, production, imports by source, producing oil and gas wells, refining capacity, natural gas reserves, drilling costs, and much more.

United States. Department of Energy. Energy Information Administration (EIA) <http://www.eia.doe.gov/> (accessed May 18, 2005). Created in 1977 by the U.S. Congress, the Energy Information Administration provides extensive, detailed current and historical information, statistics and data by geographic area, fuel, sector, and price. Petroleum and natural gas are included. Some files are available in text, PDF, and Excel format, which can be exported for further use by the searcher. The EIA also publishes some of its statistics in data in print titles, such as the *Petroleum Supply Annual*, and the *Annual Energy Review*. Among its annual publications are ones covering the following: coal, energy outlook, energy review, international energy annual, international energy outlook, natural gas, and petroleum supply.

For reasons unknown, certain publications covering statistics relating to the petroleum industry are no longer published. These include:

Guide to Petroleum Statistical Sources (1997–1998) (10th edn), New York: API EnCompass. The Guide was divided into three sections: statistical databases, recurring statistical information in print, and related publications of interest. Use of the guide helped to simplify and expedite the search for a wide variety of petroleum industry statistical information.

Natural Gas Statistics Sourcebook (2001) (7th edn), Tulsa, OK: PennWell. No volume published since the seventh edition. Contained monthly, quarterly, and annual data for the important parameters of segments of the U.S. and worldwide natural gas industry, including reserves, exploration and drilling, production, imports and exports, prices, processing, demand, and consumption

Refining Statistics Sourcebook (1999) (6th edn), Tulsa, OK: Pennwell. No volume published since the sixth edition. Contained monthly and annual data for the important parameters of segments of the U.S. and worldwide petroleum refining industry, including petroleum product demand, capacity and inputs, refining production, imports and exports, crude oil and petroleum products stocks, prices, and transportation and petroleum movement.

Data for both sourcebooks were collected from the Oil and Gas Journal Energy Database. Current data are available from the Oil and Gas Journal Online Research Center.

Statistical Annual (2000) London: Petroleum Economist. Last known volume is 2000. Primary focus was on production statistics for oil, natural gas, coal, hydro power, and nuclear power. Data were presented in graphs and tables only.

Yearbooks and almanacs

Canadian Oilfield Service and Supply Directory – Forecast and Almanac (2004–) Edmonton, AB: JuneWarren Publishing (0835–1740). Provides a review of the history and a forecast of the future of the petroleum industry in Canada. Also features data, statistics and information on many different aspects of the industry, including conventional gas, heavy oil, oil sands, natural gas, by-products, pipelines, refining, revenue, personnel, and more.

International Oil and Gas Development Yearbook (annual) Austin, TX: International Oil Scouts Association. An annual two-volume set covering exploration and production. Coverage includes exploratory wells and dry holes, discoveries, new pays, extensions, wells drilled and producing geology, and annual and cumulative production of oil and gas fields. Available on CD in pdf form, since the 1997 editions.

References

American Petroleum Institute. *All About Petroleum* <http://api-ec.api.org/filelibrary/AllAboutPetroleum.pdf> (accessed November 1, 2004).

American Petroleum Institute (1961) *History of Petroleum Engineering*, New York: American Petroleum Institute.

Burdick, D.L. and Leffler, W.L. (2001) *Petrochemicals in NonTechnical Language* (3rd edn), Tulsa, OK: PennWell.

Cronin, F. (1955) "North America's Father of Oil," *Imperial Oil Review* (April): 15–20.

Giebelhaus, A.W. (1996) "The Emergence of the Discipline of Petroleum Engineering: An International Comparison," *ICON: Journal of the International Committee for the History of Technology* 2: 108–122.

Gray, F. (1995) *Petroleum Production in Nontechnical Language*, Tulsa, OK: PennWell Books.

IEEE <http://thinkstandards.net/index.html> (accessed November 12, 2004).

Johnston, D.E. and Pile, K.E. (2002) *Well Logging in Nontechnical Language* (2nd edn), Tulsa, OK: PennWell.

Leffler, W.L., Pattarozzi, R., and Sterling, G. (2003) *Deepwater Petroleum Exploration and Production: A Nontechnical Guide*, Tulsa, OK: PennWell.

Lord, C.R. (2000) *Guide to Information Sources in Engineering*, Englewood, CO: Libraries Unlimited.

Macleod, R.A. and Corlett, J. (eds) (2005) *Information Sources in Engineering* (4th edn), London: Bowker-Saur.

Meuckenheim, J.K. (ed.) (2005) *Gale Directory of Databases. Volume 1: Online Databases*, Detroit: Thompson Gale.

Morritt, H. (1993) *Rivers of Oil: The Founding of North America's Petroleum Industry*, Kingston, ON: Quarry Press.

Oil Museum of Canada http://collections.ic.gc.ca/blackgold/oil_museum/oilmuseum.html (accessed November 2, 2004).

OPEC Annual Statistical Bulletin (2003), p.10 <http://www.opec.org/Publications/AB/pdf/AB002003.pdf> (accessed November 1, 2004).

Owen, E.W. and Doth, R.H. (1975) *Trek of the Oilfinders: A History of Exploration for Petroleum*, Tulsa, OK: American Association of Petroleum Geologists.

PATEX. About Patents and Patent Search <http://www.patex.ca/about_patents.html> (accessed November 12, 2004)

Pearson, B.C. and Ellwood. K.B. (1987) *Guide to the Petroleum Reference Literature*, Littleton, CO: Libraries Unlimited.

Spitz, P.H. (1988) *Petrochemicals: The Rise of an Industry*, New York: Wiley.

20 Transportation engineering

Rita Evans

Introduction

Transportation engineering is a subdiscipline of civil engineering. Beginning in the mid-nineteenth century, industrialization and urbanization generated a need for better infrastructure and an improved transportation system. Waterway and canal development was followed by the growth of rail transportation. The advent of motorized vehicles and the tremendous demand for roads and related facilities made the need for engineers specializing in transportation more urgent.

The dynamics of traffic flow was one of the earliest areas of research in transportation engineering, with research activities underway by the 1950s. A catalyst for the emergence of transportation engineering was the establishment in the United States of the federal Department of Transportation in 1967, and the move to multi-modalism on the part of state departments of transportation.

Today, transportation engineering encompasses a wide range of activities. Transportation engineers design, construct, and maintain facilities. They are involved in traffic engineering and operations, and logistics. They may focus on a particular mode, such as highways, but intermodal operations are increasingly important. Transportation engineering does not exist in a vacuum, and issues related to city and regional planning overlap with the more technical aspects.

The literature of transportation engineering

The literature of transportation engineering has developed with the discipline. In 1922, the Advisory Board on Highway Research of the National Research Council began publishing the proceedings of its annual meeting in the *Bulletin of the National Research Council*. In 1963, the Highway Research Board began publishing *Highway Research Record* (renamed the *Transportation Research Record* in 1974). The American Society of Civil Engineers published the *Journal of the Highway Division* beginning in 1957; this became *Transportation Engineering Journal* in 1968 and *Journal of Transportation Engineering* in 1983. The American Association of State Highway Officials published works on the design of roads and highways in the mid-1950s, material used in the design of the Interstate Highway System.

The Highway Research Information Service (HRIS) began operations in 1967. This early effort by the Transportation Research Board (TRB) to provide access to transportation engineering information involved acquisition, abstracting and indexing, record storage, file storage, and batch and online retrieval. In the mid-1970s, *TRIS* (*Transportation Research Information Services*) was introduced; the Transportation Division of the Special Libraries Association worked closely with TRB to develop and implement this database.

Today, websites are an important source of information from government agencies, professional organizations, and commercial suppliers. Statistics, standards, patents, and other information that used to be difficult to identify and locate are now often just a few clicks away. Many journals are available in electronic editions, and the *TRIS* and *Transport* databases are accessed online. However, print resources remain important; as with any engineering discipline, handbooks, manuals, and textbooks are still critical information sources.

This chapter outlines some of the more important resources for transportation engineering. It is not comprehensive but does highlight a selected list of resources that encompass the range of types of materials and information.

Searching the library catalog: keywords, LC Subject Headings, LC call numbers

Searching the library catalog for information on transportation engineering frequently involves searching by mode, such as highway, rail, or marine transportation. Some key Library of Congress Subject Headings include:

Automobiles
Bridges
  Design and Construction
  Foundations and Piers
Container Ports
Electronic Traffic Controls
Elevated Highways
Express Highways
Ferries
Geographic Information Systems
Harbors
Hazardous Substances
  Transportation
High Speed Trains
Highway Capacity
Highway Communications

Highway Engineering
Highway–Railroad Grade Crossings
Intelligent Vehicle Highway Systems
Intelligent Transportation Systems
Local Transit
Locomotives
Logistics
Marine Accidents
Marine Engineering
Motor Vehicles
Naval Architecture
Pavements
  Design and Construction
Pavements, Asphalt
Pavements, Concrete
Pedestrians
Ports
Railroad Bridges
Railroad Engineering
Railroad Tunnels
Railroads
  Cars
  Safety Measures
  Signaling
  Track
Roads
  Design and Construction
  Maintenance and Repair
Street-Railroads
Streets
Subways
Traffic Accidents
Traffic Congestion
Traffic Engineering
Traffic Flows
Traffic Safety
Traffic Signs and Signals
Transportation
  Safety Measures
Transportation Engineering
Travel Time (Traffic Engineering)
Trucking
Trucks
Tunneling
Tunnels
  Design and Construction

LC terminology often lags behind new areas of research and is slow to adopt new terminology. "Street-railroads," for example, is used rather than the much more common term "light rail." Identifying the correct LC Subject Headings will result in more useful catalog search results.

The Library of Congress classification system places most transportation engineering-related information in the T schedule, although a few topics appear in the H schedule.

Among the important areas are:

HE331–380 – Traffic engineering. Roads and highways. Streets
TA501–625 – Surveying
TA800–820 – Tunneling. Tunnels
TA1001–1280 – Transportation engineering
TF1–1602 – Railroad engineering
TE1–450 – Highway engineering. Roads and pavements.
TG1–470 – Bridge engineering
TL1–484 – Motor Vehicles. Cycles
TP315–360 – Fuel
TP690–692.5 – Petroleum refining. Petroleum products
VM1–989 – Naval architecture. Shipbuilding. Marine engineering

In addition to LC Subject Headings, research in transportation engineering can be facilitated by the use of two specialized thesauri:

The ITS Thesaurus <http://www4.nationalacademies.org/trb/tris.nsf/web/its_thesaurus?OpenDocument>) (accessed May 18, 2005). This is a controlled, hierarchical vocabulary for indexing materials relating to intelligent transportation systems and is based on the *TRT*.

The Transportation Research Thesaurus (*TRT*) <http://www4.trb.org/trb/ tris.nsf/web/trt> (accessed May 18, 2005). This is designed for indexing and retrieval of transportation information. Covers all modes and aspects of transportation and users may access terms in alphabetical, hierarchical, or KWOC displays. *TRIS*, the primary bibliographic database for transportation, uses the *TRT*, which allows for much more focused subject searching than LC's Subject Headings.

Another catalog of interest to transportation researchers is *TLCat* <http://ntl.bts.gov/link.cfm> (accessed May 18, 2005), a union catalog of transportation holdings in more than twenty government and academic libraries created from catalog records in OCLC's *WorldCat* database; searches can be limited to government or university libraries. The National Transportation Library provides a guest view of the catalog for users who are not OCLC subscribers.

Indexes and abstracts

Journal literature, conference proceedings, and technical reports are primary sources of transportation engineering information, and there are several indexing and abstracting services that provide access to this literature. Particularly useful are the indexes listed below, but other indexes such as *Web of Science*, *Compendex*, *Inspec*, and *PsycInfo* cover aspects of transportation engineering and should be used as a complement to the subject-specific indexes.

The *Transport Database* is the most comprehensive source of bibliographic information in transportation. Transport is composed of *TRIS* and the publications part of the ITRD database. Available as either a web-based or CD-ROM subscription from Ovid, *Transport*'s user interface allows complex searches, and search output can be customized. More than 650,000 records from 1968 through to the present cover transportation systems, roads and highways, traffic, urban transportation, safety, intermodal transportation, and environmental effects. Sources include articles, technical reports, conference proceedings, software, books, and summaries of research in progress. Most records include abstracts.

TRIS (*Transportation Research Information Services*) is produced by the Transportation Research Board and contains more than 600,000 records on all modes and disciplines in the field of transportation. Records are indexed using the *Transportation Research Thesaurus* and most records have abstracts. It includes articles from more than 470 serial scholarly and trade publications and technical reports. *TRIS* is available free as TRIS Online <http://trisonline.bts.gov/> (accessed May 18, 2005). TRIS Online provides electronic links to the full text of many documents and to document suppliers. TRIS is available commercially from Dialog Information Services.

The *International Transport Research Documentation* (http://www.itrd.org/database.htm) (*ITRD*) database contains 350,000 citations to technical literature, current research projects, and software from twenty-three countries including most of the European countries, Australia, Latin America, Canada, China, and Japan. Subjects covered include design, construction and maintenance of roads, bridges and tunnels; traffic and transport; pavements; vehicles; and safety. Sources include more than 500 journals, books, reports, dissertations, patents, standards and specifications, and conference proceedings. Most records include abstracts. Produced by TRL Ltd. for the Organization for Economic Cooperation and Development (OECD), ITRD, formerly known as International Road Research Documentation, is available as a subscription from STN International.

The Transportation Research Board's *TRB Publications Index* <http://www4.trb.org/trb/onlinepubs.nsf/web/index> (accessed May 18, 2005) contains 26,000 records with abstracts for all TRB and Strategic Highway Research Program (SHRP) publications from the mid-1970s until the present. All NCHRP and TCRP publications, special reports, and records are included. Each individual paper in the *Transportation Research Records*, Conference proceedings, circulars, and *TR News* is indexed.

The Transportation Research Board's *Research in Progress* (RiP) <http://rip.trb.org/> (accessed May 18, 2005) database tracks current and recently completed government-funded transportation research projects in the United States and Canada. Updated monthly. Topics include highways, traffic, materials, construction, safety, and public transportation.

The *SAE Digital Library* <http://www.elecpubs.sae.org/> (accessed May 18, 2005) from the Society of Automotive Engineers provides access to thousands of SAE technical papers and standards. The database covers all aspects of engineering for ground vehicles and associated manufacturing technologies.

The *PATH Database* (*Partners for Advanced Transit and Highways*) <http://www4.trb.org/trb/tris.nsf/web/path> (accessed May 18, 2005) provides access to information on intelligent transportation systems. Includes monographs, journal articles, conference papers, technical reports, theses, websites, and selected media. Contains over 28,000 records; limited updating done after 2004. Provides full bibliographic information with abstracts and includes online links. Produced by the Harmer E. Davis Transportation Library, University of California, Berkeley.

Bibliographies and guides to the literature

Guides produced by the Transportation Division of the Special Libraries Association have been the most comprehensive sources available for transportation engineering. Many bibliographies have been produced for specific topics by librarians, academics, and government agencies; these are best identified through focused searches in library catalogs.

Compiled by members of the Transportation Division of the Special Libraries Association, the fifth edition (2001) of *Sources of Information in Transportation* <http://www.ntl.bts.gov/ref/biblio/> (accessed May 18, 2005) is the first online edition of this guide. Individual bibliographies cover general transportation, aviation, highways, inland waterways, intelligent transportation systems, maritime transportation, trucking, pipelines, and hazardous materials transportation. The fourth edition (1990) also covered railroads, urban transportation, and shipping. The bibliographies include basic references, statistical sources, directories, periodicals, conferences, indexes and abstracts; the fifth edition includes electronic resources.

The Institute of Transportation Engineers (Washington, DC) produced *Basic References for the Transportation Engineer* in 1991. This nine-page guide for practicing transportation professionals includes textbooks, manuals, and reports.

The U.S. Department of Transportation Library has produced several transportation bibliographies that may be found at <http://dotlibrary.dot.gov/bibliographies/intro.htm> (accessed May 18, 2005). Of particular interest are *Transportation and Modal Statistics* and *Transportation and Environment Bibliography.*

Databases, datasets

Most countries gather, analyze, and disseminate statistics regarding transportation of people and goods, and the data are often available electronically. Some organizations and commercial publishers are also sources for statistics.

The Bureau of Transportation Statistics (BTS) <http://www.bts.gov/> (accessed May 18, 2005) in the U.S. Department of Transportation was established in 1992 for data collection, analysis, and reporting and to ensure the most cost-effective use of transportation-monitoring resources. BTS produces *TranStats*, the intermodal transportation database; access is by mode, subject, or agency. BTS also produces annual statistical reports. Two of its more important publications are:

National Transportation Statistics <http://www.bts.gov/publications/national_transportation_statistics/> (accessed May 18, 2005) presents information on the U.S. transportation system, including its physical components, safety record, economic performance, energy use, and environmental impacts. Beginning with the 2003 edition, individual sections are updated as needed and may be downloaded individually; the entire report can no longer be downloaded and is no longer available in print.

Transportation Statistics Annual Report. U.S. Department of Transportation. Bureau of Transportation Statistics. Annual. 2005 edition 332pp. <http://www.bts.gov/publications/transportation_statistics_annual_report/> (accessed May 18, 2005) presents transportation statistics on travel, travel time, vehicles, travel behavior, costs, mass transit, safety, emissions and economics, with many charts and other graphical representations. Includes discussion of topics related to transportation indicators. Editions from 1994 available online.

The Energy Information Administration in the U.S. Department of Energy produces several annual series of data on domestic and international production and consumption; all of the following are available online from 1995:

Annual Energy Review, Washington, DC: U.S. Department of Energy. Energy Information Administration. Annual. 2003 edition <http://www.eia.doe.gov/emeu/aer/> (all accessed May 18, 2005). This annual is the primary source of historical U.S. energy statistics. Covers total energy production, consumption, and trade; and provides overviews of petroleum, natural gas, coal, electricity, nuclear energy, renewable energy, and international energy. Many series date back to 1949. For the most current data in these series, consult the *Monthly Energy Review* (http://www.eia.doe.gov/emeu/mer/contents.html). *Annual Energy Outlook 2004 with Projections to 2025*. U.S. Department of Energy. Energy Information Administration. Annual. 2005 edition 233pp. http://www.eia.doe.gov/oiaf/aeo/). This publication presents a mid-term forecast and analysis of U.S. energy supply, demand, and prices through 2025. *International Energy Annual* (http://www.eia.doe.gov/iea/contents.html) presents an overview of key international energy trends for production, consumption, imports, and exports of primary energy commodities in more than 220 countries. Includes prices for crude oil and petroleum products in selected countries and some renewable energy sources.

Also providing transportation-related energy statistics is the *Transportation Energy Data Book* <http://www-cta.ornl.gov/data/> (accessed May 18, 2005), produced by Oak Ridge National Laboratory for the U.S. Department of Energy, Office of Energy Efficiency and Renewable Energy. This annual compilation of statistics from a variety of sources includes data on petroleum and energy, vehicles, non-highway modes, and emissions. The current edition is only available online; individual chapters or the entire report may be downloaded in pdf format. Many data tables may be downloaded as Excel files. Print versions are available, but publication lags the posting of the electronic version.

The Office of Highway Policy Information <http://www.fhwa.dot.gov/policy/ohpi/> (accessed May 18, 2005) in the Federal Highway Administration, U.S. Department of Transportation, serves as a repository of highway information and statistics. Two particularly important sources are:

Highway Statistics series <http://www.fhwa.dot.gov/policy/ohpi/hss/index.htm> (accessed May 18, 2005). This consists of annual reports containing statistical data on motor fuel, motor vehicles, driver licensing, highway-user taxation, state and local government highway finance, highway mileage, and Federal aid for highways. Data are presented in tabular format and have been published each year since 1945. A data compilation, *Summary of Highway Statistics*, is published every ten years, with the last edition published in 1995. Electronic versions of *Highway Statistics* are available beginning with 1992. Data may be downloaded in html format, and most tables are also available for download as Excel or pdf files. The annual and summary volumes are available in paper or microfiche from NTIS. Ordering information for editions going back to 1969 is available online at <http://www.fhwa.dot.gov/policy/ohim/hs03/preface.htm> (accessed May 18, 2005).

National Household Travel Survey (NHTS) <http://www.fhwa.dot.gov/policy/ohpi/nhts/index.htm> (accessed May 18, 2005) provides data on personal travel behavior including the purpose of the trip, means of transportation, trip length, day of the week and month of the year, number of people on the trip, and other trip-making characteristics. Begun in 1969, the 2001 NHTS is the sixth survey in the data series. Data may be downloaded in SAS, Dbase, and ASCII formats. Some publications based on the survey are available in print from BTS.

The *Federal Transit Administration's National Transit Database* is a primary source for statistics on public transportation in the U.S. Its publications (all accessed May 18, 2005) (http://www.ntdprogram.com/NTD/ntdhome.nsf/Docs/NTDPublications?OpenDocument#) include the data tables for the National Transit Database (http://www.ntdprogram.com/NTD/NTDData.nsf/DataTableInformation?OpenForm&2001) which has detailed summaries of annual operating and financial data submitted to FTA by transit agencies. *The National Transit Summaries and Trends* (http://www.ntdprogram.com/NTD/NTST.nsf/Web/NTST?OpenDocument) presents aggregate transit operating statistics by mode, including bus, light rail, heavy rail, commuter rail, demand response, and vanpool. For information about individual agencies, consult *Transit Profiles* (http://www.ntdprogram.com/NTD/Profiles.nsf/ProfileInformation?OpenForm&2001&All), also known as Section 15. These contain service, financial, and modal data and performance indicators for more than 500 transit agencies operating at least ten vehicles. *Railroad Facts* from the Association of American Railroads provides a statistical history of Class I (eight largest freight railroads) railroad services in the United States in a pocket-sized format. Includes statistics on financial performance, traffic, operating averages and plant and equipment, and employment and wages for the industry. Figures are also given individually for the eight Class I railroads. Previous title: *Yearbook of Railroad Facts*.

European Transport Statistics <http://transtat.agderforskning.no/> (accessed May 18, 2005), sponsored by the European Commission, provides transportation statistics by mode for all countries in Europe. *Europa Transport* http://europa.eu.int/comm/transport/index_en.html, also from the EC, has links to information on transport sectors.

The International Road Federation <http://www.irfnet.org> (accessed May 18, 2005) publishes *World Road Statistics*. The 2004 edition has worldwide data for 1998 to 2002 on roads and vehicles, and presents data by country for road networks; traffic volumes; motor vehicle production, export/import and registration; fuels; safety; taxation and expenditures. Published in hard copy, on CD-ROM with Excel files, and electronically as pdf file.

The National Safety Council <http://secure.nsc.org/index.htm> (accessed May 18, 2005) compiles and publishes statistics and reports, including *Injury Facts*. *Injury Facts* compiles annual data on fatal and non-fatal unintentional injuries, including those on streets and highways. This highly regarded source has been published since 1927 and was previously known as *Accident Facts*. Available in hard copy and on CD-ROM.

Ward's Automotive Yearbook from Ward's Communications compiles extensive statistics on the global and U.S. automobile industries. Provides an overview of the industries in various parts of the world. U.S. statistics include detailed data on car and light truck platforms, dealers, vehicle production, sales, registrations, marketing and vehicle specifications. The annual *Ward's Motor Vehicles Facts and Figures* provides statistics on the U.S. automobile industry including production, sales, and registrations. Ward's compiles and provides online access to additional industry statistics through a paid subscription to WardsAuto.com <http://www.wardsauto.com/> (accessed May 18, 2005).

Directories

Print directories are no longer the essential tools they were just a few years ago, thanks to the ease with which this type of information can be found on the Web. Many professional societies and organizations have membership directories available online, as do most academic and research institutions and government agencies. A search engine such as Google can yield information on individuals.

Directories still perform a useful function, particularly those that compile information from disparate sources and those that point to vendors and equipment manufacturers; some noteworthy ones are:

ATA's Truck Fleet Directory: Profiles of For-Hire and Private Fleets in the U.S. (annual) Alexandria, VA: American Trucking Associations. One of the few sources of truck fleet information, this directory lists 56,000 fleets organized by state. Entries include contact information, fleet size, trailer types, and commodities hauled. The CD-ROM version allows searching.

Containerisation International Yearbook (annual) London: National Magazine Co. Ltd. Lists port facilities, terminals, and traffic statistics; contains manufacturers, leasing companies, repair companies, insurance companies, and ship-brokers.

Directory of Transportation Libraries and Information Centers <http://ntl.bts.gov/tldir/> (accessed May 18, 2005) (7th edn) (2002), compiled by Susan C. Dresley; prepared for the Transportation Division of the Special Libraries Association, Washington, DC: National Transportation Library, Bureau of Transportation Statistics, US Department of Transportation. Lists more than 130 libraries and information centers in the United States, Canada, and other parts of the world. Entries include contact information, collection size and scope, and a link to the library's website.

Jane's Information Group <http://www2.janes.com/public/trans_news_analysis.html> (accessed May 18, 2005) publishes a number of authoritative, annual directories for the transportation industry; all are available in print and electronic versions. Among these are the following:

High-speed Marine Transportation provides detailed specifications for all types of high-speed marine craft including component and propulsion systems. Includes a directory of manufacturers.

Marine Propulsion provides detailed specifications and images of transmissions, propellers, and related systems for all diesel and gas turbine engine and propulsion systems. Includes a directory of manufacturers.

Pipeline and Gas Journal Annual Directory of Pipelines and Equipment (annual) Houston, TX: Oildom Publishing Co. of Texas. Includes company information and industry statistics.

Select transportation associations, societies, and unions <http://ntl.bts.gov/associations.html> (accessed May 18, 2005). Compiled by the National Transportation Library. Alphabetical directory of more than 150 organizations with links to their websites.

Urban Transport Systems profiles transportation systems in more than 400 cities worldwide. Gives a brief description of the system and statistics on fleet size, passenger journeys, and route length. Other operational statistics may be provided; some entries include route maps. A directory of rail, bus, components, signaling, control, and other equipment manufacturers is included.

World Railways, edited by Ken Harris. Profiles more than 450 railway systems in over 120 countries. Entries describe each system's history and financial status and provide information on passenger, freight, and intermodal operations; fare collection and reservation systems; and traction and rolling-stock. Includes a directory of more than 1,500 manufacturers, suppliers, and service companies.

Encyclopedias and dictionaries

Encyclopedias focused on transportation engineering are rare. Listed below are select encyclopedias that focus on a specific aspect of the discipline.

Kratville, W.W. (ed.) (1997) *The Car and Locomotive Cyclopedia of American Practices* (6th edn), Omaha, NB: Simmons-Boardman Books. This compilation of railroad mechanical technical information on rolling-stock in use in North America covers freight cars, intermodal cars, passenger cars, locomotives, and high-speed train sets. It includes information on components, research and development, maintenance and repair, and technical training. Contains statistics, figures, photographs, diagrams, tables, and a dictionary.

Papageorgiou, M. (ed.) (1991) *Concise Encyclopedia of Traffic and Transportation Systems*, Oxford; New York: Pergamon Press. Although somewhat dated, this compilation of 118 articles on models, control methods, and practical aspects provides a useful starting point for information on road, rail, air, and maritime transportation systems. Most contributors are European, and there is a distinct international flavor to the entries. Many articles provide an historical context for the topic and address issues related to the interrelationship among transportation modes. Mathematical models and figures are used extensively, as are cross-references; each entry includes a bibliography.

Tufnell, R.M. (ed.); Westwood, J. (revised and updated) (2000) *The New Illustrated Encyclopedia of Railways*, Edison, NJ: Chartwell Books. This guide traces engineering developments in railway motive power and includes detailed descriptions of more than 1,000 engines manufactured throughout the world. Lavishly illustrated with hundreds of photographs, most in color, it covers steam tender, tank, articulated, electric, diesel, and gas turbine locomotives.

TDM Encyclopedia <http://www.vtpi.org/tdm/> (accessed May 18, 2005) Victoria Transport Policy Institute. Provides detailed information on dozens of travel demand management (TDM) strategies such as improved transportation options, incentives to use alternative modes, parking and land-use management, and program support. Includes chapters on planning and evaluation techniques.

Wikipedia <http://en.wikipedia.org/wiki/> (accessed May 18, 2005) is an online encyclopedia. Its many useful entries related to transportation have an international flavor. The "Highway" article, for example, provides an overview of highway types and nomenclature for nine countries and has links to information on highway history and types throughout the world.

Dictionaries are also somewhat rare. Manuals and handbooks often include glossaries specific to some part of transportation engineering. A few notable dictionaries include:

Cavinato, J.L. (2000) *Supply Chain and Transportation Dictionary* (4th edn), Norwell, MA: Kluwer Academic. Covers terms in supply chain management, transportation, distribution, logistics, material, and purchasing.

Dinkel, J. (2000) *Road and Track Illustrated Automotive Dictionary*, Cambridge, MA: Robert Bentley Publishers. Defines basic items as well as complex systems. Many illustrations.

Glossary of Transportation Terms/Glosario de Terminos Transportes (1994), U.S. Department of Transportation, Federal Highway Administration, Washington, DC: Transportation Research Board. English/Spanish version of the Urban Public Transportation Glossary published by the Transportation Research Board in 1989. Organized alphabetically and extensively cross-referenced.

Goodsell, D. (1995) *Dictionary of Automotive Engineering*, Warrendale, PA: Society of Automotive Engineers. Defines more than 3,000 terms used worldwide.

O'Leary, P. and Corey, J. (1996) *Transportation Expressions* <http://www.bts.gov/btsprod/expr/> (accessed May 18, 2005) U.S. Department of Transportation, Bureau of Transportation Statistics, Washington, DC: BTS. Compiles definitions of terms used in U.S. DOT and other government agencies. Entries include definitions and source citations. The online version can be browsed and searched.

Paasch, H. (1997) *Paasch's Illustrated Marine Dictionary*, New York: Lyons & Burford Press. Defines the main types of steam and sailing vessels; wooden and iron hulls; propulsion machinery; anchors and related equipment; masts and spars; standing and running rigging, as well as sails, tackle, blocks and ropes.

Railway Age's Comprehensive Railroad Dictionary (2002) (2nd edn) Omaha, NB: Simmons-Boardman Books. Defines more than 4,000 terms and includes 240 illustrations.

World Road Association/Association Mondiale de la Route. PIARC Technical

Dictionaries, various dates and pagination. PIARC publishes a number of multi-language dictionaries that cover concepts in highway transportation. Equivalent terms are shown for two or more languages; some terms are also defined. The seventh edition of the French/English dictionary was published in 1997; various editions of this title have been translated into as many as fifteen different languages.

Handbooks and manuals

Handbooks and manuals are essential components of the transportation engineering literature. Facts, formulas, and basic information for specific aspects of transportation engineering are compiled in a manner that facilitates quick access. Most handbooks and manuals are limited to one mode of transportation, and listed below are key handbooks and manuals grouped by mode.

General

Hall, R.W. (ed.) (2003) *Handbook of Transportation Science* (2nd edn), Boston, MA: Kluwer Academic. Focuses on the properties and characteristics common to all modes of transportation. Organized by subject, chapters written by different authors address several broad topics: the human element of transportation, including discrete choice, travel demand and vehicle operation; flows and congestion, including traffic control and system interactions; spatial models, used in network analysis and design and network assignment; and routing and network models. Extensive use of tables, figures and mathematical models; references at the end of each chapter. The index and table of contents lack detail.

Myer Kutz, M. (ed.) (2004) *Handbook of Transportation Engineering*, New York: McGraw-Hill. Provides design techniques, examples of applications, and guidelines. The thirty-eight chapters are organized in five parts: networks and systems; traffic, streets and highways; safety, noise and air quality; non-automobile transportation; and operations and economics. While multi-modal, the focus is on highway and traffic engineering and automobile transportation.

Highway and traffic engineering

American Association of State Highway and Transportation Officials (AASHTO) (1993) *AASHTO Guide for Design of Pavement Structures*, Volume 1, Washington, DC: AASHTO. Covers the range of design from construction of new pavements to rehabilitation and reconstruction of existing pavements and includes mechanistic-empirical design procedures. Provides material on overlay design and rehabilitation. Supplement to the *Guide for Design of Pavement Structures* (1998), includes alternative design procedures.

American Association of State Highway and Transportation Officials (1999) *AASHTO Maintenance Manual: The Maintenance and Management of Roadways and Bridges*, Washington, DC: AASHTO. Provides basic information regarding the processes, methods, and materials used in maintaining roadways and bridges.

American Association of State Highway and Transportation Officials (2000) *Hot Mix Asphalt Paving Handbook* (2nd edn), Washington, DC: Transportation

Research Board. Covers paving, plant operations, materials transportation, surface preparation, and quality control.

American Association of State Highway and Transportation Officials (2000) *Manual for Condition Evaluation of Bridges* (2nd edn) (2003) (revisions), Washington, DC: AASHTO. Describes how to determine the physical condition, maintenance needs, and load capacity of highway bridges in the U.S. Provides inspection procedures and load rating practices that meet the National Bridge Inspection Standards (NBIS). Covers bridge files (records), bridge management systems, field inspections, test methods, analysis, load rating methods, and field evaluation.

American Association of State Highway and Transportation Officials (2002) *Roadside Design Guide* (3rd. edn), Washington, DC: AASHTO. Compiles current information and operating practices focused on safety treatments to minimize injury severity when vehicles leave the roadway.

American Association of State Highway and Transportation Officials (2004) *A Policy on Geometric Design of Highways and Streets* (5th edn) Washington, DC: AASHTO. Known as the "Green Book," this is the standard guide for engineers designing the physical layout and dimensions of streets and roads. The first three chapters cover design controls and criteria, the elements of design, and cross-section elements. Each functional highway type (local roads and streets, collectors, arterials, and freeways) is addressed in separate chapters. The final two chapters deal with intersections and grade separations and interchanges. Uses both U.S. customary units and metric units throughout the text and figures. Available in print and on CD-ROM.

American Concrete Institute (2004) *ACI Manual of Concrete Practice*, Farmington Hills, MI: The Institute. Contains concrete and masonry codes, specifications, requirements, guides, and reports used in design of structures and pavements and for maintenance and rehabilitation. Available as six-volume set or on CD-ROM.

Brockenbrough, R.L. and Boedecker, Jr, K.J. (2003) *Highway Engineering Handbook* (2nd edn), London: McGraw-Hill. Aimed at practitioners in highway design and construction. Begins with environmental issues and the regulatory framework, and moves on to highway location and geometric and cross-section design. Pavement design incorporates AASHTO specifications and covers preventive maintenance, pavement management and rehabilitation. Bridge engineering covers bridge types, materials, deck design, and construction and corrosion and coatings. Some chapters include references.

Chen, K. and Miles, J.C. (eds) (1999) *ITS Handbook 2000: Recommendations from the World Road Association (PIARC)*, Boston, MA: Artech House. Reviews implementations of intelligent transportation systems around the world.

Chen, W-F. and Duane, L. (ed.) (2003) *Bridge Engineering*, Boca Raton, FL: CRC Press. Covers all major areas of bridge design, construction, maintenance and inspection. Discusses superstructures, substructures, and seismic design. Includes all types of bridges: concrete, steel, box, truss, suspension. Contains many tables and formulas; each chapter has references.

Concrete Reinforced Steel Institute (2002) *Design of Reinforced Concrete Structures* (9th edn), Schaumberg, IL: The Institute; also available on CD-ROM. Covers the design of reinforced concrete structural elements such as columns, beams, footings, pile caps, retaining walls, and floor systems.

Currin, T.R. (2001) *Introduction to Traffic Engineering: A Manual for Data Collection and Analysis*, Pacific Grove, CA: Brooks/Cole. Describes step-by-step techniques for observing and analyzing spot speeds, turning movements, saturation flow rates, control delay, parking, trip generation, and platoon ratio. Provides sample data collection and data analysis forms. An appendix includes additional information on statistical analysis.

Euler, G. (1999) "Intelligent Transportation Systems", in *Traffic Engineering Handbook* (5th edn), Washington, DC: Institute of Transportation Engineers. This book provides an excellent overview of ITS. Reviews history, describes ITS concepts, and discusses major considerations in technology implementation.

Highway Capacity Manual (HCM) (2000) Washington, DC: Transportation Research Board. The standard reference manual for estimating the capacity and determining service levels of highways and other transportation facilities, it covers highways and other roadways, intersections, transit facilities, and facilities for cyclists and pedestrians. The thirty-one chapters are organized into five parts: overview; concepts; methodologies; corridor and areawide analyses; and simulations and other models. The CD-ROM version includes tutorials and video clips. *The Highway Capacity Manual Applications Guidebook* from TRB (2003) contains case studies illustrating applications of the HCM to analysis real-world traffic situations.

Institute of Transportation Engineers (1997) *Traffic Signing Handbook*, Washington, DC: ITE. Aimed at practicing traffic engineers and technicians, the sixteen chapters contain information on types of signs and sign design; materials, fabrication, installation, maintenance, and inspection; and inventory systems. Describes engineering studies and appropriate processes for establishing the need for a sign. Covers special urban considerations, human factors, and dealing with the political aspects of citizen requests. Appendices contain English and metric editions of FHWA's *Standard Alphabets for Highway Signs and Pavement Markings.*

Institute of Transportation Engineers (2000) *Intelligent Transportation Primer*, Washington, DC: Institute of Transportation Engineers. Presents a comprehensive review of ITS, with a range of topics, such as traffic and vehicle control, standards, system architecture, telecommunications, and traveler information. Appendices contain a glossary and a list of acronyms.

Institute of Transportation Engineers (2003) *Trip Generation* (7th edn), 3 vols, Washington, DC: Institute of Transportation Engineers. Provides statistics used in calculating the forecasted trip generation rate for a wide variety of land uses, from housing to schools to multi-use developments. Data is based on more than 4,000 trip generation studies, but caution is urged when dealing with data from small sample sizes. Many municipalities specify use of ITE's figures in their planning processes. *Trip Generation Handbook: An ITE Recommended Practice* (2nd edn) provides assistance in choosing independent variables and time periods for analysis and methods for conducting studies. Also useful when estimating trip generation for multi-use developments.

Lay, M.G. (1998) *Handbook of Road Technology* (3rd edn) (2 vols), Melbourne: Gordon & Breach. Provides background and practical information on roadway planning, pavements, traffic, and transport. Written from an Australian perspective.

Manual of Bridge Engineering (2000) London: Thomas Telford. Presents a comprehensive overview of concept, analysis, design, construction, and maintenance of many types of bridges from a British perspective. Chapters on load distribution, structural analysis, design of specific types of bridges, technical advancements, substructures, maintenance, and monitoring are written by a variety of experts. Makes extensive use of figures and photographs; most chapters have bibliographies.

National ITS Architecture <http://www.iteris.com/itsarch/index.htm> (accessed May 18, 2005) U.S. Department of Transportation. Provides access to all architecture documents which constitute a common framework for intelligent transportation systems. Describes functions, subsystems and data flow.

Pine, J.L. (ed.) (2001) *Traffic Control Devices Handbook*, Washington, DC: Institute of Transportation Engineers. Augments the *Manual of Uniform Traffic Control Devices for Streets and Highways (MUTCD)*. Provides basic information and criteria on traffic control devices to enable small jurisdictions to handle *MUTCD* requirements. Includes numerous tables and figures, with many taken from the *MUTCD*, and others from sources such as ITE, AASHTO, and TRB. Some chapters include extensive references.

Traffic Signs Manual: Chapter 5, "Road markings," and Chapter 7, "the design of traffic signs" (2003), London: TSO. Chapter 5 describes the design and use of markings on the surface of the road for the control, warning, guidance, or information of road users in the UK. Chapter 7 describes sign design rules and principles; advance direction and direction signs; and specific types of signs.

U.S. Federal Highway Administration (2003) *Manual on Uniform Traffic Control Devices (MUTCD)*, Washington, DC: The Administration. Defines the standards used in the U.S. for the installation and maintenance of traffic control devices on all streets and highways. *MUTCD* covers signs, markings, signals, temporary traffic control, and highway-rail and light transit grade crossings.

U.S. Federal Highway Administration. Flexibility in Highway Design <http://www.fhwa.dot.gov/environment/flex/> (accessed May 18, 2005) (1997). Provides guidance in balancing the need for highway improvements with the need to safely integrate designs into the surrounding human and natural environment. Outlines the options available to state and local officials in using flexibility built into state standards, such as lowering design speeds.

In the U.S., every state transportation department publishes manuals and other information related to highway and pavement engineering with specifications and regulations applicable in that particular state. FHWA provides links <http://www.fhwa.dot.gov/webstate.htm> (accessed May 18, 2005) to all state transportation websites.

Non-motorized transportation

American Association of State Highway and Transportation Officials (1999) *Guide for the Development of Bicycle Facilities* (3rd edn), Washington, DC: The Association. Available in print and on CD-ROM. Provides information on planning, design, construction, and maintenance of facilities to enhance safe bicycle travel.

American Association of State Highway and Transportation Officials (2004) *AASHTO Guide for the Planning, Design and Operation of Pedestrian Facilities*, Washington, DC: The Association. Describes effective measures for providing pedestrian facilities on public rights-of-way along streets and highways. Also examines the connection between pedestrian mobility and site design.

Institute of Transportation Engineers (1998) *Design and Safety of Pedestrian Facilities: A Recommended Practice*, Washington, DC: The Institute. Provides guidelines for designs to allow pedestrians safe and efficient opportunities to walk near streets and highways.

Pipelines

McAllister, E.W. (2005) *Pipeline Rules of Thumb Handbook: A Manual of Quick, Accurate Solutions to Everyday Pipeline Engineering Problems* (6th edn), Boston, MA: Elsevier Gulf Professional Publishing. Provides practical techniques and quick methods for pipeline design, engineering, and construction.

Railroad engineering

American Railway Engineering and Maintenance Association (annual) *Manual for Railway Engineering*, Landover, MD: AREMA. Loose-leaf. Compiles engineering reference material in four volumes: *Track, Structures, Infrastructure*, and *Passenger and Systems Management*. Thousands of pages document design, specifications, definitions, procedures, and other aspects of railroad engineers. Makes extensive use of diagrams, tables, and figures and includes a detailed index.

American Railway Engineering and Maintenance Association (annual) *Portfolio of Trackwork Plans*, Landover, MD: AREMA. Loose-leaf. Companion volume to the *Manual for Railway Engineering* contains plans and specifications for switches, frogs, turnouts and crossovers, crossings, rail and special trackwork.

American Railway Engineering and Maintenance Association (annual) *Communications and Signals Manual of Recommended Practice*, Landover, MD: AREMA. Contains recommendations and instructions for all aspects of railroad communications and signals including signal systems, crossing warnings, yard systems, relays, track circuits, power supplies, mechanical systems, electrical systems, data transmissions, and quality.

American Railway Engineering and Maintenance Association (2003) *Practical Guide to Railway Engineering*, Landover, MD: AREMA. Book and CD-ROM. Covers railway fundamentals. Provides an overview of railway engineering history and details on how a railroad operates, right-of-way management issues, track design and layout, and railway bridge design concepts; includes a primer on signals and communication systems. Includes chapters and course modules on basic track, drainage, environmental permitting, structures, electric traction, and passenger transit.

Vehicle engineering

Automotive Handbook (2000) (5th edn), Stuttgart: Bosch. Provides facts and figures for passenger and commercial vehicles. Covers processes, systems, and techniques.

Fitch, J.W. (1993) *Motor Truck Engineering Handbook* (4th edn), Warrendale, PA: SAE International. Incorporates the latest technology, including electronic applications to power trains and operations.

Garrett, T.K. (2001) *The Motor Vehicle* (13th edn), Warrendale, PA: SAE International. Provides current information on vehicle technology with sections on engines, transmissions, and the carriage unit.

Gillespie, T.D. (1992) *Fundamentals of Vehicle Dynamics*, Warrendale, PA: Society of Automotive Engineers. Explains the performance of an automotive vehicle, with chapters focusing on acceleration performance; braking performance; aerodynamics and rolling resistance; ride; tires; steady-state cornering; suspensions; steering systems; and roll-over.

Jurgen, R.K. (ed.) (1999) *Automotive Electronics Handbook* (2nd edn), New York:

McGraw-Hill. Covers electronic components and systems including sensors and actuators; control systems; displays and information systems; and emerging technologies.

Seiffert, U.W. and Wech, L. (2003) *Automotive Safety Handbook*, Warrendale, PA: SAE International. Covers both active and passive safety systems and addresses accident avoidance, occupant protection, biomechanics, and the interrelationships among the occupant, the vehicle, and the restraint system.

Monographs and textbooks

Textbooks and monographs provide in-depth discussion of design, control mechanisms, operations, and other aspects of transportation.

Antaki, G.A. (2003) *Piping and Pipeline Engineering: Design, Construction, Maintenance, Integrity, and Repair*, Boca Raton, FL: CRC Press. Covers technical principles in materials, design, construction, inspection, testing, and maintenance. Includes codes and standards, design analysis, welding and inspection, corrosion mechanisms, fitness-for-service, and failure analysis, and an overview of valve selection and application.

Banks, J.H. (2002) *Introduction to Transportation Engineering* (2nd edn), Boston, MA: McGraw-Hill. Textbook. Focuses on highway and road design and traffic flow, control and capacity. Road and highway sections include facilities design, geometric design and surfaces. Addresses social, political, and economic contexts including environmental impacts, travel demand, planning, and project evaluation. Extensive use of figures, tables, exercises, and references.

Castro, C. and Horberry, T. (eds) (2004) *Human Factors of Transport Signs*, Boca Raton, FL: CRC Press. Examines sign research and new technologies for improving signaling.

Daganzo, C.F. (1997) *Fundamentals of Transportation and Traffic Operations*, New York: Elsevier Science. Introduces the basics of traffic operations, and covers tools and their applications.

Daganzo, C.F. (1999) *Logistic Systems Analysis* (3rd edn), New York: Springer-Verlag. Examines logistics systems, including operation and organization. Focuses on finding reasonable, rational solutions. Extensive references.

Fricker, J.D. and Whitford, R.K. (2004) *Fundamentals of Transportation Engineering: A Multimodal Systems Approach*, Upper Saddle River, NJ: Pearson Prentice Hall. Begins with an examination of traffic flow and highway design and moves into planning, safety, pavement design, and freight movement. Makes extensive use of scenarios, figures, illustrations, and includes exercises, references, and glossaries for each chapter.

Garber, N.J. and Hoel, L.A. (1997) *Traffic and Highway Engineering* (3rd edn), Boston, MA: PWS Publishing. Presents material at an introductory level. Covers standard topics of traffic engineering, safety, traffic flow, intersections, capacity, design, drainage, materials, and pavements.

Hay, W.H. (1982) *Railroad Engineering*, New York: Wiley. Classic textbook. It covers the fundamentals for locating, constructing, operating, and maintaining a railroad. Includes problem examples and an index.

Hiltscher, G., Mühlthaler, W., and Smits, J. (2003) *Industrial Pigging Technology: Fundamentals, Components, Applications*, Weinheim: Wiley-VCH. Describes equipment used for planning and designing pigging units and includes many practical examples.

Homburger, W.S., Hall, J.W., Reilly, W.R., and Sullivan, E.C. (2001) *Fundamentals of Traffic Engineering* (15th edn), Berkeley, CA: Institute of Transportation Studies, University of California. Serves as an introduction to the field of traffic engineering; designed for use in a one-week course for practicing professionals, but appropriate for university engineering courses. Addresses traffic studies; capacity; demand; traffic control devices; geometric design; control; operations; and traffic management. Extensive references.

IVHS America (1992) *Strategic Plan for Intelligent Vehicle-highway Systems in the United States*, Washington, DC: IVHS America. Provides blueprint for the development and deployment of intelligent vehicle-highway systems.

Kennedy, J.L. (1993) *Oil and Gas Pipeline Fundamentals* (2nd edn), Houston, TX: Gulf Publishing. Provides basic information on types of pipelines, design, construction practices, welding techniques and equipment, operations, maintenance, repair, inspection, and safety.

Larock, B.E., Jeppson, R.W., and Watters, G.Z. (1999) *Hydraulics of Pipeline Systems*, Boca Raton, FL: CRC Press. Provides comprehensive treatment of pipeline hydraulics. Includes CD with software for performing complex computations.

Liu, H. (2003) *Pipeline Engineering*, Boca Raton, FL: CRC Press, Covers the essential aspects and types of pipeline engineering. Arranged in two parts: Part 1 covers pipe flows and engineering considerations; Part 2 summarizes equipment and methods for pipeline planning, design, construction, operation, and maintenance.

Mannering, F.L. and Kilareski, W.P. (1998) *Principles of Highway Engineering and Traffic Analysis* (2nd edn), New York: John Wiley. Covers fundamental issues in highway engineering and traffic analysis. Makes extensive use of examples and problems.

McCubbin, R.R., Staples, B.L., and Mercier, M.R. (2003) *Intelligent Transportation Systems Benefits and Costs: 2003 Update*, Washington, DC: U.S. Department of Transportation. Reports on the impact that intelligent transportation system projects have on the operation of the surface transportation network. Includes cost information for representative ITS deployments.

Mohitpour, M., Murray, A., and Golshan, H. (2003) *Pipeline Design and Construction* (2nd edn), New York: ASME International. Contains practical techniques for the design, construction, commissioning, and assessment of pipelines and related facilities.

Muller, G. (1999) *Intermodal Freight Transportation* (4th edn), Washington, DC: Eno Transportation Foundation. Provides comprehensive overview and discusses terminal operations, transport equipment, containerization, and technology trends.

Murphy, P.R. and Wood, D.F. (2003) *Contemporary Logistics* (8th edn), Upper Saddle River, NJ: Prentice Hall. Provides overview of logistics, elements of logistics systems, materials handling, transportation management, inventory, warehousing and supply management, and implementing a logistics system. Stresses transportation as a critical element in the physical system.

Najafi, M. (2004) *Trenchless Technology: Pipeline and Utility Design, Construction, and Renewal*, New York: McGraw-Hill. Covers design guidelines, construction, environmental considerations, and equipment, methods, and materials for minimal surface disruption through the use of trenchless technologies.

Papacostas, C.S. and Prevedouros, P.D. (2001) *Transportation Engineering and Planning* (3rd edn), Upper Saddle River, NJ: Prentice Hall. Undergraduate textbook. It addresses design and operations; transportation systems, including modes, planning and forecasting; transportation impacts; and supporting elements, including queuing and simulation. Includes exercises.

Smith, P. (2004) *Piping Materials Guide*, New York: Elsevier Gulf Professional Publishing. Covers the selection of piping materials, materials and fitting, troubleshooting techniques for corrosion control, and inspections.

Stone, R. and Ball, J.K. (2004) *Automotive Engineering Fundamentals*, Warrendale: SAE International. Covers the principles involved in designing a vehicle; includes engines, transmissions, steering systems, brakes, tires, and aerodynamics.

Sussman, J. (2004) *Introduction to Transportation Systems*, Boston, MA: Artech House. Provides a graduate-level introduction to transportation systems. The chapter on intelligent transportation systems examines the functional areas of ITS, institutional issues, and future visions.

Wright, P.H. (1996) *Highway Engineering* (6th edn), New York: John Wiley. Textbook covering the design of roadways and mass transit facilities. Addresses traffic engineering, materials, drainage, pavements, and maintenance.

Wright, P.H. (ed.) (1998) *Transportation Engineering: Planning and Design* (4th edn), New York: Wiley. Explores transportation engineering planning on a multimodal systems basis and examines the design of facilities that serve individual transportation modes.

Journals

Journal literature is an essential form of information in transportation engineering. There are a number of publications that address the discipline as a whole, but as is the case with much of the information in transportation engineering, most sources address only one mode. Many of the publications are peer-reviewed journals; a few trade publications have been included. General civil engineering titles such as the *ASCE Journal* are not included. Listed below are some of the key journals:

General

Accident Analysis and Prevention. Elsevier (0001–4575) bimonthly. Publishes papers dealing primarily with medical, legal, economic, educational, behavioral, theoretical, or empirical aspects of transportation accidents.

Journal of Navigation (0373–4633) <http://journals.cambridge.org> London: Royal Institute of Navigation. Publishes papers on the science of navigation.

Journal of the Transportation Research Forum (1046–1469) Fargo, ND: Upper Great Plains Transportation Institute; biannual. Publishes papers presented at TRF Annual Forum.

Journal of Transportation and Statistics (1094–8848) <http://www.bts.gov/publications/journal_of_transportation_and_statistics/> (accessed May 18, 2005) Washington, DC: U.S. Department of Transportation, Bureau of Transportation Statistics; three issues per year. Focuses on the latest developments in transportation information and data, theory, concepts, and methods of analysis relevant to all aspects of the transportation system.

Journal of Transportation Engineering (0733–947X) http://scitation.aip.org/teo/. American Society of Civil Engineers; bimonthly. Contains technical articles on planning, design, construction, maintenance, and operation of air, highway, and urban transportation, as well as pipeline facilities.

Transportation Journal (0041–1612) <http://www.astl.org/tj.htm> (accessed May 18, 2005) American Society of Transportation and Logistics; quarterly. Covers the supply chain and logistics management.

Transportation Research <http://www.elsevier.com/wps/find/journal_browse.cws_home> (accessed May 18, 2005) published in six parts.

Transportation Research, Part A (0965–8564) Policy and practice covers all aspects of transportation from various disciplines, including engineering.

Transportation Research, Part B (0191–2615) Covers methodological aspects of transportation, particularly those such as traffic flow, analysis of transportation networks, and queuing theory that require mathematical analysis.

Transportation Research, Part C (0968–090X) Focuses on the implications of emerging technologies on the planning, design, control, implementation, management, and rehabilitation of transportation systems.

Transportation Research, Part D (1361–9209) Covers the environmental impact of transportation and implications for the design and management of transportation systems.

Transportation Research, Part E (1366–5545) Focuses on topics such as logistics and supply chain management and logistics and operations models.

Transportation Research, Part F (1369–8478) Covers behavioral and psychological aspects of traffic and transportation.

Transportation Research Record (0361–1981) Transportation Research Board. The official publication of TRB, the *Record* contains peer-reviewed technical papers presented at TRB conferences on all aspects and modes of transportation. Each record contains multiple papers focused on a single topic such as soil mechanics, pavement maintenance, highway capacity, or rail-highway crossings. The *Records* are available on CD-ROM but not online.

Highway and traffic engineering

Asphalt Technology News (1083–687X) <http://www.eng.auburn.edu/center/ncat/> (accessed May 18, 2005) National Center for Asphalt Technology, biannual. Newsletter aimed at transferring asphalt research findings to practitioners.

Better Roads (0006–0208) <http://www.betterroads.com> (accessed May 18, 2005) James Information Media; monthly. Practical articles aimed at highway contractors and state and local government agencies involved in road and bridge engineering and maintenance.

Focus (1060–6637) <http://www.tfhrc.gov/focus/focus.htm> (accessed May 18, 2005) FHWA, Turner-Fairbank Highway Research Center; monthly. Covers the implementation of innovative technologies including bridge technology, pavement technology, asset management, program administration, winter maintenance, and work zone safety.

Highways <http://www.highways-mag.co.uk/highways/> (accessed May 18, 2005) London: Institution of Civil Engineers; bimonthly. Covers traffic management and highway maintenance in the U.K.

HMAT: Hot Mix Asphalt Technology (not supplied) National Asphalt Pavement Association; monthly. Promotes the use of asphalt pavements through success stories, buyers' guides, and information on bituminous materials, mixtures, applications, and maintenance.

IMSA Journal (1064–2560) <http://www.imsasafety.org/journal/journal.htm> (accessed May 18, 2005) Newark, NY: International Municipal Signal Association; bimonthly. Focuses on traffic signals and traffic control.

Intellimotion <http://www-path.eecs.berkeley.edu/PATH/Intellimotion/> (accessed May 18, 2005) Richmond, CA: California PATH Publications; quarterly. Reports on research in intelligent transportation systems at the California PATH (Partners for Advanced Transit and Highways) program.

ITE Journal (0162–8178) <http://www.ite.org/itejournal/index.asp> (accessed May 18, 2005) Institute of Transportation Engineers; monthly. Reports on research involving surface transportation systems, particularly traffic engineering.

ITS International (1463–6344) <http://www.itsinternational.com/> (accessed May 18, 2005) Swanley, Kent: Route One Publishing; bimonthly. Focuses on the deployment of technology for intelligent transportation systems and covers the full range of applications, including mass transit and light rail systems.

Journal of Bridge Engineering (1084–0702) <http://scitation.aip.org/beo/> (accessed May 18, 2005) American Society of Civil Engineers; bimonthly. Publishes research about the practice and profession of bridge engineering, and covers projects, design, construction, inspection, safety, repair, and rehabilitation.

NCHRP Report, *NCHRP Synthesis of Highway Practice*, and *NCHRP Summary of Progress* <http://gulliver.trb.org/bookstore/> (accessed May 18, 2005) Transportation Research Board. The National Cooperative Highway Research Program conducts research in areas that affect highway planning, design, construction, operation, and maintenance nationwide. Results are disseminated through the *Report*, *Synthesis*, and *Summary* publications.

Public Roads (0033–3735) <http://www.tfhrc.gov/pubrds/pubrds.htm> (accessed May 18, 2005) FHWA, Turner-Fairbank Research Center; bimonthly. Reports on advances and innovations in highway and traffic research and technology, with an emphasis on developments in federal highway programs.

Roads and Bridges (8750–9229) <http://www.roadsbridges.com/rb/> (accessed May 18, 2005). Scranton Gillette Communications; monthly. Provides information on advancements in the road and bridge industry, including reviews of products and construction and maintenance equipment. Features many small, successful projects.

Routes = Roads (1011–1891) PIARC/World Road Association; quarterly. In French and English. Publishes technical articles on road technology, safety and vehicles. European perspective.

Synthesis of Highway Practice (0547–5570) <http://trb.org/news/blurb_browse.asp?id=5> (accessed May 18, 2005)

Washington, DC: Transportation Research Board; irregular. Focuses on practical solutions and measures for resolving specific problems.

Traffic Engineering and Control (0041–0683) <http://www.tecmagazine.com/> (accessed May 18, 2005); eleven issues per year. Covers traffic control, traffic planning, and transportation management from a British perspective.

Traffic Injury Prevention (1538–9588) <http://www.tandf.co.uk/journals/titles/15389588.asp> (accessed May 18, 2005) Taylor & Francis; quarterly. Focuses on research, interventions, and evaluations within the areas of traffic safety, crash causation, injury prevention, and treatment.

Traffic Technology International (1356–9252) Dorking, Surrey: UK & International Press; bimonthly. Covers advanced traffic management systems including enforcement, intelligent parking, road pricing, electronic tolls collection, and detection and incident management.

Tunneling and Underground Space Technology (incorporating *Trenchless Technology*) (0886–7798) http://www.elsevier.com/wps/find/journaldescription.cws_home/799/description#description Elsevier Science; quarterly. Focuses on technical improvements and cost-effective methods for the planning, design, construction, operation and maintenance of underground structures. Publishes reports and papers from the International Tunnelling Association.

Tunnels and Tunnelling International (1369–3999) <http://www.tunnelsonline.info/> (accessed May 18, 2005) Polygon Media; monthly. International in coverage, T&T publishes technical profiles of ongoing and completed tunnels, including rail, vehicular and pipeline projects. Covers techniques and methods, equipment, geologic conditions, safety, and environmental issues.

World Highways/Routes du Monde (0964–4598) <http://www.worldhighways.com> (accessed May 18, 2005) Swanley, Kent: Route One Publishing; bimonthly. Covers all aspects of road infrastructure worldwide, including construction, toll systems, and traffic management. Official publication of the International Road Federation.

Maritime

Fast Ferry International (0954–3988) <http://www.fastferryinfo.com> Kent: Fast Ferry International; ten issues per year. Features articles on new vessels and designs.

Journal of Ship Production (8756–1417) <http://www.sname.org/jsp_description.htm> (accessed May 18, 2005) Jersey City, NJ: Society of Naval Architects and Marine Engineers; quarterly. Covers technical aspects of shipyards and the production of merchant and naval ships.

Journal of Ship Research (0022–4502) <http://www.sname.org/jsr_description.htm> (accessed May 18, 2005) Jersey City, NJ: Society of Naval Architects and Marine Engineers; quarterly. Publishes technical papers on applied research in hydrodynamics, propulsion, ship motions, structures, and vibrations.

Journal of Waterway, Port, Coastal and Ocean Engineering (0733–950X) <http://scitation.aip.org/wwo/> (accessed May 18, 2005) American Society of Civil Engineers; bimonthly. Covers engineering aspects of dredging, floods, pollution, and sediment transport, and the development and operation of ports, harbors, and offshore facilities.

Marine Technology Society Journal (0025–3324) <http://www.mtsociety.org/publications/journal.cfm> (accessed May 18, 2005) Columbia, MD: The Society; quarterly. Peer-reviewed publication covers advances in ocean science, engineering, and marine technology.

Maritime Engineering. (1542–0566) <http://www.sname.org/mtonline_description.htm> (accessed May 18, 2005) Jersey City, NJ: Society of Naval Architects and Marine Engineers; quarterly. Publishes technical papers on the advances, trends, concepts, and discoveries in the marine industry.

Sea Technology (0093–3651) <http://www.sea-technology.com/> (accessed May 18, 2005) Arlington, VA: Compass Publications; monthly. Covers the design, engineering, and application of equipment and services for the marine industry.

Motor vehicles

Automobile Abstracts. (0309–0817) <http://www.mira.co.uk> (accessed May 18, 2005) Motor Industry Research Association (MIRA). Automobile Abstracts, Warwickshire: MIRA; monthly. Electronic journal of abstracts of technical articles from worldwide automotive literature. Covers all aspects of vehicle design and performance, as well as fuel, lubricants, materials, production, and environmental aspects.

Automotive Engineering International (1543–849X) <http://www.sae.org/automag/> (accessed May 18, 2005) SAE International; monthly. Covers new products and technology for the global automotive industry. Focuses on applied vehicle engineering and design concepts.

IEEE Transactions on Vehicular Technology (0018–9545) <http://ieeexplore.ieee.org/xpl/RecentIssue.jsp?puNumber=25> (accessed May 18, 2005) New York: Institute of Electrical and Electronic Engineers; bimonthly. Focuses on electrical and electronic technology in vehicles and vehicle systems.

International Journal of Vehicle Autonomous Systems (1471–0226) <https://www.inderscience.com/browse/index.php?journalID=30> (accessed May 18, 2005) Geneva, Switzerland: Inderscience; quarterly. Reports on driver assistance systems, intelligent vehicle systems, collision avoidance, by-wire systems, and new electrical and electronic systems.

International Journal of Vehicle Design: The Journal of Vehicle Engineering and Components (0143–3369) <http://www.inderscience.com/browse/index.php> (accessed May 18, 2005) Geneva, Switzerland: Inderscience Publishers; monthly. Publishes articles on engineering design and research into all types of self-propelled vehicles and components.

Vehicle System Dynamics: International Journal of Vehicle Mechanics and Mobility (0042–3114) <http://www.ingentaconnect.com/content/tandf/vesd> (accessed May 18, 2005) Lisse, The Netherlands: Taylor & Francis; monthly. Publishes research on vehicle dynamics, intelligent highway systems, and related topics.

World Automotive Manufacturing (1463–1857) <http://www.awknowledge.com/WAM/> (accessed May 18, 2005) London: Automotive World; monthly. Covers the design process, materials handling, equipment, logistics, and the manufacturing process.

Pipelines

Pipeline and Gas Journal (0032–0188) <http://www.pipelineandgasjournalonline.com> (accessed May 18, 2005) Houston, TX: Oildom Publishing Co. of Texas; monthly. Covers engineering, construction, and operations of pipelines transporting crude oil products and natural gas.

Pipeline and Gas Technology (not supplied) <http://www.pipelineandgastechnology.com/> (accessed May 18, 2005) Houston, TX: Hart Energy Publishing, LP; monthly. Covers processes and technological advances in pipeline operating maintenance and construction.

Pipeline World <http://www.pipemag.com/> (accessed May 18, 2005) Beaconsfield, Bucks: Scientific Surveys Ltd; bimonthly. Provides technical information on pipes and pipelines research with a European focus. Formed by the merger of *Pipeline World* and *Pipes and Pipelines International*.

Railroad engineering

International Railway Journal (not supplied) <http://www.railjournal.com/> (accessed May 18, 2005) Omaha, NB: Simmons-Boardman Books; monthly. Covers international developments in high speed rail, systems, signaling, and wheel sets.

Japanese Railway Engineering (0448–8938) Published by Japan Railway Engineers' Association; biannual. Publishes articles in English on railway engineering in Japan.

Quarterly Report of RTRI (0033–9008) <http://www.rtri.or.jp/> (accessed May 18, 2005) (hard copy in English) and *RTRI Report* (full papers in Japanese). Railway Technical Research Institute. English abstracts are available online (<http://www.rtri.or.jp/infoce/tic_E.html> (accessed May 18, 2005). Covers technological developments in the Japanese railroad industry.

Railway Gazette International (0373–5346) <http://www.railwaygazette.com> (accessed May 18, 2005) Reed Business Information; monthly. Features technical developments involving rolling-stock, track, switches and signals, and fare systems.

Railway Technical Review: The International Journal for Railway Engineers (0079–9548) Hestra-Verlag; quarterly. Publishes technical papers on all aspects of railway engineering.

Railway Track and Structures (RT&S) (0033–9016) Omaha, NB: Simmons-Boardman Books; monthly. Focuses on rail engineering, maintenance-of-way, and communications and signals in the U.S. Reports on railroad testing done at the Transportation Technology Center, Inc. (TTCI).

Proceedings

Many societies publish proceedings of meetings, seminars, and other gatherings. Some are quite specialized and proceedings dealing with detailed technical aspects of any mode can be identified using catalogs and indexes described previously. Some of the more important published proceedings include the following:

American Association for the Advancement of Automotive Medicine <http://www.carcrash.org/publications_proceedings.htm> (accessed May 18, 2005) *Proceedings of the Annual Conference.* Des Plaines, IL: The Association. Papers focus on motor vehicle crash injury prevention and control from a multi-disciplinary perspective. Available in hard copy and on CD-ROM.

American Railway Engineering and Maintenance Association <http://www.arema.org/> (accessed May 18, 2005) *Proceedings.* Beginning in 1998, these proceedings superseded the *AREA Proceedings.* Produced by the Communications and Signals and the Tracks and Structures Committees, they contain technical papers and committee reports. The American Railway Engineering Association (AREA) published the *Bulletin* five times a year from 1900 to 1997; in 1997 AREA became part of AREMA. Four issues of the *Bulletin* contain technical papers, proposed manual changes, and committee reports that were then bound together to form AREA's annual *Proceedings.* The fifth issue is a *Membership and Committee Directory.*

ARRB Transport Research Ltd <http://www.arrb.com.au/> (accessed May 18, 2005) Proceedings. Vermont South, VA, Australia; biennial. Includes papers from the conference on road-related issues such as pavement and asphalt mix design, road safety, traffic engineering, and environmental practice. Formerly known as Australian Road Research Board.

Association of Asphalt Paving Technologists <http://www.asphalttechnology.org/> (accessed May 18, 2005) Asphalt Paving Technology. The Association; annual.

Covers all aspects of asphalt pavements from a global perspective: materials, mix design, testing, construction, and performance. Published irregularly from 1974 to 1986; annual since 1988.

Institute of Electronic and Electrical Engineers (IEEE) <http://ieeexplore.ieee.org/xpl/conferences.jsp> (accessed May 18, 2005) Intelligent Transportation Systems Council (ITSC) Conference Proceedings and Intelligent Vehicles Symposium Proceedings, Piscataway, NJ: The Institute; annual. Focuses on advanced technologies for vehicles.

Institute of Transportation Engineers (ITE) <http://www.ite.org/bookstore/index.asp> (accessed May 18, 2005) *Annual Meeting Compendium of Papers* (title varies), Washington, DC. Papers on technical topics including traffic engineering, traffic operations, traffic management, safety, mobility, and transit. Proceedings issued by the Institute of Traffic Engineers from 1930 to 1976. Beginning in 1997, available only on CD-ROM.

International Congress on Noise Control Engineering. *Inter-noise Proceedings*. Place of publication varies: The International Institute of Noise Control Engineering (INCE); annual. Covers research on traffic and pavement noise and vibration engineering. INCE/USA sponsors the annual Noise-Con Conference.

IEEE/ASME Joint Rail Conference <http://ieeexplore.ieee.org/xpl/conferences.jsp> (accessed May 18, 2005) The Vehicular Technology Section, Land Transportation Division, Institute of Electrical and Electronic Engineers and Rail Transportation Division, American Society of Mechanical Engineers; annual. Focuses on innovations in railway mechanical and electrical engineering. Beginning with 2002, available on CD-ROM; available online to members and subscribers from 1989 to the present.

Planning Transport Research and Computation (PTRC) Ltd. European Transport Conference: PTRC Education and Research Services <http://www.aetransport.co.uk/lc_cms/page_view.asp?id=667> (accessed May 18, 2005) annual. Papers on topics such as transportation management, modeling, public transportation, road maintenance, and intelligent transportation systems. Content is primarily British with some international coverage. Formerly known as the PTRC Summer Annual Meeting. Separate volumes published for each seminar topic. Beginning with 2001, available only on CD-ROM.

Society of Automotive Engineers (SAE) <http://www.sae.org/events/> (accessed May 18, 2005). SAE sponsors numerous conferences and symposia on all aspects of the design, manufacture, and total life-cycle technology for ground vehicles; see SAE's website for details.

Stapp Car Crash Journal. Proceedings. Warrendale, PA: Society of Automotive Engineers; annual. Presents research in impact biomechanics, human injury tolerance, automotive safety measures and appliances, and related fields concerned with land-vehicle crash injury protection. Previously known as the Stapp Car Crash Conference.

Transportation Association of Canada. Proceedings of TAC Annual Conference. Ottawa. Covers topics such as geometric design, pavement, road maintenance, and traffic control and engineering. Published on CD-ROM.

Transportation Research Board. Preprints. Washington, DC: National Academy of Sciences; annual. A primary source for research in all aspects of transportation engineering. Papers are published in series by subject. Covers the full range of multi-modal transportation topics: pavements, structures, transportation management, traffic engineering, freight transportation, safety, rail transit, security, aviation, logistics, modeling, and so on. Most preprints are revised and published in the *Transportation Research Records* series. Beginning in 1998, available only on CD-ROM. Prior to 1962, published as *Proceedings of the Annual Meeting of the Highway Research Board.*

World Congress on Intelligent Transport Systems; annual. Title varies. Sponsored by ITS America, VERTIS, and ERTICO. Recent proceedings available on CD-ROM, with papers accessible by author, subject, session, paper number, and paper type.

World Conference on Transport Research. *Proceedings*; triennial. Papers on transportation modeling, operation, planning, and control.

Patents and standards

Patents and standards can be key resources for transportation engineers involved in the design of new equipment and processes, particularly in the area of motor vehicles. Locating this material used to be rather difficult, but the development of electronic databases and online resources has greatly enhanced access. Key patent databases and portals are included in Chapter 2 of this book.

Standards and specifications

American Association of State Highway and Transportation Officials (1999) *Guide Specifications for Seismic Isolation Design*, Washington, DC: AASHTO. Provides fundamental requirements for seismic isolation design.

American Association of State Highway and Transportation Officials. (2002) *Standard Specifications for Highway Bridges* (17th edn), Washington, DC: AASHTO. Known as "The Bridge Book," this is the basic source of structural design standards for public and private sector bridge engineers. Includes CD-ROM.

American Association of State Highway and Transportation Officials (2004) *AASHTO LRFD Bridge Construction Specifications* (2nd edn), Washington, DC: The Association. Companion volume to *LFRD Bridge Design Specifications* focused on construction.

American Association of State Highway and Transportation Officials (2004) *AASHTO LRFD Bridge Design Specifications* (3rd edn), Washington, DC: The Association. Based on the load-and-resistance factor design (LRFD) philosophy. Available in U.S. (standard), SI (metric) or standard/metric. Includes CD-ROM with both editions and additional features.

American Association of State Highway and Transportation Officials (2004) *Standard Specifications for Transportation Materials and Methods of Sampling and Testing* (24th edn), and *AASHTO Provisional Standards* (2004), Washington, DC: AASHTO. Also available on CD-ROM. Contains more than 400 materials and specifications and test methods commonly used in highway construction; volumes organized by materials and testing. Includes relevant ASTM specifications.

American Society of Testing and Materials (2004) *Annual Book of ASTM Standards. Section 4, Construction; Volume 4.02, Concrete and Aggregates; Volume 4.03, Road and Paving Materials and Vehicle-Pavement Systems*, Philadelphia, PA: ASTM. Covers road and paving materials. Provides specifications, tests and practices for field measurements, traffic monitoring, and vehicle roadside communication.

Association of American Railroads. *Manual of Standards and Recommended Practices*, Washington, DC: ARA. Contains about two dozen sections which are updated on an irregular basis. Includes regularly adopted specifications, standards, and recommended practices for freight cars, locomotives, and their components.

Code of Federal Regulations. Title 49 Transportation <http://www.access.gpo.gov/nara/cfr/cfr-table-search.html> (accessed May 18, 2005) annual. Washington, DC: Office of the Federal Register, National Archives and Records Service, General Services Administration. Contains the general and permanent rules published in the Federal Register by the Department of Transportation, Federal Railroad Administration, National Highway Traffic Safety Administration, Federal Transit Administration, Surface Transportation Board, and other federal departments and agencies.

Institute of Transportation Engineers (1999) *Management and Operations of Intelligent Transportation Systems*, Washington, DC: Institute of Transportation Engineers. Recommended practice. Contains guidelines for intelligent transportation systems operations.

ITS Standards <http://www.standards.its.dot.gov/standards.htm> (accessed May 18, 2005). Provides access to publications from various organizations that contain standards for intelligent transportation systems.

SAE Handbook (annual) Warrendale, PA: Society of Automotive Engineers. Available in hard copy (three volumes) and on CD-ROM. Contains SAE's ground vehicle standards.

U.S. Federal Highway Administration (1996) *Standard Specifications for Construction of Roads and Bridges on Federal Highway Projects*, Washington, DC: FHWA. Covers the process used for road and bridge construction on federal highway projects from bidding through construction.

Important websites and portals

As more information becomes available online, websites and portals are essential resources. There are hundreds of sites with relevant transportation engineering information; this is a selection of the more important ones.

The U.S. Department of Transportation is an excellent source of transportation engineering information. Some of the department's more useful sites include the following (all accessed May 18, 2005):

The Federal Highway Administration <http://www.fhwa.dot.gov/> (FHWA) is a primary source for information on highway engineering for projects receiving federal funding. Extensive online information is available for bridges, pavements, environmental design, and safety; a list of FHWA sites is available at <http://www.fhwa.dot.gov/fhwaweb.htm>. Among the important FHWA sites are:

FHWA's Office of Infrastructure (<http://www.fhwa.dot.gov/infrastructure/>) provides technical assistance and program assistance in Federal-aid highway programs, asset management (<http://www.fhwa.dot.gov/infrastructure/asstmgmt/index.htm>), pavements and bridges. An extensive amount of information and publications related to geotechnical engineering, structures, hydraulics, asphalt, concrete, pavement design, and asset management are available.

In the FHWA Office of Bridge Technology (<http://www.fhwa.dot.gov/bridge/index.htm>), the National Bridge Inspection Program includes the National Bridge Inspection Standards (<http://www.fhwa.dot.gov/bridge/nbis.htm>) and the National Bridge Inventory (<http://www.fhwa.dot.gov/bridge/nbi.htm>) of deficient bridges, listed by state.

The FHWA Office of Pavement Technology (<http://www.fhwa.dot.gov/pavement/index.htm>) includes publications, conferences, training programs, contacts, and links to web-based information on asphalt and concrete pavements.

The Asphalt Technology Team (<http://www.fhwa.dot.gov/pavement/ ashome.htm>) promotes implementation of the Superpave system and works to improve pavement smoothness (<http://www.fhwa.dot.gov/pavement/pshome. htm>).

The Concrete Technology Team (<http://www.fhwa.dot.gov/pavement/conhome.htm>) site includes current projects and links to a design guide.

The Construction and Maintenance website (<http://www.fhwa.dot.gov/construction/>) in FHWA's Office of Asset Management links to National Highway Specifications (<http://fhwapap04.fhwa.dot.gov/nhswp/index.jsp>) with a searchable library of highway specifications from throughout the U.S. It also links to the Construction Program Guide (<http://www.fhwa.dot.gov/construction/cqit/index.htm>) where there are links to topics such as design-build, safety and state preference for materials. The Generic Construction Related Review Guidelines (<http://www.fhwa.dot.gov/construction/reviews.htm>) provides examples of process and in-depth reviews performed by field offices on subjects such as pavements, structures, environmental protection, and work zone safety.

In FHWA's Office of Operations (<http://www.ops.fhwa.dot.gov/traffic/>), the Workzone Mobility and Safety Program (<http://www.ops.fhwa.dot.gov/wz/about/about-us.htm>) website links to a set of practitioner's tools including best practices, fact sheets, reports, technologies, and decision support tools.

FHWA's Office of International Programs publishes International Guide to Highway Transportation Information (<http://ntl.bts.gov/DOCS/finalrep.html>). The guide identifies key highway transportation libraries and research centers; websites; document suppliers; databases; and associations, organizations and societies. Separate volumes are published for each type of information and data are presented for thirty-three countries. The first four volumes are also available in pdf; the first four volumes are also available in pdf:

<http://international.fhwa.dot.gov/pdfs/ightivol1.pdf>
<http://international.fhwa.dot.gov/pdfs/ightivol2.pdf>
<http://international.fhwa.dot.gov/pdfs/ightivol3.pdf> and
<http://international.fhwa.dot.gov/Pdfs/Vol4.pdf>

FHWA's Transportation and Highway Related Web Sites <http://www.fhwa.dot.gov/webtrans.htm> is an excellent list of agencies, organizations and research centers.

Other Department of Transportation and government sites of particular interest for highway and traffic engineering are as follows:

The National Transportation Library (NTL) <http://ntl.bts.gov/> (accessed May 18, 2005), part of the Bureau of Transportation Statistics, supports transportation policy, research, operations, and technology transfer activities by providing access to relevant information. The NTL's digital collection serves as a repository of technical, research, and policy documents from public, academic, and private organizations. Users can search the collection or they may browse it by selecting one of sixteen topics. The NTL's reference staff assists customers in finding statistics, reports, and technical experts; they also answer questions about *TRIS Online* and TRANSTATS databases.

The Local Technical Assistance Program (LTAP) Clearinghouse <http://www.ltapt2.org/> (accessed May 18, 2005) connects to LTAP centers in each state and Tribal Technical Assistance Program (TTAP) centers. Centers provide employees of local agencies with cost-effective access to training, information clearinghouses, and technology updates.

The Intelligent Transportation Systems (ITS) Program <http://www.its.dot.gov/> (accessed May 18, 2005) guides policy in the development of advanced technologies to improve transportation efficiencies and safety. Information is available on Standards, (<http://www.standards.its.dot.gov/standards.htm>), the Intelligent Vehicle Initiative (<http://www.its.dot.gov/ivi/ivi.htm>), and the Travel Management Program (<http://www.its.dot.gov/TravelManagement/travel.htm>).

ITS Benefits and Unit Costs Database http://www.benefitcost.its.dot.gov/ United States. Joint Program Office for Intelligent Transportation Systems. Tracks data on the impact of intelligent transportation systems projects. Contains estimates of ITS costs data. The Turner-Fairbank Highway Research Center (<http://www.tfhrc.gov/>) conducts research to find solutions to complex technical problems in transportation. The Long Term Pavement Performance (<http://www.tfhrc.gov/pavement/ltpp/ltpp.htm>) program is a twenty-year project to study in-service asphalt, concrete, and portland cement pavements. Software, reports, manuals, and product information briefs and applications notes are available. The Safety Research program (<http://www.tfhrc.gov/safety/safety.htm>) focuses on intersections, pedestrian and bicyclists, roadsides, run-off-road accidents and speed management, and provides access to technical reports, articles, and other resources.

Operations and Intelligent Transportation Systems <(http://www.tfhrc.gov/its/its.htm>) has information on enabling technologies and travel management products and services as well as links to technical reports and articles. (all accessed May 18, 2005).

The National Highway Traffic Safety Administration <http://www.nhtsa.dot.gov/people/crash/Index.html> (accessed May 18, 2005) provides data on crashes through databases such as the Fatality Analysis Reporting System (FARS) and the Crashworthiness Data System.

The National Transportation Safety Board <http://www.ntsb.gov/> (accessed May 18, 2005) publishes accident reports for highways, marine transportation, pipelines, and railroads.

The Office of Pipeline Safety <http://ops.dot.gov/index.htm> (accessed May 18, 2005) in the Pipeline and Hazardous Materials Safety Administration includes accident and incident statistics; regulations; and documents on risk management.

Other highway and traffic engineering-related sites of interest are as follows:

The Institute of Transportation Engineers (ITE) (<http://www.ite.org>) develops intelligent transportation system (ITS) standards (<http://www.ite.org/standards/index.asp>) in a cooperative agreement with the U.S. Department of Transportation. ITE's site also has technical information (<http://www.ite.org/library/

index.asp>) and links on topics such as traffic calming, road safety audits, safety and signal timing (site updating is somewhat spotty) (all accessed May 18, 2005).

Other sites with transportation engineering information are as follows:

American Association of Port Authorities (AAPA) <http://www.aapa-ports.org/> (accessed May 18, 2005). Provides links to sites with port information for North America.

BIMCO <http://www.bimco.dk/> (accessed May 18, 2005) (Baltic and International Maritime Council) provides members with access to databases of information on technical aspects of shipping, cargoes, safety, environmental issues, and so on.

DOLPHIN: Directory of Online Ports and Harbor Information <http://dolphin.tamu.edu/> (accessed May 18, 2005). Portal for marine transportation system information with links organized by subject, such as operations and modes.

Lloyd's Marine Intelligence Unit <http://www.lloydsmiu.com> (accessed May 18, 2005). Provides members with access to detailed databases on vessels, shipping, casualties, and shipping companies.

National Truck Equipment Association (NTEA) <http://www.ntea.com/> (accessed May 18, 2005). Provides access to technical reports and articles, safety information, and a hoist database.

Office of Coast Survey, U.S. National Ocean Service, National Oceanic and Atmospheric Administration <http://chartmaker.ncd.noaa.gov/> (accessed May 18, 2005). Provides access to databases, charts, surveys, and other navigational products.

Physical Oceanographic Real-Time Systems (PORTS) <http://co-ops.nos.noaa.gov/d_ports.html> (accessed May 18, 2005). From U.S. National Ocean Service, National Oceanic and Atmospheric Administration, supplies shipmasters and pilots with water levels, currents and other data to avoid groundings and collisions.

Railway Track and Structures magazine's Links to Important Railroad Industry Information <http://www.rtands.com/links.html> (accessed May 18, 2005) provides quick access to the sites of Class I, regional and short line railroads; associations; regulatory agencies; and information resources.

TrafficLinq Search Directory <http://www.trafficlinq.com/> (accessed May 18, 2005). Produced by Dutch traffic and transportation engineers, this has an extensive directory of links covering transportation, road traffic, intelligent transportation systems, modeling, and safety. Covers about 1,000 transportation websites worldwide, and the TrafficLinq search engine allows searching of all the sites with one query.

Transport Canada <http://www.tc.gc.ca/en/menu.htm> (accessed May 18, 2005). This site provides access to statistics, policies, regulations, and other information about all modes of transportation.

U.S. Dept of Transportation Maritime Administration <http://www.marad.dot.gov/> (accessed May 18, 2005). Links to information on the Marine Transportation System initiative, security, shipbuilding, and domestic operation of ports. Provides access to publications and forms.

University Transportation Centers Program <http://utc.dot.gov/> (accessed May 18, 2005). Focuses on education, research, and technology transfer at thirty-three university-based centers of excellence.

Associations, organizations, and societies

For transportation engineering, entities such as professional associations and organizations are important sources of information. Most deal with just one mode, but some, such as the Transportation Research Board, are multi-modal in focus.

The American Association of State Highway and Transportation Officials (AASHTO) <http://www.transportation.org/> (accessed May 18, 2005). This association fosters the development, operation, and maintenance of an integrated national transportation system. Publishes essential handbooks including the *Manual on Uniform Traffic Control Devices* (with FHWA and ITE) and *A Policy on Geometric Design of Highways and Streets*. Lobbies for increased government funding for highways and public transit. AASHTO has Standing Committees on Aviation, the Environment, Highway Traffic Safety, Highways, Public Transportation, Rail Transportation, and Water Transportation.

American Public Transportation Association (APTA) <http://www.apta.com/> (accessed May 18, 2005). An advocacy group for bus, rapid transit, and commuter rail systems, APTA sponsors conferences on bus and rail transit and applications of intelligent transportation systems. It publishes proceedings, reports, and statistics. Formerly known as the American Public Transit Association and the American Transit Association.

American Railway Engineering and Maintenance Association <www.arema.org/eseries/scriptcontent/index.cfm> (accessed May 18, 2005). This association develops and publishes technical and practical knowledge and recommended practices for the design, construction, and maintenance of railway infrastructure. Formed from the 1997 merger of the American Railway Bridge and Building Association, the American Railway Engineering Association, and the Roadmasters and Maintenance of Way Association, along with functions of the Communications and Signal Division of the Association of American Railroads.

American Road and Transportation Builders Association (ARBTA) <http://www.artba.org/> (accessed May 18, 2005). This association represents the U.S. transportation construction industry to the federal government. It lobbies for increased funding for transportation projects, conducts public information campaigns, and publishes intelligence reports.

American Society of Civil Engineers (ASCE) <http://www.asce.org/> (accessed May 18, 2005). This is the world's largest publisher of civil engineering information, including technical and professional journals, and proceedings. It sponsors conferences and continuing education programs, and produce policies and standards.

American Society of Mechanical Engineers (ASME) <http://www.asme.org/> (accessed May 18, 2005). ASME's Rail Transportation Division (RTD) focuses on intercity freight and passenger railroads as well as rail rapid transit. Along with the Land Transportation Division of the IEEE Vehicular Technology Society, RTD is a partner in the annual IEEE/ASME Joint Rail Conference which focuses on innovation in railway mechanical and electrical engineering. RTD publishes *Rail Transportation* as part of the ASME International Mechanical Engineering Congress proceedings.

American Society of Naval Engineers (ASNE) <http://www.navalengineers.org> (accessed May 18, 2005). ASNE promotes naval engineering through publications and sponsorship of conferences.

Association of American Railroads (AAR) <http://www.aar.org/> (accessed May 18, 2005). AAR represents the major freight railroads in North America and Amtrak. Publishes technical standards (<http://www.aar.com/aartech/index.htm>). A subsidiary, Transportation Technology Center, Inc. (http://www.aar.com/ttci/index.htm), conducts railroad and transit research on rolling-stock, track components, structures, signals, and safety devices

CODATU (Cooperation for Urban Mobility in the Developing World) Association <http://www.codatu.org/index2.htm> (accessed May 18, 2005). An international group, CODATU sponsors the biennial Urban Transport Conference to further the development of urban transportation in developing and industrialized countries.

Institute of Transportation Engineers (ITE) <http://www.ite.org/> (accessed May 18, 2005). ITE offers education programs and provides certification for professional traffic operations engineers. Eleven areas-of-interest councils develop standards and recommended practices, organize conferences and seminars, and issue briefings and informational reports. ITE's publications include standard works such as *Transportation Engineering Handbook*, *Trip Generation* and *Parking Generation*.

Intermodal Association of North America (IANA) <http://www.intermodal.org/index.html> (accessed May 18, 2005). IANA is an association of railroads, truckers, port authorities, suppliers, and manufacturers, and promotes intermodal transportation.

International Road Federation (IRF) <http://www.irfnet.org/> (accessed May 18, 2005). IRF promotes road planning, development, construction, and usage through a wide range of conferences. Compiles and publishes *World Road Statistics*.

International Union of Railways (IUR) <http://www.uic.asso.fr/> (accessed May 18, 2005). IUR, primarily a European organization, works to facilitate international railway transportation. It produces technical standards, regulations and recommendations, and publishes leaflets, reports and technical documents, and *International Railway Journal*.

Japan is a world leader in railroad engineering, and the Railway Technical Research Institute (RTRI) <http://www.rtri.or.jp/> (accessed May 18, 2005). RTRI, the research center for the seven Japan Railway companies, develops basic technology and research applications such as the Maglev system, promotes technology transfer, and studies safety. RTRI publishes *Quarterly Report of RTRI, RTRI Report*, and a calendar of international railway conferences.

The National Asphalt Paving Association (NAPA) <http://www.hotmix.org/> (accessed May 18, 2005). NAPA supports an active research program to address environmental issues and improve the quality of asphalt pavement mixes and paving techniques. Produces technical, marketing, and educational materials.

Society of Automotive Engineers (SAE) <http://www.sae.org/servlets/index> (accessed May 18, 2005). SAE is an international membership organization that sponsors meetings, conferences, and symposia on the technologies related to the design, manufacture, and life cycle of the automotive, truck, bus, and other mobility industries. In addition to its published proceedings, it is well known for developing and publishing more than 8,300 standards for ground vehicles and aerospace.

Transportation Association of Canada (TAC) <http://www.tac-atc.ca/english/index.cfm> (accessed May 18, 2005). TAC promotes knowledge on technical guidelines and best practices for roadways and urban transportation systems.

The Transportation Research Board (TRB) <http://gulliver.trb.org/> (accessed May 18, 2005). TRB is the primary transportation research organization in the U.S. It is charged with promoting transportation innovation through research; areas of interest include all modes of transportation and cover technical and policy aspects. It facilitates information sharing through its annual meeting and an extensive program of publication of reports and peer-reviewed technical papers, including its journal, *Transportation Research Record.* TRB manages research programs, including cooperative programs in highway and transit research. It produces *TRIS*, the online database of transportation information. A division of the National Research Council, TRB is supported by state departments of transportation, the U.S. Department of Transportation, and other federal agencies.

Transport Research Laboratory (TRL) <http://www.trl.co.uk/1024/mainpage.asp> (accessed May 18, 2005). TRL is an independent organization in the U.K. that provides research and testing for all aspects of transportation and publishes research reports.

The U.S. Department of Transportation's Federal Highway Administration (FHWA) <http://www.fhwa.dot.gov/> (all accessed May 18, 2005). FHWA provides finance and technical support to state, local, and tribal governments for the construction, improvement, and maintenance of the highway system. Its many websites (<http://www.fhwa.dot.gov/fhwaweb.htm>) provide extensive information on a wide variety of topics related to highway, structure, and pavement engineering.

Acknowledgments

I am deeply indebted to my colleagues in the Transportation Division of the SLA who contributed to *Sources of Information in Transportation*, particularly Susan Dresley, series editor of this excellent resource.

My special thanks to Bonnie Osif for giving me the wonderful challenge and opportunity of writing this resource guide to transportation engineering.

References

Sinha, K.C. (ed.) (2002) "Development of Transportation Engineering Research, Education, and Practice in a Changing Civil Engineering World," *Journal of Transportation Engineering*, 128 (4): 301–313.

Sources of Information in Transportation (2001) (5th edn) compiled by members of the Transportation Division, Special Libraries Association; Series Editor, Susan C. Dresley, Washington, DC: National Transportation Library, Bureau of Transportation Statistics, U.S. Department of Transportation <http://www.ntl.bts.gov/ref/biblio/>.

Sources of Information in Transportation (1990) (4th edn) compiled by members of the Transportation Division, Special Libraries Association, Monticello, IL: Vance Bibliographies.

Index

eBooks – at www.eBookstore.tandf.co.uk

A library at your fingertips!

eBooks are electronic versions of printed books. You can store them on your PC/laptop or browse them online.

They have advantages for anyone needing rapid access to a wide variety of published, copyright information.

eBooks can help your research by enabling you to bookmark chapters, annotate text and use instant searches to find specific words or phrases. Several eBook files would fit on even a small laptop or PDA.

NEW: Save money by eSubscribing: cheap, online access to any eBook for as long as you need it.

Annual subscription packages

We now offer special low-cost bulk subscriptions to packages of eBooks in certain subject areas. These are available to libraries or to individuals.

For more information please contact webmaster.ebooks@tandf.co.uk

We're continually developing the eBook concept, so keep up to date by visiting the website.

www.eBookstore.tandf.co.uk

FOR REFERENCE ONLY
NOT TO BE TAKEN FROM LIBRARY

CONCORDIA UNIVERSITY LIBRARIES
SIR GEORGE WILLIAMS CAMPUS
WEBSTER LIBRARY